AF324612

CLIMATE CHANGE

CLIMATE CHANGE

AN ENCYCLOPEDIA OF SCIENCE, SOCIETY, AND SOLUTIONS

Volume 3
Human Impact and Primary Documents

Bruce E. Johansen

An Imprint of ABC-CLIO, LLC
Santa Barbara, California • Denver, Colorado

Library of Congress Cataloging-in-Publication Data
Names: Johansen, Bruce E. (Bruce Elliott), 1950–
Title: Climate change : an encyclopedia of science, society, and solutions /
    [edited by] Bruce E. Johansen.
Description: Santa Barbara, California : ABC-CLIO, LLC, 2017. |
    Includes bibliographical references and index.
Identifiers: LCCN 2016059526 (print) | LCCN 2016059775 (ebook) |
    ISBN 9781440840852 (hardcopy : alk. paper : set) | ISBN 9781440848698
    (vol. 1) | ISBN 9781440848704 (vol. 2) | ISBN 9781440848711 (vol. 3) |
    ISBN 9781440840869 (ebook)
Subjects: LCSH: Climatic changes—Encyclopedias.
Classification: LCC QC902.92 .C54 2017 (print) | LCC QC902.92 (ebook) |
    DDC 577.2/2—dc23
LC record available at https://lccn.loc.gov/2016059526

ISBN: 978-1-4408-4085-2 (set)
        978-1-4408-4869-8 (vol. 1)
        978-1-4408-4870-4 (vol. 2)
        978-1-4408-4871-1 (vol. 3)
EISBN: 978-1-4408-4086-9

21 20 19 18 17   1 2 3 4 5

This book is also available as an eBook.

ABC-CLIO
An Imprint of ABC-CLIO, LLC

ABC-CLIO, LLC
130 Cremona Drive, P.O. Box 1911
Santa Barbara, California 93116-1911
www.abc-clio.com

This book is printed on acid-free paper ∞

Manufactured in the United States of America

Contents

VOLUME II: WEATHER AND GLOBAL WARMING

Weather

Global Warming

VOLUME III: HUMAN IMPACT AND PRIMARY DOCUMENTS

Agriculture and Food

Human Health

Indigenous Peoples and Cultures

Greenhouse Gases and Energy

Solutions and Actions

AGRICULTURE AND FOOD

OVERVIEW

Every plant is keyed to a habitat, an important part of which is temperature tolerance. Soil and precipitation also are vital. Given the restraints of habitat, those who propose that we cope with a warming world simply by moving North America's breadbasket to the Yukon and Northwest Territories are displaying an unusual amount of hubris—even for human beings. What, one may ask, about the suitability of soils scraped clear by millennia of glaciation? Such proposals rank with those who tell polar bears to turn brown, scavenge human garbage dumps, and get with the program.

Adaptation to a climate that changes more swiftly, and in malign ways, has become increasingly important for those who grow our food. As an editorial in *Nature* phrased agriculture's challenge ("Prepare Farms" 2015): "Farmers must prepare for, and adapt to, a changed climate that is likely to feature more erratic rainfall, temperature extremes, drought, soil erosion, invasive weeds and durable pests. . . . But if adaptation is to work, climate scientists, agricultural researchers, farmers and government officials must work closely together."

This section contains two types of entries. One describes what rising greenhouse gas emissions have done and will do to the agricultural activities that provide the food we eat. Those who believe that rising levels of carbon dioxide in the air will be beneficial for domesticated plants tend to downplay the damage that heat inflicts and dwell on the obvious, if rather simplistically analyzed, biological fact that plants breathe carbon dioxide and expel oxygen. More carbon dioxide, it is assumed by nonreaders of agricultural science journals, will deliver bigger, greener, and more robust harvests.

A body of evidence indicates that although "enhanced" levels of carbon dioxide may indeed accelerate the growth of some plants, these are not usually the types we will eat unless we become extremely hungry. Weeds tend to flourish under elevated levels of CO_2, whereas edible plants suffer damage from both exceptional heat and added greenhouse gases. In some cases, foliage increases, but seeds and fruits do not.

Hartwell Allen, a researcher with the University of Florida and the U.S. Department of Agriculture, has been growing rice, soybeans, and peanuts under controlled conditions, including varying temperatures, humidity, and carbon dioxide levels. Allen and his colleagues have found that although higher temperatures (to a point) and CO$_2$ levels stimulate lusher and faster growth, they are deadly during flowering and pollinating stages. At temperatures above 36°C during pollination, peanut yields dropped 6 percent per degree C of temperature. John Sheehy of the International Rice Research Institute in Manila found that damage to the world's major grain crops begins during flowering at approximately 30°C. Around 40°C, yields for these crops essentially fall to zero. Crops grown under elevated levels of carbon dioxide also may be deficient in vital nutrients such as zinc and iron, which already pose a substantial public health problem around the world.

Even modest temperature increases that are anticipated by some climate models could be enough to reduce rice yields significantly over the next century. Researchers at the University of Florida tested several varieties of rice, growing them in chambers that simulated various temperature-change situations. They found that although the rice plants flourished no matter what the temperature, yields of grains declined precipitously as temperatures increased. The more modest temperature increases, the researchers say, could reduce rice yields by 20 percent to 40 percent by 2100, whereas the larger increases predicted by more dire forecasts could cut rice production entirely.

Climate-Ready Crops

In some quarters, adaptation has become a hot ticket. Agriculture contributes approximately 20 percent of human greenhouse gas emissions. Agricultural research is underway to reduce this figure, including such projects as breeding low-flatulence cattle (beef animals are a surprisingly large source of atmospheric methane, as is rice farming). Low-till or zero-till farming can help keep carbon in the soil, and more selective use of nitrogen fertilizer may inhibit emission of nitrous oxide, a greenhouse gas 310 times more potent than the main greenhouse gas carbon dioxide. In 2008, the National Farmers Union paid farmers in the United States $5.8 million to adopt environmentally constructive practices in a "carbon credit" program, many of which (such as no-till farming and rotational grazing) capture carbon dioxide. Agencies that aid farmers already are developing new varieties of corn, wheat, rice, and sorghum as well as programs to encourage more efficient use of water and soil resources and to develop new practices to reduce greenhouse gas emissions from farming.

Large multinationals already have been racing to dominate so-called climate-ready crops. By 2008, three companies—Germany's BASF, the Chinese multinational Syngenta, and U.S. agribusiness giant Monsanto—had filed applications that would allow them to control two-thirds of gene families in worldwide patent office filings, according to ETC Group of Ottawa, Canada, which advocates causes that benefit subsistence farmers. These new crops are being bred to resist not only heat and drought but also saltwater inundation, flooding, and increasing ultraviolet radiation (associated with depletion of stratospheric ozone).

Farming with an eye to carbon sequestration utilizes soil restoration and woodland regeneration, no-till farming, cover crops, nutrient management, manuring and sludge application, improved grazing, water conservation, efficient irrigation, agroforestry practices, and the growth of energy crops on spare lands.

In the meantime, people are being encouraged to eat with an eye toward reducing carbon dioxide and methane emissions—more vegetables and less meat. Local production reduces use of fossil fuels in transport. Food production in the United States today requires an average of six calories of energy to produce one calorie of food. Much of this energy is consumed in transport and in energy-intensive cultivation of factory farms, often because of a popular preference for energy-intensive types of food such as meats (most notably beef).

Further Reading

"Prepare Farms for the Future: Scientists Must Work Closely with Farmers to Ensure That Agriculture Can Stand Up to the Ravages of Climate Change." *Nature* 523 (July 23, 2015): 381. http://www.nature.com/news/prepare-farms-for-the-future-1.18018.

AGRICULTURE, FUTURE OF

Agriculture contributes approximately 20 percent of human greenhouse gas emissions. Low-till or zero-till farming can help keep carbon in the soil, and more selective use of nitrogen fertilizer may inhibit emission of nitrous oxide, a greenhouse gas 310 times more potent than the main greenhouse gas, carbon dioxide. In 2008, the National Farmers Union paid farmers in the United States $5.8 million to adopt environmentally constructive practices in a "carbon credit" program, many of which—such as no-till farming and rotational grazing—capture carbon dioxide. Agencies that aid farmers already are developing new varieties of corn, wheat, rice, and sorghum as well as programs to encourage more efficient use of water and soil resources and to develop new practices to reduce greenhouse gas emissions from farming.

Climate-Ready Crops

Large multinationals have been racing to dominate so-called climate-ready crops. By 2008, three companies—Germany's BASF, the Chinese multinational Syngenta, and U.S. agribusiness giant Monsanto—by 2008 had filed applications that would allow them to control two-thirds of gene families in worldwide patent office filings, according to ETC Group of Ottawa, Canada, which advocates causes that benefit subsistence farmers. These new crops are being bred to resist not only heat and drought but also saltwater inundation, flooding, and increasing ultraviolet radiation (associated with depletion of stratospheric ozone).

Although the ETC Group maintains that the companies are engaged in "an intellectual-property grab," the companies assert that "gene-altered plants will be crucial to solving world hunger but will never be developed without patent protections" (Weiss 2008). Patenting genes will prevent farmers in poor countries from saving seeds for future harvests and require them to purchase new seeds from the companies. The rush to patent new seeds also may prevent distribution by public-sector agencies affiliated with the United Nations and the World Bank. "When a market is dominated by a handful of large multinational companies, the research agenda gets biased toward proprietary products," said Hope Shand, ETC's research director. "Monopoly control of plant genes is a bad idea under any circumstance. During a global food crisis, it is unacceptable and has to be challenged" (Weiss 2008).

Speaking for Monsanto, Ranjana Smetacek said these companies should be appreciated for developing crops that will survive adverse environmental conditions. "I think everyone recognizes that the old traditional ways just aren't able to address these new challenges. The problems in Africa are pretty severe" (Weiss 2008). Monsanto maintains that not all of its work is profit oriented. It has joined with BASF, for example, to support the Bill and Melinda Gates Foundation's development of drought-resistant corn free of royalties to farmers in four southern African countries. "We aim to be at once generous and also cognizant of our obligation to shareholders who have paid for our research," Smetacek said (Weiss 2008). The patents may be applied to a wide variety of crops, including "maize, wheat, rye, oat, triticale, rice, barley, soybean, peanut, cotton, rapeseed, canola, pepper, sunflowers, potato, tobacco, eggplant, tomato, peas, alfalfa, coffee, cacao, tea, Salix, oil palm, coconut, perennial grass, and forage crop plants" (Weiss 2008).

Fog and Fruit Trees

One side effect of intense and enduring drought that affects California agriculture has been a notable reduction in coastal and valley fog. This "Tule fog," as it is called locally, decreased 46 percent between 1981 and 2013, according to reports from the NASA Earth Observatory. In 2014, the observatory called the decrease "bad news for California's fruit and nut farmers" ("Winter Fog" 2014).

University of California–Berkeley scientists Dennis Baldocchi and Eric Waller used satellite photography to compile a record of "fog days." They explained via the NASA Earth Observatory:

> Fog forms on cold winter nights after rain, when the air near the ground is moist with water evaporating out of the soil. When temperatures dip low enough, the moist air condenses into fog. Fewer fog days occur when temperatures are warmer or conditions are drier. The fog is important to California's crops because fruit and nut trees, like people, need sufficient rest before they can be their most productive. They get that rest in the winter when cold temperatures—between [0 and 7°C (32 and 45°F)]—bring on a dormant period. Fewer fog days corresponds with fewer cold days. . . . The trees are now being exposed to hundreds fewer cold hours compared to 1982. The fog also shields the trees from direct sunlight during the winter. Direct sunlight can warm the buds even when surrounding air temperatures are cool. The warming decreases the number of cold-rest hours the tree gets during the winter, which decreases its productivity. ("Winter Fog" 2014; Baldocchi and Waller 2014)

Reduction of cool winters required by fruit and nut trees in California already is imperiling some crops, according to a joint study by the University of

California–Davis (UC Davis) and the University of Washington. "Depending on the pace of winter chill decline, the consequences for California's fruit and nut industries could be devastating," said Minghua Zhang, a professor of environmental and resource science at UC Davis ("Warming Climate Jeopardizes" 2009). By 2000, winter temperatures already were too high in most of California's Central Valley to sustain apples, cherries, and pears. By 2100, the same area may no longer sustain walnuts, pistachios, peaches, apricots, and plums. Lack of cold weather affects the plants' flowering time.

"Our findings suggest that California's fruit and nut industry will need to develop new tree cultivars with reduced chilling requirements and new management strategies for breaking dormancy in years of insufficient winter chill," said Eike Luedeling, a postdoctoral fellow in UC Davis' department of plant sciences ("Warming Climate Jeopardizes" 2009).

Twenty-First Century Projections

Temperatures by the end of the 21st century will rise to such a level that Europe's extreme heat of August 2003 (which killed 30,000 people) will become common and devastate world agriculture and provoke a "perpetual food crisis," including crop failures in many regions. That was the conclusion reached in a study published in *Science* January 9, 2009, that had been conducted by scientists at the University of Washington and Stanford University. Yields of staples such as wheat, corn, and rice may be reduced 20 percent to 40 percent. The study's lead author, University of Washington climate researcher David Battisti, said that effects will be most intense in the tropics and subtropics were many people already live at the margin of survival (Mittelstaedt 2009). The scientists used data from 23 global climate models to show a high probability that more than 90 percent of growing-season temperatures in the tropics and subtropics by the end of the 21st century will exceed the most extreme seasonal temperatures recorded from 1900 to 2006 (Battisti and Naylor 2009, 240).

How will agriculture fare in a warmer world? The MINK Study (an acronym for Missouri, Iowa, Nebraska, Kansas) surveyed potential climate change in the central United States, North America's agricultural heartland. Under certain circumstances, the authors found, higher levels of carbon dioxide might enhance the growth of some crops, but as a whole the productivity of the region's agriculture would be significantly diminished. Agriculture in current farming regions will be severely affected not only by heat stress but also by reduced surface-water supplies because most global climate models predict that the interiors of continents will become not only hotter but also drier, especially during the growing season, as the atmosphere warms. An additional problem facing farmers in Nebraska and Kansas is depletion and salinization of aquifers that already support a large part of agricultural production in both states, especially their drier western areas.

Earth's population by 2050 is expected to increase by nearly half, or 3.5 billion people. Roughly 75 percent of poor people still will depend on agriculture. Hotter, drier weather combined with explosive bursts of precipitation may shorten growing seasons and threaten production in some areas (notably in Africa and India) where agricultural production is limited by availability of water rather than the onset of cold weather. Hundreds of millions of people who already live at the margin may find their survival threatened.

Various modes of adaptation such as breeding crops that can resist more heat, flood, drought, or insect infestations may provide some help in the short range, but their ability to mitigate the destructive nature of warming will probably decline as temperatures rise. Accelerating climate change will provide farmers with an ever-changing environment that is bereft of old environmental benchmarks. A research report by the Consultative Group on International Agricultural Research (CGIAR) in Washington, D.C., a worldwide network of agricultural research centers, is already developing new crop strains that can withstand rising temperatures, drier climates, and increasing soil salinity. CGIAR also researches measures to reduce the carbon footprint of farming (Zeller 2006).

In the National Climate Assessment, which was published in January 2013, major agricultural areas in the middle of the United States were warned to anticipate more heavy spring rains of intensities that will ruin planting seasons and fewer summer rains that help relieve drought and replenish corn and other crops. Loss of topsoil from occasional but extremely heavy rains in a "drought or deluge" scenario is also anticipated. Higher nighttime temperatures will interfere with the pollination of many crops, which will suffer heat damage as weeds, plant diseases, and pests flourish during warmer growing seasons. Runoff from wide swings in snowpacks will become more erratic in irrigated areas as hotter weather strains general resources. Wider swings in both temperatures and rainfall also will play havoc with agriculture (Gaarder 2013, D1).

Warming and Canadian Grain Production

Climate-change skeptics often advance a simplistic solution to agricultural problems caused by warming temperatures and a more explosive hydrological cycle: move it all northward. Following their climatic logic, one can almost imagine corn sprouting along the shores of Hudson Bay. This assumes, however, that the soils will be right for growing crops. Even today, wheat, soy and canola are grown almost to the 60th parallel in the Peace River valley in Northern Alberta and British Columbia. Some varieties of grain, rye, flax, and canola mature in 120 days. Given enough rain (a problem in some northern latitudes) warmer weather could benefit Canadian agriculture. Grain-growing areas in Russia also may move northward in areas where soil is suitable.

The northern areas of Alberta and Saskatchewan are already producing hardy varieties of wheat. In northern Ontario and part of western Quebec, the clay belts are currently being farmed. Many of the Canadian Shield's soils do not support agriculture, mainly because the soil is poor, although some areas are fertile. Warmer temperatures, especially in August and September, would allow for increased farming in Canada. The fertile eastern townships of Quebec in the St. Lawrence Lowlands have been farmed since at least the early 1700s. Most of these areas are former lake or sea bottoms and have deep, rich soils. The limiting factor has been the growing season. The boreal forests of northern Canada include tens of millions of acres with abundant water, in which trees have grown, died, rotted, and regrown for millennia. Add warmth and some of these areas might become productive farmlands.

Eastern Canada north of the Great Lakes and the St. Lawrence River valley is not generally suitable for grain. The western plains from western Lake Superior to the Rocky Mountains contain excellent grain-growing soils, however. Grain from Manitoba, Saskatchewan, and Alberta fed Great Britain from 1939 to 1941 during the early days of World War II.

Global-Scale Issues

Temperature increases and shifts in rainfall patterns probably will reduce growing periods in sub-Saharan Africa by more than 20 percent, with some of the world's poorest nations in eastern and central Africa at greatest risk. A warming climate probably will reduce wheat production in India's breadbasket. Production may decline by around 50 percent by 2050, a decrease that could put as many as 200 million people at greater risk of chronic hunger.

Although some studies anticipate rising food production in some areas during the early stages of global warming, others anticipate "negative surprises" in world agriculture, especially during the last half of the 21st century. Poor (i.e., "developing") nations may lose 334 million acres of prime farmland during the next half century as temperatures rise and storms intensify, according to three studies in the December 11, 2007, edition of the *Proceedings of the National Academy of Sciences*: Tubiello et al., Howden et al., and Schmidhuber and Tubiello. By the last half of the 21st century, even cooler regions that may benefit from earlier temperature rises could experience declines in productivity. The authors of these studies argue that extremes of heat will join with other factors such as the spread of weeds and diseases to compound agricultural problems. These problems will inhibit increases in food production necessary to feed rising populations. These studies project that as many as 170 million people may be "at risk of hunger" by 2080 (Schmidhuber and Tubiello 2007, 19703).

"Many people assume that we will never have a problem with food production on a global scale. But there is a strong potential for negative surprises," said Francesco Tubiello, a physicist and agricultural expert at NASA's Goddard Institute for Space Studies, who coauthored the three PNAS reports published. The PNAS study authors said that much previous research work is oversimplified; as a consequence, the potential for bigger and more rapid problems remains unexplored. Heat waves and extreme storms could have their greatest effects on crops at crucial critical germination or flowering times. Tubiello says this is already happening on smaller scales.

Redrawing the Maps

During 2008, the U.S. Department of Agriculture issued a revised climatic-zone map for gardeners. This map assigns zones to various plants for winter survival. In the 18 years since the map had been revised in 1990, growing zones for many plants have moved northward. Southern magnolias, for example, once restricted to coastal Virginia southward, now thrive into Pennsylvania. In 1990, kiwis died north of Oklahoma. In 2008, they fruited in St. Louis, Missouri.

The Arbor Day Foundation has made similar adjustments in its maps, which were last revised in 2005. New drafts of U.S. Department of Agriculture maps may be difficult to find because they were delayed by political bickering within the agency between climate scientists and skeptics, as well as between owners of nurseries. Nursery owners feared losing money if they sold plants with money-back guarantees and the plants died in the new zones, but other sellers want to expand their marketing northward (Weise 2008).

Further Reading

Baldocchi, D., and E. Waller. "Winter Fog Is Decreasing in the Fruit Growing Region of the Central Valley of California." *Geophysical Research Letters* 41(9) (May 15, 2014): 3251–3256. http://onlinelibrary.wiley.com/doi/10.1002/2014GL060018/abstract.

Battisti, David. S., and Rosamond L. Naylor. "Historical Warnings of Future Food Insecurity with Unprecedented Seasonal Heat." *Science* 323 (January 9, 2009): 240–244.

Gaarder, Nancy. "An Ag 'Battleground:' Federal Climate Report Says Midlands at Risk of Experiencing More Extremes." *Omaha World-Herald*, February 4, 2013, D1, D4.

Howden, S. Mark, et al. "Adapting Agriculture to Climate Change." *Proceedings of the National Academy of Sciences* 104 (December 11, 2007): 19691–19696.

Mittelstaedt, Martin. "World Faces Perpetual Food Crisis: Study." *Toronto Globe and Mail*, January 8, 2009. http://www.theglobeandmail.com/servlet/story/RTGAM.20090108 .wclimate0108/BNStory/International/home (no longer available).

National Climate Assessment. Washington, DC: U.S. Government Printing Office, January 2013. http://nca2014.globalchange.gov/.

Schmidhuber, Josef, and Francesco Tubiello. "Global Food Security under Climate Change." *Proceedings of the National Academy of Sciences* 104 (December 11, 2007): 19703–19708.

Tubiello, Francesco, Jean Francois Soussana, and S. Mark Howden. "Crop and Pasture Response to Climate Change." *Proceedings of the National Academy of Sciences* 104 (December 11, 2007): 19686–19690.

"Warming Climate Jeopardizes California Fruit and Nut Crops." Environment News Service, July 23, 2009. http://www.ens-newswire.com/ens/jul2009/2009-07-23-095.html.

Weise, Elizabeth. "Warming Climate Makes Gardeners' Map Out-of-date." *USA Today*, April 24, 2008, A1.

Weiss, Rick. "Firms Seek Patents on 'Climate Ready' Altered Crops." *Washington Post*, May 13, 2008. http://www.washingtonpost.com/wp-dyn/content/article/2008/05/12/AR20080 51202919_pf.html.

"Winter Fog Becoming Rare in California." NASA Earth Observatory, June 5, 2014. http:// earthobservatory.nasa.gov/IOTD/view.php?id=83806&src=eoa-iotd.

Zeller, Tom, Jr. "America's Breadbasket Moves to Canada?" *The New York Times*, December 5, 2006. http://thelede.blogs.nytimes.com/2006/12/05/americas-breadbasket-moves-to -canada.

See also: Carbon Dioxide Enrichment; Drought, United States; Drought, Worldwide; Extreme Weather; Floods; Greenhouse Gas Emissions; Growing Seasons, Europe and Asia; Rice Yields, Decline in; Temperatures, Global; Wine Grapes and Warming

CARBON DIOXIDE ENRICHMENT

Because plants require carbon dioxide to grow, folk wisdom holds that increasing the level of this trace gas in the atmosphere will accelerate plant growth. Unfortunately, climate-change skeptics who rely on the idea that higher carbon dioxide levels will benefit plants ignore the possibility that heat stress, drought, or deluge will cause problems. In addition, scientific support for the idea that elevated carbon dioxide levels aid plant growth is far from convincing. Although a few studies indicate that higher carbon dioxide levels and warmer temperatures may help "green" the Earth in the short run, longer-run studies cast considerable doubt on this assumption.

Pim Martens, director of the International Centre for Integrated Assessment, which is based in the Netherlands, contends that increased growth may be counterbalanced by increases in molds and other parasites that thrive in hot and humid weather. "It is generally believed that a climate change will have negative effects for global food production," Martens writes (Martens 1999, 540). To cite one of many examples, the Mediterranean fruit fly could expand into northern Europe in the next century given the degree of global warming projected by the U.N.'s Intergovernmental Panel on Climate Change.

Simplistic Assumptions

The case for atmospheric carbon dioxide as agricultural bromide has been advanced by some global warming skeptics with absolute and often simplistic statements about its assumed benefits. One such statement came from Sylvan Wittner, director of the Michigan Agricultural Experiment Station and chairman of the agricultural board of the National Research Council:

> Flowers, trees, and food crops love carbon dioxide, and the more they get of it, the more they love it. Carbon dioxide is the basic raw material that plants use in photosynthesis to convert solar energy into food, fiber, and other forms of biomass. Voluminous scientific evidence shows that if CO_2 were to rise above its current ambient level of 360 parts per million, most plants would grow faster and larger because of more efficient photosynthesis and a reduction of water loss. (Klug 1997)

High Carbon Dioxide Degrades Crops' Nutritional Value

Crops grown under elevated levels of carbon dioxide may be deficient in vital nutrients, such as zinc and iron, lack of which already pose a substantial public health problem around the world. Samuel S. Myers led a team that investigated this possibility, and reported the following in *Nature* (2014, 139):

> Dietary deficiencies of zinc and iron are a substantial global public health problem. An estimated two billion people suffer these deficiencies causing a loss of 63 million life-years annually. Most of these people depend on C_3 grains and legumes as their primary dietary source of zinc and iron. Here we report that [these] grains and legumes have lower concentrations of zinc and iron when grown under field conditions at the elevated atmospheric CO_2 concentration predicted for the middle of this century. C_3 crops other than legumes also have lower concentrations of protein, whereas C_4 crops seem to be less affected. Differences between cultivars of a single crop suggest that breeding for decreased sensitivity to atmospheric CO_2 concentration could partly address these new challenges to global health.

Other studies have found that elevated carbon dioxide levels in field tests degrade protein and nitrogen levels in several food crops, including wheat. As Arnold J. Bloom reported in *Nature Climate Change* (2014),

> plants under elevated CO_2 grow larger, diluting the protein within their tissues that carbohydrates accumulate within leaves, down-regulating the amount of the most prevalent protein Rubisco that carbon enrichment of the rhizosphere leads to progressively greater limitations of the nitrogen available to plants and that elevated CO_2 directly inhibits plant nitrogen metabolism, especially the assimilation of nitrate into proteins in leaves of C_3 plants. Recently, several meta-analyses have indicated that CO_2 inhibition of nitrate assimilation is the explanation most consistent with observations. Here, we present the first direct field test of this explanation.

> Bloom led a team that analyzed wheat (*Triticum aestivum* L.) grown under elevated, ambient carbon dioxide concentrations. "In leaf tissue," they wrote, "the ratio of nitrate to total nitrogen concentration and the stable isotope ratios of organic nitrogen and free nitrate showed that nitrate assimilation was slower under elevated rather than ambient CO_2. These findings imply that food quality will suffer under the CO_2 levels anticipated during this century unless more sophisticated approaches to nitrogen fertilization are employed" (Bloom et al. 2014).

> **Further Reading**
>
> Bloom, Arnold J., et al. "Nitrate Assimilation Is Inhibited by Elevated CO2 in Field-Grown Wheat." *Nature Climate Change*. April 6, 2014. http://www.nature.com/nclimate/journal/vaop/ncurrent/full/nclimate2183.html.
> Myers, Samuel S., et al. "Increasing CO_2 Threatens Human Nutrition." *Nature* 510 (June 5, 2014): 139–142.

Like Wittner, Sherwood Idso asserts that Earth's ecosystem is starved for carbon dioxide. Idso paints a wondrous picture of giant plants nourished by carbon dioxide levels "in the thousands of parts per million" (Klug 1997). Idso calls such levels "a breath of fresh air for the planet's close-to-suffocating vegetation" (Klug 1997). The word *rejuvenation* is prominent in Idso's case, describing his vision of a carbon-enriched world atmosphere. Idso does not tell his audience that such carbon dioxide levels were once experienced when global temperatures over much of Earth's surface were insufferable during summers for warm-blooded animals.

According to Idso,

> The whole face of the planet will likely be radically transformed—rejuvenated as it were—as the atmospheric CO_2 content reverses its long history of decline and returns, in significant measure, to conditions much closer to those characteristic of the Earth at the time when the basic properties of plant processes were originally established. (Idso 1982, 9)

Research by Fakhri A. Bazzaz and Eric D. Fajer has long cast doubt on the contrarians' assertions that an atmosphere "enriched" by carbon dioxide will lead to more plant growth and greater agricultural yields.

> Studies have shown that an isolated case of a plant's positive response to increased CO_2 levels does not necessarily translate into increased growth for entire plant communities. . . . [P]hotosynthetic rates are not always greatest in CO_2-enriched environments. Often plants growing under such conditions initially show increased photosynthesis, but over time this rate falls and approaches that of plants growing under today's carbon dioxide levels. . . . When nutrient, water, or light levels are low, many plants show only a slight CO_2 fertilization effect. (Bazzaz and Fajer 1992, 68–71)

"Enhanced" Carbon May Stunt Plant Growth

A slight rise in temperatures and carbon dioxide levels would stimulate the growth of many plants, but any gardener knows that each plant has its own heat limit. Thus, studies by several researchers (see, for example, Schimel 2006, 1889–1890; and Long et al. 2006) reveal a considerably more complex relationship between plant growth and "enhanced" carbon dioxide. These levels vary considerably between species. Every plant has a level of temperature at which the rate of respiration surpasses the rate of photosynthesis, causing the organism to eventually die. Every plant also requires moisture in certain amounts. Too little (or too much) moisture will kill many plants. Nighttime temperatures also play a crucial role in relieving moisture stress for some species. If nighttime temperatures are too high, some trees will die.

Levels of carbon dioxide above a certain level may actually stunt the growth of many plants. Many insects are extremely sensitive to temperature and humidity, and even a small rise in temperature can cause massive outbreaks, as witnessed by insect population explosion in New Orleans during 1995, which had experienced five years without a killing frost.

Researchers Roseanne D'Arrigo and Gordon Jacoby of the Lamont–Doherty Earth Observatory studied tree rings near the timberline in boreal forests in northern and central Alaska. They found that growth increased with modest increases in warming during the 1930s and 1940s. After that, increased warming produced heat and drying that stressed the trees and caused, along with increasing insect activity, tree growth to slow most years.

Growth was higher than usual during years when moisture was plentiful but below average during relatively warm and dry years. In the meantime, insects usually flourish in warmer weather. The bark beetle, a major pest in Alaska's forests, has shortened its reproductive cycle by 100 percent because of warming, according to Jacoby and D'Arrigo (Gelbspan 1997, 141). David Deming has measured soil temperatures in northern Alaska. His records indicate that the soil there has warmed by 2 to 5°C (3.6 to 9°F) during the 20th century.

Warming Harms Crop Yields

Fakhri A. Bazzaz and Eric D. Fajer (1992, 68) are not convinced of CO_2's benefits. "[W]e do not expect that agricultural yields will necessarily improve in a CO_2-rich future." William R. Cline (1992, 91) added that scarcity of water (which is forecast for many continental interiors as the atmosphere warms) also may reduce agricultural yields. Global warming is damaging production of the world's major food crops, according to scientists at Lawrence Livermore National Laboratory and the Carnegie Institution at Stanford University. Ecologists there reported that yields of corn, wheat and barley have declined by about 40 million tons every year since 1981 from what farms worldwide should have produced. The annual value of those lost crops is roughly $5 billion.

Horticulturalist N. C. Bhattacharya has found that although enriched carbon dioxide accelerates growth in most plants, others respond negatively. Increasing the growth rate of plants also tends to accelerate depletion of soils, which stunts later growth. Heat also may be detrimental to some plants even as their growth is being stimulated by rising carbon dioxide levels in the atmosphere (Bhattacharya 1993).

Climatologist Y. A. Izrael summarizes the challenge of agriculture in a warmer world: "Estimates of the impact of doubled CO_2 on crop potential have shown that in the northern midlatitudes summer droughts will reduce potential production by 10 percent to 30 percent. The impact of climate change on agriculture in all, or most, food-exporting regions will entail an average cost to world agricultural production [of] no less than 10 percent" (Izrael 1991, 83). Parry elaborated: "While global levels of food production can probably be maintained in the face of climate change, the cost of this could be substantial" (Parry 1990, 279). Gains in production at higher latitudes are unlikely to balance reductions in the hotter mid-latitudes, which are major grain exporters today (Parry, 1990, 279).

Although crop yields continue to improve, they would have been better during the last 20 years without warming temperatures, according to Livermore climate scientists David Lobell and Christopher Field, head of the Carnegie Institution's Department of Global Ecology at Stanford University (Hoffman 2007). "At least for wheat, corn, and barley, temperature trends in the last few decades have been in the direction of holding yields down," Field said. "They're still increasing, but if temperatures hadn't been warming, they would have been increasing more" (Hoffman 2007).

According to a report in the magazine *World Watch*, "For wheat and corn alone, the annual global losses are equal to the wheat and corn production of Argentina. The researchers likened the effect of global warming on crops to driving a car with the parking brake on. As a result, they said, farmers and plant scientists will have to work harder at meeting rising food demand—and harder still if farming for energy as well" (Hoffman 2007). The same fungus that caused the Irish potato famine has migrated 4,000 feet up slopes in the Peruvian Andes, because of warmer, wetter weather. Potato breeders are trying to outwit the fungus by developing new breeds of tubers (Halweil 2005, 18).

Studies of Specific Plants and Pests

Hartwell Allen, a researcher with the University of Florida and the U.S. Department of Agriculture, has been growing rice, soybeans, and peanuts under controlled conditions and varying temperatures, humidity, and carbon dioxide levels. Allen and his colleagues have found that although higher temperatures (to a point) and CO_2 levels stimulate lusher and faster growth, they are deadly at the flowering and pollination stages. At temperatures above 36°C during pollination, peanut yields dropped 6 percent per degree Celsius. John Sheehy of the International Rice Research Institute in Manila found that damage to the world's major grain crops begins during flowering near 30°C. Around 40°C, yields fall to zero. At his center, the average temperature has risen 2.5°C in 50 years, frequently reaching damaging levels. In rice, wheat, and maize, yields fall 10 percent for every degree above 30°C. Higher nighttime temperatures in particular inhibits plants' ability to respire and saps their energy (Halweil 2005, 19–20). Planting shade trees among crops may help mitigate heat in the short range.

Studies by Martin Parry (1990; Parry and Jiachen 1991) estimate that the European corn borer could move 165 to 200 kilometers northward (in the Northern Hemisphere) with each one-degree Celsius rise in temperature. The potato leaf-hopper, a major pest for soybeans, currently spends its winters along the Gulf Coast. Global warming could move this range northward. The range of the horn-fly, which caused about $700 million a year in damage to beef and diary cattle across the United States during the late 1980s, could be similarly affected.

In *Global Warming: Signs to Watch For* (1993), Harold W. Bernard described global warming's anticipated role in the spread of wheat rust, a fungus that thrives in dry heat. Wheat rust destroyed millions of tons of wheat in North America during the Dust Bowl decade of the 1930s. At the same time, the pale western cutworm, another pest that favors hot and dry conditions, damaged thousands of acres of wheat in Canada and Montana (Bernard 1993, 47).

William R. Cline, an economist who specializes in global warming, modeled global warming three centuries into the future. By that time, he expects that agricultural production in many contemporary breadbaskets will have been devastated. Estimating an average July maximum for Iowa between 100 and 108°F, Cline scoffs at the idea that higher carbon dioxide levels in the atmosphere will enhance agricultural yields. By 2275, Cline anticipates that temperatures may become so hot that most staple grain crops in present-day Iowa (and surrounding states) may die of heat stress. Wheat, barley, oats, and rye simply will not grow in the climate that Cline projects for the U.S. Midwest roughly three centuries from today. In Eurasia, rice, corn, and sorghum will be pressed to their limits by such temperatures as far north as Moscow. By 2275, according to Cline, the atmosphere's carbon dioxide load may be eight times the levels of the 1990s (Cline 1992).

According to Thomas R. Karl et al. (1997), increases in minimum temperatures are important because of their effects on agriculture. Observations over land areas during the latter half of the 20th century indicate that average minimum temperatures have increased at a rate more than 50 percent greater than that of maximums. The rise in minimum temperatures has lengthened the growing (frost-free) season in many parts of the United States. In the Northeast, for example, Karl et al. write that the frost-free season began an average of 11 days earlier during the 1990s than during the 1950s. The compression of daily high and low temperatures may be related to increasing cloud cover and evaporative cooling in many areas, Karl and associates propose. Clouds depress daytime temperatures because they reflect sunlight, and they warm nighttime temperatures by inhibiting loss of heat from the surface. "Greater amounts of moisture in the soil from additional precipitation and cloudiness inhibit daytime temperature increases because part of the solar energy goes into evaporating this moisture," they wrote (Karl et al. 1997).

Effects of Droughts and Deluges

In a warmer world, both erosion caused by deluges *and* persistent drought probably will pose dangers to agriculture. This is not the paradox that it seems because, according to Karl and colleagues, "not only will a warmer world be likely to have more precipitation, but the average precipitation event is likely to be heavier" (Karl et al. 1997). Karl and colleagues cited data indicating that heavy precipitation events increased roughly 25 percent from the beginning of the 20th century to its end. Karl describes how a generally wetter world also will be a place in which drought may threaten flora and fauna more often:

> As incredible as it may seem with all this precipitation, the soil in North America, southern Europe and in several other places is actually expected to become drier in the coming decades. Dry soil is of particular concern because of its far-reaching effects, for instance, on crop yields, groundwater resources, lake and river ecosystems. . . . Several models now project significant increases in the severity of drought. (Karl et al. 1997)

Karl et al. temper this statement by citing studies indicating that increased cloud cover may reduce evaporation in some of the areas that are expected to become drier. All in all, however, most of the world's flora and fauna will suffer significantly in a significantly warmer world. Here, as elsewhere, an accurate forecast is nuanced and complex.

Further Reading

Bazzaz, Fakhri A., and Eric D. Fajer. "Plant Life in a CO_2-Rich World." *Scientific American,* January 1992, 68–74.

Bernard, Harold W., Jr. *Global Warming: Signs to Watch For.* Bloomington: Indiana University Press, 1993.

Bhattacharya, N. C. "Prospects of Agriculture in a Carbon-Dioxide-Enriched Environment." Pp. 487–505 in Richard A. Geyer, ed., *A Global Warming Forum: Scientific, Economic, and Legal Overview.* Boca Raton, FL: CRC Press, 1993.

Cline, William R. *The Economics of Global Warming.* Washington, D.C.: Institute for International Economics, 1992.

Gelbspan, Ross. *The Heat Is On: The High Stakes Battle over Earth's Threatened Climate.* Reading, MA: Addison-Wesley, 1997.

Halweil, Brian. "The Irony of Climate." *World Watch,* March–April 2005, 18–23.

Hoffman, Ian. "Global Warming Killing Crops." *Daily Democrat,* March 17, 2007. http://www.dailydemocrat.com/article/ZZ/20070317/NEWS/703179724.

Idso, Sherwood B. *Carbon Dioxide: Friend or Foe?* Tempe, AZ: IBR Press, 1982.

Izrael, Yu. A. "Climate Change Impact Studies: The IPCC Working Group II Report." Pp. 83–86 in J. Jager and H. L. Ferguson, eds., *Climate Change: Science, Impacts, and Policy.* Proceedings of the Second World Climate Conference. Cambridge, UK: Cambridge University Press, 1991.

Karl, Thomas R., Neville Nicholls, and Jonathan Gregory. "The Coming Climate: Meteorological Records and Computer Models Permit Insights into Some of the Broad Weather Patterns of a Warmer World." *Scientific American* 276 (1997): 79–83. http://www.scientificamerican.com/0597issue/0597karl.htm (no longer available).

Klug, Edward C. "Global Warming: Melting Down the Facts about This Overheated Myth." CFACT Briefing Paper Number 105, November 1997. http://www.cfact.org/IssueArchive/greenhouse.bp.n97.txt (no longer available).

Long, Stephen P., et al. "Food for Thought: Lower-Than-Expected Crop Yield Stimulation with Rising CO_2 Concentrations." *Science* 312 (June 30, 2006): 1918–1921.

Martens, Pim. "How Will Climate Change Affect Human Health?" *American Scientist* 87(6) (November–December 1999): 534–541.

Parry, Martin. *Climate Change and World Agriculture*. London: Earthscan, 1990.

Parry, Martin, and Zhang Jiachen. "The Potential Effect of Climate Changes on Agriculture." Pp. 279–289 in J. Jager and H. L. Ferguson, eds., *Climate Change: Science, Impacts, and Policy.* Proceedings of the Second World Climate Conference. Cambridge, UK: Cambridge University Press, 1991.

Schimel, David. "Climate Change and Crop Yields: Beyond Cassandra." *Science* 312 (June 30, 2006): 1889–1890.

See also: Agriculture, Future of

DISASTER RELIEF

International disaster aid will not be able to keep up with the impact of global warming, the Red Cross has said, as it reported a sharp increase in weather-related disasters during the late 1990s. In its annual "World Disasters Report," the International Federation of Red Cross and Red Crescent Societies said that floods, storms, landslides, and droughts numbered around 200 a year before 1996. The number rose to 392 in 2000. Recurrent disasters "are sweeping away development gains and calling into question the possibility of recovery," the report said (Capella 2001). Roger Bracke, Red Cross director of disaster-relief operations, blamed the trend on global warming, warning that it "is probable that these kind of disasters will increase even more spectacularly [in the future]" (Capella 2001).

Floods have accounted for more than two-thirds of the average 211 million people a year affected by natural disasters during the 1990s, according to the Red Cross. Famine caused by drought was responsible for about 42 percent of the deaths caused by natural disasters, according to this report, which said that the poor are most vulnerable to disasters. Eighty-eight percent of those affected and two-thirds of those killed in the past 10 years lived in the most impoverished countries (Capella 2001).

Environmental Refugees

During the 21st century, the term *environmental refugee* is becoming more familiar around the world. Lowland residents who could be forced out of their homes by a three-foot rise in sea levels during in this century include 26 million people in Bangladesh, 70 million to 100 million in China, 20 million in India, and 12 million in Egypt's Nile River delta (Gelbspan 1997, 162). In Egypt, a one-meter sea-level rise could cost 15 percent of the country's gross national product, including much of its agricultural base (Edgerton 1991, 72–73). A 14-inch rise in sea levels could flood 40 percent of the mudflats that ring Puget Sound, obliterating a significant habitat for shellfish and waterfowl (Gough 1999, 48). These locations are only a few of a great many examples, because Earth's junctures of seas and rivers have been important crossroads for human trade (as well as fertile farming areas) throughout human history. Many river deltas are densely populated and extremely vulnerable to even a small amount of sea level rise.

The World Bank released a report in 2013 listing the 10 cities in the world that may incur the highest levels of damage from rising sea levels: Miami, New York City, New Orleans, Tampa, and Boston (all in the United States), as well as Guangzhou and Shenzen in China, Nagoya and Osaka in Japan, and Mumbai in India. The World Bank surveyed prospective damage in 136 large coastal cities and concluded that damage could rise to $1 trillion a year. Directed by World Bank economist Stephane Hallegatte and consulting with the Organization for Economic Cooperation and Development, the study warns that coastal cities "face a high risk from increasingly costly flooding as sea levels rise amid climate change. Their current defenses will not be enough as the water level rises" ("10 Coastal Cities" 2013). Many of these cities are at risk not only from rising seas but also from subsiding land.

Further Reading

Edgerton, Lynne T. *The Rising Tide: Global Warming and World Sea Levels.* Washington, DC: Island Press, 1991.

Gelbspan, Ross. *The Heat Is On: The High Stakes Battle over Earth's Threatened Climate.* Reading, MA: Addison-Wesley, 1997.

Gough, Robert. "Stress on Stress: Global Warming and Aquatic Resource Depletion." *Native Americas*, 16(3–4) (Fall–Winter 1999): 46–48. http://nativeamericas.aip .cornell.edu (no longer available).

"10 Coastal Cities at Greatest Flood Risk as Sea Levels Rise." Environment News Service, September 3, 2013. http://ens-newswire.com/2013/09/03/10-coastal-cities-at -greatest-flood-risk-as-sea-levels-rise/.

The same report was critical of the international charity infrastructure, asserting that emergency international aid to the poorest countries declined during the late 1990s and the amount sent elsewhere rose. The report complained that many donors focus on high-profile projects to rebuild infrastructure, not people's daily livelihoods. A large amount of aid also is often spent on studies rather than actual street-level aid. For example, the report said that nearly two-thirds of the funds spent on a flood-action plan in Bangladesh from 1990 through 1995 left the country to pay foreign-aid consultants, "thereby undermining the local economy" (Capella 2001).

Anticipated climatic changes related to global warming will probably deepen the gap between the richest and poorest nations, especially in parts of Africa, South Asia, and South America, according to assessments of the anticipated effects of global warming on food production. Nations in tropical climates such as India, Brazil, and much of sub-Saharan Africa will probably experience the largest proportional losses in food production (McFarling 2001). Most of the world's poorest peoples live in the tropics, where the effects of warming on agriculture may be the most catastrophic, leaving widespread starvation and malnutrition, according to the report.

By contrast, people who live in cooler climates may experience gains in crop yields as higher temperatures lengthen growing seasons (McFarling 2001). For the poorest nations, "there is no margin for loss," said Mahendra Shah, one of the report's authors and a United Nations advisor and expert on land use from Austria's International Institute for Applied Systems Analysis (McFarling 2001).

Although poor nations would bear the heaviest burden as a group, some of the largest "developing" nations (China, Indonesia, Mexico, Chile, Congo, and Kenya are examples) would probably see increased production. Some developed countries, including Britain, the Netherlands, and Australia, could see crop yields decline as warmer, wetter weather increases diseases and pests (McFarling 2001).

Further Reading

Capella, Peter. "Disasters Will Outstrip Aid Effort as World Heats Up." *The Guardian* (U.K.), June 29, 2001, 15.

McFarling, Usha Lee. "Warmer World Will Starve Many, Report Says." *Los Angeles Times*, July 11, 2001, A3.

See also: Drought, Worldwide; El Niño and La Niña; Temperatures, Global

GREENHOUSE GAS EMISSIONS

People who do not farm for a living often have a stereotyped view of agriculture as a family affair, a builder of character, and a style of employment that evokes, for tillers of the soil, a basic sense of enjoyment from communion with nature. During the 20th century, however, agriculture became progressively more mechanized on a massive scale suited to large, worldwide markets. Like other modes of production in our machine culture, agriculture has come to demand less human labor and an increasing amount of fossil fuel energy. Industrial-scale agriculture also requires copious amounts of synthetic fertilizers. For all except a comparatively few remaining (and often struggling) family farmers, agriculture has become as industrialized as factory work.

Ecologist Barry Commoner described how the American farm has changed:

> Between 1950 and 1970, the total U.S. crop output increased by 38 percent, although the acreage decreased by 4 percent and the labor [number of people employed] fell by 58 percent. This sharp increase in productivity was accomplished by an 18 percent increase in the use of machinery and a 295 percent increase in the application of synthetic pesticides and fertilizer. (Commoner 1990, 49)

Industrial Agriculture's Emissions

Like many industries, modern agriculture emits copious greenhouse gases: methane from cattle (from their breath as well as bodily wastes), nitrous oxides from fertilizer (23 times as potent a greenhouse gas as carbon dioxide), carbon dioxide released to the atmosphere when soil is tilled, CO_2 emitted by farm machinery and trucks used to transport produce to market, and so forth. "Altogether," wrote Mark Schapiro in *Carbon Shock* (2014, 24), "farmers are in the unenviable, though not totally unavoidable, position of being at the front lines of climate shifts that they are helping to trigger." Agriculture in the United States is an enormous industry, with more than 2 million farms on 900 million acres harvesting some $350 billion worth of crops that are shipped all over the world.

Worldwide, industrial agriculture accounts for roughly one-third of human greenhouse gas emissions. "The industrialized, meat-heavy food system of the United States takes a heavy toll on the atmosphere," wrote Mark Hertsgaard, author of *Hot: Living through the Next Fifty Years on Earth,* in *The New York Times* (2012). "It takes an enormous amount of fossil fuel to run farm equipment and harvest the mountains of corn that fatten livestock. And most fertilizers contain nitrous oxide, a greenhouse gas 298 times more potent than carbon dioxide over a century."

In March 2014, the U.N. Intergovernmental Panel on Climate Change (IPCC) addressed the issue. "Throughout the 21st century, climate-change impacts are projected to further erode food security—particularly in urban areas and emerging hot spots of hunger. Climate change affects many aspects of food security, including food access, utilization and price stability. . . . If these challenges are not addressed, consumers will need to be prepared for higher food prices and potential food shortages" (Rinne 2014). Rising temperatures and extreme precipitation events imperil farmers perhaps more than any other group.

Writing in *Native Americas*, Craig Benjamin described how large-scale monocultural farms make themselves vulnerable to a risking risk of pest attack in a warmer, more humid world.

> This impressive vulnerability of industrial agriculture is key to understanding how climate change will likely have an impact on global agriculture and on the relationship between industrial agriculture and indigenous farming communities. Faced with rapid and dramatic climate change, the impressively vulnerable industrial farm can conceivably continue to use large-scale irrigation and artificial fertilizers to counter the effects of changing temperature and precipitation. (Benjamin 1999, 80)

Some people believe that global warming will be ameliorated by such adaptations as forecasting systems that advise farmers to switch crops or change the timing of planting. Crops are already being modified to survive heat and drought. These may buy only a few decades, however, as warming continues. "After that," said Francesco Tubiello, "all the bets are off" ("Warming Climate" 2007).

Global Warming and International Stability

"In many cases, we are not prepared for the climate-related risks that we already face," said Vicente Barros, co-chair of the panel and professor emeritus of climatology at the University of Buenos Aires. "This is a wake-up call for the agriculture sector," said Craig Hanson, director of food, forests, and water programs at the World Resources Institute. "Climate change is a food-security issue. It's not just an environmental issue" (Porter 2014).

Accelerated photosynthesis stoked by elevated carbon dioxide levels may produce more plant mass, but this actually favors weeds over cereal crops because high temperatures impede the growth of food plants. At the same time, Earth's population is expected to rise past 9 billion by 2050; then climatic feedback loops will probably cause warming and all of its associated effects to accelerate sharply.

"A world with a more unstable food supply is likely to be a more volatile place. And those most exposed, of course, will be the world's poor," wrote Eduardo Porter in *The New York Times* (2014). Recent experience suggests that the productivity of farmland will not decline gradually as the world grows warmer. World food prices stopped their long secular decline around 2007 and have been on a roller coaster ever since. More volatile weather patterns promise to bring sharp disruptions to agricultural production that can cause spikes in food prices.

"There is a rigorous correlation between food price spikes and urban unrest," said Andrew Holland, who studies climate change at the American Security Project, a research group in Washington, D.C. "There was a food price spike in 2008, and you can see unrest spread throughout Africa. And there's a relatively clear line that leads from the food price spike in 2010 to unrest in the Middle East and the Arab Spring" (Porter 2014).

Ever since Thomas Malthus warned the world in the late 18th century that population growth would outstrip food supply and lead to widespread famine, agriculture has adapted to produce higher yields and bring marginal land into cultivation, averting global episodes of famine, even as population rose to levels that Malthus could not have imagined. Today, research is underway into heat-resistant crops and other measures, including more efficient use of food stocks in a world where one-quarter of production never reaches consumers.

Changes in demand and logistics could also help cope with scarcer food. Hanson pointed out that one-quarter of food produced in the world today is wasted by the poor storage and transport infrastructure in developing countries or wasteful consumers in the rich world. "For all the evidence of humankind's ability to adapt to its environmental constraints, it would be reckless to assume that ingenuity will arrive just in time to pull us from the brink," commented Porter (2014).

The summer of 2012 provided a glimpse of the climatic future in the farming areas of the U.S. Midwest. That year brought the hottest July on record (higher than the Dust Bowl years of the 1930s) and the worst drought in half a century. The intense heat wave killed large parts of the corn harvest as temperatures rose too high to allow pollination in many nights in July. That summer reminded observers of another climatic foretaste—Europe's exceptionally hot summer of 2003, which killed more than 30,000 people and also reduced harvests of some crops by 30 percent or more. Corn pollination suffers in the heat. So does pollination of common garden fare such as tomatoes, peppers, green beans, and squash. At temperatures above 85°F at night, pollen dies in flowers that would have matured into fruits and vegetables (Gaarder 2011).

"The negative impacts of global climate change on agriculture are only expected to get worse," said researchers at the London School of Economics and the Information Technology and Innovation Foundation (in Washington, D.C.). They are promoting the development of "more resilient crops and agricultural production systems than we currently possess in today's world." Concern arose in 2007 and 2008 as prices for major grains more than doubled after a string of weather disasters in important growing regions, provoking panicked buying and food riots in at least 30 countries (Gillis 2013, November 11). The situation was complicated by the diversion of corn to produce ethanol. All of this collided with rising demand, thus providing a window on the future as well.

By 2013, reports from the IPCC warned of the sensitivity of important crops to heat waves and reduced production as demand continues to rise. The agricultural risks "are greatest for tropical countries, given projected impacts that exceed adaptive capacity and higher poverty rates compared with temperate regions," the IPCC said (Gillis 2013, November 1). In addition, placing new land into production to meet demand for food could entail deforestation that would remove large numbers of trees that consume carbon dioxide.

Warming a Positive in North Dakota

Since the 1960s, parts of North Dakota have seen barley and corn, both cool-weather crops, replaced by corn and soybeans, crops that favor a warmer climate. From 1900 to 2010, Langdon, North Dakota's growing season has added 21 days, part of a trend across the area toward generally longer and hotter summers with more variations in rain and snowfall (both dry and wet, often alternating) and winters with less of that once-familiar Northern Plains Arctic bite.

"In the town of Rugby, N.D., 50 miles south of the Canadian border, climate change is written in the fields. North Dakota's average temperature has increased 2.7 degrees F during the last century. . . . Where once wheat was king, field after field is now full of feed corn. At the beginning of September, farmers are hustling to get combines out to cut the golden wheat but green fields of corn are everywhere—and still a month from harvest," wrote Elizabeth Weise in *USA Today* (2013). North Dakota is one area where climatic warming has been an advantage—so far. Aided by new varieties of seeds as well as a warmer, longer growing season, area harvests are of sizes that would have been unimaginable years ago, said Mike Ostlie, an agronomist at the Carrington Research Extension Center of North Dakota State University, some 150 miles northwest of Fargo. "Used to be, every three to five years there was a crop failure. Now I don't know when that last happened" (Weise 2013).

Other crops are migrating northward in the United States, too. New York state, once too cold for soybeans, harvested 320,000 acres of them in 2013. The Finger Lakes region has become wine grape country. However, California almonds (the state's single most valuable crop harvest), cherries, and apricots no longer receive optimal winter chill in their dormant periods to produce abundant flowers and fruit.

Further Reading

Weise, Elizabeth. "Some Crops Migrate North with Warmer Temperatures." *USA Today*, September 17, 2013. http://www.usatoday.com/story/news/nation/2013/09/17/climate-change-agriculture-crops/2784561/.

U.S. Agricultural Policy Adapts

In the meantime, U.S. federal agriculture policy encourages greenhouse gas emissions. "Instead of helping farmers take common-sense measures to limit their land's vulnerability to extreme weather," wrote Mark Hertsgaard in *The New York Times*, federal subsidies "spend billions more on crop insurance—sticking taxpayers with the bill. 'It's like giving a homeowner cut-rate fire insurance but not requiring fire extinguishers,' Jim Kleinschmit of the Institute for Agriculture and Trade Policy said" (Hertsgaard 2012). Hertsgaard wrote that even though U.S. farm policies and subsidies favor huge monocultures of single major crops (corn, soybeans, wheat, for example), "Farmers can best boost resilience to climate change, scientists say, by improving their soil's fertility and capacity to retain moisture. That means cutting back on chemical fertilizers, which kill many of the microorganisms that ventilate soil, and shifting to compost and manure fertilizers and crop rotations" (2014).

At the same time, new varieties of crops are being bred to cope with heat and extreme weather. As Justin Gillis reported in *The New York Times*:

> At the end of a dirt road in northeastern India, nestled between two streams, lies the remote village of Samhauta. Anand Kumar Singh, a farmer there, recently related a story that he could scarcely believe himself. Last June, he planted 10 acres of a new variety of rice. On Aug. 23, the area was struck by a severe flood that submerged his field for 10 days. In years past, such a flood would have destroyed his crop. But the new variety sprang back to life, yielding a robust harvest. "That was a miracle," Mr. Singh said. (Gillis 2011)

The new strain of rice took several years to develop, plus a $20 million grant from the Bill & Melinda Gates Foundation.

Food Production and Population

By 2011, the growth in farm output that was characteristic of the late 20th century had slowed and was no longer keeping up with rising populations and affluence. In fact, it is now failing to keep up with the demand for food driven by population increases and rising affluence in once-poor countries. Consumption of corn, wheat, rice, and soybeans was outstripping their production, exhausting surpluses and raising prices. Speculative buying also has driven some markets from time to time.

Price rises have been "destabilizing politics in scores of countries, from Mexico to Uzbekistan to Yemen. The Haitian government was ousted in 2008 amid food riots, and anger over high prices has played a role in the recent Arab uprisings" (Gillis 2011). "The success of agriculture has been astounding," said Cynthia Rosenzweig, a researcher at NASA who helped pioneer the study of climate change and agriculture. "But I think there's starting to be premonitions that it may not continue forever" (Gillis 2011).

Millions of people, mainly across large parts of Asia, have added meat and dairy products to their diets, increasing demand for grains. The conversion of U.S. corn crops to ethanol has added to demand as well, just as weather disasters—heat waves, droughts, floods—pinch production in many areas. Many of the disasters have signatures that can be tied to global warming such as the 2003 heat wave in Europe, which cut grain production as much as 30 percent. A long drought in Australia, broken partly in 2015 by flooding rains in some areas, also reduced wheat and rice production.

As Justin Gillis also wrote in *The New York Times*: "A broad body of research suggesting that extra carbon dioxide does act as plant fertilizer, but that the benefits are less than previously believed—and probably less than needed to avert food shortages." "One of the things that we're starting to believe is that the positives of CO_2 are unlikely to outweigh the negatives of the other factors," said Andrew D. B. Leakey, another of the Illinois researchers. "Other recent evidence suggests that longstanding assumptions about food production on a warming planet may have been too optimistic" (Gillis 2011).

Economists Michael J. Roberts of North Carolina State University and Wolfram Schlenker of Columbia University compared crop yields and temperature. They found that yields fall sharply (30 percent or more) at a threshold for each type of grain—for corn around 84°F, and for soybeans around 86°F—no matter what the atmospheric CO_2 level may be. Research by David B. Lobell of Stanford University suggests that rising temperatures may already be suppressing crop yields in Russia, France, China, and other countries. "I think there's been an under-recognition of just how sensitive crops are to heat, and how fast heat exposure is increasing," Lobell said (Gillis 2011).

Further Reading

Benjamin, Craig. "The Machu Picchu Model: Climate Change and Agricultural Diversity." *Native Americas* 16(3–4) (Summer–Fall 1999): 76–81.

Commoner, Barry. *Making Peace with the Planet.* New York: Pantheon, 1990.

Gaarder, Nancy. "Harvests Will Take Hit from Heat's Effect on Pollen." *Omaha World-Herald*, July 25, 2011, A1, A2.

Gillis, Justin. "A Warming Planet Struggles to Feed Itself." *The New York Times*, June 4, 2011. http://www.nytimes.com/2011/06/05/science/earth/05harvest.html.

Gillis, Justin. "Climate Change Seen Posing Risk to Food Supplies." *The New York Times*, November 1, 2013. http://www.nytimes.com/2013/11/02/science/earth/science-panel-warns-of-risks-to-food-supply-from-climate-change.html?_r=0.

Gillis, Justin. "A Jolt to Complacency on Food Supply." *The New York Times*, November 11, 2013. http://www.nytimes.com/2013/11/12/science/earth/warning-on-global-food-supply.html.

Hertsgaard, Mark. "Harvesting a Climate Disaster." *The New York Times*, September 12, 2012. http://www.nytimes.com/2012/09/13/opinion/the-farm-bill-should-help-the-planet -not-just-crops.html.

Porter, Eduardo. "Old Forecast of Famine May Yet Come True." *The New York Times*, April 1, 2014. https://www.nytimes.com/2014/04/02/business/energy-environment/a-200-year -old-forecast-for-food-scarcity-may-yet-come-true.html.

Rinne, Tim. "Local View: City Dwellers Can Grow Food in a Risky Climate." *Lincoln Journal Star* (Nebraska), November 30, 2014. http://journalstar.com/news/opinion/editorial /columnists/local-view-city-dwellers-can-grow-food-in-a-risky/article_b5198714-8021 -54c8-aa89-9cb09d42b186.html.

Schapiro, Mark. *Carbon Shock: A Tale of Risk and Calculus on the Front Lines of the Disrupted Global Economy; How Carbon Is Changing the Cost of Everything.* White River Junction, VT: Chelsea Green Publishing, 2014.

"Warming Climate Jeopardizes California Fruit and Nut Crops." Environment News Service, July 23, 2009. http://www.ens-newswire.com/ens/jul2009/2009-07-23-095.html.

See also: Agriculture, Future of; Carbon Dioxide Enrichment; Drought, United States; Drought, Worldwide; Extreme Weather; Floods; Growing Seasons, Europe and Asia; Rice Yields, Decline in; Temperatures, Global; Wine Grapes and Warming

GROWING SEASONS, EUROPE AND ASIA

A network of phenological observers monitors gardens stretching from the north of Norway southeastward to Macedonia, as well as from from Valentia Observatory's phenological garden in County Kerry, Ireland, eastward to Finland, using clones of a single sample of each species bred in Germany when the first gardens were established. Phenology, "the timing of seasonal activities of animals and plants" (Walther et al. 2002, 389), has been used to assemble a 40-year record that was analyzed by two German scientists, Annette Menzel and Peter Fabian, from the University of Munich. Observations gathered with this system indicate that spring events, including leaf unfolding, have advanced by some six days during the period, whereas autumn events such as leaf coloring now occur 4.8 days later on average. Therefore, the average growing season in Europe has lengthened by 10.8 days since the early 1960s (McWilliams 2001).

These changes can be attributed almost entirely to changes in average temperatures, especially in winter. Other variables—including soil composition, water supply, biological factors, and general surroundings—have as far as possible been kept identical. The results agree with similar studies carried out in the United States. That study concluded that during three decades the average first leaf date had moved progressively earlier each year, advancing from around May 4 in 1965 to April 23 by the early 1990s. According to one observer, "Both studies confirm what we already know—that average winter temperatures in both regions have been rising slightly, but significantly, throughout the period concerned. But more importantly, the results establish phenology as an important tool for the monitoring of global warming" (McWilliams 2001).

In 2001 and 2002, British phenologists gathered information at various locations in the British Isles, "showing that the arrival of spring is no longer constrained to its traditional March slot and the end of autumn now continues well beyond late October" (Hann 2002). Collectively, the phenologists' studies indicated that "higher-than-average temperatures from January to April [of 2002] led to almost every characteristic of this year's spring occurring up to three weeks earlier than in 2001. On average, insects such as bees and butterflies, were three weeks early, while plants flowered two weeks ahead and many birds, including the turtle dove, arrived a week earlier than usual" (Hann 2002).

According to these studies, other notable early spring signs included:

the call of the cuckoo—five days earlier;
hazel flowering—23 days earlier;
emergence of snowdrops—seven days earlier;
the first shimmering bluebell carpets—16 days earlier;
alder leafing—13 days earlier;
hawthorn leafing—17 days earlier (Hann 2002).

Changes in the autumn included:

beech leaf coloring—12 days later;
oak leaf coloring—nine days later;
beech leaf fall—12 days later;
oak leaf fall—5 days later;
and, "some people in milder parts of the United Kingdom were condemned
    to cutting their lawns all year-round" (Hann 2002).

Great Britain's Woodland Trust recruited 13,500 gardeners and wildlife watchers throughout the United Kingdom to monitor the first signs of spring early in 2004. A volunteer in Lochgilphead, Scotland, reported the first frogspawn on February 1, six weeks before its usual expected occurrence. The report also confirmed the earliest recorded sighting of a bumblebee in Scotland on February 15. Ten more sightings were recorded during the same month. Blackbirds were observed building nests on January 8 near Kircudbright, two or three months ahead of usual. In London, during the 1920s and 1930s, bumblebees were never sighted before the first week of February. In 2003, one was sighted in Isleworth, West London, on December 23. Another was noted on Christmas Eve in Devon, lured out by mild temperatures. During the second week of February 2004, however, many of these "early wakers" were killed by a hard freeze, snow, and ice (Derbyshire 2004).

As the phenologists concluded,

Interconnected and often complex relationships between trees, insects, and animals in woods are also affected by early spring arrivals. For example, results from this spring [2002] show that some synchronous relationships between birds, insects, and plants could become disturbed. Crucially, for many species, this could have serious implications for their future survival. As ancient woods become even more fragmented and isolated some characteristic plants and animals may face further threats as the climate changes. (Hann 2002)

Researchers at Boston University and the American Geophysical Union have described dramatic changes in the timing of both the appearance and fall of leaves (as recorded by weather satellites since the 1980s), according to data published in the September 16, 2000, issue of the *Journal of Geophysical Research*. The growing season is now almost 18 days longer in Eurasia, with spring arriving a week early and autumn delayed by 10 days. In North America, the growing season appears to be as much as 12 days longer (Dube 2001). Although a longer growing season may seem like cause for celebration, researcher Ranga Myneni, an associate professor in the department of geography at Boston University, argued the news is cause for concern. "What is good for plants is not necessarily good for the planet" (Dube 2001).

Wolfgang Lucht and colleagues documented a two-decade-long trend toward earlier spring bud burst and increased leaf cover in northern latitudes that was briefly interrupted only following the 1991 eruption of Mount Pinatubo in the Philippines in 1992 and 1993 (Lucht et al. 2002, 1687). "We conclude," they wrote, "that there has been a greening trend in the high northern latitudes, associated with the gradual lengthening of the growing season . . . and reduced snow-cover extent" (Lucht et al. 2002, 1688).

Further Reading

Derbyshire, David. "Baffled Bumble Bee Lured Out Early by Changing Climate." *Daily Telegraph* (London), March 12, 2004, 15.

Dube, Francine. "North America's Growing Season 12 Days Longer." *National Post* (Canada), September 5, 2001. http://www.nationalpost.com (no longer available).

Hann, Judith. "Spring Wakes Early, but Will Autumn Lie in Late Again? What Will Tomorrow's World Look Like?" United Kingdom Woodland Trust, 2002. http://www.woodland -trust.org.uk/news/subindex.asp?aid=328 (no longer available).

Lucht, Wolfgang, et al. "Climatic Control of the High-Latitude Vegetation Greening Trend and Pinatubo Effect." *Science* 296 (May 31, 2002): 1687–1689.

McWilliams, Brendan. "Study of Plants Confirms Global Warming." *Irish Times*, November 1, 2001, 26.

Walther, Gian-Reto, et al. "Ecological Responses to Recent Climate Change." *Nature* 416 (March 28, 2002): 389–395.

See also: Agriculture, Future of; Drought, United States; Drought, Worldwide; Extreme Weather; Floods; Temperatures, Global; Wine Grapes and Warming

LOW-CARBON DIET

Today food production in the United States requires an average of six calories of energy to produce one calorie of food. Much of this energy is consumed not only in transport but also in energy-intensive cultivation of factory farms, as well as by the preference of many people for energy-intensive types of food, such as meats (most notably beef) (Hillman and Fawcett 2007, 60).

Advocates of localized farming in the U.S. Midwest have pointed out that these areas once produced for home consumption. "More than 60 years ago," however, "even [people in] a renowned farm state like Nebraska gave up growing food for [their] own diets. . . . As much as anyone living in New York City and Los Angeles, we depend on the global food system to stock our pantries and dish up our meals."

> Like the rest of America, we're getting half of our produce—including 70 percent of the lettuce—from the California Central Valley (which is, incidentally, mired in a record-breaking 500-year drought). Even more telling, $4 billion of the $4.4 billion we Nebraskans annually spend on food is leaving the state. We're not buying food that's from here. Instead, we're blithely counting on some faceless, anonymous source to supply all our meals and snacks. (Rinne 2014)

Midwest advocates of localized agricultural economies now say, "Just as eating is a 'local' act (stuffing our mouths is about as local as you can get), we need a food supply that's locally based as well. As consumers, we need to be supporting our local farmers and ranchers and building the market for locally produced food" (Rinne 2014). Urban dwellers are also being advised to raise some of their own food.

Tim Rinne, state coordinator of Nebraskans for Peace, has called on everyone to become the most local of farmers and suggests joining with neighbors in Lincoln, Nebraska to start a community garden:

> We can grow the perishable items (particularly the lettuce greens) that are the hardest to keep on the grocery store shelf. And because they're grown and harvested right where we live, they're fresher and more nutritious. All over town, from the grass front lawn to the sidewalk space in the city right-of-way, there's room for beds of lettuce and spinach, carrots and peppers, onions, tomatoes and potatoes. And with the onset of climate change and the threat of food shortages, it's none too soon to be trying our hands at a little gardening and learning something about our food. (Rinne 2014)

By the end of 2009, labels listing CO_2 emissions incurred in the production of various foods were appearing in some Swedish grocery stores and on restaurant menus. Swedes are being advised to favor foods that consume less energy in production: carrots, for example, which are grown outside there, as opposed to hot-house tomatoes, and chicken over beef. They are advised to reduce fish consumption because European stocks have been depleted.

A large hamburger produces 1.7 kilograms of carbon dioxide emissions, whereas a chicken sandwich has 0.4 kilograms (Rosenthal 2009). Roughly 25 percent of emissions in industrialized nations result from the foods people eat. Sweden is aiming to eliminate the use of fossil fuels to produce electricity by 2020; gasoline-powered cars are supposed to be obsolete by 2030.

Far-Fetched Food

Food items sold in U.S. grocery stores travel an average of 1,500 miles to reach consumers, according to David Pimentel, professor of ecology and agricultural science at Cornell University. The food industry burns almost one-fifth of all the petroleum consumed in the United States—about as much as automobiles, according to Michael Pollan's *Omnivore's Dilemma* (Knoblauch 2007, 46). The average American eats more than 200 pounds of red meat, poultry, and fish per year, an increase of 23 pounds compared to 1970.

A walk through the average U.S. supermarket is a tour of the world, a cornucopia of food choices. Carbon-conscious consumers sometimes take pride in eating low on the food chain, substituting vegetables and fruits for meat. How carbon conscious, however, are vegetables that come from Chile, Nicaragua, or New Zealand out of season? In that case, most of their carbon calories come in the form of transportation, most of it powered by oil. The choices are beguiling. How long has it been since we could only indulge our taste for watermelon during a few summer months?

A tour of a typical U.S. supermarket also reveals that modern transport and various free-trade agreements have rendered distances nearly irrelevant to our food supply and the idea of "seasonal" produce nearly meaningless. During winter in the United States, many fruits and vegetables are now routinely imported from the Southern Hemisphere. Geography (and energy expenditure) has become so irrelevant in today's American grocery stores that some salmon is raised in the Pacific Northwest, shipped to China to be cut (where labor is cheap), and shipped back to North America for sale.

Latino and Asian immigration in the United States has given rise to all manner of imports (packaged goods, mainly) from these countries—especially Mexico southward, China, Korea, Vietnam, and India. A typical city of a half-million people (Omaha, Nebraska, in this case) includes at least a half-dozen East Asian grocery stores, with goods from China and nearby countries, and at least two specializing in goods from India and Bangladesh. Another sells groceries from the Philippines and Vietnam. Mainline grocery stores now also sell extensive offerings from East Asia and Latin America. Omaha has three Spanish-language newspapers that carry advertising from groceries. Many of these goods are shipped from thousands of miles away, usually by using oil-based fuels.

The energy intensity of food is a worldwide phenomenon. Cod caught off Norway is shipped to China, carved and turned into filets, and shipped back to Norway for sale. (The cost in fuel, not counting greenhouse gas emissions, is less than the difference in labor). Argentine lemons fill supermarket shelves in Spain as lemons grown nearby rot on the ground. Half of Europe's peas in 2008 were being grown in Kenya.

The World Meets at the Supermarket

Cruising the aisles of most supermarkets in the United States yields thousands of food choices from dozens of countries. In the following examples, the first figure shows miles and the second kilometers. Distances are in air miles from New York City. For surface miles via ocean transport add 10 percent. Some of these goods probably have been shipped by air, but most have reached the stores via surface, ships, trucks, and trains, which involve longer distances but are far more energy efficient.

Alaska—salmon (Anchorage: 3,371/5,425)
Belgium—banana nectar (3,600, 5840)
Chile—grapes and other winter fruits (Santiago: 5,107/8,219)
China—beer (Beijing: 6,843/11,012)
Costa Rica—pineapples and bananas (San Jose: 2,509/4,039)
Ecuador—cut flowers (Quito: 2,834/4,560)
Germany—beer (Frankfurt: 3,858/6,208)
Honduras—shrimp (Tegucigalpa: 2,004/3,225)
Hong Kong—jasmine rice (8,059/12,968)
Italy—premium pasta (Rome: 4,283/6,891)
Kenya—tilapia (fish), shrimp (Nairobi: 7,358/11,842)
Mexico—mangoes and papayas, watermelon (all seasons), cantaloupe, winter strawberries (Mexico City: 2,090/3,363)
Netherlands—flower seeds (Amsterdam: 3,652/5,877
Nicaragua—winter watermelon, shrimp (Managua: 2,310 miles/3,719 kilometers)
Peru—winter asparagus (Lima: 3,640/5,857)
Philippines—mangoes (Manila: 8,509/13,693)
Russia—caviar (Moscow: 4,668/7,511)
Spain—olives (Madrid: 3,591/5,779)
United Kingdom—Coleman's mustard (London: 3,463/5,772)
Venezuela—winter watermelon (Caracas: 2,149/3,459)

Within the United States, many fruits and vegetables are shipped nationwide from California (Los Angeles to New York City—2,451 miles or 3,944 kilometers), at what would be international distances in many other countries. In summer, however, produce is more localized. The area around Omaha produces some of the best watermelons on Earth, but for only two to three months (August through early October).

Some U.S. food chains that specialize in organic foods, such as Whole Foods Market, make an effort to purchase locally grown fruits and vegetables. Local grocery chains also purchase locally in season some of the time in areas with extensive agricultural bases (corn in Nebraska and Iowa, for example). Of course, as in Britain, United States businesses have many central-distribution systems that defy geography.

To shorten the grocery food chain, more urban residents are raising food on vacant lots. By 2007, 2,000 community gardens were operating in U.S. urban areas.

Omaha, Seattle, New York City, Boston, and many other urban U.S. locales also host seasonal farmers' markets. In Omaha, at least, some of the best-tasting produce in the world comes from small fruit and vegetables stands that pop up around the city every summer and fall

The Carbon Footprint of Meat

Animal agriculture is a major source of carbon dioxide. The production of animal protein requires some 10 times the fossil fuel input (producing 10 times the carbon dioxide) compared to the production of edible plants. The feeding, killing, and processing of meat is enormously energy intensive. The world meat industry produces 18 percent of the world's greenhouse gas emissions, more than transportation. Nitrous oxide in manure (its warming effect is 296 times greater than that of carbon dioxide) and methane from animal flatulence (23 times greater) mean that "a 16-oz. T-bone is like a Hummer on a plate" (Will 2007). Billions of chickens, cattle, turkeys, pigs, and cows being raised in factory farms produce methane from digestion and feces. In fact, raising animals for human consumption is the single largest source of methane emissions in the United States, and a methane molecule is 20 times more effective at retaining heart than a carbon dioxide molecule.

Many people are now going vegetarian as part of a personal climate-change strategy. A study by University of Chicago professors Gidon Eshel and Pamela Martinargues that a completely vegetarian diet (avoiding eggs and dairy products as well as meat, fish, and foul) can remove as much carbon dioxide and methane from the air (or more in some cases) than driving a hybrid car (compared to one with an internal-combustion engine). They calculated that the difference between a meat-centered diet and vegetarianism had the same effect on greenhouse gas production as switching from a sport utility vehicle to a standard sedan.

Some overtly "green" businesses are not. Consider Ben & Jerry's ice cream, invented by eco-talking Vermont hippies, a gallon of which "requires electricity-guzzling refrigeration from manufacture to table, as well as four gallons of milk from cows that, at the same time, produce eight gallons of manure and eight gallons' worth of methane flatulence. The cows eat grain and hay cultivated with tractor fuel, chemical fertilizers, herbicides and insecticides, transported by trains and trucks" (Will 2007).

The Green Restaurant Movement

Customers at American Flatbread, a pizzeria in the Virginia city of Ashburn, a Washington, D.C., suburb, drink beer that is brewed nearby. Leaves of their iced tea are grown and packaged on a local farm. Buyers acquire scallions, spinach, cheese, and corn within a few miles. On the restaurant's back wall, a hand-painted map of Loudoun County shows the farms and dairies that the restaurant uses to assemble its menu. Its wood-fueled oven is made of Virginia red clay (Lazo 2007). Several other restaurants in the same area have found that using local foods is tasty, popular, and profitable, suiting an area of mixed suburbs and farms that have been struggling to survive urban sprawl. Local farms supply local groceries (including burgeoning organic food markets) as well as restaurants with corn, melons, tomatoes, peppers, squash, eggplant, and cucumbers—and even lamb. American Flatbread's frozen pizzas also are sold at local grocery stores.

At the Habana Outpost in Brooklyn, a solar-panel awning shades the outdoor dining area while providing the energy for the first solar-powered restaurant in New York City. A bicycle crank provides the power that blends fruit smoothies. "I wanted to make being environmentally responsible fun and friendly, without being didactic," said the eatery's owner, Sean Meenan (Stukin 2007, 51). Cell phones and computers are solar powered. Toilets use rainwater, kitchen waste is composted, and utensils and drinking cups are made of cornstarch. Meenan, who also owns two other eco-eateries in New York City, also buys his organic food as close to home as possible.

National interest in green restaurants prompted the formation of a Green Restaurant Association (GRA) in the United States. By 2015, the 25-year-old group had certified several hundred eating establishments. Certification is earned by buying locally, avoiding Styrofoam, making recycling a priority, using environmentally friendly cleaners, avoiding plastic, using energy-efficient appliances, and collecting used frying oil for uses as biodiesel fuel. The GRA's publicity points out that the 1 million restaurants in the United States consume more electricity than any other type of retail business. The average restaurant turns out 50,000 pounds of garbage (most of it organic) and uses 300,000 gallons of water per year (Stukin 2007, 52).

Further Reading

Hillman, Mayer, and Tina Fawcett. *The Suicidal Planet: How to Prevent Global Climate Catastrophe.* New York: St. Martin's Press/Thomas Dunne Books, 2007.

Knoblauch, Jessica A. "Have It Your (the Sustainable) Way." *EJ* (*Environmental Journalism*), Spring 2007, 28–30, 46.

Lazo, Alejandro. "A Shorter Link between the Farm and Dinner Plate." *Washington Post,* July 29, 2007, A1. http://www.washingtonpost.com/wp-dyn/content/article/2007/07/28/AR2007072801255.html?wpisrc=newsletter.

Rinne, Tim. "City Dwellers Can Grow Food in a Risky Climate." *Lincoln Journal-Star* (Nebraska), November 30, 2014. http://journalstar.com/news/opinion/editorial/columnists/local-view-city-dwellers-can-grow-food-in-a-risky/article_b5198714-8021-54c8-aa89-9cb09d42b186.html.

Rosenthal, Elisabeth. "To Cut Global Warming, Swedes Study Their Plates." *The New York Times*, October 23. 2009. http://www.nytimes.com/2009/10/23/world/europe/23degrees.html.

Stukin, Stacie. "The Lean, Green Kitchen." *Vegetarian Times*, September 2007, 51–54.

Will, George F. "Fuzzy Climate Math." *Washington Post*, April 12, 2007, A27. http://www.washingtonpost.com/wp-dyn/content/article/2007/04/11/AR2007041102109_pf.html.

See also: Agriculture, Future of

RICE YIELDS, DECLINE IN

Even modest temperature increases that are anticipated by some climate models could be enough to reduce rice yields significantly over the next century. Researchers at the University of Florida tested several varieties of rice, growing them in chambers that simulated various temperature-change situations. They found that although the rice plants flourished no matter what the temperature, yields of grains declined precipitously as temperatures increased. The more modest temperature increases, the researchers say, could reduce rice yields by 20 percent to 40 percent by 2100, and the larger increases predicted by more dire forecasts could cut rice production essentially to zero (Fountain 2000).

In another study, researchers from China, the United States, and the Philippines analyzed weather data from the International Rice Research Institute Farm in Los Baños, Philippines (near Manila), from 1979 to 2003. They compared these records with rice yields at the same location from 1992 to 2003. They found that annual mean maximum and minimum temperature at the location increased by 0.35°C and 1.13°C, respectively, between 1979 and 2003. Grain yield declined by 10 percent for each 1°C rise in growing-season minimum temperatures during the dry cropping season (January through April). "This report," they wrote in the *Proceedings of the National Academy of Sciences*, "provides a direct evidence of decreased rice yields from increased nighttime temperature associated with global warming" (Peng et al. 2004, 9971). Globally, nighttime temperatures have increased faster than daytime readings because greenhouse gases tend to trap more heat radiating from the ground and thus reduce cooling (Fountain 2004).

Kenneth G. Cassman of the University of Nebraska, who participated in this study, said that researchers are working to determine the cause of the yield reduction, but they speculate that it is because hotter nights make the plants work harder just to maintain themselves, diverting energy from growth. "If you think about it, world records for the marathon occur at cooler temperatures because it takes much more energy to maintain yourself when running at high temperatures. A similar phenomenon occurs with plants," he said (Schmid 2004).

Tim L. Setter, a professor of soil, crop, and atmospheric science at Cornell University, did not take part in this study but commented that higher nighttime temperatures "could consume carbohydrates in a nonproductive way, and by reducing the reserves of carbohydrates, particularly at time of flowering and early grain filling, would decrease the number of kernels that would be set" (Schmid 2004).

Further Reading

Fountain, Henry. "Observatory: Threat to Rice Crops." *The New York Times*, December 12, 2000, F5.

Fountain, Henry. "Observatory: Rice and Warm Weather." *The New York Times*, June 29, 2004, F1.

Peng, Shaobing, et al. "Rice Yields Decline with Higher Night Temperature from Global Warming." *Proceedings of the National Academy of Sciences* 101(27) (July 6, 2004): 9971–9975.

Schmid, Randolph E. "Warming Climate Reduces Yield for Rice, One of World's Most Important Crops." Associated Press, June 28, 2004 (LEXIS).

See also: Greenhouse Gas Emissions

WINE GRAPES AND WARMING

The British invented sparkling wine in 1728, but the French and Italians capitalized on it with champagne, partly because summers in the United Kingdom are dark and damp and yield substandard grapes. With a warming climate, however, British bubbly is improving, so much so that in January 2010 a wine from the Nyetimber Estate won a title as world's best sparkling wine at a prestigious tasting contest in Italy. Warmer weather is producing better grapes in Britain, and they are improving winemaking techniques to make the most of them. The grape-growing season has extended itself by two weeks in many parts of Britain, allowing grapes to be brought in before rainy, cool October weather. English wines also have become less acidic and sweeter (Naik 2010, A1). Even so, France still produced 320 million bottles of champagne in 2009 compared to 1.4 million in Britain (Naik 2010, A18).

Problems for the U.S. Grape Industry

Significant warming could pose major problems for the United States wine grape industry, reducing the area suitable for growing premium grapes by 50 percent to 80 percent by the end of the 21st century, according to a study published in the *Proceedings of the National Academy of Sciences* (PNAS) (White et al. 2006). The article's authors anticipate that weather problems will intensify for grapes in such areas as California's Napa and Sonoma Valleys, including more extremely hot days, something Noah Diffenbaugh of the department of earth and atmospheric sciences at Purdue University agrees with (Schmid 2006).

Using a climate model from the Intergovernmental Panel on Climate Change, the authors of the PNAS study assert that "potential premium wine grape production area in the conterminous United States could decline by up to 81 percent by the end of the 21st century" (White et al. 2006, 11217). An increasing number of days with daily maximum temperatures above 95°F could eliminate wine grape production in many of today's growing areas. Grape production probably will be restricted by century's end to a narrow band along the West Coast and cooler areas of the Northeast and Northwest, many of which face problems related to excess moisture.

At temperatures above 95°F, many wine grapes cannot maintain photosynthesis, so sugars in the grapes break down, according to Purdue University's Diffenbaugh (Schmid 2006). "We have very long-term studies of how this biological system (of vineyards) responds to climate," said Diffenbaugh, and that gives the researchers confidence in their projection. Diffenbaugh is a coauthor of the paper (Schmid 2006). James A. Kennedy, a professor of food science and technology at

Oregon State University, said he was shocked by the report on the potential effects on wine grapes. "The lion's share of the industry is in California, so it's a huge concern from a wine quality standpoint," he said. For people in the industry, "this paper is going to be a bit of a shocker" (Schmid 2006).

French Grape Harvest Moves Up 40 Days in 50 Years

Grapes that were once harvested in the French Alsace district in mid-October were ripening in mid-September by 1998. In 2007, the harvest began on August 24, dramatic evidence that warming temperatures gave changed growing seasons. "I noticed the harvest was getting earlier before anybody had a name for it," said 59-year-old Ren Mur, the 11th generation of his family to produce wine from the clay and limestone slopes of the Vosges Mountains near the German border. "When I was young, we were harvesting in October with snow on the mountaintops. Today we're harvesting in August" (Moore 2007).

Grapes are extremely sensitive to temperature and other weather conditions, and detailed records have been kept on them. "The link of wine to global warming is unique because the quality of wine is very dependent on the climate," said Bernard Seguin, an authority on the impact of global warming and viniculture at the French National Agronomy Institute. "For me, it is the ultimate expression of the consequences of climate change" (Moore 2007).

Provence and other southern regions of France have become too dry and hot for grapes. Pests also move north with warmer temperatures, including the leafhopper, which is spreading yellow leaf disease in Alsatian vineyards for the first time. In some Mediterranean vineyards, weather has become so warm that grapes are harvested at night. Vineyards are being relocated to higher elevations. Some French growers are planning to move to England, where the winemaking industry is reemerging. In France, vintners no longer add sugar to the wines to improve their flavors and alcohol content. Warmer summers do that now naturally. Temperatures in Alsace have risen 3.5°F in 30 years.

British Columbia Wines

In the Western Hemisphere, by 2007 southern British Columbia was developing a wine grape industry for the first time near Tappen, some 200 miles northeast of Vancouver and adjacent to the Okanagan Valley, although the occasional early frost still caused problems at harvest time in the middle of October. The grapes grown there are increasingly showing up in some of the world's most expensive chardonnay and cabernet sauvignon wines. The ancient French names stem from regions where warming temperatures have begun to ruin temperature-sensitive harvests.

Before 1990, Canadian wine was a joke, along the lines of the Nebraska Navy, but with longer and warmer summers, wine grape production has become a serious business as farmers abandon dairy cows for grapes. Temperatures have risen 5°F over 60 years, and the growing season has extended itself almost two weeks. Winter temperatures have risen substantially, enough to enable survival of grape vines (Belkin 2007). By 2007, British Columbia hosted 136 wineries, and the province's wines were winning international competitions.

Further Reading

Belkin, Douglas. "Northern Vintage: Canada's Wines Rise with Mercury." *Wall Street Journal*, October 15, 2007, A1, A20.

Moore, Molly. "In Northern France, Warming Presses Fall Grape Harvest into Summertime." *Washington Post*, September 2, 2007, A1. http://www.washingtonpost.com/wp-dyn/content/article/2007/09/01/AR2007090101360_pf.html.

Naik, Gautam. "Warmer Climate Gives Cheer to Makers of British Bubbly." *Wall Street Journal*, May 11, 2010, A1, A18.

Schmid, Randolph E. "Climate Change Could Devastate U.S. Wineries." Associated Press, July 10, 2006 (LEXIS).

White, M. A., et al. "Extreme Heat Reduces and Shifts United States Premium Wine Production in the 21st Century." *Proceedings of the National Academy of Sciences* 103(30) (July 25, 2006): 11217–11222.

See also: Agriculture, Future of; Temperatures, Global

HUMAN HEALTH

OVERVIEW

Britain's foremost medical journal, *The Lancet*, has been projecting the anticipated toll of accelerating climate change on human health for many years. Its 2015 report updated its first such worldwide collaboration, which was compiled in 2009. The new report raised the projected toll dramatically, anticipating that more people than previously thought will suffer health effects from heat waves, floods, and droughts provoked by a warming climate. "Everything that was predicted in 2009 is already happening," said Nick Watts, a public health expert at the Institute for Global Health at University College London, leader of more than 40 scientists from China, Africa, and Europe who coauthored the report. "Now we need to take a further step forward. The science has substantially moved on" (Tavernise 2015; Watts et al. 2015).

According to the 2015 report in *The Lancet*, using anticipated population increases and trends in greenhouse gas generation, compared to the 1990s, the number of people exposed to extreme rainfall events by the end of the 21st century will more than quadruple and affect about 2 billion people a year. At the same time, the number exposed to debilitating drought will increase threefold, and the exposure of elderly people—the most vulnerable group to health-compromising heat waves—will increase by a factor of 12 (affecting 3 billion people a year in all age groups), according to Peter Cox, one of the authors, who is a professor of climate-system dynamics at the University of Exeter in Britain (Tavernise 2015).

"We are saying, let's look at climate change from the perspective of what people are going to experience, rather than as averages across the globe," Cox said. "We have to move away from thinking of this as a problem in atmospheric physics. It is a problem for people" (Tavernise 2015). Even the 2009 report called climate change "the biggest global health threat of the 21st century." Six years later, without any binding international agreement to control emissions, *The Lancet* said that worldwide carbon emissions and their anticipated impact on human health "have risen above the worst-case scenarios used in 2009 [as] projections of mortality and illness, and other effects, like famine, have worsened," a situation that should "persuade people that the topic is urgent, not simply a distant matter for scientists" (Tavernise 2015).

"Coal Is by Far the Worst"

Burning fossil fuels creates health problems that surpass warming the atmosphere. Black soot from coal combustion aggravates lung diseases. Nitrogen oxides are also released by burning fossil fuels and contribute to the creation of ground-level ozone, haze, and smog that intensifies allergies, asthma, chronic bronchitis, and chronic obstructive pulmonary disease. Asthma afflicts 300 million people worldwide, and many of its sufferers are children. Asthma rates worldwide doubled between 1980 and 1995 before roughly stabilizing (Epstein and Ferber 2011, 88).

The combustion of coal in particular provokes many health problems above and beyond those delivered by rising temperatures because it contributes more carbon dioxide (as well as atmospheric pollution) per unit of energy generated than oil or natural gas. These include "particulates, nitrates, sulfates, and mercury, which pollute soils and water, including the aquatic food chain, where they concentrate in fish. . . . The resulting air pollution results in heart attacks and lung cancer" (Epstein and Ferber 2011, 233). Alan Lockwood, a senior scientist with Physicians for Social Responsibility, estimated in 2015 that coal-fired power plants were contributing to 50,000 deaths a year in the United States. Combustion of coal, he says, contributes to many of the country's most prevalent causes of death: cancer, heart disease, stroke, and chronic respiratory diseases. "Though other energy sources pose health risks, in terms of health effects, coal is by far the worst," Lockwood said (Szabo 2015).

Rising Temperatures and Disease Vectors

Rising temperatures also provide hospitable environments for insects, most notably mosquitoes, which spread several crippling and sometimes fatal diseases. When climate-change scientists and diplomats met in Buenos Aires in 1998, they were greeted by news that mosquitoes carrying dengue fever had invaded more than a third of the homes in Argentina's most populous province with its 14 million people. The *Aedes aegypti* mosquito first appeared in Argentina in 1986 but within 12 years was found in 36 percent of homes in Buenos Aires province (Webb 1998). Periodic outbreaks of dengue recurred in northern Argentina in 2009 and 2015–2016, the last also accompanied by Zika, a viral disease carried by the same mosquitos (Gilbert 2016). The worldwide incidence of dengue fever increased by a factor of 10 between 1975 and 2015, with rising minimum temperatures an important factor ("Human Health" 2016).

Ancient Influenza Strains in Old Ice

Ancient strains of influenza that have not afflicted humans for many centuries could be released by melting ice and migratory birds, according to Professor Scott Rogers, chair of biology at Bowling Green State University. "We've found viral RNA in the ice in Siberia, and it's along the major flight paths of migrating waterfowl." These same pathways take the birds to North America, Asia, and Australia and interconnect with other migratory paths to Europe and Africa, Rogers said. The virus that Rogers and his collaborators found resembles a strain that emerged between 1933 and 1938 and again in the 1960s ("Melting Ice" 2006). The research was published in the December 2006 edition of *Journal of Virology*.

These viruses may be preserved in ice and released after human immunity has lapsed. Survivors of the worldwide flu pandemic of 1918 had immunity to the H1N1 strain, for example, but their descendants did not. A recurrence "could take hold as an epidemic," said Rogers, who explained that no one yet knows whether long-frozen viruses retain their ability to infect people. But "we think they can survive a long time" in ice, he said, noting that tomato mosaic virus has been found in ice 140,000 years old in Greenland ("Melting Ice" 2006).

Further Reading

"Melting Ice May Release Frozen Influenza Viruses." Environment News Service, November 27, 2006. http://www.ens-newswire.com/ens/nov2006/2006-11-27-09 .asp#anchor3 (no longer available).

In 1995, an explosion of termites, mosquitoes, and cockroaches hit New Orleans, following an unprecedented five years without a killing frost. Mild winters with a lack of freezing conditions allow many disease-carrying insects to expand their ranges. John T. Houghton, author of *Global Warming: The Complete Briefing* (1997), believes that global warming will accelerate the spread of many diseases from the tropics to the middle latitudes. Malaria could increase from its current level, Houghton warned. "Other diseases that are likely to spread for the same reason are yellow fever, dengue fever, and viral encephalitis," he wrote (Houghton 1997, 132). After 1980, "[malaria], dengue, yellow fever, cholera, and a number of rodent-borne viruses are also appearing with increased frequency," Paul R. Epstein of Harvard University reported (Epstein 1998). In 1995, mortality from infectious diseases attributed to causes other than HIV and AIDS rose 22 percent above the levels of 15 years earlier in the United States. Adding deaths complicated by HIV and AIDS, deaths from infectious diseases had risen 58 percent in 15 years (Epstein 1998). Several allergies, including hay fever, also have been spreading and becoming more intense as temperatures rise. Aside from insect-vector diseases, warmer and more humid weather may aggravate urban air pollution (especially tropospheric ozone), which is a factor in many human maladies, including asthma. Pim Martens points to several studies that indicate that air pollution's effects on human health increase with temperature: "Simultaneous exposure to heat and pollution appears too be more harmful than the sum of the individual effects," one observer wrote (Martens 1999, 535). Allergenic pollens and spores are more readily dispersed during hot, dry summers.

Asthma is aggravated by heat—which increases the pollen production of many plants—as well as by air pollution. According to the American Lung Association, more than 5,600 people died of asthma in the United States during 1995, a

45.3-percent increase in mortality over 10 years and a 75-percent increase since 1980. Roughly one-third of those cases were children under age 18. Since 1980, children under age five have experienced a 160-percent increase in asthma.

Dengue fever is one of several mosquito-vector diseases that have been increasing their coverage in many areas of Earth as global temperatures have warmed, climbing higher altitudes in the tropics and increasing their latitudes in temperate zones. Rising temperatures and humidity increase the range of many illnesses spread by insects, including mosquitoes, warm-weather insects that die at temperatures below 50 to 61°F, depending on species. By 2014, the number of mosquitoes bearing dengue had increased 30-fold in 50 years, mainly in Asia ("Health Risks" 2015, 97).

The Spread of Malaria

According to the Intergovernmental Panel on Climate Change's projections for human health, a rise in global average temperatures of 3 to 5°C by 2100 could lead to 50 million to 80 million additional annual cases of malaria worldwide, primarily in tropical, subtropical, and poorly protected temperate-zone populations. Italy experienced a brief outbreak of malaria in 1997. Researchers with the Hadley Centre for Climate Prediction and Research expect the same disease to reach the Baltic states by 2050. In parts of the world where malaria is now unknown, most people have no immunity. The World Health Organization projects that warmer weather will cause tens of millions additional cases of malaria and other infectious diseases. The Dutch health ministry anticipates that more than a million people may die annually as a result of the impact of global warming on malaria transmission in North America and Northern Europe (Epstein 1999, 7).

Countering the majority view that a warmer world will spread malaria, David J. Rogers and Sarah E. Randolph (2000), using their own models, wrote in *Science* that even extreme rises in temperature will not spread the disease. They argue that the spread of malaria is too poorly understood to base a forecast several decades into the future on one variable—temperature. For example, the Dengie Marshes of Essex, England, a breeding ground for malaria-carrying mosquitoes in the 17th century, have dried up, making an increase in temperatures not a factor vis à vis malaria's transmission.

Malaria is not a new disease in the temperate zones. It was common throughout the time of the Roman Empire and later. A British invasion of the Netherlands in 1806 failed to drive out French troops because large numbers of the Britons became ill with malaria. Malaria was a public health problem in most of the eastern United States during warm, humid summers before medications were developed for it late in the 19th century. Even so, the resurgence of malaria (as well as other diseases) in a warming world is a matter of utmost concern to everyone, as well as scientists.

Further Reading

Epstein, Paul R. *Climate, Ecology, and Human Health.* December 18, 1998. http:// www.iitap .iastate.edu/gccourse/issues/health/health.html (no longer available).

Epstein, Paul R. "Profound Consequences: Climate Disruption, Contagious Disease, and Public Health." *Native Americas* 16(3–4) (Fall–Winter 1999): 64–67.

Epstein, Paul R., and Dan Ferber. *Changing Planet, Changing Health: How the Climate Crisis Threatens Our Health and What We Can Do about It.* Berkeley: University of California Press, 2011.

Gilbert, Jonathan. "Argentina Struggles to Control a Major Outbreak of Dengue." *The New York Times*, February 18, 2016, A10.

"Health Risks." *National Geographic,* November 2015, 96–97.

Houghton, John. *Global Warming: The Complete Briefing.* Cambridge, UK: Cambridge University Press, 1997.

"Human Health." *National Geographic,* January 2016, n.p. https://www.scribd.com/doc /293386356/National-Geographic-USA-January-2016.

Martens, Pim. "How Will Climate Change Affect Human Health?" *American Scientist* 87(6) (November–December 1999): 534–541.

Rogers, David J., and Sarah E. Randolph. "The Global Spread of Malaria in a Future, Warmer World." *Science* 289 (September 8, 2000): 1763–1766.

Szabo, Liz. "Health Groups Say Obama Plan on Coal Will Save Lives." *USA Today*, August 4, 2015, 2A.

Tavernise, Sabrina. "Risk of Extreme Weather from Climate Change to Rise over Next Century, Report Says." *The New York Times*, June 22, 2015. http://www.nytimes.com/ 2015/06/23/science/risk-of-extreme-weather-from-climate-change-to-rise-over-next- century-report-says.html.

Watts, Nick, et al. "Health and Climate Change: Policy Responses to Protect Public Health." *The Lancet*, June 23, 2015. http://www.thelancet.com/commissions/climate-change.

Webb, Jason. "Mosquito Invasion as Argentina Warms." Reuters, 1998. http://ces.iisc.ernet .in/hpg/envis/doc98html/enarg117.html.

ALLERGIES

Scientific evidence has been accumulating that warming temperatures "turbocharge" the growth of plants that release allergy-causing pollens, that also provoke asthma, which has been on the rise in many countries. A warmer world finds more people itching their eyes and sneezing as rising temperatures enhance the growth of weeds that produce pollen and mold spores. Conventional allergy medications that used to produce relief now offer little help. Hay fever, asthma, and allergic reactions have been growing as well. The season when allergies are prevalent also has become longer as winters shorten. "Allergic disorders represent an important group of chronic diseases in the United States, with estimated costs at approximately $21 billion per year," wrote U.S. Department of Agriculture weed ecologist Lewis Ziska who, with colleagues, researches the link between rising carbon dioxide levels in the atmosphere and rising allergy rates (2011, 4248).

> Aeroallergen exposure is associated with two principal allergic diseases: allergic rhinitis (hay fever) and asthma. For much of geographic North America, there are three distinct plant-based aeroallergen seasons; tree pollen in the spring; grass pollen in the early summer, and weed pollen, including ragweed

(*Ambrosia* spp.) in the summer and fall. Pollen from the genus *Ambrosia*, which includes *A. artemisiifolia* (short or common ragweed), *A. trifida* (giant ragweed), *A. psilostachya* (western ragweed), and *A. bidentata* (lanceleaf ragweed) has long been acknowledged to be a significant cause of allergic disease. An extensive skin test survey demonstrated that at least 10 percent of the U.S. population is ragweed sensitive; the prevalence of ragweed sensitivity among atopic individuals was 27 percent in two large case series. It has been reported that *Ambrosia* may cause more seasonal allergic rhinitis than all other plants combined.

Allergies and asthma are closely linked. More than 70 percent of people with asthma also have allergies (Naik 2007, A1). In its 2007 report, the Intergovernmental Panel on Climate Change says that spring, the most intense season for allergies, now arrives 10 to 15 days earlier than in 1975 (Naik 2007, A1). By 2007, 35 million people in the United States suffered from allergies and 20 million from asthma; the percentage of people in the United States with asthma has doubled since 1980 (Naik 2007, A13).

Rising humidity and more rainfall in some areas also enhances the growth of ragweed and other prolific pollen producers that have already enjoyed enhanced growth from warming temperatures. "We've definitely seen a big increase in patients," said allergist Brian Rotskoff, who runs the Clarity Allergy Center in Chicago, a hot spot for ragweed. He has been seeing children at younger ages than he used to, and many adults' symptoms have become worse. Ragweed releases its sperm into the air, to which people react by producing histamine. "The severity is affecting their quality of life," he said, to the point that the allergy, which produces coughing, shortness of breath, itchy eyes, a runny nose, and sneezing impedes sleep and ability to focus while awake (Koch 2013). Sometimes ragweed allergy develops into respiratory problems as well.

Humidity and Heat Boost Ragweed

For the 10 percent of Americans who suffer from an allergy to ragweed, the fall season has become more tortuous. For the first time, a study has documented that the season for ragweed, the most debilitating of plant allergens, is lengthening in North America. Ten sites providing data for a study of ragweed conditions between 1995 and 2009 indicated that the fall ragweed season has lengthened by 11 days.

"We were surprised . . . by how much change had been seen over such a short period," said Kim Knowlton, one of the lead authors and a health scientist at Columbia University and senior scientist at the Natural Resources Defense Council (Gaarder 2011).

Starting in 2001, plant physiologist Lewis Ziska with the U.S. Department of Agriculture's Crop Systems and Global Change Laboratory in Beltsville, Maryland, planted ragweed in the autumn across urban, suburban, and rural settings throughout the Baltimore area. He used the same seeds and same soil and watered the plants the same way. Soon, the urban plants were growing much faster than the rural crop and producing five times as much pollen (Naik 2007, A1).

Thomas Casale, chief of allergy and immunology at Omaha's Creighton University, both said he has also seen the season extend to later in the fall. "Our falls are warmer than they used to be, our frosts have been delayed. If you think about it, you shouldn't be surprised, because we've lived it," said one report coauthor (Gaarder 2011). Change is more noticeable traveling northward. In Oklahoma City, the allergy season was one day longer in 2009 than it was in 1995. The allergy season grew by an average of 16 days in Minneapolis and by 27 days during the same period in Saskatoon, Canada.

Urban ragweed's pollen was more potent as well, all fed by warmer temperatures and 20 percent more carbon dioxide, mainly from auto traffic. Ziska himself has allergies and asthma. He found that ragweed loves warmth. The urban plants in the previously noted experiment around Baltimore emerged three to four days earlier at the urban site than at the rural one 40 miles away; the urban plants were almost twice as large at the urban site as the rural location (Naik 2007, A13). The allergen that causes reactions in humans is also more potent in the city-grown plants than the rural ones. Ziska has found that temperatures between the two sites can vary by 5 to 10°F, and when the carbon dioxide level is less than 400 parts per million (ppm) at the rural site it can be more than 470 ppm in the city.

Ziska's data indicate a robust relationship between carbon dioxide levels and prevalence of pollen. Pollen production doubled as the CO_2 level rose from around 280 parts per million in 1970 to 370 ppm in 2000. By 2015, it was above 400 ppm. Leonard Bielory, an allergy and immunology expert at Rutgers University's Center for Environmental Prediction, has detected the same change. "There's clear evidence that pollen season is lengthening and total pollen is increasing," says George Luber, associate director for climate change at the Centers for Disease Control and Prevention (CDC) in Atlanta. "It's one of the ways climate change is already affecting your community" (Koch 2013).

The CDC has used allergy skin tests to detect increased human sensitivity to ragweed, as well as ryegrass, oak, Bermuda grass, and mold pollens across the United States. Federal data reported a 17-percent increase in U.S. asthma prevalence from 2001 through mid-2012 (Koch 2013). Asthma is now being called an epidemic affecting one in every 12 people. Rutgers' Bielory is also investigating whether food allergies are related to atmospheric carbon dioxide levels. "There are so many pieces of the puzzle, and it's not a flat puzzle. It's a 3-D puzzle," he said. Climate change is one factor. Others may be toxic chemicals and a compulsion to be too clean. The "hygiene hypothesis" holds that people who are not exposed to germs and allergens early in life have underdeveloped immune systems that lack the ability to respond. Simply put, the immune system needs work to do.

In only 15 years, the pollen season of one common allergen has lengthened by as much as 27 days in some parts of North America, according to research by a team led by Lewis Ziska at the U.S. Department of Agriculture in Beltsville, Maryland, which compared readings of ragweed pollen since 1995 at 10 stations across North America with changes in temperature and first frost. They found a clear link between recent warming and the length of the pollen season. What's more, the farther north they looked, the greater the extension to the season—so allergy-prone Canadians should consider buying tissues in bulk.

Ragweed was nearly unknown in Europe until the 1990s. It has now spread through Eastern Europe, as well as France, Italy, the Netherlands, and parts of the Nordic countries. Harvard's Paul Epstein, associate director of the Center for Health and the Global Environment, says that doubling the carbon dioxide level in a test plot of ragweed increases its pollen production by 61 percent (Naik 2007, A13).

Shorter Winters and Hay Fever

Warmer temperatures and shorter winters have been related to rising incidence of hay fever. The epidemic seems to be worldwide. Hay fever, otherwise known as *seasonal allergic rhinitis*, is caused by an allergy to pollens and fungal spores. Writing in the *The Guardian* (London), summarized the world hay-fever situation:

> They are sneezing in Stockholm, throats are itchy in India, and Irish eyes are streaming. From Algeria to Iceland and Hong Kong to Aberdeen, record numbers of people are suffering the misery of hay fever—and it's getting worse. Everything from global warming to air pollution is conspiring to make hay fever the number-one global irritant. The figures are truly remarkable. The number of people rubbing their eyes in British doctor's surgeries [offices] has risen fivefold since the 1950s, and about a quarter of people in the United Kingdom are now believed to be hay-fever sufferers. (Adam 2003)

Similar trends have been observed across Europe, Adam wrote. Cases of hay fever doubled or even trebled in Sweden and Finland during the 1970s and 1980s, and Swiss surveys show approximately one in 10 are affected, an increase from fewer than one in 100 when a similar count was made in 1926. Some 40 percent of the people in Australia and the United States said they suffered from hay fever or similar allergies by 2002. Areas such as West Africa, where hay fever was once all but unknown, have also been reporting it (Adam 2003).

English scientists (as well as street-level observers) have maintained that the allergy season is arriving earlier because of global warming. Many trees and grasses are flowering sooner and for extended periods, creating more of the pollen that is the main trigger of hay fever. In 2002, the hay fever season began as early as January 30, according to research by Tim Sparks of the Centre for Ecology and Hydrology. The same research also indicated that varieties of common grass were flowering as many as 13 days earlier than in 2001. Hazel and birch trees also were causing allergic reactions in many hay fever sufferers (Chapman 2003).

These findings were part of the world's largest phenological survey by the center and the Woodland Trust. By early February 2003, said Sparks, hazel trees were in flower. According to Jean Emberlin, of the British National Pollen Research Unit, "Last year, the grass pollen season was exceptionally long because the weather was wet and warm. The season extended into August instead of ending in July" (Chapman 2003). Young people appear to be worst hit, with 36 percent suffering from it. The figure in the wider population is 15 percent to 25 percent, with rates doubling since 1965 (Chapman 2003).

In the past, London hay-fever sufferers usually had not stocked up on remedies in earnest until the beginning of May. Pharmacies in 2003 reported a 30-percent increase in sales each week between mid-March and mid-April. Muriel Simmonds, chief executive of the medical charity Allergy U.K., cited "definite evidence" that hay fever was returning to the British Isles earlier year over year (Galloway and Rhodes 2003). Simmonds' organization has moved National Allergy Week from June to May because of the earlier onset of hay fever. "This is the earliest I've ever seen it in London," she said (Galloway and Rhodes 2003).

Asthma and Wildfires

A Harvard study released in 2004 said that the asthma among preschool children rose 160 percent between 1980 and 1994, double the rate in the overall population. The study linked the rise in asthma to increases "an epidemic in inner-city youth" to increases in pollen and fungal growth, both spurred by a warmer and often wetter climate (Naik 2007, A13). Dust mites, mold, and obesity also may play a role. A lack of infectious agents in childhood (which stimulate the immune system) also may be a factor. "I do think that climate change contributes," said Christopher Randolph, an allergist and clinical professor at Yale University, "but it's not the only answer" (Naik 2007, A13).

In California's Central Valley, drought, heat, and sooty air from wildfires combined by 2015 to produce a hell on Earth, as hospitals were inundated with wheezing asthma patients. After five years of record heat and drought, the state's largest fire scorched more than 160 square miles of hills near Fresno in the San Joaquin Valley, turning the air a murky brown as temperatures reached 106°F, sealing bad air into the valley even after the fire. Thus, California, even with the toughest air regulations in the United States, could not control the effects of geography and climate. Already, more than 20 percent of the children in the Central Valley have asthma, according to the American Lung Association, and soot from wildfires only makes the situation worse (Lovett and Medina 2015).

"At local asthma clinics," reported *The New York Times*, "doctors have had an explosion in business since the drought began. Dr. Malik N. Baz said his asthma and allergy practice had grown at least 20 percent each of the past three years, and he has opened five new clinics around the Valley in the past two years—all of them overflowing with patients" (Lovett and Medina 2015).

Dr. Vipul Jain, a pulmonologist who runs a chronic lung disease program in Fresno, told *The New York Times:* "It's kind of a worst-nightmare situation," said Dr. Jain, an associate professor at the University of California–San Francisco. "And the worst is still yet to come. We can see it, but anyone with a lung problem, they feel it, and there is no way to prevent an exacerbation of their problems. People have to go outside and work, and they will suffer" (Lovett and Medina 2015).

Further Reading

Adam, David. "Hatchoooooh! Record Numbers of People Are Complaining of Hay Fever." *The Guardian* (London), June 18, 2003, 4.

Chapman, James. "Early Spring Misery for 12 Million Hay Fever Sufferers." *The Daily Mail* (London), February 4, 2003, 23.

Gaarder, Nancy. "Come Autumn, Ragweed Will Stick around Longer." *Omaha World-Herald,* March 6, 2011. http://www.livewellnebraska.com/apps/pbcs.dll/article?AID=/20110306/ LIVEWELL01/703069909/1161 (no longer available).

Galloway, Elaine, and Chloe Rhodes. "Warm Spell Brings Early Start to Hay-Fever Misery." *Evening Standard* (London), April 14, 2003, 16.

Koch, Wendy. "Climate Change Linked to More Pollen, Allergies, Asthma." *USA Today,* May 31, 2013. http://www.usatoday.com/story/news/nation/2013/05/30/climate-change -allergies-asthma/2163893/.

Lovett, Ian, and Jennifer Medina. "Fires in West Have Residents Gasping on the Soot Left Behind." *The New York Times*, September 9, 2015. http://www.nytimes.com/2015/09/10/ us/fires-in-west-leave-residents-gasping-on-the-soot-left-behind.html.

Naik, Gautam. "Global Warming May Be Spurring Allergy, Asthma." *Wall Street Journal*, May 3, 2007, A1, A13. https://www.wsj.com/articles/SB117815689129890415.

Ziska, Lewis, et al. "Recent Warming by Latitude Associated with Increased Length of Ragweed Pollen Season in Central North America." *Proceedings of the National Academy of Sciences* 108(10) (February 22, 2011): 4248–4251.

See also: Australia, Heat and Drought in; Drought, United States; Drought, Worldwide; Greenhouse Gases, Effects of

GREENHOUSE GASES, EFFECTS OF

In Bangladesh, dengue fever appeared in 2000 and redeveloped each year since as the *Aedes aegypti* mosquito, which carries the disease, breeds more frequently as warmer temperatures decrease the size of its larvae, increasing incidence of the disease. The incubation period of dengue type 2 virus lasts 12 days at 30°C but seven days at 32 to 35°C (McKibben 2010, 73). Thus, a few degrees of temperature rise can increase transmission of dengue by 300 percent (McKibben 2010, 73). Dengue has no cure, and the mosquitoes feed during daylight hours, making bed netting useless.

Mosquitoes, Termites, Cockroaches

Dengue fever is only one of several mosquito-vector diseases that have been increasing their coverage in many areas of Earth as global temperatures have warmed by increasing its altitude in the tropics and escalating its latitude in temperate zones. Rising temperatures and humidity increase the range of many illnesses spread by insects, including mosquitoes, which are warm-weather insects that die at temperatures below 50 to 61°F, depending on the species. A common disease in tropical regions, dengue is a prolonged, flulike viral infection that can cause internal bleeding, fever, and sometimes death. Also sometimes called *breakbone fever*, dengue may be accompanied by headache, rash, and severe joint pain. The World Health Organization lists dengue fever as the tenth deadliest disease worldwide. It is increasing faster than any other disease. Brazil, for example, reported nearly 1.6 million cases of dengue in 2015, compared to 569,000 the previous year. At least 839 people died of dengue there in 2015, an 80-percent increase in one year (Romero 2015).

Temperature, Humidity, Air Pollution, and Human Maladies

During 1995, an explosion of termites, mosquitoes, and cockroaches hit New Orleans after an unprecedented five years without a killing frost. "Termites are everywhere. The city is totally, completely, inundated with them," said Ed Bordees, a New Orleans health official, who added, "The number of mosquitoes laying eggs has increased tenfold" (Gelbspan 1997, 15). The situation in New Orleans was aggravated not only by unusual warmth but also by above-average rainfall totaling 80 inches the previous year. Some of the 200-year-old oaks along New Orleans's St. Charles Avenue were found to have been eaten alive from the inside by billions of tiny, blind, Formosan termites. The same year, dengue fever spread from Mexico across the border into Texas for the first time since records have been kept. Dengue fever, like malaria, is carried by a mosquito with a range that is defined by temperature. At the same time, Colombia was enduring swarms of mosquitoes and outbreaks of the diseases they carry, including dengue fever and encephalitis, triggered by a record heat wave followed by heavy rains.

Mild winters with a lack of freezing conditions allow many disease-carrying insects to expand their ranges. "Indeed," commented Paul Epstein of Harvard Medical School's Center for Health and the Global Environment, "Fossil records indicate that when changes in climate occur, insects shift their range far more rapidly than do grasses, shrubs, and forests, and move to more favorable latitudes and elevations hundreds of years before larger animals do. 'Beetles,' concluded one climatologist, 'are better paleo-thermometers than bears'" (Epstein 1998).

Epstein identified three tendencies in global climate change, and related each to an increasingly virile environment for infectious diseases. The three indicators are:

1. increased air temperatures at altitudes of two to four miles above the surface in the Southern Hemisphere;
2. a disproportionate rise in minimum temperatures in either daily or seasonally averaged readings; and
3. an increase in extreme weather events such as droughts and sudden heavy rains (Epstein 1998).

A Sierra Club study indicated that frequent and lengthy El Niño episodes provide an indication of the sensitivity of some diseases to changes in climate. This study cited evidence that warming waters in the Pacific Ocean contributed to a severe outbreak of cholera that led to thousands of deaths in Latin American countries during the 1990s. According to health experts quoted by the Sierra Club study, "The current outbreak [of dengue fever], with its proximity to Texas, is at least a reminder of the risks that a warming climate might pose" (Sierra Club 1999). The study concluded, "While it is difficult to prove that any particular outbreak was caused or exacerbated by global warming, such incidents provide a hint of what might occur as global warming escalates" (Sierra Club 1999).

Aside from insect-vector diseases, warmer, more humid weather may aggravate urban air pollution (especially tropospheric ozone), which is a factor in many human maladies such as asthma. Pim Martens of the Netherlands' Center for Integrative

Studies (and a senior scientist at the University of Maastricht) pointed to several studies that indicate that air pollution's effects on human health rise as temperatures increase: "Simultaneous exposure to heat and pollution appears too be more harmful than the sum of the individual effects" (Martens 1999, 535). Allergenic pollens and spores are more readily dispersed during hot, dry summers, Martens said.

Jonathan Patz, associate professor of environmental studies and population health sciences at the University of Wisconsin at Madison and a lead author of the 2007 Intergovernmental Panel on Climate Change assessment of North America, said currently projected warming alone will probably mean that by 2050 the U.S. Northeast will experience 68 percent more "red ozone alert" days, which indicates the air is unhealthy to breathe (Eilperin 2007).

According to Dr. Joel Schwartz, an epidemiologist at Harvard University, present-day air-pollution concentrations are responsible for 70,000 early deaths per year and more than 100,000 excess hospitalizations for heart and lung disease in the United States. This could increase 10 percent to 20 percent in the United States as a result of global warming, with significantly greater increases in countries that are more polluted to begin with, according to Schwartz (Sierra Club 1999).

In addition to aggravation of specific diseases, persistent heat and humidity has a general debilitating influence on most warm-blooded animals, including human beings. Farm animals are especially adversely affected when the air temperature remains higher than usual throughout the night. During the latter half of the 20th century, according to Epstein, nighttime minimum temperatures over land areas have risen at a rate of 1.86°C per 100 years, and maximum temperatures have risen at 0.88°C during the same period (Epstein 1998).

A warming world with alternating droughts and deluges is increasingly bringing tropical weather to larger parts of Texas and other parts of the U.S. South, where infrequent freezes no longer kill disease-carrying mosquitoes and other pests. As weather warms, mosquitoes reproduce more often and become more active.

Robert Haley, director of epidemiology at the University of Texas Southwestern Medical Center, said that warm winters with few freezing days and periods of drought punctuated by major rainstorms fill culverts force mosquitoes and birds to congregate and are followed by warm temperatures and a hot, early summer. "Those are perfect conditions. . . . Climate change is broadening the tropical latitudes, and Texas is going to be tropical eventually," Haley said (Warrick 2015). Diseases such as dengue fever and the chikungunya virus, both carried by mosquitoes, which had rarely been seen on the U.S. mainland, have become more common, joining the now endemic West Nile virus. All appear earlier in the spring and linger into mid- and late autumn.

Large cities, including Dallas, restarted community-wide pesticide spraying against mosquitoes and other pests for the first time in almost 50 years. Joby Warrick wrote in the *Washington Post* (2015) that Chagas disease, a sometimes fatal parasitic infection spread *Rhodnius prolixus*, a tropical disease also known as the "kissing bug," also has been moving northward into Texas. Its carriers have been sighted in Indiana and Ohio, according to the Centers for Disease Control and Prevention.

"The warmer the planet gets, the more pathogens and vectors from the tropics and subtropics are going to move into the temperate zones," said Daniel R. Brooks, an evolutionary biologist and senior research fellow at the University of Nebraska's Manter Laboratory of Parasitology. "Countries such as the United States tend to have a false sense of security, but vectors and pathogens don't understand international boundaries. You can't just put up a fence to keep them out" (Warrick 2015).

Greenhouse Gas Reduction and Health

As Luis Cifuentes and colleagues observed in *Science*, "The same actions that can reduce the long-term buildup of greenhouse gases—reductions in burning of fossil fuels—can also yield powerful, immediate benefits to public health by reducing the adverse effects of local air pollution" (Cifuentes et al. 2001, 1257). Their study cited estimates that 18,700 premature deaths per year could be avoided in the United States by reducing emissions from older coal-fired electrical generating plants (Cifuentes et al. 2001, 1257). Deaths from bronchial problems, heart disease, and other ailments would be reduced substantially. Another of the surveyed studies indicated that air pollution from traffic causes more deaths than traffic accidents.

The World Health Organization estimated that 460,000 avoidable deaths occurred annually worldwide by 1995 "as a result of suspended particulate matter, largely from outdoor urban exposures" (Cifuentes et al. 2001, 1257). The authors noted that several types of pollution rise with temperatures. According to estimates developed by Cifuentes and colleagues, reducing fossil fuel air pollution in four of the world's largest cities—New York, Mexico City, São Paulo, and Santiago, Chile— could have prevented 64,000 premature deaths between 1980 and 2000 and reduced rates of infant mortality, asthma, cardiovascular problems, and respiratory ailments.

"The benefits of lowering emissions are immediate" because many of the gases emitted when fuels are burned are also pollutants, said George Thurston, one of the review's authors, who is an associate professor of environmental medicine at the New York University School of Medicine. "Universal studies have shown when air pollution levels go up, you get an increase in the numbers of deaths and hospital admissions, missed days at work and school, and other adverse effects," Thurston said (Surendran 2001).

Kidney Stones

Writing in *The Proceedings of the National Academy of Sciences* July 15, 2008, several scientists proposed a connection between kidney stones and a warming environment. Tom H. Brikowski, Margaret S. Pearle, and Yair Lotan wrote that patients in the U.S. South have a 50-percent higher incidence of kidney stones than in the Northeast. The researchers expect that the fraction of the U.S. population living in high-risk stone zones is predicted to grow from 40 percent in 2000 to 56 percent by 2050, provoking as much as a 30-percent increase (Brikowski et al. 2008, 9841), resulting in 1.6 million new cases in the United States by 2050.

Incidence of kidney stones increases with dehydration, which aggravates formation of stones from salts that crystallize in the kidneys, a condition that increases with a warmer climate. The scientists asserted that similar conditions may also increase kidney stone incidence in large parts of Africa, the Middle East, and South Asia.

Margaret S. Pearle, of the University of Texas Southwestern Medical School in Dallas, said, "We see a relationship between kidney stones and temperatures everywhere. Even in places with air conditioning, warmer temperatures mean more stones" (Vergano 2008). By 2007, kidney stones were afflicting 12 percent of men and 7 percent of women in the United States.

Further Reading

Brikowski, Tom H., Yair Lotan, and Margaret S. Pearle. "Climate-Related Increase in the Prevalence of Urolithiasis in the United States." *Proceedings of the National Academy of Sciences* 105(28) (July 15, 2008): 9841–9846.
Vergano, "Kidney Stone Cases Could Heat Up." *USA Today*, July 15, 2008, A1.

Another study reported that alternative transportation policies initiated during the busy 1996 Summer Olympics in Atlanta "not only reduced vehicle exhaust and air pollutants such as ozone by about 30 percent [but] also decreased the number of acute asthma attacks by 40 percent and pediatric emergency admissions by about 19 percent" (Surendran 2001).

Warming and West Nile Virus

Until 1999, the West Nile virus had never even been detected in North America. No one knows exactly how the virus reached the United States, but once it did it spread rapidly across the continent. By 2001, it had infected 29 species of mosquitoes, 100 species of birds, and many mammals, including humans. By the summer of 2002, West Nile had reached 36 states, as well as the southern regions of eastern Canada (Grady 2002). By 2003, most of the continental United States was reporting incidences of West Nile virus. Global warming may be a contributing factor because of warm winters and pervasive summer droughts that seem to favor the spread of the mosquito-borne virus.

The disease initially brings fever, aches, and profound fatigue, sometimes followed by paralysis and other neurological complications, including meningitis and encephalitis, which can leave a victim physically disabled and brain damaged. Between 1999 and 2004, the West Nile virus infected more than 16,000 people in the United States, killing more than 600, and afflicting 6,500 others with severe neurological problems (Chase 2004, A1).

Epstein said that drought helps the mosquito species *Culex pipiens*, which plays a major role in spreading West Nile (Grady 2002). Epstein added that drought also may wipe out darning needles, dragonflies, and amphibians, which destroy

Cholera's Climatic Connection

Scientists are only beginning to describe the many ways in which a warming climate affects human health. One example is the connection between climate and the spread of cholera.

In January 1991, an epidemic of the disease began in Chancay on Peru's coast near Lima. It was the first time that cholera had appeared in the Western Hemisphere in more than 100 years, the beginning of an epidemic that within 15 months would afflict 500,000 people, killing almost 500. The epidemic was researched by Rita Colwell, who was looking for the vector that was spreading the disease, as part of her doctoral dissertation research at the University of Washington. She found ancient accounts associating the start of cholera epidemics with the arrival of ships in seaports. Her idea that seawater was acting as a "reservoir" for cholera was revolutionary in the 1990s—and unsettling to commercial fishing interests when she directed it in Chesapeake Bay. The spread of the disease in contaminated freshwater already was well known.

Colwell, however, was the first to associate the spread of cholera with the onset of warming water provoked by El Niño conditions on the Pacific Coast of South America (and, by implication, the disease's presence in warmer water worldwide). She established a relationship between cholera and seaborne plankton. "In one fell swoop, Colwell had solved one of the great mysteries of this ancient disease: the location of the *V. cholerae* reservoir. . . . Global warming could threaten human health in ways that scientists in ways that scientists had yet to fully anticipate," wrote Paul Epstein and Dan Ferber in *Changing Planet, Changing Health* (2011).

Further Reading

Epstein, Paul R., and Dan Ferber. *Changing Planet, Changing Health: How the Climate Crisis Threatens Our Health and What We Can Do about It*. Berkeley: University of California Press. 2011.

mosquitoes. Drought also may aid the spread of infection by drawing thirsty birds to the pools and puddles where mosquitoes breed. "Hot weather plays a role, too," Epstein said. "Warmth increases the rate at which pathogens mature inside mosquitoes" (Grady 2002).

Heat and Human Sex Ratios

We know that a warming climate changes the sex ratio of some turtles. It also may be changing the sex ratio of human males and females by producing more girls than boys. In 2014, a team of Japanese scientists published a study associating environmental warming in that country with an increasing number of female births since the 1970s. Most of the decline came in pregnancy, as a higher number of male fetuses died during temperature extremes. The phenomenon occurred during both extremely hot and extremely cold weather: After an extremely hot summer in 2010, fewer male babies were born nine months later, and the same happened after a severely cold winter in 2011.

Published in the journal *Fertility and Sterility* (September 2014), the study compared spontaneous miscarriages and birth records with temperature records between 1968 and 2012. "The recent temperature fluctuations in Japan seem to be linked to a lower male–female sex ratio of newborn infants, partly via increased male fetal deaths," the authors wrote. "Male *concepti* seem to be especially vulnerable to external stress factors, including climate changes" (Fukuda et al. 2014). The scientists did not venture an explanation as to why this is the case, but loggerhead sea turtles have been similarly affected.

As Fukuda et al. wrote,

A statistically significant positive association was found between yearly temperature differences and sex ratios of fetal deaths; a statistically significant negative association was found between temperature differences and sex ratios of newborn infants from 1968 to 2012, and between sex ratios of births and of fetal deaths. The sex ratios of fetal deaths have been increasing steadily along with temperature differences, whereas the sex ratios of newborn infants have been decreasing since the 1970s. (Fukuda et al. 2014)

Further Reading

Chase, Marilyn. "As Virus Spreads, Views of West Nile Grow Even Darker." *Wall Street Journal*, October 14, 2004, A1, A10.

Cifuentes, Luis, et al. "Hidden Health Benefits of Greenhouse Gas Mitigation." *Science* 252 (August 17, 2001): 1257–1259.

Eilperin, Juliet. "U.S., China Got Climate Warnings Toned Down." *Washington Post*, April 7, 2007, A5. http://www.washingtonpost.com/wp-dyn/content/article/2007/04/06/AR200 7040600291.html

Epstein, Paul R. *Climate, Ecology, and Human Health.* December 18, 1998. http://www.iitap .iastate.edu/gccourse/issues/health/health.html (no longer available).

Fukuda, Misao, et al. "Climate Change Is Associated with Male:Female Ratios of Fetal Deaths and Newborn Infants in Japan." *Fertility and Sterility,* September 2014. http://www.fert stert.org/article/S0015-0282%2814%2901840-8/fulltext.

Gelbspan, Ross. *The Heat Is On: The High Stakes Battle over Earth's Threatened Climate.* Reading, MA: Addison-Wesley, 1997.

Grady, Denise. "Managing Planet Earth: On an Altered Planet, New Diseases Emerge as Old Ones Re-Emerge." *The New York Times*, August 20, 2002, F2.

Martens, Pim. "How Will Climate Change Affect Human Health?" *American Scientist* 87(6) (November–December 1999): 534–541.

McKibben, Bill. *Eaarth: Making a Life on a Tough New Planet.* New York: Times Books, 2010.

Romero, Simon. "Alarm Spreads in Brazil over a Virus and a Surge in Malformed Infants." *The New York Times*, December 30, 2015. http://www.nytimes.com/2015/12/31/world/ americas/alarm-spreads-in-brazil-over-a-virus-and-a-surge-in-malformed-infants.html.

Sierra Club. "Global Warming: The High Costs of Inaction." 1999. http://www.sierraclub. org/global-warming/resources/innactio.htm (no longer available).

Surendran, Aparna. "Fossil Fuel Cuts Would Reduce Early Deaths, Illness, Study Says." *Los Angeles Times*, August 17, 2001, A20.

Warrick, Joby. "A Warming Globe Brings Invisible Threat: Disease" *Washington Post*, November 27, 2015. http://www.washingtonpost.com/sf/national/2015/11/27/disease/?wpmm=1&wpisrc=nl_headlines.

See also: Allergies; Drought, United States; Drought, Worldwide; Insect Bites; Malaria; Temperatures, Global; Zika Virus

INSECT BITES

Warming climate has favored the spread of many disease-transmitting insects that were previously kept under control by seasonal frosts and freezes. Spreading insect infestations have been notable in many midlatitude urban areas, among them London, England, and Boston, Massachusetts. According to James Meek, writing in *The Guardian* (London):

> They're chomping in Chelsea, dining al fresco in Fulham and more than pecking at their food in Pimlico. Despite their fancy taste in London addresses they are neither posh nor particularly fussy. They are alien vine weevils, and they want to eat your plants. Two species of vine weevil previously unable to survive Britain's cold winters have been discovered in southwest London. One of the species has also been detected in Surrey and Cardiff and as far north as Edinburgh. Two new species, *Otiorhynchus armadillo* and *Otiorhynchus salicicola*, not previously known north of Switzerland, will add to gardeners' woes. (Meek 2002)

"This is probably the most serious new garden pest in recent memory," said Max Barclay, curator of beetles at the Natural History Museum in London, who has been following waves of insect infestations novel to Britain. Classed as beetles, the immigrating species are now prevalent in Chelsea, Victoria, Pimlico, and Fulham, causing significant damage to gardens and in some cases defoliating garden plants almost entirely. "It's very likely these weevils have been introduced to Britain through imported ornamental plants from Italy," said Barclay. "It looks like they're here to stay" (Meek 2002). Earlier springs and later, milder winters have drawn other new pests to Britain. "Of course, the fact [that] Mediterranean species are doing well in Britain invites speculation about climate amelioration," said Barclay (Meek 2002).

Urban Wildlife in Boston

After one of the city's mildest winters on record in 2001–2002, Boston residents and exterminators reported "one of the largest explosions of urban wildlife in recent memory" (Wilmsen 2002). Rats, ants, wasps, snakes, squirrels, rats, skunks, and other animals used the mild winter to procreate at times when the rigors of a New England winter usually would have killed many and kept the rest sexually dormant. Many small mammals had litters early and probably would have one or more extra litters of young during the warm months of early 2002. (Many of the same creatures, however, experienced the traditional perils of a New England winter during 2002–2003, 2003–2004, and 2014–2015 when snow and cold returned with a vengeance.)

"I expect sharp increases in rat populations all over New England," said Bruce Colvin, a Lynnfield, Massachusetts, ecologist who is an expert on them. "When you have a mild winter, there's more food and warm places for them to live, which means less stress. Instead of killing each other in the competition for food, they're all getting along and reproducing." Rat pups born during the winter were already mature enough by early spring to have litters of their own (Wilmsen 2002, B1). Larger-than-usual numbers of carpenter ants and other insects, including bees and other wasps, were emerging in Boston by late March 2002, well ahead of their usual seasonal introduction in June.

Also in Boston, by the first week of June 2002, pollen levels were at levels not usually observed until the end of the month. Paul Epstein, a professor of medicine at Harvard University, said that if carbon dioxide levels double, ragweed pollen levels will rise 61 percent. He anticipates that level will be reached by 2050. Epstein wrote in the *Annals of Allergy, Asthma, and Immunology*, as reported in the *Boston Globe* (Smith 2002; Harvard 2002). The record mild weather of the 2001–2002 winter was followed in New England by two severe winters that played a role in reducing pest populations to more normal levels.

Further Reading

"Harvard Medical School: Pollen Production—and Allergies—May Rise Significantly over Next 50 Years." *ScienceDaily*, March 22, 2002. http://www.sciencedaily.com/releases/2002/03/020322075343.htm.

Smith, Stephen. "Comin' Ah-Choo: Tepid Temperatures Speeding Allergy Season." *Boston Globe*, April 10, 2002, A1.

Wilmsen, Steven. "Critters Enjoy a Baby Boom; Mild Winter's Downside Is Proliferation of Vermin." *Boston Globe*, March 30, 2002, B1.

Rats, Spiders, and Scorpions Thrive in Great Britain

Great Britain's rat population rose by nearly a third between 1998 and 2002. According to a report described in the *Daily Mail* (U.K.), by 2002 an estimated 60 million brown rats inhabited Great Britain's buildings, streets, sewers, and waterways. As the newspaper reported, "The rodents narrowly outnumber the human population" (Utton 2003). Even Buckingham Palace called in exterminators after a rat infestation in its kitchens in 2002. The 29-percent four-year rise in rat populations was blamed on milder, wetter winters as well as fast-food litter on which the rats feast. The amount of rubbish on London streets had grown by an estimated 80 percent in 35 years, according to the report. Rats breed quickly, reaching sexual maturity in just eight weeks. They can have sex 20 times a day and can give birth every four weeks. A single pair of rats can produce as many as 2,000 offspring a year (Utton 2003).

Barrie Sheard of the British National Pest Technicians' Association, which conducted the study, pointed to a rapid increase in rat numbers during the summer months, with warming a probable factor. Other factors also are at work. For example, many sewer systems built during the late 19th century have decayed, allowing more rats to reach the surface. To save money, many British town councils started billing private property owners for rat control—and this led many residents to avoid reporting infestations.

"Until these last four to five years, our summer months were always recognized as a period of the year when rat complaints were always at a far-reduced level," Sheard said (Utton 2003). Peter Gibson, speaking for Keep Britain Tidy, said, "The worrying thing is not just their numbers, but their behavior. Rats are not just in the sewers any more. They are coming out into the open, on to the streets, because they know that's where the most food is" (Utton 2003).

Spiders, wasps, and similar insect species also have been moving northward though Britain at unprecedented rates as temperatures rise. Insects with historic ranges in the northern reaches of England are moving into Scotland. Those previously resident in southern England have arrived in the north, with European spiders increasingly seen invading Britain from the south and west. Global warming is being held responsible (Browne and Simons 2002).

With no natural predators, scorpions have been spreading in England as temperatures have risen. Two-inch-long, European yellow-tailed scorpions probably were brought to the island from Italy or southern France during 1860 on ships docking at Sheerness in Kent. For many decades, a colony of 1,000 hid in the brickwork of the dockyard. As temperatures warmed after the late 1980s, the scorpion colony grew to around 3,000 and spread into neighboring residential areas, where people have been advised to follow the practices typically used in tropical areas: turn shoes upside down and shake them before wearing. Paul Hillyard, curator of arachnids at the Natural History Museum in London, said: "In the warmer temperatures these scorpions can flourish and people could start transporting them throughout Britain" (Ingham 2004). European yellow-tailed scorpions also have been sighted in Harwich and Pinner and at the Ongar underground station in Essex.

According to one English observer, "The magnificent yellow and black wasp spider, *Argiope bruennichi*, which grows up to 50 millimeters (two inches) long, arrived in Hampshire from the Continent and now occurs as far north as Derbyshire" (Browne and Simons 2002). Peter Harvey, head of the Spider Recording Scheme for the British Arachnological Society, said the wasp spider had spread rapidly during the five years ending in 2002. "It is almost certainly because of warmer and longer summers and autumns. It lays its eggs in autumn, so cold autumns are very bad for it" (Browne and Simons 2002).

Examples of insect migrations are plentiful in the British Isles:

The bee wolf, a wasp that feeds honeybees to its offspring, had until recently been confined to the Isle of Wight and East Anglia. However, it has spread rapidly in recent years and is now found as far north as Yorkshire. Arachnophobes in Scotland will be anxious to learn that the giant house spider, *Tegenaria gigantea*, which was once confined to England, has now reached Ullapool in the northwest Highlands. A species of jumping spider entirely new to Britain was recently discovered in London, having arrived from the Continent. A survey of Mile End Park in the East End of London has revealed a thriving colony of *Macaroeris nidicolens*, which is usually found in continental Europe, where it lives on small trees and bushes. Exactly how the spider arrived in Britain is unclear, although the young spiderlings could have crossed the Channel using thread of gossamer blown on the wind. (Browne and Simons 2002)

Further Reading

Browne, Anthony, and Paul Simons. "Euro-Spiders Invade as Temperature Creeps up." *The Times* (London), December 24, 2002, 8.

Ingham, John. "Stingers Thrive as the Country Gets Warmer; Invasion of the Scorpions." *Daily Express* (U.K.), June 18, 2004, 40.

Meek, James. "Global Warming Gives Pests Taste for Life in London." *The Guardian* (U.K.), October 8, 2002, 6.

Utton, Tim. "The Rat Rampage." *Daily Mail* (U.K.), January 22, 2003, n.p. (LEXIS).

See also: Allergies; Drought, United States; Drought, Worldwide; Malaria; Temperatures, Global; Zika Virus

MALARIA

Malaria is the most pervasive severe disease on Earth, with some 300 million new cases every year—one of every 25 people. It produces high fever, shaking, chills, and debilitating weakness. Even after initial recovery, victims may suffer anemia, fevers, and weakness for years. Temperature directly affects the speed at which the most dangerous malaria parasite (*Plasmodium falciparum*) matures inside a host mosquito: 56 days at 18°C (64°F), 19 days at 22°C (72°F), and eight days at 30°C (86°F). A typical host mosquito (such as the female *Anopheles gambiae*, which is widespread in Kenya) lives only two to three weeks and usually dies before malaria matures at 64°C. At higher temperatures, transmission to humans is much more likely (Epstein and Ferber 2011, 44).

Writing in the *Bulletin of the American Meteorological Society* (March 1998), Paul Epstein and colleagues described the spread of malaria and dengue fever to higher altitudes in tropical areas of Earth because of warmer temperatures. Rising winter temperatures also have allowed disease-bearing insects to survive in areas previously closed to them. According to Epstein et al., more intense precipitation and frequent flooding associated with warmer temperatures also promotes the growth of fungus and provides excellent breeding grounds for large numbers of mosquitoes. The flooding caused by Hurricane Floyd and other storms in North Carolina in 1999 are cited by some people as a real-world example of global warming promoting conditions ideal for the spread of diseases imported from the tropics (Epstein et al. 1998).

Temperatures Rise, Malaria Spreads

According to projections for human health by the Intergovernmental Panel on Climate Change (IPCC), a rise in global average temperatures of 3 to 5°C by 2100 could lead to 50 million to 80 million additional cases of malaria worldwide per year, primarily in tropical, subtropical, and less well-protected temperate-zone populations. Britain's Hadley Centre for Climate Prediction and Research expects the same disease to reach the Baltic states by 2050. In parts of the world where malaria is now unknown, most people have no immunity (Epstein 1999). The World Health Organization also projected that warmer weather will cause tens of millions additional cases of malaria and other infectious diseases. The Dutch health ministry anticipates that more than a million people may die annually as a result of the impact of global warming on malaria transmission in North America and northern Europe (Epstein 1999, 67).

Malaria could return to Britain as an endemic disease, warned scientists at the University of Durham as they announced a plan to produce a "risk map" showing which areas were most likely to suffer an outbreak (Connor 2001). With millions of tourists visiting malaria-infested regions of the world, the risk of the disease making a comeback is further increased by global warming, which expands mosquito habitats in the United Kingdom, said the university's Rob Hutchinson, an entomologist speaking at the annual meeting of the Royal Entomological Society in Aberdeen. He said that of the 25 million overseas visitors who came to Britain in 1999, some 260,000 came from Turkey and the countries of the former Soviet Union, where vivax malaria was endemic and health care was poor (Connor 2001).

In North America, by the late 1990s, malaria had been transmitted by mosquitoes as far north as Toronto, Canada, according to Epstein. "The extreme events we are seeing today in Nicaragua and Honduras as a result of Hurricane Mitch in [1998] are spawning outbreaks of cholera and dengue fever with new breeding sites for mosquitoes and increased water-borne diseases," Epstein said (Webb 1998). In northerly latitudes, nighttime and winter temperatures have warmed twice as fast as overall global temperatures since 1950, Epstein said, meaning that fewer pests are being killed by frost in the southern reaches of the temperate zones. Humidity also has increased in many regions, including much of the eastern United States, helping mosquitoes to breed. Disease-carrying mosquitoes usually required a certain level of temperature *and* humidity to survive.

In May 1995, researchers in the Netherlands and England estimated the increase in malaria's geographic range that could occur if the IPCC.'s projections for global warming prove correct. These researchers concluded that the epidemic potential of the mosquito population would double in tropical regions. In temperate climates, according to these projections, the epidemic potential could increase 100 times. Furthermore, this study said, "There is a real risk of reintroducing malaria into nonmalarial areas, including parts of Australia, the United States, and southern Europe" ("Rachel's" 1995).

By the late 1970s, dwindling investments in public health programs, growing resistance to insecticides among some species, and prevalent environmental changes (such as deforestation) also contributed to a widespread resurgence of malaria, according to Epstein. "By the late 1980s," he reported, large epidemics of malaria were being associated with warm, wet weather. Between 1993 and 1998, worldwide incidence of malaria quadrupled, according to Epstein (1998).

Global Warming and Disease Vectors

John T. Houghton, author of *Global Warming: The Complete Briefing* (1997), believes that global warming will accelerate the spread of many diseases from the tropics to the middle latitudes. Malaria could increase from its current levels, Houghton warns. "Other diseases [that] are likely to spread for the same reason are yellow fever, dengue fever, and . . . viral encephalitis," he wrote (Houghton 1997, 132). After 1980, small outbreaks of locally transmitted malaria occurred in Texas, Georgia, Florida, Michigan, New Jersey, New York, and California, usually during hot and wet spells. Worldwide, according to Epstein, between 1.5 million and 3 million people die of malaria each year, mostly children. Mosquitoes and parasites that carry the disease have evolved immunities to many drugs.

According to Epstein, "If tropical weather is expanding it means that tropical diseases will expand. We're seeing malaria in Houston, Texas" (Glick 1998). Epstein suggested that a resurgence of infectious disease may be one result of global warming. Although warming may appear beneficial at first, he said, with some plants benefiting from additional warmth and moisture, an earlier spring, and more carbon dioxide and nitrogen in the air, "warming and increased CO_2 can also stimulate microbes and their carriers" (Epstein 1998).

Since 1976, Epstein reported, 30 diseases have emerged that are new to medicine. Old ones such as drug-resistant tuberculosis have been given new life by new diseases (such as HIV and AIDS) that compromise the human immune system. By 1998, tuberculosis was claiming 3 million lives annually around the world. "Malaria, dengue, yellow fever, cholera, and a number of rodent-borne viruses are also appearing with increased frequency," Epstein reported (1998). Warming temperatures tend to accelerate the reproductive cycles of several disease-carrying pests, especially mosquitoes, different types of which specialize in various maladies. The increasing concentration of humanity into ever-larger urban areas, especially in tropical areas where mosquitoes thrive, increases their threat. The IPCC included a chapter on public health in an update of its 1990 assessment with the conclusion, "climate change is likely to have wide-ranging and mostly adverse impacts on human health, with significant loss of life" (Taubes 1997).

Writing in *Climatic Change*, Willem Martens et al. (1997) attempted to sketch how a warmer, wetter climate would affect transmission of three vector-borne diseases: malaria, schistosomiasis, and dengue fever. Martens and colleagues anticipated that the periphery of the currently endemic areas "will expand with global warming, with diseases notable at higher elevations in the tropics, an expectation that has been borne out by several observers." Martens and colleagues expected that the "increase in epidemic potential of malaria and dengue transmission may be estimated at 12 [percent] to 27 percent and 31 [percent] to 47 percent respectively" (Martens et al. 1997, 145). In contrast, they forecast that the transmission potential of schistosomiasis may decrease 11 percent to 17 percent.

Diarrhea is another disease that may become more common in the temperate zones of a warmer world; during the 1990s, diarrhea killed more than 3 million children a year worldwide, mainly in the tropics of Asia, Africa, and the Americas. The bacteria that cause diarrhea thrive in warm weather, especially after heavy rainfall.

Warming and Excess Deaths Worldwide

According to three U.N. organizations in 2003, at least 150,000 people die each year as a direct result of global warming. Warming already has been a factor in increases in malnutrition as well as outbreaks of diarrhea and malaria, the three largest killers in the poorest countries of the world, according to these reports. Diarmid Campbell-Lendrum, a World Health Organization scientist, said that estimates of deaths were extremely conservative. Furthermore, he said, the number of deaths attributable to rapid warming is expected to double by 2035. "People may say that this is a small total compared with the totals who die anyway, but these are needless deaths. We must do our best to take preventative measures," he said (Brown 2003).

The report produced by the World Health Organization, the U.N. Environment Program, and the World Meteorological Program detailed how increased warmth has intensified the spread of diseases. Diarrhea spreads by bacteria, mostly through unclean water and food, because bacteria develop and diffuse more quickly in warmer temperatures. Dirty water is the largest killer of children younger than five years of age. In Lima, Peru, a six-year study at a clinic set up to treat diarrheal complaints showed a 12-percent increase in cases for every 1°C rise in temperature during cooler months and a 4-percent increase in warmer months (Brown 2003). Similar results were found in a survey of 18 Pacific islands. The problem is made worse by high rainfall or drought when water supplies often become contaminated.

Many diseases spread by rats and insects also become more common in warmer weather. Malaria, dengue fever, and Lyme disease are all on the increase. Many threats can be curtailed by dispensing preventive medicine and providing clean water and sanitation. Climate change makes these issues more urgent, the report said. The combined effects of increased warmth and the greater volume of standing water brought by storms create malaria epidemics by providing breeding sites and an accelerated life cycle for mosquitoes. In Africa, where the death toll from malaria is highest, mosquitoes carrying the disease are spreading into mountain areas previously too cool for them to thrive, according to the report.

Health Benefits Despite Warming?

Some studies have questioned the assumption that a warming world will be beset by more malaria. Control measures have trumped increasing temperatures, according to a study by Gething et al. and published in *Nature* (2010):

> A comparison of a recently published evidence-based map of the distribution of the *Plasmodium falciparum* malaria parasite with data from 1900, before the introduction of major malaria control measures, suggests that rising temperatures are a threat to malaria control efforts are misplaced. During a century when increases in global temperature have been unequivocal, the range and intensity of malaria has diminished dramatically. The postulated effect of warming is at least an order of magnitude smaller than the effects of control measures, suggesting that the success or failure of any anti-malaria program is likely to be determined by factors other than climate. . . . The range of malaria has contracted through a century of economic development and disease control. Here, for the first time, we quantify this contraction and the global decreases in malaria endemicity since approximately 1900. (Gething et al. 2010, 342; Ledford 2010)

One climate contrarians' journal, *The World Climate Report*, replied to an editorial in the prominent British medical journal *Lancet* that asserted that malaria and other mosquito-borne diseases will spread into the temperate zones with warming temperatures. According to the contrarians, malaria's spread has little to do with temperature or humidity and more to do with medical technology and availability of air conditioning. The contrarians assert that epidemics of malaria were common in most of the United States before the 1950s. In 1878, according to their argument, about 100,000 people were infected in the United States, and one-quarter of them died.

Pim Martens has written that although the overall impact of global warming on human health is expected to be markedly negative, human beings may experience a few positive outcomes. Some diseases that thrive in cold weather (such as influenza) may find their ranges and effects reduced in a warmer world. The elderly might die less frequently of cardiovascular and pulmonary ailments, which peak during cold weather. "Whether the milder winters could offset the mortality during the summer heat waves is one of the questions that demands further research," Martens wrote (Martens 1999, 535).

Countering the views of Epstein and others, some health researchers contend that global warming will do little to increase the incidence of tropical diseases. "For mosquito-borne diseases such as dengue, yellow fever, and malaria, the assumption that warming will foster the spread of the vector is simplistic," contended Bob Zimmerman, an entomologist with the Pan American Health Organization. Zimmerman pointed out that in the Amazon River basin, more than 20 species of *Anopheles* mosquitoes can transmit malaria, and each is adapted to a different habitat: "All of these are going to be impacted by rainfall, temperature, and humidity in different ways. There could actually be decreases in malaria in certain regions, depending on what happens" (Taubes 1997).

Virologist Barry Beaty of Colorado State University in Fort Collins, Colorado, agreed with Zimmerman: "You don't have to be a rocket scientist to say we've got a problem," he said. "But global warming is not the current problem. It is a collapse in public-health measures, an increase in drug resistance in parasites, and an increase in pesticide resistance in vector populations. Mosquitoes and parasites are efficiently exploiting these problems" (Taubes 1997).

Countering the majority view that a warmer world will cause malaria to spread, David J. Rogers and Sarah E. Randolph, using their own models, wrote in *Science* that even extreme rises in temperature will not spread the disease. They argued that the spread of malaria is too poorly understood to base a forecast several decades into the future on temperature as a singular variable. For example, the Dengue marshes of Essex, in England, a breeding ground for malaria-carrying mosquitoes in the 17th century, dried up, making an increase in temperatures not a factor vis à vis malaria's transmission. Malaria is not a new disease in the temperate zones. It was common in the Roman Empire. A British invasion of the Netherlands in 1806 failed to drive out French troops because large numbers of British became ill with malaria. Malaria was also a public health problem in most of the eastern United States during warm, humid summers before medications were developed for it at the turn of the 20th century.

Paul Reiter, a dengue expert with the Puerto Rico office of the U.S. Centers for Disease Control and Prevention, has argued against the relative importance of climate in human disease by pointing to periods in the past when malaria and other tropical diseases pervaded cooler regions. He argued that the spread of malaria is more closely linked to deforestation, agricultural practices, human migration, poor public health services, civil war, and natural disasters. "Claims that malaria resurgence is due to climate change ignore these realities and disregard history," he wrote in an article about malaria's spread through England during the Little Ice Age, which began about 1450 and lasted for several hundred years in a climate that was cooler than today's (McFarling 2002).

S. I. Hay and colleagues investigated long-term meteorological trends in four high-altitude sites in East Africa where increases in malaria have been reported since the mid-1980s. "Here we show that temperature, rainfall, vapor pressure, and the number of months suitable for *P. falciparum* transmission have not changed significantly during the past century or during the period of reported malaria resurgence." As a result, they find that associations between resurgence of malaria and climate change at high altitudes in these areas "overly simplistic" (Hay et al. 2002, 905).

Further Reading

Brown, Paul. "Global Warming Kills 150,000 a Year; Disease and Malnutrition the Biggest Threats, UN Organisations Warn at Talks on Kyoto." *The Guardian* (U.K.), December 12, 2003, 19.

Connor, Steve. "Malaria Could Become Endemic Disease in U.K." *The Independent* (London), September 12, 2001, 14.

Epstein, Paul R. "Climate, Ecology, and Human Health." December 18, 1998. http:// www .iitap.iastate.edu/gccourse/issues/health/ health.html (no longer available).

Epstein, Paul R., "Profound Consequences: Climate Disruption, Contagious Disease, and Public Health." *Native Americas* 16(3–4) (Fall–Winter 1999): 64–67.

Epstein, Paul R., and Dan Ferber. *Changing Planet, Changing Health: How the Climate Crisis Threatens Our Health and What We Can Do about It.* Berkeley: University of California Press, 2011.

Epstein, Paul R., et al. "Biological and Physical Signs of Climate Change: Focus on Mosquito-Borne Diseases." *Bulletin of the American Meteorological Society* 79, Part 1 (1998): 409–417.

Gething, Peter W., et al. "Climate Change and the Global Malaria Recession." *Nature* 465 (May 20, 2010): 342–345.

Glick, Patricia. *Global Warming: The High Costs of Inaction.* San Francisco: Sierra Club, 1998.

Hay, S. I., et al. "Climate Change and the Resurgence of Malaria in the East African Highlands." *Nature* 425 (February 21, 2002): 905–909.

Houghton, John. *Global Warming: The Complete Briefing.* Cambridge, UK: Cambridge University Press, 1997.

Ledford, Heidi. "Malaria May Not Rise as World Warms." *Nature* 465 (May 20, 2010): 280.

Martens, Pim. "How Will Climate Change Affect Human Health?" *American Scientist* 87(6) (November–December 1999): 534–541.

Martens, Willem J. M., Theo H. Jetten, and Dana A. Focks. "Sensitivity of Malaria, Schistosomiasis, and Dengue to Global Warming." *Climatic Change* 35 (1997): 145–156.

McFarling, Usha Lee. "Study Links Warming to Epidemics; the Survey Lists Species Hit by Outbreaks and Suggests That Humans Are Also in Peril." *Los Angeles Times,* June 21, 2002, A7.

"Rachel's No. 466: Climate and Infectious Disease." Rachel's Democracy and Health News. November 2, 1995. Environmental Research Foundation, Annapolis, Maryland. https:// rachelsdemocracyandhealthnews.wordpress.com/2013/10/27/rachels-466-climate-and -infectious-disease-part-1/.

Taubes, Gary. "Apocalypse Not." 1997. http://www.junkscience.com/news/taubes2.html (no longer available).

Webb, Jason. "Mosquito Invasion as Argentina Warms." Reuters, 1998. http://ces.iisc.ernet .in/hpg/envis/doc98html/enarg117.html.

See also: Allergies; Drought, United States; Drought, Worldwide; Insect Bites; Temperatures, Global; Zika Virus

ZIKA VIRUS

One of the major species to benefit from a warming planet is mosquitoes, a major disease vector. In 2015 and 2016, public health authorities in several mainly tropical countries were confronted with a new threat—the Zika virus, which is spread by the same species of mosquito (*Aedes aegypti*) that is responsible for transmitting yellow fever and an exploding epidemic of dengue fever in many of the same places. As temperatures rise, the mosquito's reproductive cycle accelerates and its range expands. Also in 2015 and 2016, Earth's temperatures stoked by an intense El Niño cycle as well as underlying warming from greenhouse gases in the atmosphere, had risen to record levels.

In 2016, a global health emergency was declared for Zika, which plays a major role in the birth of deformed babies to women who have been afflicted by the virus. At the same time, the number of people afflicted by dengue had risen to 100 million a year, affecting populations from India to Argentina, and leaving several thousand people dead. Neither Zika nor dengue are curable, but they are preventable if anti-mosquito hygiene measures are strictly followed.

Zika and Dengue Spread as Climate Warms

Health authorities warn that without a vaccine or a cure, both dengue fever and Zika may spread into more temperate countries, including the United States, as the global climate continues to warm. Given rising population and temperatures, according to one analysis in *The New York Times*, in a worst-case scenario, high levels of global emissions couple with fast population growth to double the number of people exposed to the principal mosquito from roughly 4 billion today to as many as 8 billion or 9 billion before 2100. At the same time, scientists have been finding more ways in which the Zika virus can spread. The latest, described in June 2016 in the *New England Journal of Medicine* is oral sex—even kissing (D'Ortenzio et al. 2016, 2195).

A warming climate is not the only causal factor here—it never is. As Justin Gillis wrote in *The New York Times* (2016): "In interviews, experts noted that no epidemic was ever the result of a single variable. Instead, epidemics always involve interactions among genes, ecology, climate and human behavior, presenting profound difficulties for scientists trying to tease apart the contributing factors. 'The complexity is enormous,' said Walter J. Tabachnick, a professor with the Florida Medical Entomology Laboratory, a unit of the University of Florida in Vero Beach."

The *Aedes aegypti* mosquito has adapted over thousands of years to life in human company, breeding in cities' stagnant water and carrying an appetite for human blood. Tropical urban areas have been growing more quickly than those in any other part of the world; Lagos (Nigeria), Karachi (Pakistan), and many other large urban areas now host populations that exceed those of New York City and London. The mosquitoes have spread as well, especially in cities in which many people who lack piped-in water store it cisterns on roofs, providing mosquito habitat. Such methods of open-air water storage near homes is common in many areas where Zika has been spreading, including Recife and Salvador in northeastern Brazil and São Paulo, Brazil's largest city, where dengue fever increased rapidly in 2015. Worldwide, dengue killed 20,000 people in 2015, including more than 800 people in Brazil, 40 percent more than in 2014.

Warming Accelerates Mosquito Breeding

"As we get continued warming, it's going to become more difficult to control mosquitoes," said Andrew Monaghan, who studies climate and health at the National Center for Atmospheric Research in Boulder, Colorado. "The warmer it is, the faster they can develop from egg to adult, and the faster they can incubate viruses" (Gillis

2016). Increasing international air travel also acts as a disease vector, as the occasional mosquito hitches a ride.

Higher temperatures provide more time for a virus to reproduce inside mosquitos as they look for humans to bite. Warmth also accelerates viral incubation because mosquitos mature more quickly. "You're actually speeding up the whole reproductive cycle of the mosquitoes," said Charles B. Beard, who heads a unit in Fort Collins, Colorado, that is studying insect-borne diseases for the Centers for Disease Control and Prevention in Atlanta. "You get larger populations, with more generations of mosquitoes, in a warmer, wetter climate. You have this kind of amplification of the risk" (Gillis 2016).

One case study of warming and mosquito infestation is provided by Mexico City, the largest metropolitan area in the Americas with 21 million people. Although mosquitos and the diseases they transmit are common in Mexico's lowlands, the city has thus far escaped them because it is at much higher elevation and thus cooler. However, as temperatures warm, mosquitoes have been emerging closer to the city, as they work their way to higher elevations. "Literally," said Monaghan, "The mosquitos are just down the hill" (Gillis 2016).

Further Reading

D'Ortenzio, Eric, et al. "Evidence of Sexual Transmission of Zika Virus." *New England Journal of Medicine* 374 (June 2, 2016): 2195–2198.

Gillis, Justin. "Zika Outbreak Could Be an Omen of the Global Warming Threat." *The New York Times*, February 19, 2016. http://www.nytimes.com/2016/02/19/science/zika-outbreak-could-be-an-omen-of-the-global-warming-threat.html.

See also: Allergies; Drought, United States; Drought, Worldwide; Insect Bites; Malaria; Temperatures, Global

INDIGENOUS PEOPLES AND CULTURES

OVERVIEW

Climate change induced by human consumption of fossil fuels affects everyone on Earth and has become the signature environmental issue of our time. Native peoples of North America, with their historically close philosophical connection to Earth and subsistence styles of life, are among the first to be significantly affected by a rapidly changing climate. This is most evident in the Arctic, which is the swiftest warming area on Earth, where an Inuit world built on ice is melting away. Alaskan Native communities also face climate-induced change, including relocation of entire coastal villages. Elsewhere in North America, Native water resources and food sources already have been damaged by a warming climate.

American Indians Suffer More Intensely

The report of the U.S. National Climate Assessment in 2014 from a team of more than 300 experts included a chapter, "Indigenous Peoples, Lands, and Resources," that read, "Key vulnerabilities include the loss of traditional knowledge in the face of rapidly changing ecological conditions, increased food insecurity due to reduced availability of traditional foods, changing water availability, Arctic sea ice loss, permafrost thaw, and relocation from historic homelands" ("Indigenous Peoples" 2014).

A National Wildlife Federation report, *Facing the Storm: Indian Tribes, Climate-Induced Weather Extremes and the Future for Indian Country*, suggests that American Indians suffer more intensely than other ethnic groups from climate change. "The high dependence of tribes upon their lands and natural resources to sustain their economic, cultural, and spiritual practices, the relatively poor state of their infrastructure and the great need for financial and technical resources to recover from such events all contribute to the disproportionate impact on tribes," the report said ("American Indians" 2011).

Across North America, changing climate influences the availability of traditional foods, including beans, corn, squash, seals, shellfish, finned fish, bison, bear, caribou, walrus, deer, moose, and wild rice. Medicinal and food plants sometimes move out of historical (and often treaty-circumscribed) ranges as weather warms. "For example," noted the National Climate Assessment:

Climate change and other environmental stressors are affecting the range, quality, and quantity of berry resources for the Wabanaki tribes in the Northeast. The Karuk people in California have experienced a near elimination of both salmonids and acorns, which comprise 50 percent of a traditional Karuk diet. In the Great Lakes region, wild rice is unable to grow in its traditional range due to warming winters and changing water levels, affecting the Anishinaabe peoples' culture, health, and well-being. ("Indigenous Peoples" 2014)

"We need to be worried about climate change because it's clearly already affecting our region in ways that impact many areas—we're seeing landscapes burning, dying because of heat and dryness," Jonathan Overpeck of the Southwest Climate Science Center told Indian Country Today Media Network. "We're seeing reservoirs that were full just 10 years ago now only half full on the Colorado [River]. These are visible harbingers of what might come. What we need to do as a society is talk about it and figure out how to deal with these challenges" (Allen 2012).

"In a very real sense . . . indigenous peoples are the advance guard of climate change. They are the first communities to observe climate and environmental changes firsthand, and are already using their traditional knowledge and survival skills—the heart of their cultural resilience—to trial adaptive responses" (McLean 2010). Many of these cultures and peoples are under stress not only from climate change imposed from the outside but also from many other forms of pollution imposed by industrial-scale resource development (Johansen 2016). "Ironically," said the U.N. report, "indigenous peoples, whose livelihood activities are most respectful of nature and the environment, suffer immediately, directly, and disproportionally from climate change and its consequences" (McLean 2010). Infrastructure, notably housing, is vulnerable because houses are often of marginal quality to begin. Reservation lands located in drought-prone areas are suffering desertification (the Navajo being one major example) as precipitation declines below previously minimal amounts. Record numbers of wildfires have become the devastating twin sister of drought, which has been made more intense by insect infestation such as from the bark beetles that have devastated large areas of pine forests in western Canada and the United States.

"Winter Counts" as Weather Records

Plains American Indian peoples—notably Lakota (Sioux), Kiowa, Blackfeet, Mandan, and others—kept pictorial accounts of important events on buffalo hides, called *waniyetu wówapi*, in Lakota, in which "Waniyetu" referred to winter, and "Wówapi" to anything that was noted, read, or counted. These may have been battles, the deaths of leaders, or famines, as well as distinctive natural events—"such as meteor storms, eclipses, and unusual weather and climate" (Therrell and Trotter 2011, 583). These accounts are now being used by climate scientists and have been described in *The Bulletin of the American*

Meteorological Society, which notes, "An analysis of the winter count records in conjunction with observational and proxy climate records and other historical documentation suggests that the winter counts preserve a unique record of some of the most unusual and severe climate events of the early American period and provide valuable insight into the impacts upon people and their perceptions of such events in the ethnographically important region of the Great Plains" (Therrell and Trotter 2011).

A winter count for a seasonal cycle usually began with the first snow of fall and extended to the next season's initial snowfall. An image became a mnemonic device that jogged human memory to recall an extensive historical narrative. Many winter counts have been studied in detail and compared with European–American historical records such as diaries, newspaper accounts, tree rings, and logs kept at forts. Winter counts have been useful in confirming records of extreme weather in the modern states of Colorado, the Dakotas, Iowa, western Minnesota, Montana, Nebraska, Oklahoma, and Wyoming, as well as the southern parts of Canadian provinces north of these states.

Many records described intense cold and deep snow; heat waves and droughts were depicted much less often. Winter counts in 1788 and 1789 depicted many Crow people dying in the cold. No European–American records existed in this area at that time, but the winter was extremely severe in New England. George Washington described cold weather during February 1789 at his planation in northern Virginia. Thomas Jefferson, American minister to France at the time, described the conditions there as "a winter of such severe cold, as was without example in the memory of man, or in the written records of history" (Koch and Peden 1944, 91).

Severe flooding was recorded on winter counts in 1825 and 1826 in which several hundred people drowned after an ice dam broke up on the Missouri River in March 1826. Fort post journals noted related flooding on the Red and Assiniboine Rivers, where ice broke up in early May 1826 and carried away 47 homes (Therrell and Trotter 2011, 588). Heavy snow was recorded during the winter of 18927 and 1828, and again in 1852, a trend that was confirmed by Anglo-American immigrants from Minnesota to the West Coast. In 1865 and 1866, cold and snow returned, killing many horses and provoking famine among the Mandans.

Further Reading

Koch, A., and W. Peden, eds. *The Life and Selected Writings of Thomas Jefferson.* New York: Random House, 1944.

Therrell, Matthew D., and Makayla J. Trotter. "Waniyetu Wówapi: Native American Records of Weather and Climate." *Bulletin of the American Meteorological Society* 92(5) (May 2011): 583–592. http://journals.ametsoc.org/doi/pdf/10.1175/2011 BAMS3146.1.

Native Peoples' Paradox

Native peoples around the world are confronted with a paradox. Even as their traditional knowledge is used for foods, drugs, cosmetics, and other things by invasive majority cultures, they are under continual assault by the industrial effluvia of these same cultures. The languages that hold the lockboxes of memory for traditional knowledge are disappearing as well. Global warming is thus a relatively new stanza in a very old song. "As foods, medicines, fuels, and habitats are disappearing," according to this U.N. compendium, "small communities are suffering particular hardships, and indigenous cultures, traditions, and languages are facing major challenges to their existence" (McLean 2010). Witness Alaskan Native villages washing onto the ocean as Navajos, their land swept by sand dunes, give up their sheep. Inuit hunters fall through thin ice as mothers do not dare breastfeed their babies, their milk laced with toxins. Some small villages (mostly, thus far, along the coasts of Alaska and Louisiana) have been threatened with removal—displacement, a reminder that a changing climate makes some homelands uninhabitable.

Scientists are using Native accounts to augment their observations. "Winter counts" of the Plains peoples have been used to confirm events in meteorological history. The observations and descriptions by Crow elders of rising temperatures and diminishing snowpack in southeastern Montana were compared to weather observations and fleshed out with their effects on people. Arctic peoples and scientists have been communicating about their melting ice world for many years.

In Crow country, berries often bud out earlier and become vulnerable to sudden spring freezes that kill the blossoms. Longer autumns mean that buffalo berries dry out before the first frost, making them worthless. Because of warmer winters, trees may emerge from dormancy and then die in sudden subsequent freezes. Increasingly, hot late springs and summers have made fasting for a sun dance in late May or June risky for peoples' health. The average temperature in Hardin, Montana (Crow country), has risen from 45.6°F in the 1950s to 50.1°F after 2000. Snow used to melt slowly and thus ensured summer water supplies. Now shorter winters and hotter summers threaten water supplies by the hot season.

Persistent drought in the U.S. Southwest is forcing Navajos who have no indoor plumbing to travel several miles for water as their wells run dry. They are also being forced to sell their livestock early when formerly scanty pastures turn to bare dirt, "perhaps among the worst of those impacts," wrote Terri Hansen in the Indian Country Today Media Network (2014).

This may resemble the future faced by the majority culture in a century.

Further Reading

Allen, Lee. "Southwest Tribes Struggle with Climate Change Fallout." Indian Country Today Media Network, June 14, 2012. http://indiancountrytodaymedianetwork.com/article/southwest-tribes-struggle-with-climate-change-fallout-118386.

"American Indians Feel the Effects of Climate Change at Higher Rate Than Other Groups." Indian Country Today Media Network. August 9, 2011. http://indiancountrytodaymedia network.com/article/american-indians-feel-the-effects-of-climate-change-at-higher-rate-than-other-groups-46365.

Hansen, Terri. "Climate Disruptions Hitting More and More Tribal Nations." Indian Country Today Media Network, May 7, 2014. http://indiancountrytodaymedianetwork. com/2014/05/07/climate-disruptions-hitting-more-and-more-tribal-nations-154747.
"Indigenous Peoples, Lands, and Resources." Chapter 12, *National Climate Assessment*, U.S. Global Change Research Program, 2014. http://nca2014.globalchange.gov/report/sectors/ indigenous-peoples.
Johansen, Bruce E. *Resource Exploitation in Native North America: A Plague upon the Peoples.* Santa Barbara, CA: Praeger, 2016.
McLean, Kirsty Galloway. *Advance Guard: Climate Change Impacts, Adaptation, Mitigation and Indigenous Peoples: A Compendium of Case Studies.* U.N. University Institute of Advanced Studies. Traditional Knowledge Initiative. Traditional Knowledge Initiative. Darwin, Australia, 2010. http://archive.ias.unu.edu/resource_centre/UNU_Advance_Guard_Com pendium_2010_final_web.pdf.

ARCTIC HUNTERS

"Talk to hunters across the North and they will tell you the same story—the weather is increasingly unpredictable," said Sheila Watt-Cloutier, Inuit activist and president of the Inuit Circumpolar Conference in 2006.

> The look and feel of the land is different. The sea ice is changing. Hunters are having difficulty navigating and traveling safely. We have even lost experienced hunters through the ice in areas that, traditionally, were safe! Our Premier, Paul Okalik, lost his nephew when he was swept away by a torrent that used to be a small stream. . . . The melting of our glaciers in summer is now such that it is dangerous for us to get to many of our traditional hunting and harvesting places. (Pegg 2004)

The Inuits' ancient connection to their hunting culture may disappear within her grandson's lifetime, Watt-Cloutier said. "My Arctic homeland is now the health barometer for the planet" (Pegg 2004).

For years, Inuit hunters and elders have reported changes to the environment that are now supported by American, British, and European computer models that have concluded that climate change is amplified at high latitudes. Inuit hunters above the Arctic Circle in Alaska say they are definitely seeing a trend: the ice regularly comes a month later than it did in 1980 and roughly two months later than in 1970. Ice also breaks up earlier than previously, so hunting seasons that require transport over ice are becoming shorter. Steven Kooneeliusie and the other Inuit who hunt caribou, seal, and other animals say the signs of a gradual increase in temperatures are everywhere around them. "When I went hunting years ago, I used to wear a full-length caribou skin coat, but now I just wear a light parka. It is so hot these days my snowmobile often overheats," Kooneeliusie said in the small town of Pangnirtung, 100 miles north of Iqaluit, astride the Arctic Circle (Ljunggren 2000).

Hunters as the Ultimate Weather Watchers

Arctic hunters are attuned to their environment because they must be. The natural world is a tough master, and they must know how animals behave.

They are the ultimate weather watchers, observing the

> formation of seasonal ice crusts, changes in snow forms used as navigational aids, animal behavior and sea-ice conditions to conclude that the wind was becoming less persistent and predictable. . . . monitoring of sea-ice thickness, extent and other climate-related phenomena . . . [for] hunters who were most interested in changes in snow conditions and wind, any shift [of wind direction] can mean the difference between life and death if the weather signs are misinterpreted. (Weatherhead et al. 2010)

The ability to thrive in the Arctic depends in large part on the ability to anticipate and respond to dangers, risks, opportunities, and change. Knowing where caribou are likely to congregate is as important as knowing how to stalk them. Sensing when sea ice is safe for travel is an essential part of bringing home a ringed seal (nattiq).

The accuracy and reliability of this knowledge has been repeatedly subjected to the harshest test because people's lives depend on decisions made on the basis of their ecological knowledge. Mistakes can lead to death, even for those with great experience. Thus, information of particular relevance to survival has been valued and refined through countless generations as individuals combine the lessons of their elders with their own personal experience.

The dependency of people on each other is the first lesson of a demanding environment. "According to ancient Inuit philosophy," said one informed observer

> sharing among all beings makes survival in the Arctic possible. A real Inuk would never brag about his hunt. Nor would he decline to share the hunt's reward, for to do so would be to contravene the basic laws of respect among all creatures, who exist as equals. To be disrespectful in this way would offend the seals and encourage them to disappear. (Pelly 2001, 106)

A rapidly warming climate seems to have offended nature in basic and profound ways, causing animals to lose their instincts. For example, Native peoples in parts of the Arctic have described ringed seal pups being abandoned by their disoriented mothers as ice melts sooner now than it used to and thus impeding breeding cycles. These pups often are too young to fend for themselves, not having been properly weaned, so they have died in large numbers. Spotted seals have had similar problems, but bearded seals, for unknown reasons, seemed to be in relatively good condition, at least until 2010. Walruses have been forced to travel farther as sea ice melts earlier, and they have been losing weight. Because of violent weather and early ice melt, polar bears have been less able to find shelter. All of this puts pressure on Native Alaskan hunters, who must range far afield of their homes to find seals and walrus and must negotiate unfamiliar and often treacherous conditions on ice that breaks up under their feet (Voggesser 2010).

The Arctic's rapid thaw has also made hunting—never a safe or easy way of life—even more difficult and dangerous. Hunters in and around Iqaluit say that the weather has been seriously out of order since roughly the middle 1990s. One Inuit hunter, Simon Nattaq, was hunting alone when he fell through unusually thin ice and became mired in icy water long enough to lose both his legs to hypothermia, one of several injuries and deaths reported around the Arctic because of thinning ice.

Pitseolak Alainga, another Iqaluit-based hunter, said that climate change compels caution. One must never hunt alone, he said. Before venturing onto ice in fall or spring, hunters should test its stability with a harpoon, he said. Such tests are especially important when a hunter travels in a snowmobile. In older times, sled dogs could sense thin ice, something a machine cannot do. Alainga knows the value of safety on the water. His father and five other men died in late October 1994 after an unexpected ice storm swamped their hunting boat. The younger Alainga and one other companion barely escaped death in the same storm. He believes that more hunters are suffering injuries not only because of climate change but also because basic survival skills are not being passed from generation to generation as in years past when most Inuit lived off the land.

An Ice World Melts

Ice is fundamental to the entire Arctic food chain, as described by Susan McGrath in the *National Geographic* (2011):

> Sea ice is the foundation of the Arctic marine environment. Vital organisms live underneath and within the ice itself, which is not solid, but pierced with channels and tunnels large, small, and smaller. Trillions of diatoms, zooplankton, and crustaceans pepper the ice column. In spring, sunlight penetrates the ice, triggering algal blooms. The algae sink to the bottom, and in shallow continental shelf areas they sustain a food web that include clams, sea stars, arctic cod, seals, walruses—and polar bears (McGrath, 2011, 70).

The swiftly changing Inuit world revolves around the ice, which is melting rapidly year by year. The traditional economy and its spiritual bonds are tethered to the ice. The pronounced thinning of Arctic sea ice has made the ice pack more brittle and susceptible to wind drift. The volume of Arctic sea ice decreased by one-third during 2007–2011 compared with the 1979–2006 mean.

In the Eskimo village of Kaktovik, Alaska, on the Arctic Ocean roughly 250 miles north of the Arctic Circle, a robin built a nest in town during 2003—not an unusual event in more temperate latitudes but quite a departure from the usual in a place where, in the Inupiat Eskimo language, no name exists for robins. Some residents of Baker Lake, Nunavut, 1,330 kilometers east of Iqaluit, spotted magpies flitting around town in May 2006. These scavengers, relatives of crows, had never been seen in Nunavut before. The magpies are not expected to become permanent residents, however, even if the climate warms, because they roost in trees. The tundra has no trees—not yet. However, in the Okpilak River valley on Alaska's North Slope, which heretofore has been too cold and dry for willows, they are now sprouting profusely. Never mind the fact that in the Inupiat language *Okpilak* means "river with no willows." In addition, three kinds of salmon have been caught in nearby waters in places where they were once unknown.

Correspondent Jerry Bowen, reporting for the CBS Morning News from Barrow, Alaska, the northernmost town in Alaska, on August 29, 2002, quoted native elder Simeon Patkotak as saying residents there had just witnessed their first mosquitoes. Ice cellars carved out of permafrost were melting as well, forcing local Native people to borrow space in electric freezers for the first time to store whale meat. Average temperatures in Barrow have risen 4°F during the past 30 years (Bowen 2002). The average date at which the last snow melts at Barrow in the spring or summer receded some 40 days between 1940 and the years after 2000—up to mid- or late May from early July (Wohlforth 2004, 27).

Warming has extended to all seasons. In Iqaluit, for example, thunder and lightning used to be rarities. The world of the Arctic is changing rapidly before indigenous peoples' eyes. Lightning strikes, for example, ignite fires on the tundra just as in forests; more than 20,000 were recorded on Alaska's North Slope in 2007, a period of record high temperatures and widespread drought in the area. One strike ignited a fire near the Anaktuvuk River. By the time its embers were snuffed out by falling snow in early October, it had scorched an area the size of New York City (Qiu, 2009, 34). According to the U.S. Bureau of Land Management, the frequency of lightning strikes on the North Slope has increased 10-fold in 10 years (Qiu, 2009, 34). Warming brings more lightning, and lightning-sparked fires add carbon dioxide to the air (Qiu, 2009, 34). Fires also accelerate the thawing of permafrost, adding even more carbon dioxide and methane to the atmosphere.

Rapid warming has extended into midwinter. During the first week of January 2011, record warmth in Iqaluit drove temperatures above freezing, an extreme rarity in midwinter that created a debilitating muddy mess. After that, everything flash froze into dangerous sheets of ice. The temperature at one point was 68°F above average. In New York City, a similar departure from average would have produced a 100°F reading on New Year's Eve. In Iqaluit, a high temperature of 35°F in the first week of January was an all-time record. Nearby, the village of Pangnirtung hit 8°C (46°F) on January 4. "The normal around this time of year is around −22°C (−8°F)," Yvonne Bilan-Wallace, a meteorologist with Environment Canada, told the *Nunatsiaq News* in Iqaluit. "So yeah, you're way above normal" (Windeyer 2011). "Accustomed to blizzards, Iqaluit road crews are finding it challenging to keep the streets passable," reported the Indian Country Today Media Network. "When rain covered the previous day's sand, then froze, they couldn't put down more sand until they had chipped the ice away with a grader" (Nunavut Capital, 2011).

Hunters Attempt to Adapt

Rapid changes in weather and climate complicate hunters' finely tuned relationship with their environment, including the timing and speed of freezes and thaws, thickness of ice, positions of floes and their edges, movements of sea ice, and migrations of wildlife. Hunters report that the ringed seal harvest has been delayed as much as a month due to later freezing (Laidler et al. 2009, 376). Even then, ice tends to be thinner, making travel even more hazardous than usual. Ice now often forms, breaks up, and refreezes, further complicating travel and hunting. Hunters also report that winds have become more variable, increasing the roughness of ice surfaces as they freeze, also impeding transportation and hunting (Laidler 2009, 377).

Inuit hunters have done their best to adapt to rapidly changing conditions. Attentiveness to the weather has always been valued in the Arctic, as in any place where the natural environment can be lethal. In addition, modern technology has been put to work. It is not unusual to see hunters consult the Global Positioning System and satellite images to check ice conditions before leaving town with a dog team. Contact is maintained with people at home via VHF radio, and immersion suits are sometime worn during periods of unstable ice to provide buoyancy and warmth to hunters who fall through the ice. Sharing traditional foods (especially for those who cannot hunt themselves within extended families) is considered an affirmation of Inuit identity, especially when food is scarce, even at a time when more Inuit are buying imported food.

Weather watchers in the Arctic say that anticipating conditions, especially regarding wind speed, direction, and variability, has become more difficult. Inuit have always experienced changeable weather—the major reason some act as watchers to anticipate changes—but starting in the 1960s, and most notably during the 1990s, experience has exceeded natural bounds. Shari Gearheard lives in the Inuit hamlet of Clyde River, which includes some 800 people on the shore of Baffin Bay along the northwestern spine of Baffin Island. In *Climatic Change* (2010, 274) she wrote, "What is unprecedented is how quickly conditions change from one to the next and the persistence of this unpredictability over time."

Further Reading

Gearheard, S., et al. "Linking Inuit Knowledge and Meteorological Station Observations to Understand Changing Wind Patterns at Clyde River, Nunavut." *Climatic Change*, 100 (2010), 267–294. doi: 10.1007/s10584-009-9587-1.

Laidler, G. J., et al. "Travelling and Hunting in a Changing Arctic: Assessing Inuit Vulnerability to Sea Ice Change in Igloolik, Nunavut." *Climatic Change* 94 (2009): 363–397. doi: 10.1007/s10584-008-9512-z./

Ljunggren, David. "Effects of Global Warming Clear in Canada Arctic." Environmental News Network, April 20, 2000. http://www.enn.com/enn-subsciber-news-archive/2000/04/04202000/reu_arctwarm_12170.asp (no longer available).

McGrath, Susan. "On Thin Ice." *National Geographic*, July, 2011, 64–75.

McLean, Kirsty Galloway. *Advance Guard: Climate Change Impacts, Adaptation, Mitigation and Indigenous Peoples: A Compendium of Case Studies.* U.N. University Institute of Advanced Studies. Traditional Knowledge Initiative. Traditional Knowledge Initiative. Darwin, Australia, 2010. http://archive.ias.unu.edu/resource_centre/UNU_Advance_Guard_Compendium_2010_final_web.pdf.

"Nunavut Capital Battles Heat Wave." Indian Country Today Media Network January 5, 2011. http://indiancountrytodaymedianetwork.com/article/nunavut-capital-battles-heat-wave-9443.

Pegg, J. R. "The Earth Is Melting, Arctic Native Leader Warns." Environment News Service, September 16, 2004.

Pelly, David F. *The Sacred Hunt: A Portrait of the Relationship between Seals and Inuit.* Vancouver, BC: Greystone, 2001.

Qiu, Jane. "Arctic Ecology: Tundra's Burning." *Nature* 461 (September 3, 2009): 34–36.

Voggesser, Garrit. "The Tribal Path Forward: Confronting Climate Change and Conserving Nature." *The Wildlife Professional,* Winter 2010.

Weatherhead, E., S. Gearheard, and R. G. Barr. "Changes in Weather Persistence: Insight from Inuit Knowledge." *Global Environmental Change* 20 (2010): 523–528.

Wohlforth, Charles. *The Whale and the Supercomputer: On the Northern Front of Climate Change.* New York: North Point Press/Farrar, Strauss and Giroux, 2004.

See also: Adaptation, Animals, Lizards, and Warming Habitats; Animal Life, Arctic; Climate Change, Abrupt Nature of; Ice Melt, Arctic; Sea Ice, Arctic; Summer Ice, Arctic

INUIT

On southern Baffin Island, July 2001 was extremely warm in Iqaluit (pronounced "Ee-ha-loo-eet," meaning "many fish" and "fishing place"), the capital of the semisovereign Inuit nation of Nunavut in the Canadian Arctic. The bizarre weather was the talk of the town. The urgency of global warming was on everyone's lips. Even as U.S. President George W. Bush faulted the concept of global warming as lacking "sound science," after several warm days, the temperature hit 25°C on July 28, in a community that nudges the Arctic Circle. It was the warmest summer anyone in the area could remember. Travelers joked about forgetting their shorts, sunscreen, and mosquito repellant—all now necessary equipment for a globally warmed Arctic summer.

A Natural World Turned Upside Down

In Iqaluit, a warm, desiccating northwesterly wind raised whitecaps on nearby Frobisher Bay and rustled carpets of purple saxifrage flowers on bay-side bluffs as people emerged from their overheated houses (which were built to absorb every scrap of passive solar energy), swabbing ice cubes wrapped in hand towels across their foreheads. The high temperature at 78°F was 30°F above the July average of 48°F and comparable to a 110 to 115°F day in New York City or Chicago. The wind raised eddies of dust on Iqaluit's gravel roads as residents swatted slow, corpulent mosquitoes.

The bay on which Iqaluit sits is named for Martin Frobisher, an early explorer who sought the legendary Northwest Passage from Europe to Asia. With the polar ice cap in the Canadian Arctic melting, the fabled Northwest Passage is now opening. Commercial shippers are eyeing the prospect of cutting more than 4,000 miles off their usual trip through the Panama Canal, bringing one more finger of the industrial south to the Arctic.

Inuit people speak of a natural world that is being turned upside down. "We have never seen anything like this. It's scary, *very* scary" said Ben Kovic, Nunavut's chief wildlife manager. "It's not every summer that we run around in our T-shirts for weeks at a time" (Johansen 2001, 18). At 11:30 a.m. on a Saturday in late July, Kovic was sitting in his backyard and repairing his fishing boat, wearing a T-shirt and blue jeans in the warm wind, with many hours of Baffin's 18-hour July daylight remaining. On a nearby beach, Inuit children were building sandcastles with plastic shovels and buckets, occasionally dipping their toes in the still-frigid seawater.

Ice Melts Even in Winter

Warming in winter brought mud at unfamiliar times. Although snow never brings Iqaluit to a stop, winter mud can paralyze the town. During the first week of January 2011, record warmth in Iqaluit drove temperatures above the freezing point, a rarity in midwinter, which created a debilitating, muddy mess. After that, the thaw froze into dangerous sheets of ice. The temperature at one point was 68°F above average. In New York City, a similar departure from average would have produced a 100°F New Year's Eve. A high of 35°F in the first week of January was an all-time record. Nearby, the village of Pangnirtung experienced 8°C (46°F) on January 4. "The normal around this time of year is around –22 [°C (–8°F)]," Yvonne Bilan-Wallace, a meteorologist with Environment Canada, told the *Nunatsiaq News* in Iqaluit. "So yeah, you're way above normal" (Windeyer 2011).

Pangnirtung, north of Iqaluit, is the main gateway to Auyuittuq National Park—an Inuit word, now out of date, that means "the land that never melts," the most northerly national park in North America. Established in 1972, the park covers 19,500 square kilometers of deep mountain valleys, dramatic fjords, ancient glaciers, and spiny peaks. Auyuittuq ("Ow-you-ee-tuk") is famed for its enormous glaciers. Local people say that even these glaciers are slowly melting. "The glaciers have receded over the last 10 years, and the ice is much worse," according to hunter Solomon Nakoolak (Ljunggren 2000).

In Inuktitut, the word *sila* means "the environment" and is also a synonym for "weather." The weather is both changing (*silaup aulaninga*) and becoming "mixed up" (*uggianaqtuq*)—that is, seemingly turned on its head (Wright 2014, 236). In some areas, not enough snow falls anymore to build *igluit* (igloos), the iconic ice houses of the Inuit (Wright 2014, 237). At the same time, "Satellite phones have become essential equipment, as being able to make contact with search and rescue may mean life and death. Because elders can no longer predict the weather based on clouds, wind direction, sky, the northern lights and even stars, their knowledge is no longer relevant and is not always treated with the same respect. This has an impact on the cultural role of elders" (Wright 2014, 239).

Just how dramatic is *silaup aulaninga* (more roughly meaning "climate change" these days)? According to the Arctic Climate Impact Assessment,

> Change has been so dramatic that during the coldest month of the year . . . [in December 2001], torrential rains have fallen in the Thule region so much that there appeared a thick layer of solid ice on top of the sea ice and the surface of the land. . . . The snow that usually covers the sea ice became *nilak* (freshwater ice), and the lower level became *pukak* (crystalized ice), which was very bad for the paws of our sled dogs. (Arctic Climate Impact Assessment 2004, 84)

Late Freezes, Early Thaws

Thunderstorms are now far more common across the Arctic. The same day that Iqaluit had a high of 78°F, the forecast high for Yellowknife, the capital of the Northwest Territories, was 85°F with scattered thunderstorms. During the summer of 2004, highs in the 80s became routine in Fairbanks, Alaska. The winters of 2000–2001 and 2001–2002 in Iqaluit were notable for liquid precipitation (freezing rain) in December. Tromso, in far northern Norway, had almost no snow during the winter of 2001–2002, along with extremely mild weather—as warm as plus 15°C at one point shortly after Christmas. There was almost no snow at the time. Around the Tromso airport near the ocean, residents could mow their lawns. A few years later, thunderstorms bearing heavy rain were reported during brief temperature spikes even in February across Alaska's North Slope, over and near Iqaluit, and along the southwestern coast of Greenland.

Early in January 2004, Sheila Watt-Cloutier wrote from Iqaluit on Baffin Island that Frobisher Bay had just frozen over for the season at a record late date:

> We are finally into very "brrrrrr" seasonal weather and the Bay is finally freezing straight across. At Christmas time the Bay was still open and as a result of the floe edge being so close we had a family of Polar bears come to visit the town a couple of times. Also in Pangnirtung [north of Iqaluit] families from the outpost camps came into town for Christmas by boat! Imagine that the ice was not frozen at all there by Christmas. But this week our temperatures with the wind-chill reached minus 52 so we are pleased. . . . (Watt-Cloutier 2005)

Snows in Iqaluit also now tend to be heavier and wetter than ever before. Winter cold spells, which still occur, have generally become shorter, according to longtime residents of the area.

Climate change has been rapid and easily detectable within a single human lifetime. "The main climate change impacts reported by Nunavik residents are earlier springs, longer and hotter summers than before (30 to 40 years ago) with less precipitation, colder and shorter winters with less snow, thinner ice covering on lakes and rivers, lower lake levels, decreased river flows, and the drying up of small rivers and small lakes," wrote one informed observer (McLean 2010). "When I was a child," said Watt-Cloutier, "we never swam in the river [Kuujjuaq] where I was born [in Nunavik, northern Quebec]. Now kids swim there all the time in the summer." Cloutier does not remember even having worn short pants as a child (Johansen 2001, 19).

Dirty Glaciers and Land-Dwelling Polar Bears

Some of the rivers to which Arctic char return to spawn have dried up, according to Kovic. The summer of 2001 brought drought as well as record warmth to Iqaluit and its hinterland. Flying above glaciers in the area, Kovic has noticed that their coloration was changing. "The glaciers are turning brown," he said, speculating that melting ice may be exposing debris, and that air pollution from southern latitudes may be a factor. Some ringed seals with little or no hair have been caught, said Kovic. Asked why seals have been losing their hair, Kovic answered, "That is a big question that someone has to answer" (Johansen 2001, 19).

As a wildlife officer, Kovic sees changes that alarm him. Polar bears, for example, now often come onshore rather than stay on the ice, sometimes with dire consequences for unwary tourists. The harbor ice at Iqaluit did not form in 2000 and 2001 until late November or December, five or six weeks later than usual. The ice also breaks up earlier in the spring—sometimes in May in places that once were icebound into early July. In Resolute Bay, bears attracted by the smell of seal meat have been known to chase children on their way to school. Some local Inuit hunters have turned profits in the thousands of dollars by leading tourists on hunts for land-bound, hungry bears that local people have come to regard as a dangerous nuisance.

In Iqaluit, warmer weather is imperiling the safety of Arctic drinking water. Chlorine-treated water is delivered daily in tank trucks in some parts of Nunavut, but large proportion of Inuit still depend on untreated water, which poses health risks in an area with large numbers of migratory animals at a time when temperatures are rising. Studies began in 2003 and 2004 to evaluate the risk of gastroenteric diseases related to climate change. Factors influencing water quality include

> major shoreline erosion on certain rivers, higher turbidity of running water, and a deterioration of raw water quality. Collection of untreated water is a traditional activity among Inuit, particularly elders, who have spent a large part of their lives outdoors. For most of the people interviewed, gastrointestinal symptoms were usually associated with tap water. When they started having problems with the tap water, they would look for water outdoors. However, they mentioned that in spring, when the ice was starting to break up, the water

obtained outdoors tasted different, and cases of diarrhea were more frequent. (McLean 2010)

Alaskan Native Villages Fall to Encroaching Seas

In the 227 federally recognized Alaskan Native communities and villages, many residents obtain sustenance by hunting walrus, caribou, and other land animals as well as fishing for salmon and other species. Climate change poses serious problems for Alaska Native populations. Thinning sea ice can trap people many miles from home as rapidly changing snow and ice endanger hunters. Other problems include "malnutrition and food insecurity from lack of access to subsistence food; contamination of food and water; increasing economic, mental, and social problems from loss of culture and traditional livelihood; increases in infectious diseases; and the loss of buildings and infrastructure from permafrost erosion and thawing, resulting in the relocation of entire communities" ("Indigenous Peoples" 2014). Permafrost, subsoil that used to remain frozen, now sometimes dissolves into a melting, muddy mess and is no longer the "glue" that provides a base for building.

Coastal erosion is provoking health problems, including "loss of clean water for drinking and hygiene, saltwater intrusion, and sewage contamination that could cause respiratory and gastrointestinal infections, pneumonia, and skin infections" ("Indigenous Peoples" 2014). Permafrost thaw also threatens Native food supplies as it destroys ice cellars used to store meat acquired by hunters. Food may be spoiled. To prevent the risk of contamination and illness, many Native people rely on imported food that is less healthy and more expensive than anything caught by hunters.

In Alaska, these climate-induced changes threaten village infrastructure, water supplies, health, and safety. Habitat changes associated with loss of sea ice and landscape drying and extensive wildfires occur with a frequency and ferocity unknown before—to the point of altering the hunting landscape. A simple example are lichens, which provide food for caribou in winter. They do not recover from fire for 70 years.

Several coastal Alaskan Native villages have been declared disaster areas following several severe storms. By 2013, more than 30 villages were experiencing or facing, relocation because of climate change, including Shishmaref and Newtok. "Alaska is seeing all these things the rest of the country hasn't seen yet," said Dr. Jerome Montague, Native Affairs and Natural Resources Advisor for the Alaskan Command Joint Task Force (Jessepe 2012).

If the U.S. government was as good at arranging relocations as issuing reports on them, new locations would have been found for villages such as Newtok and Kivalina long ago. A growing stack of government reports has described flooding and erosion threatening four villages—Kivalina, Koyukuk, Newtok, and Shishmaref—whose residents have all agreed they must move.

Kivalina's Steady Erosion

Kivalina, a village of about 400 primarily Inupiat residents 80 miles north of the Arctic Circle, spreads across a thin, barrier reef island that was first used as a hunting ground by peoples who have lived in the area for thousands of years. In 1905, Native parents were required to enroll their children in school there by the federal government or face imprisonment. The parents moved in with the children and began a settlement around the school. Today, Kivalina hangs precariously onto a slender, eroding peninsula. The U.S. Army Corps of Engineers projects that the sea will wipe it off the map by 2025.

For many years, sea ice routinely formed around Kivalina early each fall. The hard freeze provided a hard surface, and the ice protected the settlement against storms. As temperatures have risen in recent decades, ice forms in late November and even December, exposing the village to rough seas and erosion. The strip of land under Kivalina, which lies 120 miles north of the Arctic Circle, shrunk from 55 acres of livable space in 1953 to 27 acres in 2003 as the Chukchi Sea eroded the shoreline from the west and Kivalina lagoon at the mouth of the Wulik and Kivalina Rivers undercut it from the east (Shearer 2011, 13–14, 50). The people of Kivalina decided that their location was unsustainable by 1990 and voted to move first in 1992 but found that no aid was available.

Over the years, Kivalina's problem has come to be shared by as many as 200 coastal villages in Alaska, of which at least 30 face relocation, according to a Government Accountability Office report issued in 2005. In 2006, the Army Corps of Engineers estimated that the village of Kivalina, as well as Newtok and Shishmaref, would be lost to erosion in 10 to 15 years, with an estimated relocation cost of $80 million to $200 million for each village. In 2004 and 2005, storms of unusual ferocity struck Kivalina anew, eroding between 70 and 80 feet of coastline, exposing the permafrost, and endangering several houses. In 2008 and 2009, the Corps of Engineers built a rock revetment that it said would buy the village 10 to 15 years—at most and barring further episodic flooding.

By roughly 2000, the sea was not freezing by the time autumn storms arrived. By 2004, what used to be an annual, bearable cycle of fall storms "became a life-threatening event." The island, said Swan, "was falling apart into the Chukchi Sea before the very eyes of its inhabitants," who struggled to buttress the crumbling shoreline. "Every object placed along the edges," said Swan, "was being sucked into the angry sea." The violence of the storms during that and following years made evacuation by sea impossible. "There was nowhere to go, and nothing the volunteers did worked to keep the island together—the people were trapped!" (Shearer 2011, 100).

In Kivalina, erosion and flooding has poured sediment into the community water system, requiring filtration to prevent ingestion of unsanitary water. People In Kivalina could take only sponge baths and limited access to laundry for the winter. "As hand-washing and bathing decreased, respiratory and skin diseases increased," health aides said ("Climate Change" 2011).

Newtok's Life Runs Out to Sea

In far western Alaska, the Yup'ik Eskimos—the *Qaluyaarmiut* or "dip net people"—of Newtok on the Ninglick River adjacent to the Bering Sea in far western Alaska have lived along the Bering Sea coast for at least 2,000 years. Until recently, Newtok was home to some 400 people in 63 houses (before several of them moved away), few of which are well insulated, and some of which have been sinking into thawing permafrost.

Storm tides flooded the village water supply, causing raw sewage to spread through the community, displacing residents from homes, destroying food storage, and shutting down essential utilities. Public infrastructure has been significantly damaged or destroyed because of the combination of extreme weather events and ongoing erosion. A major storm in 2005 took out the Ninglick River barge landing, so fuel and other supplies could no longer be delivered by sea during the annual thaw.

The beginnings of infrastructure have been built at a new site: a barge landing, six houses, and a foundation for what might become an emergency evacuation center. However, no sewage, water, or electric utilities were available as of 2013. Every step has encountered a governmental gauntlet. In the meantime, angry seas and melting permafrost are slowly turning old Newtok into a climate-change memorial.

Shishmaref's Demise

Six hundred people in the Inuit village of Shishmaref on a barrier island just north of the Bering Strait (and called Kigiktaq by the Inupiag in the village) on the Chukchi Sea, some 60 miles north of Nome, have been watching their village erode into the sea. The site had been occupied for almost 400 years when increasing storm surges began to erode more than 23 feet a year off its shoreline late in the 20th century. Within a few years, the new climate regime in which ice no longer held back the increasingly turbulent sea was threatening the old village with destruction (Cordalis and Suagee 2008, 47).

The permafrost that had reinforced its coast is thawing. "We stand on the island's edge and see the remains of houses fallen into the sea," wrote Anton Antonowicz in the *Daily Mirror*. "They are the homes of poor people. Half-torn rooms with few luxuries. A few photographs, some abandoned cooking pots. Some battered suitcases" (Antonowicz 2000, 8). Percy Nayokpuk, a village elder, runs the local store, which now perches dangerously close to the edge of the advancing sea. "When I was a teenager, the beach stretched at least 50 yards further out," said Percy. "As each year passes, the sea's approach seems faster" (Antonowicz 2000, 8). Five houses have washed into the sea; the U.S. Army Corps of Engineers has moved or jacked up others. The villagers have been told they will soon have to move.

Year by year, the hunting season, which depends on the arrival of the ice, starts later and ends earlier. "Instead of dog mushing, we have dog slushing," said Clifford Weyiouanna, 58, a reindeer herder near Shishmaref (Antonowicz 2000, 8). Villagers have been catching fish, such as flounder, which are usually associated with warmer water.

Because of retreating ice, Shishmaref hunters are being forced to search as far as 200 miles from town for walrus. They also now use boats to hunt seals they used to track over ice. "This year the ice was thinner, and most of the year at least part of the ice was open. We don't normally see open water in December," said Edwin Weyiouanna, an artist who has lived most of his life on the Chukchi Sea (Murphy 2001). In earlier years, the sea was more likely to be frozen during much of the stormy winter season. With warming, the erosive, wind-whipped ocean corrodes Shishmaref's waterfront. Now the town's residents have come to fear the full moon, which brings unusually high tides.

Old photos indicate a time when the village fronted on wide, sandy beaches. Today, each major storm carves off chunks of the island. Then, as now, most of Shishmaref's houses had no running water or plumbing, so the town captured rain and snow for reuse because most residents shower and wash their clothes at a public facility. As its island shrinks and houses crumble into the sea, people crowd into what remains. Several plans to move have come and gone, and by 2014 many people in the village had given up on moving because of the $179 million price tag, according to a study by the Army Corps of Engineers in 2004.

According to village elders, "Fourteen houses on Shishmaref's north side had to be put on skids and dragged down to the opposite end of the island after a major storm in October 1997. Another big storm in October 2001 sloughed off huge chunks of the northern shoreline" (Sheppard 2014). In the meantime, money from the U.S. federal government dried up after Sen. Ted Stevens was defeated in 2008. Seawalls have been built, and the sea has swallowed them. Talk of relocation did little except stall life on the current site. "It stopped all investment in this community," said Percy Nayokpuk, owner of the island's general store, of the relocation plan. "There's been a lot of projects lost because of the vote to move. That's about all that's resulted from that vote" (Sheppard 2014). Dozens of reporters and television crews came and went as the island continued to shrink over years that became decades.

Further Reading

Antonowicz, Anton. "Baking Alaska: As World Leaders Bicker, Global Warming Is Killing a Way of Life." *Daily Mirror* (U.K.), November 28, 2000, 8–9.

Arctic Climate Impact Assessment. *Impacts of a Warming Arctic.* Cambridge, UK: Cambridge University Press, 2004.

Cordalis, D., and D. B. Suagee. "The Effects of Climate Change on American Indian and Alaska Native Tribes." *Natural Resources and Environment* 22(3) (Winter 2008): 45–49.

"Climate Change Puts Health of Arctic Villagers on Thin Ice." Indian Country Today Media Network, March 07, 2011. http://indiancountrytodaymedianetwork.com/article/climate -change-puts-health-of-arctic-villagers-on-thin-ice-21391.

"Indigenous Peoples, Lands, and Resources." National Climate Assessment, U.S. Global Change Research Program, 2014. http://nca2014.globalchange.gov/report/sectors/indig enous-peoples.

Jessepe, Lorraine. "Alaskan Native Communities Facing Climate-Induced Relocation." Indian Country Today Media Network, June 21, 2012. http://indiancountrytodaymedianetwork .com/article/alaskan-native-communities-facing-climate-induced-relocation-119615.

Johansen, Bruce E. "Arctic Heat Wave." *The Progressive*, October 2001, 18–20.

Ljunggren, David. "Effects of Global Warming Clear in Canada Arctic." Environmental News Network, April 20, 2000. http://www.enn.com/enn-subsciber-news-archive/2000/04/04202000/reu_arctwarm_12170.asp (no longer available).

McLean, Kirsty Galloway. *Advance Guard: Climate Change Impacts, Adaptation, Mitigation and Indigenous Peoples: A Compendium of Case Studies.* UN University Institute of Advanced Studies. Traditional Knowledge Initiative. Traditional Knowledge Initiative. Darwin, Australia, 2010. http://archive.ias.unu.edu/resource_centre/UNU_Advance_Guard_Compendium_2010_final_web.pdf.

Murphy, Kim. "Front-Row Exposure to Global Warming; Climate: Engineers Say Alaskan Village Could Be Lost as Sea Encroaches." *Los Angeles Times*, July 8, 2001, A1.

Shearer, Christine. *Kivalina: A Climate Change Story.* Chicago: Haymarket Books, 2011.

Sheppard, Cate. "Climate Change Takes a Village as the Planet Warms, a Remote Alaskan Town Shows Just How Unprepared We Are." Huffington Post, December 14, 2014. http://www.huffingtonpost.com/2014/12/14/shishmaref-alaska-climate-change-relocation_n_6296516.html.

Watt-Cloutier, Sheila. "The Climate Change Petition by the Inuit Circumpolar Conference to the Inter-American Commission on Human Rights: Presentation by Sheila Watt-Cloutier, Chair, Inuit Circumpolar Conference [at the] 11th Conference of Parties to the U.N. Framework Convention on Climate Change," December 7, 2005. Montreal. http://inuitcircumpolar.com/index.php?ID=318&Lang=En.

Windeyer, Chris. "South Baffin Swelters in Winter Heat Wave." *Nunatsiaq News*, January 4, 2011. http://www.nunatsiaqonline.ca/stories/article/98789_south_baffin_swelters_in_winter_heat_wave/.

Wright, Shelly. *Our Ice Is Vanishing/Sikuvut Nunguliqtuq: A History of Inuit, Newcomers, and Climate Change.* Montreal: McGill-Queen's University Press, 2014.

See also: Adaptation, Animals, Lizards, and Warming Habitats; Animal Life, Arctic; Climate Change, Abrupt Nature of; Ice Melt, Arctic; Sea Ice, Arctic; Sea Level Rise; Summer Ice, Arctic

LOUISIANA NATIVES

In Louisiana, where land is subsiding as it also erodes from rising sea levels, Native tribes are losing so much land that many people have been forced to abandon their homelands. As land sinks and seas rise, saltwater intrusion has been ruining farm fields. Loss of land also has been accelerated by the extraction of oil and gas and by construction of levees and dams.

The tribes of the area include the Atakapa-Ishak living in Grand Bayou Village in Plaquemines Parish, Louisiana. The Atakapa-Ishak are a mix of various tribes and French immigrants. They have inhabited this site for some 300 years and the wider area for several more centuries. The Grand Caillou and Dulac Band (Biloxi-Chitimacha Muskogees Confederation), who live in or near several bayous in Terrebonne Parish, Louisiana, have likewise lived in the area for several hundred years.

Land Losses for Louisiana Tribes

The Isle de Jean Charles Band (Biloxi-Chitimacha Confederation of Muskogees) also live in Terrebonne Parish, on a narrow ridge of land between Bayou Pointe-aux-Chene and Montegut Bayou (called Isle de Jean Charles), on land that until 1876 was regarded as uninhabitable swamp by the state of Louisiana and before the state began selling tracts of it to immigrants. Also of mixed heritage (combining French and Choctaw, among others), the Isle de Jean Charles Band arrived in the area around 1840. Today they are battling land loss and saltwater intrusion that has rendered cultivation of food especially difficult ("Vulnerability" 2012).

In 2016, this community, then down to some 60 people, received $48 million from the U.S. government so it could be relocated. It thus became the first climate-refugee community to be moved with federal money. By that time, more than 90 percent of their homeland on the Gulf of Mexico had washed away (Davenport and Robertson 2016).

In 2012, the Pointe-au-Chien Indian Tribal Community had 680 members living in lower Pointe-au-Chien, Terrebonne Parish, Louisiana, a traditional Chitimacha village, which combines Biloxi, Acolapissa, and Atakapas ancestry with other Native peoples who survive by hunting alligators, fishing, and catching shrimp, crabs, and oysters. Like the others, this group also has historically cultivated food but saltwater intrusion and loss of land to the ocean is making agriculture difficult ("Vulnerability" 2012).

This site was not purposefully chosen as a place to live by its inhabitants; they sought refuge there to escape European–Americans who had seized their lands. To avoid being relocated by force or killed, they hid in the bayous and in dense forested swamps. Over the years they built a community as the land sank and the sea slowly rose, with both hurricanes and oil-company canals accelerating the erosion. The land also sank as dikes and levees were constructed upstream and the Mississippi River was dammed, preventing the sediment and silt that used to flow toward the coast from reaching the delta. Natural wetlands that once protected the island from coastal storms died. Between 2005 and 2013, six major storms battered the area, the best known of which were Katrina and Rita. By 2013, a hurricane was not required to flood what was left of the community. High tide on a full moon would do it ("Vulnerability" 2012):

> Many of the Isle's trees, traditional and medicinal plants, gardens, and trapping grounds are gone. There were 78 homes in 2002, but only about 25 remain by 2012. The community also continues to experience health and livelihood impacts from decades-long industrial contamination and encroaching toxic industries, chemicals from dispersants, oil spills, including the 2010 BP oil disaster, and post-storm debris contaminating the air, soil, and water. (Maldonado et al. 2013, 601)

Unlike Alaskan villages that are eroding into the sea, the official file on Isle de Jean Charles' desperate straits is extremely thin. In fact, as a tribe without federal recognition, a file does not even exist. All four of these tribes lack U.S. federal recognition status, which thus denies them funding for many programs that might provide education, housing, and utilities, as well as other services from the Federal Emergency Management Agency and the Bureau of Indian Affairs. They face their situation more or less alone.

Writing in *The New York Times Magazine* (October 5, 2014), Nathaniel Rich sketched the speed with which Louisiana is losing its coastline:

> Since 2011, the National Oceanic and Atmospheric Administration has delisted more than 30 place names from Plaquemines Parish alone. English Bay, Bay Jacquin, Cyprien Bay, Skipjack Bay, and Bay Crapaud have merged like soap bubbles into a single amorphous body of water. The lowest section of the Mississippi River delta looks like a maple leaf that has been devoured down to its veins by insects. The sea is rising along the southeast coast of Louisiana faster than it is anywhere else in the world.
>
> The land loss is swiftly reversing the process by which the state was built. As the Mississippi shifted its course over the millennia, it deposited sand and silt in a wide arc. This sediment first settled into marsh and later thickened into solid land. But what took 7,000 years to create has been nearly destroyed in the last 85. Dams built on the tributaries of the Mississippi, as far north as Montana, have reduced the sediment load by half. Levees penned the river in place, preventing the floods that are necessary to disperse sediment across the delta. The dredging of two major shipping routes, the Mississippi River Gulf Outlet and the Gulf Intracoastal Waterway, invited saltwater into the wetlands' atrophied heart.
>
> Beneath the surface, the oil and gas industry has carved more than 50,000 wells since the 1920s, creating pockets of air in the marsh that accelerate the land's subsidence. The industry has also incised 10,000 linear miles of pipelines, which connect the wells to processing facilities; and canals, which allow ships to enter the marsh from the sea.

Tribes Share Knowledge

Native communities along Louisiana's coast that are facing rapid climate changes met in January 2012 to "share knowledge, support, cultural connectivity, and adaption strategies." The National Resources Conservation Service also took part, with members of tribes, Native leaders from outside, representatives of government agencies, and others; proceedings were reflected in the 2013 National Climate Assessment report.

The Native people living on the coast and on low-lying islands near the swamps, marshes, bayous, and deltas of southern Louisiana have experienced storms of varying intensities, the rise and fall of sea levels and tides, and changes in the level and path of the Mississippi River for hundreds of years. "However," according to a report on the meeting in 2012,

> in recent years, environmental changes including subsidence, land sinking and shrinking, and sea-level rise have posed uncommon challenges to these

indigenous communities. Natural disasters, such as Hurricanes Katrina, Rita, Gustav, Ike, Lee and Isaac, have taken a significant toll. Additionally, the tribes have also had to cope with various impacts resulting from the petroleum industry, ranging from standard canal construction to large-scale disasters such as the BP oil spill [as well as [levee installations, fossil fuel infrastructure and extraction]. This profile explores the ways in which climate change may exacerbate the challenges already facing coastal Louisiana tribes. ("Vulnerability" 2012)

In addition, all of these factors have been compounded by natural subsidence of a land base that is steadily eroding into the ocean and afflicted with a rising tide of saltwater that has made food cultivation more difficult. The saltwater intrusion is also killing coastal forests. "There used to be trees and forest for as far as you could see or run. We would go out to play and there was land all around us, now there is just water," said Shirell Parfait Dardar, who is Pointe-au-Chien. Donald Dardar told the hearing, "I used to get lost walking in the trees behind my house. Now there's nothing. Bays and bayous were miles from home, now they're all around" ("Vulnerability" 2012). Even before climate change is factored in, peoples of the area told the 2012 workshop, "The land and waters that we depend on for our lives, our culture, [and] our heritage, have been abused, broken, and poisoned" ("Vulnerability" 2012).

Levees (and some canals dug out by oil companies) may protect some communities in the area, but Native areas usually end up flooded with displaced water, robbed of sediment that used to replenish their slivers of land on and near the coast. The levees and canals impede the natural flow of the Mississippi River and its sediment load. Robbed of sediment, the Native peoples' cultivated land erodes (as it sinks and sea levels rise), forcing people to purchase expensive processed food at grocery stores (yet another form of colonization). Offshore oil spills (the largest being BP's *Deepwater Horizon* in 2010 (five years after Hurricanes Katrina and Rita) have further ravaged what was left of the coastal peoples' ability to farm and gather fish and shellfish in the bayou country. The Grand Bayou Village's fishing and shrimping was fundamentally damaged by the BP spill, reducing the catch by 70 percent for a time (Faerber 2010).

Research by Bethel and colleagues in (2011, 569) combined local knowledge and geospatial mapping to illustrate how land area available to the Grand Bayou area of Plaquemines Parish had diminished over the preceding 45 years. During the 1950s, Isle de Jean Charles was 11 miles long and five miles wide. In 2012, it was no more than two miles long and one-quarter mile wide. A Public Broadcasting System "News Hour" report in 2012 described the scene:

Along the coastal bayous of Louisiana, forests are full of dead bald cypress trees, their roots ravaged by saltwater that has been creeping increasingly inland from the Gulf of Mexico. These ghost forests stand in stark contrast to the landscape of blooming trees and grazing cattle that Theresa Dardar walked across as a child to reach the burial site of her ancestors. Now as water is swallowing up her ancestors' land, she takes a boat. (PBS News Hour 2012)

Many traditional medicinal plants cannot grow in soil laced with salt. Canals built by oil companies tend to increase tidal exchange and salinity thus increases and kills plants. Once plants die, more land is lost because of increased erosion, "regular tides and storms to more easily erode existing marshland, creating a positive feedback loop of deterioration and land change. Additionally, tropical storms wreak havoc on the fragile and quickly disappearing gardens and vegetation" ("Vulnerability" 2012).

Changing Social Relationships

Dwindling harvests of homegrown foods change social relationships. People used to trade things (garden produce for freshly caught shrimp, for example), but no one trades store-bought groceries. The declining number of medicinal plants also decreases the number of natural remedies, requiring people with little income to pay for medical services. "Because of land changes and the loss of traditional foods and medicinal plants, tribal lifestyles once rooted in local ecosystems and traditional cultural practices are now disconnected and dependent on nontribal systems" ("Vulnerability" 2012).

Erosion of land and loss of forests has made what is left of costal communities more vulnerable to hurricanes. Under these conditions, even a minor hurricane now can become a major disaster. "Back when I was a child, we used to ride out the hurricane on the island and not worry about flooding. We didn't have to worry about the winds either, because there were a lot of trees. . . . And now it's basically an open field. So when a hurricane comes, it's like here we are, come and get us," said Chief Albert Naquin of Isle de Jean Charles ("Vulnerability" 2012). When hurricanes strike, disaster-relief efforts often bypass small coastal communities in favor of urban areas:

> While national and international concern focused on the destruction of the city of New Orleans during Hurricane Katrina, coastal Louisiana tribes were fighting their own battle. Tribal communities lost boats, homes, and livelihoods to the mud and floodwater that washed over the area during severe storms. In addition, the hurricanes caused the loss of many coastal wetlands, and devastated the "seafood, crabbing, oystering, shrimping, hunting, alligator, and fur processing industries" which the tribes depend on. (Archambeault 2006)

As Hurricane Katrina approached, some Grand Bayou community members evacuated by boat, realizing that roads would be impassible for weeks afterward. On boats, they were able to return quickly to begin cleanup and rebuilding. Some of them lived on boats as long as nine months after Katrina.

Louisiana's state government proposed that coastal tribes move away from the area as individuals, but that would split up tribes and end any possibility of federal recognition. A dwindling land base already has forced some people out in such numbers that they are having trouble maintaining a sense of community. The Grand

Caillou and Dulac is an example. "On the Isle de Jean Charles, 95 percent of the community has had to move away for work (leaving just the older folks and grandkids). . . . In the Grand Bayou Village, developers [who] seemingly ignore subsiding land and rising sea levels want to 'buy out' the tribes and build condominiums and fishing camps near the bayous the tribes have traditionally owned for generations" ("Vulnerability" 2012).

Given their small numbers and lack of federal recognition, the small tribes of costal Louisiana are being left alone to drown. Garret Graves, chairman of the Coastal Protection and Restoration Authority, told the PBS News Hour that "looking at the amount of money it would take [to protect the Isle de Jean Charles community], and the [number of] homes . . . you have to prioritize" (PBS News Hour). In the meantime, remaining residents struggle to survive with what resources they can muster, building bulkheads and raised gardens, and asking to be dealt into disaster planning.

Further Reading

Archambeault, William G. "Louisiana Indians: Survivors in a Post-Katrina Environment." *Indigenous Policy Journal*, 2006. Indigenous Studies Network (ISN). http://indigenouspolicy. org/Articles/VolXVIINo3/LouisianaIndians/tabid/181/Default.aspx. Accessed May 16, 2012.

Bethel, M. B., et al. 2011. "Blending Geospatial Technology and Traditional Ecological Knowledge to Enhance Restoration Decision-Support Processes in Coastal Louisiana." *Journal of Coastal Research*, 27(3) (2011): 555–571.

Davenport, Carol, and Campbell Robertson. "Resettling the First American 'Climate Refugees.'" *The New York Times*, May 2, 2016. http://www.nytimes.com/2016/05/03/us/resettling-the-first-american-climate-refugees.html.

Faerber, Fritz. 2010. *Oil Spill Threatens Native American "Water" Village.* National Geographic. Online film. http://news.nationalgeographic.com/news/2010/06/100608-us-oil-gulf-indians-video/. Accessed May 16, 2012.

Maldonado, J. K., et al. "The Impact of Climate Change on Tribal Communities in the U.S.: Displacement, Relocation, and Human Rights." *Climatic Change* 120 (2013): 601–614.

PBS News Hour. "Native Lands Wash Away as Sea Levels Rise." Louisiana Public Broadcasting (co-producer), June 1, 2012. http://www.pbs.org/newshour/updates/climate-change/jan-june12/louisianacoast_05-30.html (no longer available).

Rich, Nathaniel. "The Most Ambitious Environmental Lawsuit Ever." *The New York Times Magazine*, October 5, 2014. http://www.nytimes.com/interactive/2014/10/02/magazine/mag-oil-lawsuit.html.

"Vulnerability of Coastal Louisiana Tribes in a Climate Change Context." Natural Resources and Conservation Services (National Resources Conservation Service) Workshop. Tribes and Climate Change. Northern Arizona University. September 26, 2012. http://www4.nau.edu/tribalclimatechange/tribes/gulfcoast_lacoastal.asp.

See also: Floods; Sea Level Rise

NAVAJO, THE

Persistent drought in the U.S. Southwest is forcing Navajos who have no indoor plumbing to travel several miles for water as their wells run dry. Drought is also forcing the early sale of livestock because former scanty pastures are turning to bare dirt. "Perhaps among the worst of those impacts," wrote Terri Hansen in the Indian Country Today Media Network (2014),

> are the runaway sand dunes it has unleashed, which extend over one-third the 27,000-square-mile reservation. During the 1996–2009 drought period the extent of dune fields increased by some 70 percent. These dunes are moving at rates of approximately 35 meters per year, covering houses, burying cars and snarling traffic, degrading grazing and agricultural lands, contributing to the loss of rare and endangered native plants, and when they occur contributing to poor air quality, a serious health concern for many of the reservation's 173,667 residents.

The 25 percent to 40 percent of Navajos who haul their own water pay 20 times per volume that of non-Navajos who have piped in supplies on per capita income that is less than half the U.S. average—before adding the expense of round trips that average 28 miles (Cozzetto 2013, 569). During droughts, which are becoming more frequent, both the cost of water and the distances required to acquire it increase.

The drought also has become pervasive in other parts of the U.S. Southwest. "Our 30,000-acre reservation is pretty dry because of drought," said Lawrence Snow, Land Resources Manager for Utah's Shivwits Band of Paiutes. "Wildfires in the last decade have burned half our acreage and changed the landscape. We've got less trees, and bark beetles are trying to kill off the ones we do have. Once the fires happened and took out the ground cover, major storms brought big flooding" (Allen 2012). When Native areas in the U.S. Southwest receive rain and snowfall, it increasingly comes in flooding deluges. Arizona's Havasupai tribe between 2008 and 2010 endured several damaging floods.

From Little Water to Nearly None

Navajo and Hopi lands in Arizona have always been relatively dry, but climate change in recent years often has made matters worse. The Navajo and other Southwestern Native peoples have made a fine art of surviving on little water for centuries and made their ways of farming more efficient and their animals able to deal with scare water. There is a difference, however, between little rain and nearly none, and that is what they have been dealing with for 20 years. Cindy Dixon's sheep, for example, used to forage scrub on the desert near Farmington, New Mexico. By 2014, however, even that had died and Dixon had to turn to expensive bales of hay. "The landscape around her Navajo Reservation homestead," wrote Bobby Magill in Climate Central, "was as brown and bleak as the open-pit coal mine a few miles to the west and well within earshot" (Magill 2014).

Dixon lives without electricity or running water, but her sheep cannot eat sand. "Since it's all dry and bare and deserted—no vegetation—I have to constantly buy hay and grain to keep the sheep fed," Dixon said, looking at the land around her trailer. "This is a bad, bad area for livestock" (Magill 2014). Sometimes Dixon cuts her own grocery spending so that she can buy hay for the sheep. The lack of forage is compounded by coal dust blowing in from the mine on stiff winds that are now pushing sand dunes across the brown, desiccated land.

Migrating Sand Dunes

By 2014, sands dunes were "covering housing, causing transportation problems, and contributing to loss of endangered native plants and grazing land" (Cozzetto 2013, 569). Rainfall in some parts of the Navajo Nation fell to 3 inches a year during the latest drought (Redsteer et al. 2011a, 2011b). According to the U.S. Geological Survey (USGS), "More than one-third of Native lands on the Colorado Plateau (Navajo Nation and Hopi tribal lands) are covered with sand dunes and sand sheets" because of the enduring drought (Redsteer et al. 2011a, 2011b). Lands that once were marginally productive for sheep grazing and dry-land agriculture (a long-time practice among the Navajo and Hopi) are becoming true water-starved deserts. Or, as the USGS phrases it, "Reactivation of inactive dunes could have serious consequences on human and animal populations, agriculture, grazing, and infrastructure on the Navajo Nation and similar areas in the Southwest" (Redsteer et al. 2011a, 2011b). Wind and drought have been worst in the spring.

Dunes are migrating faster across the landscape at speeds heretofore unknown in Navajo country. In 2009, the USGS measured dune migration as fast as 112 to 157 feet per year. Some dunes moved more than 3.3 feet in a single windstorm. The Grand Falls dune field has grown in areal extent by 70 percent (laterally and downwind) in 15 years (1992 to 2007). The drought has continued since then, punctuated by rare deluges that quickly run off the cracked, parched, and increasingly sandy soil.

Streams that once were sources of water have dried up, feeding the wind with plumes of gritty, irritating sand. According to the USGS, "The formation and movement of active dunes on the downwind side of stream-bed sand sources is currently endangering housing and transportation, potentially jeopardizing native plants and grazing lands, increasing health hazards to humans and animals, and affecting regional air quality" (Redsteer et al. 2011a, 2011b).

The Navajo Nation has experienced several decades of rising temperatures, declining snowfall, and decreased stream flow (and nonexistent flow in some cases), which results in water scarcity that has "magnified the impacts of drought that began in 1996 and continues today" (Redsteer et al. 2013, 390). Stream flow data and historic information on surface-water features (such as springs, lakes, and streams) show significant changes over the past century (Redsteer et al. 2010). Historical sources as well as elders' accounts describe many watercourses that are dry today. Some began to disappear in the early to middle 20th century. "Moreover," wrote Redsteer et al. (2013, 390), "significant reductions in the number and length of stream reaches with perennial flow have occurred since 1920, and for some historic ephemeral streams, no flow during spring run-off and summer rains occurs today." Many surface-water features began to disappear in the early to mid-1900s and are now only ephemeral or entirely dry year-round.

Elders Share Observations

Traditional Navajo elders shared observations of weather patterns' impacts, including "declines in snowfall, surface water features, and water availability" (Redsteer et al. 2011b), as well as "lack of available water and changing socio-economic conditions as leading causes for the decline in the ability to grow corn and other crops. . . . Other noticeable changes reported in these accounts include the disappearance of springs and the plants and animals found near water sources or in high elevations, such as certain medicinal plants, cottonwood trees, beavers, and eagles" (Redsteer et al. 2010). The elders also noticed changes in wind speed that were changing the behavior of sand and dust storms. These changes have been impeding the ability to grow corn, the staff of life, as well as the use of corn pollen, which has a central role in a large majority of Navajo ceremonies. When ceremony is impeded, the disruption emphasizes a paradigm shift in climate going back hundreds if not thousands of years.

Navajo traditional livelihoods have always been tied to the land. With more heat, less precipitation, and expanding sand dunes, however, sheep herding has become ever more difficult. Unemployment on the Navajo reservation was 47 percent in 2011, and the poverty rate was 37 percent. The median household income was $24,000 (Magill 2014). Roughly 40 percent of Navajos on the reservation have no electricity and running water, and at least one-third of families hauled their water 10 miles or more. Amenities that are taken for granted in many urban areas in the United States and Canada do not exist for Navajos. Coal is mined in Navajo country to power electricity-generating plants that supply Tucson, Las Vegas, and Phoenix, their lines running over Navajo hogans with no power.

A Sand Dune as a Neighbor

Lester and Louise Williams, who live near Tuba City, have a sand dune as a neighbor that has worn out its welcome. Despite the dune, the Williamses say they have no plans to leave their house. Kathy Ritchie (2014) described the sand dune astride the Williams' house:

> Lester Williams—a.k.a. "Chee Willie"—and his wife, Louise, live just a few feet from a massive sand dune. The gigantic pile looks to be around 20 feet high. This is the Williamses' fifth house, and they share it with their children and grandchildren. The very same dune that looms right outside swallowed their four previous homes, along with their sheep corrals. Inside the tiny home, family photos hang from the wall, along with a calendar from Hank's Trading Post and a framed poster of the Canadian Rockies. . . . Chee Willie doesn't

speak English, but he's incredibly animated when he speaks Navajo. Through a translator, he talks about the difficulties of living in this kind of environment, where the wind whips the sand so furiously that the family can't leave the house. He says he once tried to remove the sand himself, but it came back. The sand always comes back.

Chee Willie's wife, Louise, says that blowing sand makes breathing difficult. The wind howls, and the sand swirls. Yet they stay. Home is a sacred place among Navajos. Home is a sacred bond—over and above the practical difficulty of obtaining a new lease, moving livestock, and the fact that more and more Navajo land is laced with migrating dunes.
Like heavy snowdrifts, dunes move across roads and block them with increasing regularity.

As we're packing up and preparing to leave Chee Willie's house, Tohannie tells me that yesterday's windstorm shifted a nearby dune, causing it to cover part of a road used by the handful of families in the area, including Tohannie's parents. "I got stuck with my son," he says. "We had to shovel our way out." These are not small drifts, and they can't be moved, like snow, with a plow. And, if course, they never melt. (Ritchie 2014)

Ritchie described one dune as a "magnificent sculpture shaped from mostly eroded Navajo and Entrada sandstone. . . . Many of the dunes in the area are unexpectedly tall, some measuring anywhere from 30 feet to 40 feet high. Begay tells me that near Preston Mountain, some 45 minutes north of where I'm standing, the dune field is even higher, possibly 60 feet to 80 feet in places" (Ritchie, 2014). The dominant plant species in some areas, if any survive, has become tumbleweed, which has evolved to move with the wind and anchors nothing.
Across the reservation, persistent winds have been driving sand into homes and across roads. Most of the roads are not paved, and their surfaces can become parts of the moving dunes. Even major paved roads have been blocked. On April 16, 2013, driven by winds gusting to 60 miles an hour, drifts of sand closed parts of Interstate 40, as traffic backed up 12 miles. A NASA satellite photographed the dust plume from space.

Memories of Wetter Times

Wise Navajo drivers carry shovels. "Kee Tohannie, Begay's grandfather and Huskie Tohannie's father always carries a shovel, chains and sometimes a hatchet in case he or one of his neighbors is marooned in the sand. 'I've been stuck in the sand many times—it's a lot of digging,' he says. 'You just have to know how to drive in sand. Like you learn to drive in snow'" (Ritchie 2014).
Redsteer, who is of Crow descent, was raised in their homeland near the Montana–Wyoming border, but in the 1970s she married a Navajo, moved to his homeland, and mothered three children. Many people told her how much plant life in the area

had changed over the years. Only later did she begin to associate these changes with climate change. In 1986, Redsteer and her family moved to Flagstaff, Arizona, when she was 29 years old. She studied for a doctorate at Northern Arizona University, and shortly after 2000, now employed by the U.S. Geological Survey, she switched her focus of study from volcanic deposits near Yellowstone National Park to the effects of climate change on the Navajo Nation as intensifying drought was causing sand dunes to grow and migrate there. She wrote several academic papers and had a key role in the National Climate Assessment, released by President Barack Obama in 2013. She has become known for linking elders' recollections with weather records to trace the evolution of climate change.

Bobby Magill wrote in Climate Central (2014),

> Navajo elders remember wetter times, when winter snows were knee-deep, water always ran in springs and arroyos, and the rangeland among the canyons, mesas and volcanic hills could support large herds of livestock, a mainstay of the Navajo economy. Some elders recalled a time when they were children, with moist ground until the Fourth of July, climate data on the Navajo Nation indicates a marked drying trend since the middle 1940s, and a warming of 4 [°F] in many areas since the 1960s. The decrease in snowfall has long-term implications. "Snow is like water in the bank," Redsteer said. "It takes a long time to melt. It soaks into the ground slowly." ("The Navajo" 2011)

Since the 1950s, the Southwest has experienced extremely wet and extremely dry periods, but the current drought has dominated the weather since the 1980s, with brief wet periods in 2004, 2005, and 2010 doing little to alleviate that long-term trend.

Navajo farmer Jonathan Yazzie said that for many years he grew squash, corn, zucchini, chilies, and even cantaloupe. The drought has put him out of business. "The water is just not there no more," he said. "We're down to 15 sheep. No cattle. Two horses. That's our kids' future" (Magill 2014). He has no access to irrigation he has sought from the Navajo and Hopi governments. Without it, continued drought will force his family to move. Traditional herding is dying. "I don't know a single young Navajo person today who's thinking about having their own sheep herd," Redsteer said. "Part of that is due to their own market economy. The feasibility of doing that is just impossible now. There is still a holdout group of elderly people who really don't have a choice. Their language is Navajo. Their culture is Navajo. They really don't have any other place to go" (Magill 2014). The Navajo Times remarked, "Hardly anyone is making a profit from their livestock any more; they've become expensive pets" ("The Navajo" 2011).

A technical report for the National Climate Assessment issued in 2013 (Garfin et al. 2013) said that the Four Corners area probably will continue to endure warmer weather on average during coming decades as soil continues to dry with droughts becoming more intense and frequent. The drought—and spread of sand dunes—is worst in the southwestern quarter of the Navajo reservation, where many families may be forced to move, according to Redsteer (Magill 2014), who has used the recall of elders as well as weather data to trace climatic changes. Average snowfall across the Navajo Nation has declined dropped from around 31 inches in 1930 to 11 or so inches by 2010, according to a United Nations case study. "Every tribal elder mentioned the lack of snowfall," Redsteer said. "They describe winters [in the past] where the snow was 'chest high on horses.' The snowfall shows a significant decline over the 20th century, and is still declining in recent years" (Magill 2014). Elders' memories are especially important in recent years as heat and drought have accelerated because many U.S. government weather stations were shut down during the early 1980s to save money.

By 2014, Margaret Redsteer had studied the spreading Navajo sand dunes for 14 years in collaboration with the Navajo Nation. She also works with Northern Arizona University's Tribal Environmental Education Outreach Program in a continuing effort to stabilize dunes with native vegetation. The battle is never easy, and the changing climate grants no favors. Although native plants require time to get established after precious rains, invasive tumbleweeds explode nearly overnight. Tumbleweed sucks up moisture before native plants get it, according to Redsteer. "Tumbleweed is a major blow to rangeland conditions," Redsteer said. "It is amazing how huge the areas are that are affected by tumbleweed" (Ritchie 2014).

Further Reading

Allen, Lee. "Southwest Tribes Struggle with Climate Change Fallout." Indian Country Today Media Network, June 14, 2012. http://indiancountrytodaymedianetwork.com/article/southwest-tribes-struggle-with-climate-change-fallout-118386.

Cozzetto, K., et al. "Climate Change Impacts on the Water Resources of American Indians and Alaska Natives in the U.S." *Climatic Change* 120 (2013): 569–584.

Garfin, Gregg, et al. *Assessment of Climate Change in the Southwestern United States: A Report Prepared for the National Climate Assessment.* Washington, DC: Island Press, 2013. http://swccar.org/sites/all/themes/files/SW-NCA-color-FINALweb.pdf.

Hansen, Terri. "Climate Disruptions Hitting More and More Tribal Nations." Indian Country Today Media Network, May 7, 2014. http://indiancountrytodaymedianetwork.com/2014/05/07/climate-disruptions-hitting-more-and-more-tribal-nations-154747.

Magill, Bobby. "The Navajo Nation's Shifting Sands of Climate Change." Climate Central, May 28, 2014. http://www.climatecentral.org/news/navajo-nation-climate-change-17326.

"The Navajo Memory Complements Science in Study of Climate Change." USGS Newsroom. October 21, 2011. http://www.usgs.gov/newsroom/article.asp?ID=3007&from=rss#.VAb2Wmv29vZ (no longer available).

Redsteer, M. H., et al. "Disaster Risk Assessment Case Study: Recent Drought on the Navajo Nation, Southwestern United States." Chapter 3 in *Global Assessment. Report on Disaster Risk Reduction,* 2010. http://www.preventionweb.net/english/hyogo/gar/2011/en/what/drought.html.

Redsteer, M. H., R. C. Bogle, and J. M. Vogel. "Monitoring and Analysis of Sand Dune Movement and Growth on the Navajo Nation, Southwestern United States." 2011a. Fact Sheet Number 3085. Reston, VA: U.S. Geological Survey.

Redsteer, M. H., K. B. Kelley, and H. Francis. "Increasing Vulnerability to Drought and Climate Change on the Navajo Nation." 2011b. Paper GC43B-0928, delivered at American Geophysical Union annual meeting, December 5–9, 2011, San Francisco.

Redsteer, M. H., et al. "Unique Challenges Facing Southwestern Tribes." Pp. 385–404 in G. Garfin et al., eds., *Assessment of Climate Change in the Southwest United States: A Report Prepared for the National Climate Assessment*. Washington, DC: Island Press, 2013. http://www.swcarr.arizona.edu/sites/default/files/ACCSWUS_Ch17.pdf.

Ritchie, Kathy. "Dune and Gloom." *Arizona Highways*. 2014. http://www.arizonahighways.com/extras/dune.asp (no longer available).

See also: Deforestation; Desertification; Drought, United States; Drought, Worldwide

PACIFIC NORTHWEST TRIBES

Before European–Americans arrived, the Hoh and Quileute, like many other tribes in what is now western Washington, ranged widely to hunt, gather, and fish. The Quileute and Hoh used some 900 square miles north and west of the Olympic Mountains. According to treaties signed during the 1850s, they and other tribes surrendered most of the land to the federal government but retained rights to fish that were defended legally through U.S. courts in the 1970s. Both the Quileute and Hoh today have reservations that are only a mile square. In the traditional fashion, the reservations are located at the confluence of rivers and saltwater, the Quileute on the Pacific Ocean and the Hoh on the Strait of Juan de Fuca, the best place to harvest salmon.

The Hoh and Quileute

With the advent of a warming climate, both locations have been assailed by increasingly violent rains and storm surges. Both reservations lie in rain forest (with as much as 160 inches of rain per year), compared to Seattle, a more sheltered location, which averages 36 inches annually. In addition, the Hoh River descends 7,000 feet from Mount Olympus to the strait within 50 miles. This location has become more vulnerable to both freshwater flooding and saltwater storm surges as more precipitation falls as rain rather than snow and as wind-driven waves increase coastal erosion. The Hoh already abandoned one lower village during the 1970s when storm surges rose and the river changed course.

As the 2014 National Climate Assessment reported, on the coast of Washington state, "[the] Quileute tribe in northern Washington is responding to increased winter storms and flooding connected with increased precipitation by relocating some of their village homes and buildings to higher ground within 772 acres of Olympic National Park that has been transferred to them; the Hoh tribe is also looking at similar options for relocation."

At Quilieute, many people "report that storms are increasing in intensity and frequency during the winter . . . superstorms have hit the area with fierce winds and violent rains. In 2006, waves whipped up by a superstorm threw giant driftwood up to the elementary and middle school playgrounds" for the first time in many elders' memories (Papiez 2012, 71). Later, another, even more violent storm broke through a bulkhead and "almost took the school away. . . . More trees are falling, They [storms] pack more wallop now than they ever did" (Papiez 2012, 71). Storms have grown so violent that crabbers often cannot leave port, and state regulations have been amended to make their season more flexible. Quileute has been cut off by flooding so often that local road closures have become routine. Septic tanks backed up (one person recalled that "our toilets became fountains!") as families were evacuated from their homes (Papiez 2012, 74).

The Swinomish Face Rising Waters

The same wind, rain, and storm surges that devastated the Quilieute and Hoh reservations during 2006 also inundated parts of costal Swinomish land on northern Puget Sound, provoking an effort there to use projections of the Intergovernmental Panel on Climate Change's fourth assessment to anticipate what a century of global warming might do (Intergovernmental Panel on Climate Change 2007). The Swinomish (whose name means "people by the water"), using a $400,000 federal grant, found that 1,100 acres of their land (15 percent of the reservation), including all of their agricultural land, is at risk of saltwater inundation from rising sea levels (expressed mainly as storm surges). This area included land planned for economic development as well as shorelines. In addition, 160 homes, and 18 nonresidential properties (mainly commercial structures) with a combined value of $102 million were at risk from sea level rise. Large areas (2,218 acres and more than 1,500 properties worth $518 million) could be at risk for wildfires.

Along the shore, important shellfish harvesting sites also faced inundation. "Important 'keystone' species such as shellfish and salmon are at risk of higher levels of contamination from algal blooms and other diseases that may be exacerbated by increased temperatures and other changes," the report said ("Swinomish Climate" 2012). The elderly, those with asthma, and others of all ages risk respiratory ailments from increased heat. In addition, cultural sites, including those used for ceremonial and medicinal plants, could be lost or damaged by persistent hot and dry weather.

In 2014, the Swinomish tribe received a three-year, $756,000 U.S. Environmental Protection Agency grant to further develop its study of climate-change impacts that will focus on indigenous health indicators, including cultural, family, and emotional aspects of climate change, with an emphasis on storm surges. The project will "develop a model showing projected coastal erosion due to sea-level rise, storm surge, and wave energy through Year 2100 on the shores of the Swinomish Reservation." Swinomish Tribal historic preservation officer Larry Campbell said, "We're protecting the universal resource rather than the tribal resource. We're doing a lot more for the state and the county, and then in the end the tribe benefits by taking care of the whole. We're a very aggressive tribe when it comes to our environment" ("Swinomish Study" 2014).

Writing for the Pacific Northwest Tribal Climate Change Project at the University of Oregon, Carson Viles (n.d.) described how sea level rise affects biodiversity in estuaries—that is, natural areas where freshwater rivers meet saltwater. He described the Swinomish, who harvest at the junction of saltwater and rivers, and described the damage of slowly rising seas and storm surges in such locations:

> As sea levels continue to rise, access to these areas is being lost. In addition to the loss of traditional fishing sites, rising sea levels are threatening estuaries along the coast. Estuaries are incredibly productive ecosystems that provide valuable habitat for many first foods plant and animal species, most notably salmon. The shifting state of estuaries has the potential to radically diminish important first foods populations. Damaged estuaries could mean a loss of gathering grounds for indigenous peoples along the Pacific Northwest coast, as well as a massive loss of a critically important ecosystem type. These areas represent not only access to productive fishing, but also the knowledge accumulated over time about how local populations of fish behave in particular places. Loss of these fishing places constitutes a loss of interaction between indigenous peoples and specific species populations. (Swinomish Indian Tribal Community 2010)

Further Reading

Intergovernmental Panel on Climate Change. *Climate Change 2007: Synthesis Report.* IPCC Fourth Assessment Report (AR4). 2007. https://www.ipcc.ch/publications_and_data/publications_ipcc_fourth_assessment_report_synthesis_report.htm.

Papiez, Chelsie. "Climatic Change in the Quileute and Hoh Nations of Coastal Washington." Pp. 68–88 in Zoltan Grossman and Alan Parker, eds., *Asserting Native Resilience: Pacific Rim Indigenous Nations Face the Climate Crisis.* Corvallis, OR: Oregon State University Press, 2012.

"Swinomish Climate Change Initiative." Pp. 133–145 in Zoltan Grossman and Alan Parker, eds., *Asserting Native Resilience: Pacific Rim Indigenous Nations Face the Climate Crisis.* Corvallis, OR: Oregon State University Press, 2012.

Swinomish Indian Tribal Community. *Swinomish Climate Change Initiative Climate Adaptation Action Plan.* La Conner, WA: Swinomish Indian Tribal Community Office of Planning and Community Development, 2010.

"Swinomish Study Social and Physical Impacts of Climate Change with EPA Grant." Indian Country Today Media Network, September 2, 2014. http://indiancountrytodaymedianetwork.com/2014/09/02/swinomish-study-social-and-physical-impacts-climate-change-epa-grant-156714.

Viles, Carson. "First Foods and Climate Change." Pacific Northwest Tribal Climate Change Project, University of Oregon. No date. http://tribalclimate.uoregon.edu/files/2010/11/firstfoods_climatechange_12-14-11_final.pdf. Accessed September 13, 2014.

See also: Extreme Weather; Fisheries; Floods; Ocean Acidity, North America; Ocean Acidity, Worldwide; Sea Level Rise

GREENHOUSE GASES
AND ENERGY

OVERVIEW

A universal desire of Earth's growing population for comfort, convenience, and profit requires the generation of energy for transportation and industrial production, most of which is supplied by fossil fuels that add greenhouse gases to the atmosphere and thus raise worldwide temperatures and destabilize climate. Just 200 years ago, the proportion of carbon dioxide, which makes up less than 1 percent of the atmosphere, had cycled between roughly 180 parts per million (ppm) and 280 ppm for almost a million years. As coal was combusted to fire the first steam engines, that proportion began to rise. By 2015, it had reached 400 ppm, probably the highest level since the Pliocene Epoch 2 million to 3 million years ago when Earth had almost no long-lasting ice and sea levels were 80 to 100 feet higher.

Energy—especially for transportation—is the crux of the world's fossil fuel dilemma. Aviation in 2013 accounted for some 15 percent of fossil fuels consumed in transport; surface transport accounted for 80 percent. During the 10 years before 2008, passenger air traffic increased by 60 percent. Passenger vehicles in the United States account for 40 percent of the country's oil consumption and 10 percent of the world's (Kolbert 2007, 88, 90).

According to Bill McKibben, "The average American car, driven the average American distance—ten thousand miles—in an average . . . year releases its own weight in carbon into the atmosphere" (McKibben 1989, 6). In a sense, global warming is the exhaust pipe of the "American Dream," of the bigger and better, the new and improved, the mobile life in the fast lane. Cars and trucks used in the United States burn 15 percent of the world's oil production. Transportation alone consumes one-fourth of the energy and two-thirds of oil used in the United States.

Today energy is being generated from fossil fuels at record rates as China, India, and other countries industrialize and populations rise. Coal in particular is the most carbon-intensive fuel per unit of energy provided. Ninety percent of Earth's remaining fossil fuel reserves are coal, the most dangerous fossil fuel from a greenhouse point of view, because most coals produce roughly 70 percent more carbon dioxide per unit of energy generated than natural gas and about 30 percent more than oil. Coal is also the most plentiful fossil fuel, especially in places with large populations such as China, which controls 43 percent of remaining reserves. During the

1980s, China passed the Soviet Union as the world's largest coal producer. China also built 114,000 megawatts of coal-fired power in 2006 and 95,000 more in 2007. Coal poses environmental problems other than carbon emissions. Coal mining also produces methane, and its combustion produces sulfur dioxide and nitrous oxides as well as carbon dioxide. Transport of coal also usually requires more energy than any other fossil fuel.

Air Conditioning's Climatic Irony

Air conditioning engages us in dangerous climatic irony. As global temperatures rise (and as affluence spreads), more of it is used. Because most air conditioning coolants produce greenhouse gases and increased use requires more electric power (and combustion of fossil fuels), its use aggravates climate change. Stan Cox, author of *Losing Our Cool: Uncomfortable Truths about Our Air-Conditioned World*, commented in the *Washington Post* (2016), "Over the past half-century, American cities have taken on an unstable thermodynamic form, coming to resemble collections of boxes full of cool air crowded onto concrete heat islands. Turn off the AC, and office buildings would become uninhabitable, vehicles sitting in traffic would become torture chambers, apartment buildings would become death traps."

In 1960, 12 percent of households in the United States had air conditioning. That figure rose to almost 90 percent by 2016. Between 1950 and 2006, the size of the average U.S. house tripled from 290 to 900 square feet per person. Between 1993 and 2005, consumption of electricity for air conditioning doubled, according to the U.S. Energy Information Administration (Cox 2016). Use of air conditioning, which produces waste heat, also aggravates the urban heat-island effect, exposing the poor and elderly to the worst of heat waves.

As Cox wrote, "Reducing air conditioning dependence in urban areas will require steadily shrinking parking and driving space, accompanied by improvements in public transportation; de-paving and revegetating the rescued space; retrofitting large buildings with ventilation shafts to allow air in when weather permits" (2016). The Lawrence Berkeley National Laboratory also recommends at least three shade trees per building in urban areas, along with a crash program to make all roofs and pavement in U.S. cities reflective, two practical moves that could reduce cooling demand by as much as 20 percent.

Further Reading

Cox, Stan. "Your Air Conditioner Is Making the Heat Wave Worse." *Washington Post*, July 22, 2016. https://www.washingtonpost.com/posteverything/wp/2016/07/22/your-air-conditioner-is-making-the-heat-wave-worse/?wpisrc=nl_headlines&wpmm=1.

Realizing coal's dangers, many energy producers have begun to move away from it and toward using natural gas, wind, and solar. By 2015, coal production and consumption was falling in the United States, much of Europe, and even in China. Financially, many coal-mining companies were in trouble, as were coal-transporting railroads such as the Union Pacific. Wind power has reached price parity with coal for electrical generation, and capacity is growing quickly. The state of Iowa, for example, acquired 50 percent of its electricity from wind in 2016.

Even so, more people are traveling by air, which is especially carbon intensive. One journey across the Atlantic Ocean from the United States to Europe on a jet aircraft will emit as much greenhouse gases *per person* as an average automobile commuter creates in an entire year. At the same time, efforts are underway to make air transport more energy efficient.

Alternative fuels do not provide the thrust required to lift commercial aircraft and keep them aloft; no one will ever build a solar or wind-powered passenger airplane. Some improvements in fuel efficiency have been achieved, however. Hydrogen-powered and electric cars are of little help in reducing greenhouse gas emissions as long as their power sources stem from coal, oil, and gas. Fuel efficiency has improved in this area as well, but not enough to counter the rising number of cars and trucks worldwide.

China, with its 1.4 billion people and booming economy, has become the world's largest generator of greenhouse-gas emissions, surging ahead of the United States during the last 15 years. China presents a paradox. Its policies are de-emphasizing use of coal in favor of wind and solar, not only to reduce carbon-dioxide emissions, but also to alleviate air pollution, even as new coal-consuming power plants are being built.

Although the United States long was Earth's largest generator of greenhouse gas emissions, China has recently surged ahead. Nominally, almost 75 percent of growth in global fossil fuel carbon and cement production between 2010 and 2012 occurred in China, but estimates of greenhouse gas production there are often not exact. In 2015, China admitted to having underreported its coal usage (and thus carbon dioxide emissions) by 17 percent—a billion tons of CO_2 annually. World emissions of CO_2 stabilized in 2014 at 32 billion tons. China's CO_2 emissions have risen from 3 billion tons annually in 2000 to 8 billion in 2014. China had been underreporting its coal consumption since the year 2000, with the disparity increasing over time. By 2015, the 1-billion-ton shortfall was as much coal as the United States burned in total (Buckley 2015).

Even as China has been installing more solar power than any other nation, it has been burning more coal than the United States, Europe, and Japan combined. In 2010, China overtook the United States as the world's largest consumer of energy, having used only half as much as 10 years previously. China has accelerated as a consumer of energy (mainly low-energy coal) more quickly than any other country in world history. The scale of industrial development in China in the late 20th and early 21st centuries also has no parallel in human history. To gauge the scale of the building boom (and greenhouse gas generation) in China, consider the amount of cement manufactured and used there between 2010 and 2013—6.1

gigatons. The United States made and used 4.4 gigatons of cement during the *entire* 20th century. For the first time, in 2009, the Chinese purchased more cars than residents of the United States, 12.8 million to 10.3 million. China's car sales increased 42 percent in 2009 over 2008, 72 percent of them sport utility vehicles, a year when most of the world was in recession.

China is beginning to embrace a need for an energy paradigm shift to renewable sources of energy like most of the world. The major question for energy planners (and everyone else) is, will this paradigm change fast enough to spare Earth a climate catastrophe?

Further Reading

Buckley, Chris. "China Burns Much More Coal Than Reported, Complicating Climate Talks." *The New York Times*, November 4, 2015, A1.

Kolbert, Elizabeth. "Running on Fumes: Does the Car of the Future Have a Future?" *The New Yorker*, November 5, 2007, 87–90.

McKibben, Bill. *The End of Nature.* New York: Random House, 1989.

AIR TRAVEL

By 2020, greenhouse gas emissions of international aircraft entering or departing the United States are expected to be 75 percent more than in 1990, according to testimony by Annie Petsonk, international counsel for the Environmental Defense Fund, before the U.S. Senate Committee on Commerce, Science, and Transportation on June 6, 2012 (Schapiro 2014, 3). Furthermore, nitrous oxides and water vapor expelled by jet aircraft in flight as exhaust (contrails) intensify their greenhouse effect (Schapiro 2014, 3). "The higher you go, the more destructive the impact from the wastes of burned jet fuel," wrote Mark Schapiro in *Carbon Shock* (2014, 3).

One journey across the Atlantic Ocean from the United States to Europe on a jet aircraft emits as much greenhouse gases *per person* as an average automobile commuter creates in an entire year. United States citizens, whose aviation mileage per person increased 400 percent between 1970 and 2006 (Hillman and Fawcett 2007, 55), contribute an enormous and growing carbon overload. The rest of the world is not far behind. Addressing air travel's impact on world climate is most necessary in the United States, origin point for a third of the world's commercial aviation. By 2004, air travel in the United States consumed 10 percent of all fossil fuel energy, with passenger loads expected to double between 1997 and 2017, making air travel the fastest-growing source of carbon dioxide and nitrous oxides in the economy (Flannery 2005, 282).

Airline travel contributed an estimated 5 percent to the total of global human-generated greenhouse gases in 1990, an amount that was increasing at a much faster rate than overall fossil fuel usage. Between 1969 and 1989, according to the International Energy Agency, worldwide airline passenger miles increased 400 percent (Cars and Climate 1993, 44). China's domestic air travel mileage was increasing 15 percent a year by 2008, with 186 new airports planned by 2010 (Stern 2009, 45).

Mark Jacobson of Stanford University reported at the 2009 meeting of the American Geophysical Union that emissions from commercial airline flights could contribute 15 percent to 20 percent of warming in the Arctic. Worldwide, the contribution was estimated at 4 percent to 8 percent of surface global warming, or 0.03 to 0.06°C thus far. A major part of this contribution results from black carbon (soot). This analysis indicated that aircraft emissions increased sky coverage of cirrus clouds over areas where vapor trails were most abundant such as the Arctic. In some areas, sky coverage was reduced. These estimates were calculated from models using data from 2004 to 2006 (Dalton 2009).

Air Travel's Effect on the Atmosphere

Aircraft emissions are especially damaging because much of their pollution takes place in the upper atmosphere at jet-stream level. The problem is significant. Carbon dioxide emissions from air traffic originating in the United Kingdom, for example, rose 85 percent between 1990 and 2000, according to that nation's Department of Trade and Industry.

Jet aircraft emits a toxic cocktail of gases. Atmospheric emissions from aircraft have three times the global warming potential of their carbon dioxide content alone. In addition to carbon dioxide, combusted jet fuel injects water vapor and sulfur dioxide into the stratosphere, both of which enhance ozone depletion at that level. Nitrous oxides enhance ozone at lower levels, where it is a pollutant, and deplete it in the stratosphere, where it helps guard against ultraviolet radiation. The chemicals emitted into the atmosphere by combusting aviation fuel "cause the formation of polar stratospheric clouds [that] affect markedly the aerosol composition of the atmosphere, and intensify the greenhouse effect" (Kondratyev et al. 2004, 249).

"As far as climate change is concerned," wrote George Monbiot in *The Guardian* (U.K.),

> [this] is an utter, unparalleled disaster. It's not just that aviation represents the world's fastest-growing source of carbon dioxide emissions. The burning of aircraft fuel has a "radiative forcing ratio" of around 2.7. What this means is that the total warming effect of aircraft emissions is 2.7 times as great as the effect of the carbon dioxide alone. The water vapor they produce forms ice crystals in the upper troposphere (vapor trails and cirrus clouds) that trap the earth's heat. According to calculations by the Tyndall Centre for Climate Change Research, if you added the two effects together (it urges some caution as they are not directly comparable), aviation's emissions alone would exceed the government's target for the country's entire output of greenhouse gases in 2050 by around 134 percent. (Monbiot 2006)

The government excludes international aircraft emissions from the target.

Role of Jet Contrails in Climate Change

By 1999, in some air-traffic corridors above Europe, jet contrails sometimes covered as much as 4 percent of the sky at any given time. Vapor trails form cirrus clouds that contribute to global warming in addition to the emissions from aircraft engines. Brian Hoskins, a research professor at the Royal Society and former head of meteorology at the United Kingdom's University of Reading, said that if growth of air travel continued unrestricted, airline vapor trails would cover 10 percent of the sky over Britain by 2050 (Clover and Millward 2002). In the northeastern United States, jet contrails sometimes cover as much as 6 percent of the sky. Contrail coverage over Asia was increasing so quickly that contrails could multiply by a factor of 10 in 50 to 60 years. Jet contrails have a net atmospheric warming effect similar to that of high thin ice clouds, trapping outgoing long-wave radiation. The contrails also reflect some incoming solar radiation. Given this balance, nighttime flights (December though February) at a site in southeast England "were responsible for most of the contrail radiative forcing" (Stuber et al. 2006, 864). Night flights "account for only 25 percent of daily air traffic, but contribute 60 to 80 percent of the contrail forcing" (Stuber et al. 2006, 864). Winter flights account for only 22 percent of annual air traffic but contribute half of the annual mean forcing. "These results suggest that flight rescheduling could help to minimize the climate impact of aviation" (Stuber et al. 2006, 864).

A study published in the *Journal of Climate* estimated that increasing coverage of cirrus clouds over the United States (to which air traffic is a major contribution) could increase tropospheric temperatures 0.2 to 0.3°C per decade (Minnis et al. 2004, 1671). The study by researchers at a NASA facility in Langley, Virginia, concluded that contrails contributed to a 0.27°C per decade warming trend in the United States between 1975 and 1994.

The NASA study was the first time weather observations had been used to document temperature change relating to contrails, said Patrick Minnis, a senior research scientist at Langley. "Cirrus clouds can have a net warming or net cooling effect on the Earth, depending on how thick they are," Minnis said. Contrails may stretch 1,600 kilometers and widen to 60 kilometers, depending on the weather. "Cirrus clouds from contrails tend to be thin, and the effect of thin clouds tends to be warming," he said (Schleck 2004).

Contrails from high-flying jets also may be helping to narrow the diurnal range of high and low temperatures, making days slightly cooler and nights slightly warmer, according to measurements taken during the three days after the September 11, 2001, terrorist attacks on New York City and Washington, D.C. With nearly all air travel in the United States grounded, scientists were provided a rare view of nearly contrail-free skies for the first time in almost half a century ("Contrails Linked" 2002).

"Scientists have been noticing unusual changes in diurnal temperatures for quite some time, but can't explain why," said Travis. "We're providing one possible explanation here. Maybe jet contrail coverage is one of the reasons for this shrinking temperature range" ("Contrails Linked" 2002: Travis et al. 2002, 601). David Travis, a climatologist at the University of Wisconsin–Whitewater, led a study that offered some of the first evidence for the climate-changing effects of contrails. Travis and colleagues, including Pennsylvania State University geographer Andrew Carleton and University of Wisconsin–Whitewater undergraduate Ryan Lauritsen, used satellite images to compare cloud cover from those three days to 30 years of data for mid-September. Then they reviewed daytime and nighttime surface air temperatures across North America collected from 4,000 weather stations.

Travis said the findings of this study may complicate the global warming debate because in some regions contrails may offset some of the temperature increases anticipated by global warming models. The study also underscores the point that not all influences on climate are global. Factors such as contrails can make a difference on a regional or local scale.

The Jetsons Come to Life

As scientists worry about the atmosphere's overload of greenhouse gases, commuting by air in the image of the 20th-century cartoon family in *The Jetsons* has taken on a life of its own. Beginning in the 1990s, a rising number of people in the United States were commuting to work via airlines, sometimes thousands of miles per week. *The New York Times* has carried accounts of Manhattan jobholders who fly into the city from Rochester, New York. Silicon Valley, where the price of the average house was $1 million by the year 2015, was drawing weekly commuters, according to one *Times* account, from "Arizona, Idaho, Nevada, Oregon, and Utah" (Johnston 2000). Some people who work at New York's Lincoln Center commute from Florida. The typical schedule includes long workdays from Tuesday through Thursday, travel on Monday and Friday, and a weekend at home hundreds, sometimes thousands, of miles from the office.

For those with a penchant for technology and no concern about rising greenhouse gas levels in the atmosphere, the status vehicle in elite circles of the future may not be a sports utility vehicle. According to one observer, it will be "the family plane" (Fallows 1999, 84). Such an aircraft, the Cirrus SR20, was being produced by the late 1990s, as the National Aeronautics and Space Administration were "quietly advocating an Interstate Skyway Network" (Fallows 1999, 88).

Features in the *Wall Street Journal* have promoted "flying cars," with no mention of their impact on the atmosphere. Environmentally, "air cars" make Hummers look like bicycles. By 2006, about 100 people had paid $25,000 each to reserve a $500,000 Moller International "Skycar," which was at the test stage. The "flying car" to some people was the next step beyond the private jet. Indeed, between 2000 and 2005, the number of private jets in the United States grew by 40 percent.

Skycars are expected to have an initial sticker price of about $500,000 each (Stoll 2006). The Skycar is designed to taxi from home to a designated take-off area (a "vertiport") laid out on a parking lot or field. The Skycar, which its builders say will sell for around $60,000 once in mass production and will run on methanol, ethanol, diesel, or gasoline.

Further Reading

Fallows, James. "Turn Left at Cloud 109." *New York Times Sunday Magazine*, November 21, 1999, 84–89.

Johnston, David Cay. "Some Need Hours to Start Another Day at the Office." *Omaha World-Herald*, Feb. 6, 2000, 1G.

Stoll, John D. "Visions of the Future: What Will the Car of Tomorrow Look Like? Perhaps Nothing Like the Car of Today." *Wall Street Journal*, April 17, 2006, R8.

Reducing Aviation's Carbon Footprint

Aviation is the fastest-growing mode of travel in the United States. To handle this growth, thirty-two of the nation's fifty busiest airports have expanded their facilities in two decades. Sixty of the hundred largest airports built new runways. Since the late 1990s, passenger and freight air transport mileage has doubled roughly every ten years. The number of passengers was increasing by 8 percent a year, on average, as the volume of cargo rises by about 10 percent annually.

According to a report published by the Institute of Public Policy Research, "Flying by jet plane is the least environmentally sustainable way to travel and transport goods" (Lean 2001). "So as we are flying into the Alps for our ski holiday we are contributing to their destruction," wrote one European author. "Our honeymoon flight to the Maldives is slowly sinking it under rising sea levels and destroying coral through bleaching associated with global warming; and finally our safari flight to Africa is contributing to drought, famine and disease. It's not an appetizing thought, is it?" (Francis 2006).

By 2009, international aviation bodies were pledging to cut the industry's carbon emissions by half of 2005 levels by 2050. Giovanni Bisignani, the lead officer of the International Air Transport Association, told U.N. Secretary-General Ban Ki-moon in October, "The aviation industry is serious about its climate change responsibility. We have united all the players with a clear strategy and targets that are even tougher than those our regulators are prepared to administer. No other industry is as united, ambitious or determined" ("Global Aviation" 2009). He said that the aviation industry is reducing its emissions via technology, infrastructure, and daily operations. The International Civil Aviation Organization, whose members account for 93 percent of nonmilitary airline traffic, in mid-October 2009 approved a commitment to limit aviation emissions.

By 2008, several U.S. states and environmental groups were urging the federal Environmental Protection Agency (EPA) to regulate emissions of greenhouse gases by flights that use airports in the United States. "We want the EPA to take [its] head out of the sand and actively promulgate rules to reduce greenhouse-gas emissions," said Jerry Brown, then California's attorney general. "The EPA has taken a very passive and unimaginative approach to combating global warming" (Chea 2007). California, Connecticut, New Jersey, New Mexico, Pennsylvania, and the District of Columbia said in a petition to the EPA that the Federal Aviation Administration expects aircraft emissions over the United States to increase 60 percent by 2025. The petition urged the EPA to be more aggressive in directing airlines to raise fuel efficiency, build lighter and more aerodynamic vehicles, and develop fuels that emit fewer pollutants. The Air Transport Association, the airlines' U.S. lobby group, replied that fuel efficiency already has increased 103 percent since 1978 and will rise another 30 percent by 2025 (Chea 2007).

Could airliners use hydrogen or biofuels? Mark Lynas argues probably not—for several reasons. Hydrogen is too bulky to work as a fuel, and in any case its combustion output of water, when injected high in the stratosphere, would contribute to global warming rather than reducing it. As for biofuels, they do not have the energy density of kerosene (the major constituent of standard jet fuel), and the business of producing them in large quantities is already endangering food security and boosting deforestation across the tropics. There is simply no possibility that they could be produced in the volume needed to slake the thirst of jet aircraft in the long term (Lynas 2006, 14). Lynas added that the land plowed up for Heathrow's runways once grew some of the best apples in England. Nowadays these apples are being replaced by imported produce with a much larger carbon footprint than food grown locally (Lynas 2006, 15).

During late November 2002, two official British studies called for an end to cheap flights and a ban on new airport runways. The Royal Commission on Environmental Pollution urged Britain's government to halt airport growth, raise fares, and place financial pressure on short-haul and no-frills carriers. At the same time, the British government's body of environmental advisers, the Sustainable Development Commission, said that proposals for new runways at Stansted, Heathrow, Luton, Rugby, or a new airport at Cliffe in Kent required a "fundamental rethink" (Clover and Millward 2002). The commission—a body of academics, businesspeople, and interested citizens—recommended that the price of a one-way ticket should rise by 40 pounds sterling to have any chance of mitigating climate change. Paul Ekins, an economist and a member of the commission, said, "We believe a stable climate is a good thing and worth modifying human behavior for" (Clover and Millward 2002). The commission also called for the rapid growth of high-speed trains to replace short-haul services at transport hubs such as Amsterdam's Schiphol airport (Clover and Millward 2002).

A biofuel–jet fuel mix will propel a jet, but will it fly financially? A major problem is cost. Eighty-five percent of jet fuel's cost is feedstock—that is, its source. As of 2009, a gallon of crude oil cost approximately 90 cents U.S. compared to $7 a gallon for fuel from jatropha, $3 a gallon for camelina oil, and $20 a gallon for algae. Cost aside, each of these sources emits less soot and particulate matter than crude oil, as well as roughly half the greenhouse gases (Stein 2009).

A Carbon-Neutral Airline?

Silverjet, a luxury transatlantic air carrier, promoted itself as the first fully carbon-neutral airline because it donated $28 from each round-trip ticket to a fund for projects that could be as esoteric as fertilizing the oceans with iron so that algae can pull more carbon dioxide out of the atmosphere. In theory, said representatives of Silverjet, such projects may eventually eliminate as much carbon dioxide as the airline generates, or approximately 1.2 tons per passenger per trip.

The cure was on speculation, however, while the carbon emissions remained an everyday reality. An entire consulting industry grew to calculate offsets, bill the companies, and handle the transactions with a fee attached. By early 2007, *Businessweek* magazine said that the trade in offsets was more than $100 million a year "and growing blazingly fast" (Revkin 2007). In the meantime, actual greenhouse gas emissions barely changed.

> "The worst of the carbon-offset programs resemble the Catholic Church's sale of indulgences back before the Reformation," said Denis Hayes, president of the Bullitt Foundation, an environmental grant-making group. "Instead of reducing their carbon footprints, people take private jets and stretch limos, and then think they can buy an indulgence to forgive their sins." "This whole game is badly in need of a modern Martin Luther," Hayes added. (Revkin 2007)

Michael R. Solomon, a professor at Auburn University who wrote *Consumer Behavior: Buying, Having and Being*, said he was not surprised by the appeal of offsets. "Consumers are always going to gravitate toward a more parsimonious solution that requires less behavioral change," he said. "We know that new products or ideas are more likely to be adopted if they don't require us to alter our routines very much." He sees danger ahead "if we become trained to substitute dollars for deeds—kind of an 'I gave at the office' prescription for the environment" (Revkin 2007).

Jet Fuel Efficiency: Just an Inkling of What Needs to Be Done

The U.S. aviation industry increased its fuel efficiency 23 percent from 2000 to 2006, flying 18 percent more miles and carrying 12 percent more passengers, all on roughly 5 percent less fuel, according to the U.S. Department of Transportation. According to the International Air Transport Association, jet engines by 2006 were 40 percent more fuel efficient than they were in the 1960s. The air-travel industry may have reduced the amount of fuel it burns to transport each passenger by approximately half since the mid-1970s, but the growth of the number of people taking to the air has more than extinguished these savings in fuel efficiency. The same relationship applies to nitrogen oxides and hydrocarbons.

Fuel consumption fell to 19.6 billion gallons in 2006 compared to 20.4 billion in 2000 as airliners added navigation technology to allow more direct routing, modified wings to make them more aerodynamic, and decreased weight with lighter seats and lower inventories of food and water, among other things. However,

"There is so little left you can do other than renewing the fleet or flying less," said John Heimlich, chief economist of the Air Transport Association (Frank 2008). American and Delta Air Lines jets, for example, now sometimes taxi using one engine to reduce fuel use. Both carriers promoted their efforts to modify wings so that they reduce drag and boost efficiency, as well as find ways to reduce aircraft weight by removing ovens, galleys, and water (Wilber 2007).

Discussions of airline fuel efficiency offer some promise of incremental improvements. Flying 2,000 meters lower might potentially save 6 percent of fuel. Better planning to avoid long waits on takeoff and more direct routing may save 10 percent. The Boeing Company has pitched its 787 Dreamliner as a marvel of efficiency that will stretch fuel supplies by 20 percent. There has been some talk of jet fuel from soya, which may lower greenhouse gas emissions—which raises the question, how many square miles of soybean fields would be required to get a Dreamliner from New York City to London and back? At the end of the day, however, air travel remains a bad dream in any global warming calculus. Remember that every passenger on a long-haul flight is responsible for 124 kilograms (273 pounds) of carbon dioxide per hour (McGuire 2005, 188–189).

The long life of jets restricts an airline fleet's ability to increase fuel efficiency. The Boeing 747, for example, is still flying 36 years after it was introduced. The Tyndall Centre predicts that the Airbus A380, introduced in 2006, will be flying (in incrementally modified form), in 2070. "Switching to more efficient models," wrote George Monbiot (2006), "would mean scrapping the existing fleet."

Future efficiency gains may be meager, however, because of the mature nature of jet-engine technology. With commercial aviation mileage expected to double by 2050, the search is on for solutions to an air-transport system in which one airplane flying from New York City to Stockholm emits as much carbon dioxide as an average automobile commuter in 50 years (Daviss 2007, 33). Various systems have been tried and abandoned that might make the aerodynamics of jet aircraft more efficient. Usually, systems meant to improve laminar flow do not pay for themselves over the life of an aircraft.

Richard Branson, owner of Virgin Airlines, has revealed plans to invest $3 billion to develop ecologically friendly plant-based jet fuel. At present, the use of hydrogen fuel or ethanol in place of the usual kerosene jet fuel faces formidable obstacles. Hydrogen fuel provides only 25 percent as much energy per volume as jet fuel, meaning that a hydrogen-powered aircraft would need huge fuel tanks and have to fly with a heavier load, reducing mileage. Because of the volume, fuel could not be carried in the wings, but in the body of the aircraft, increasing drag. Hydrogen would produce no carbon dioxide, but its output of water vapor at high altitudes would increase the size of contrails, which aggravate global warming. Plant-based fuel weighs two-thirds more by volume than kerosene for the same amount of thrust. It also freezes easily at high altitudes (Daviss 2007, 35).

Air Travel as a Political Issue

In May 2013, the European Commission began considering fines for Chinese and Indian airlines (including Air China and Air India) for their refusal to comply with regulations meant to reduce greenhouse gas emissions. The same airlines were ordered to comply or face exclusion from some European airports. Airlines were being issued permits to emit greenhouse gases in European airspace after having reported their emissions. The rules allow fines of violators at 100 euros (€) for each ton of carbon dioxide not covered by a permit.

The commission said the eight Chinese carriers were facing fines of €2.4 million ($3 million U.S.). More than 20 other airlines, including some based in the United States, Japan, and Russia, also protested the new requirements, which took effect January 1, 2014, on all international flights to and from Europe. The rules earlier had been applied to flights within Europe. The international airlines contend that Europe has no right to levy taxes on emissions that occur mainly outside their borders. The permits added €3 on a round-trip ticket from Brussels to Washington, and roughly €4 between Brussels and Beijing.

The reporting and permit system is part of a general European effort to require companies that produce greenhouse gases to pay to emit them. The use of a sliding scale based on emissions is believed to encourage reductions. The system began in 2005 and was initially aimed at heavy industry, because some proprietors lobbied for and received special treatment, receiving free permits during a period of economic weakness in Europe. The prices of the permits (which were bought and sold in a market) fell too low to encourage reductions in emissions. Airlines also received some 80 percent of their initial payments for free, making a mockery of the system (Kanter 2013).

Britain's Debate over Aviation's Carbon Footprint

In London, a large headline on the front page of *The Independent*, the most aggressive of Britain's newspapers on global warming, upbraided Prince Charles for emphasizing the worldwide risks of global warming while he jetted to India on a private Airbus 319. The newspaper described in great detail a round-trip by Prince Charles from London to India via Egypt and Saudi Arabia that covered 9,272 miles and was responsible for 42 tons of carbon dioxide emissions (Hickman 2006). At the same time, Prince Charles was arguing that global warming is "the greatest challenge" facing humankind. The newspaper urged him to restrict his flying schedule.

In 2007, Prince Charles disclosed that his personal carbon footprint had shrunk 9 percent in a year to 3,775 tons. He has been taking private jets fewer times (substituting trains when available), and gassing up the royal Jaguar with used cooking oil. Otherwise, Charles has gone "carbon neutral" with offsets worth $600,000 a year. His three mansions still have a carbon footprint the size of 500 average British homes, however ("Carbon Neutral" 2007).

As newspapers argued over Prince Charles's carbon footprint, flight traffic was exploding in England, with a third runway planned at London's Heathrow Airport and similar extensions at London Stansted Airport and airports in Birmingham, Edinburgh, and Glasgow. Twelve other British airports also had announced expansion plans. According to the Environmental Audit Committee of the House of Commons, the growth the government foresaw would require the equivalent of another

Heathrow-sized airport every five years (Monbiot 2006). London's Heathrow itself was planning a sixth terminal while its fifth was under construction.

London's Heathrow, with more than 67 million passengers a year (200,000 a day) has become the world's busiest airport. One reason for the huge passenger load is the fuel-squandering nature of the world's hub-and-spoke airline-routing system. A passenger in Equatorial Guinea in West Africa who wants to get to Johannesburg in South Africa connects in London—arriving at Gatwick and then leaving from Heathrow.

One-fifth of the world's international airline passengers flew to or from an airport in the United Kingdom during 2006. The number of passengers in this hub increased fivefold in the 30 years ending in 2005 as the government envisaged that the number would more than double by 2030 to 476 million a year (Monbiot 2006). Carbon dioxide emissions from air traffic originating in the United Kingdom rose 85 percent between 1990 and 2000, according to its Department of Trade and Industry (Houlder 2002).

The phenomenal success of budget airlines such as Ryanair and easyJet contributed to a 76-percent increase in traffic through Britain's airports in a decade—to 215 million passengers in 2004 (Clark 2006). Even as they talked about reining in greenhouse gas emissions, political leaders in the United Kingdom were promoting rapid increases in air travel and expansion of airports, much as automobile manufacturers in the United States talked up greenhouse gas emissions caps while opposing higher gas-mileage standards. Both were only "green" in theory. Even with then Prime Minister Tony Blair saying "climate change is, without doubt, the major long-term threat facing our planet," British air mileage continued to explode (Lynas 2006, 12).

By 2010, under a Conservative-led coalition the British government heeded the protests of aviation expansion by enacting a ban on new runway construction at London's three international airports. In May, using rising greenhouse gas emissions from aviation as his rationale, Prime Minister David Cameron, a Conservative, canceled plans to for a third runway at Heathrow Airport. He also sought a ban on new runways at Gatwick and London Stansted. Cameron said that building new runways was incompatible with Britain's goal of reducing emissions at least 34 percent by 2020 from 1990 levels.

"The emissions were a significant factor" in the decision to cancel the runway-building plans, said Teresa Villiers, then Britain's minister of state for transport. "The 220,000 or so flights that might well come with a third runway would make it difficult to meet the targets we'd set for ourselves" (Rosenthal 2010). Peder Jensen, European Environment Agency transportation specialist, said that Britain "is the only country that had made a conscious decision based on climate considerations" (Rosenthal 2010).

Flying above the Law

Because airline travel often crosses international borders, its greenhouse gas emissions have been excluded from many tallies that add them up on a nation-by-nation basis. The airline industry routinely uses its international status to avoid national constraints such as energy-efficiency targets that go along with plans to reduce greenhouse gas emissions. The Chicago Convention on International Civil Aviation of 1944, supported by 4,000 bilateral treaties, rules that no government may levy tax on aviation fuel. "The airlines," wrote British author George Monbiot, "have been bottle-fed throughout their lives" (Monbiot 2006).

Jet fuel not only is one of the few untaxed fuels in the world but also is rated as zero value added for tax purposes. The airline industry argues that it is effectively taxed because U.K. airlines pay a duty of £5 per passenger for a European flight (a fee halved from £10 per flight in 2000). However, airlines also benefit from a £9 billion tax subsidy per year in the United Kingdom (Francis 2006).

Air-pollution emissions from international flights also have been routinely excluded from treaties designed to counter global warming and ozone depletion. Airports in Britain are exempt from pollution control, from getting planning permission for many developments, and—in most cases—from statutory noise regulation. To top it all, air travel receives billions of dollars every year in subsidies from taxpayers of industrialized countries (Lean 2001).

Ecotourism's "Hidden Pollution"

Steve McCrea, editor of the *Eco-Tourist Journal*, calls air travel ecotourism's "hidden pollution" (McCrea 1996). Tourists who take the utmost ecological care when they visit exotic locales rarely give a second thought to the greenhouse gases that they generate while reaching their destinations. According to McCrea:

> One ton of carbon dioxide enters the atmosphere for every 4,000 miles that the typical eco-tourist flies. A round trip from New York to San Jose, Costa Rica (the world's leading eco-tourist destination), is 4,200 miles, so the typical eco-tourist generates roughly 2,100 pounds of carbon dioxide by traveling to a week of sleeping in the rain forest. (McCrea 1996)

To balance carbon dioxide generated by their air travel, McCrea suggests that ecotourists plant three trees for every 4,000 miles to compensate not only for the carbon dioxide but also for other greenhouse gases created by the combustion of jet fuel.

Even as the world's glaciers melt, travel agents flock to the banner of ecotourism, advising clients to hurry up and see ancient ice before it disappears. Betchart Expeditions of Cupertino, California, in 2007 offered a 12-day tour to "Warming Island" off Greenland, which had recently emerged from melting ice, "a compelling indicator of the rapid speed of global warming" (Naik 2007, A12). The cost of the full tour was $5,000 to $7,000, plus airfare. In the Greenland coastal village of Illuissat, population 5,000, some 35,000 tourists arrived, most of them on cruise ships during 2007, up from 10,000 just five years previously. Many tourists came to witness the receding Jacobshavin glacier, which had lost nine miles in five years. Once upon a time, winter temperatures routinely fell to –40°F there, but by

2007–15°F was the usual winter low. The harbor, which used to freeze solid, now remains ice free all year, allowing fisherman to pull halibut from the waters at all seasons and thus depleting their stocks (Naik 2007, A12).

A "No-Flying Movement"

So what can fliers who were concerned about ruining the atmosphere do? The elegantly simple—and perhaps, for the time being, only—effective answer was: do not fly as much. Although airline manufacturers have gained some efficiency in recent decades, no fuel other than those made from fossil fuels provides the thrust that airliners need to gain and maintain altitude and speed. No one is seriously entertaining the idea of solar or wind (or, God help us, nuclear) driven aircraft. Like no other form of transport, the modern aircraft is a hostage of the fossil fuel age and a penultimate producer of greenhouse gases.

"In researching my book about how we might achieve a 90 percent cut in carbon emissions by 2030," wrote George Monbiot (2006), "I have been discovering, greatly to my surprise, that every other source of global warming can be reduced or replaced to that degree without a serious reduction in our freedoms. But there is no means of sustaining long-distance, high-speed travel." In Britain, a "no-flying movement" has begun to take shape as many people voluntarily commit to avoid aviation at all for nonessential trips.

Newspaper travel supplements began to provide information on train or shipping alternatives. Vacationers have been advised to "[travel] to the Alps by train and . . . get a real sense of geography, of evolving culture and changing climatic zones. Arrive by air and all you see is identical airport terminals and thousands of other culture-shocked, aggravated travelers. Slow travel, like slow food, is about clawing back quality of life" (Lynas 2006, 14–15).

Further Reading

"Carbon Neutral Chic." *Wall Street Journal*, July 9, 2007, A14.

Cars and Climate Change. Paris, France: International Energy Agency, 1993.

Chea, Terence. "Environmental Groups Call for Planes to Reduce Emissions." *USA Today*, December 6, 2007, 6B.

Clover, Charles, and David Millward. "Future of Cheap Flights in Doubt; Ban New Runways and Raise Fares, Say Pollution Experts." *Daily Telegraph* (London), November 30, 2002, 1, 4.

"Contrails Linked to Temperature Changes." Environment News Service, August 8, 2002. http://ens-news.com/ens/aug2002/2002-08-08-09.asp#anchor4 (no longer available).

Dalton, Rex. "How Aircraft Emissions Contribute to Warming; Aviation Contributes Up to One-Fifth of Warming in Some Areas of the Arctic." *Nature*, December 21, 2009. http://www.nature.com/news/2009/091221/full/news.2009.1157.html.

Daviss, Bennett. "Green Sky Thinking: Could Maverick Technologies Turn Aviation into an Eco-Success Story? Yes, but Time Is Running Out." *New Scientist*, February 24, 2007, 32–38.

Flannery, Tim. *The Weather Markers: How Man Is Changing the Climate and What It Means for Life on Earth*. New York: Atlantic Monthly Press, 2005.

Francis, Justin. Responsibletravel.com. United Kingdom. Accessed April 30, 2006. http://www.responsibletravel.com/Copy/Copy101993.htm (no longer available).

Frank, Thomas. "Planes Fly More, Emit Less Greenhouse Gas." *USA Today*, May 9, 2008, 1B.

"Global Aviation Industry Vows to Halve CO_2 Emissions by 2050." Environment News Service, October 15, 2009. http://www.ens-newswire.com/ens/oct2009/2009-10-15-01.asp.

Hickman, Martin. "The Prince of Emissions." *The Independent* (London), April 1, 2006, 1.

Hillman, Mayer, and Tina Fawcett. *The Suicidal Planet: How to Prevent Global Climate Catastrophe.* New York: St. Martin's Press/Thomas Dunne Books, 2007.

Houlder, Vanessa. "Rise Predicted in Aviation Carbon Dioxide Emissions." *Financial Times* (London), December 16, 2002, 2.

Kanter, James. "E.U. Considers Emission Fines for Chinese and Indian Airlines." *The New York Times*, May 16, 2013. http://www.nytimes.com/2013/05/17/business/global/17iht-emit17.html.

Kondratyev, Kirill, Vladimir F. Krapivin, and Costas A. Varotsos. *Global Carbon Cycle and Climate Change.* Berlin: Springer/Praxis, 2004.

Lean, Geoffrey. "We Regret to Inform You That the Flight to Malaga Is Destroying the Planet; Air Travel Is Fast Becoming One of the Biggest Causes of Global Warming." *The Independent* (London), August 26, 2001, 23.

Lynas, Mark. "Fly and Be Damned." *New Statesman* (London) April 3, 2006, 12–15. http://www.newstatesman.com/node/164067.

McCrea, Steve. "Air Travel: Eco-Tourism's Hidden Pollution." *Earth Times* (San Diego), August 1996. http://www.sdearthtimes.com/et0896/et0896s13.html.

McGuire, Bill. *Surviving Armageddon: Solutions for a Threatened Planet.* New York: Oxford University Press, 2005.

Minnis, Patrick, et al. "Contrails, Cirrus Trends, and Climate." *Journal of Climate*, April 5, 2004, 1671–1685.

Monbiot, George. "We Are All Killers: Until We Stop Flying." *The Guardian* (U.K.), February 28, 2006. http://www.monbiot.com/archives/2006/02/28/we-are-all-killers/.

Naik, Gautam. "Arctic Becomes Tourism Hot Spot, but Is It Cool?" *Wall Street Journal*, September 24, 2007, A1, A12.

Revkin, Andrew. "Carbon-Neutral Is Hip, but Is It Green?" *The New York Times*, April 29, 2007. http://www.nytimes.com/2007/04/29/weekinreview/29revkin.html.

Rosenthal, Elisabeth. "Britain Curbing Airport Growth to Aid Climate." *The New York Times*, July 2, 2010. http://www.nytimes.com/2010/07/02/science/earth/02runway.html.

Schapiro, Mark. *Carbon Shock: A Tale of Risk and Calculus on the Front Lines of the Disrupted Global Economy; How Carbon Is Changing the Cost of Everything.* White River Junction, VT: Chelsea Green Publishing, 2014.

Schleck, Dave. "High Fliers May Be Creating Clouds: Global Warming May Be Worsened by Contrails from Aircraft." *Montreal Gazette*, July 18, 2004, D6.

Stein, Mara Lemos. "Ticket to Nowhere?" *Wall Street Journal*, March 9, 2009, R9.

Stern, Nicholas H. *A Blueprint for a Safer Planet: How to Manage Climate Change and Create a New Era of Progress and Prosperity.* London: The Bodley Head, 2009.

Stuber, Nicola, et al. "The Importance of the Diurnal and Annual Cycle of Air Traffic for Contrail Radiative Forcing." *Nature* 441 (June 15, 2006): 864–867.

Travis, Davis J., Andrew M. Carleton, and Ryan G. Lauritsen. "Climatology: Contrails Reduce Daily Temperature Range." *Nature* 418 (August 8, 2002): 593–594.

Wilber, Del Quentin. "U.S. Airlines Under Pressure to Fly Greener." *Washington Post*, July 28, 2007, D1. http://www.washingtonpost.com/wp-dyn/content/article/2007/07/27/AR 2007072702256_pf.html.

See also: Corporate Sustainability; Temperatures, Global; Temperatures, Greenhouse Gas Levels and; Unburnable Carbon

AUTOMOBILES

Climate scientist Stephen Schneider recalled once sharing a stage with singer John Denver in Aspen, Colorado. "John said, in introducing me to this group," Schneider recalled, "that he had been unaware of his own impact on global nature until he learned about the global warming problem. 'As soon as I learned about that,' he confessed, 'I sold my Porsche.' Everybody applauded" (Schneider 2000).

Passenger vehicles in the United States account for 40 percent of the country's oil consumption, and 10 percent of the world's (Kolbert 2007, November 5, 88, 90). According to Bill McKibben (1989, 6), "The average American car, driven the average American distance—[10,000] miles—in an average . . . year releases its own weight in carbon into the atmosphere." Global warming, in a sense, is the exhaust pipe of the "American Dream," of the bigger and better, the new and improved, the mobile life in the fast lane. Cars and trucks used in the United States burn 15 percent of the world's oil production. Transportation alone consumes one-fourth of the energy and two-thirds of oil used in the United States (Cline 1992, 200).

Times are changing, however, as people realize that combating global warming means changing the ways in which people move from place to place—and, with time, even the ways our cities and towns are laid out. By 2007, 16 million of 150 million U.S. workers telecommuted even as 76 percent of Americans drove to work alone (Vanderkam 2008).

Automobiles consume a third of the world's oil production. The number of automobiles was increasing more quickly than population, especially in countries with very large population that have just begun to experience full-scale industrial development such as India and China.

An Exemplar of Fossil Fueled Culture

The automobile is an exemplar of fossil fuel culture—easy to use, convenient, suited to the individualism of our time, and generative of greenhouse gases in its manufacture and everyday use. Each automobile brings with it a bundle of carbon dioxide, carbon monoxide, methane, nitrous oxide, and waste heat. The tailpipes of motorized vehicles emit 47 percent of North America's nitrous oxides, a primary component of ozone in the lower atmosphere. In addition to burning fossil fuels, the automobile is also a little greenhouse factory of its own whenever the windshield captures the sun's heat and cooks the air inside the auto's steel and glass casing.

Cities in the United States and elsewhere have been remade in the image of the automobile. Any U.S. city that reached maturity after the 1930s—the last decade in which the average person did not own a car—has become an energy-gobbling sprawling mass of suburbs and freeways that virtually guarantees that most people need automobiles whether they want them or not. Outside a few major eastern cities (New York and Boston being prominent examples), the automobile has become an everyday necessity for nearly everyone. Weaning ourselves from global warming in the long run is going to require major surgery on our concepts of urban land use.

In some places, such surgery is being performed on a large-scale basis. Portland, Oregon, for example, has canceled several freeway projects that once threatened to make the center city little more than a conduit to the suburbs. Even so, Portland has no shortage of downtown freeways, but it also has a well-used light-rail system that connects neighborhoods and its airport with the central core. Property values have risen in the center city as people renovate homes that are within bicycling or walking distance of downtown businesses. Even in Detroit, General Motors (GM) during 2007 entered a partnership to develop residences in once-abandoned buildings along the waterfront. One selling point for these lofts is—note that the pitch is being made by an automobile manufacturer—the fact that people who live in them will be able to walk to work at GM's headquarters (White 2006).

Some areas are turning the tables on automobiles, however. Street parking, driveways, and home garages are generally forbidden in Vauban, Germany, near an experimental new district on the outskirts of Freiburg, near the French and Swiss borders. Approximately 70 percent of Vauban's families do not own cars, and more than half of the city's 5,500 residents vote for the German Green Party. Residents live in a rectangular square mile that is small enough for walking, bicycling, and public transportation. Stores, restaurants, banks, and schools are interspersed among homes. Communal cars are rented out by Vauban's car-sharing club for hauling that requires them (Rosenthal 2009).

Congestion Charges

While he was mayor of New York City, Michael Bloomberg proposed a congestion charge for the most crowded southern half of Manhattan Island, roughly 86th Street southward. (The proposal was never adopted.) On weekdays from 6 a.m. to 6 p.m., trucks would have been charged $21 a day and cars $8 in addition to premium parking fees charged by city-owned and private lots. Currently, only 5 percent of people who work in Manhattan but do not live in there commute by car. Those traveling only within the zone would have paid half price; taxis and livery cabs would have been exempt. With uncontrolled free access to the area, studies have shown that vehicle speeds within this area average 2.5 to 3.7 miles an hour. Many times, walking is almost as fast. Anyone who wants to go anywhere with any speed in this area takes the subway. The value of time lost to congestion delays in New York City has been estimated at $5 billion a year; add wasted fuel, lost revenue, and increasing costs of doing business, and the total rises to $13 billion a year

(Kolbert 2007, May 7)., Bloomberg's plan was derailed by the New York State Assembly in July 2007.

Singapore was the first city to introduce such a fee. London introduced a congestion charge in 2003, followed by Milan, Italy. In London, vehicle speeds have since risen 37 percent and carbon dioxide emissions have fallen 15 percent. London's mayor at that time, Ken Livingstone, a major proponent of the charge, was easily reelected in 2004 and two-thirds of London residents supported the charge in 2006. In January 2007, the congestion zone was expanded westward to include most of Kensington, Chelsea, and Westminster (Kolbert 2007, May 7). By early 2007, London had reduced private car use 38 percent and carbon emissions 20 percent in the congestion charge zone.

London also improved its public transport system to offer Londoners easier alternatives; many commuters had complained that the buses were slow and expensive. By the time the London bus system was upgraded, more than 6 million people were using it everyday. The number of people commuting by bicycle in London (at no charge) soared 80 percent after the congestion charge was implemented ("Big City Mayors" 2007).

In the meantime, British Petroleum (now formally known by its initials—BP—has been taken by green publicists to mean "Beyond Petroleum") has proposed that every motorist in Great Britain sign up for a plan called "Target Neutral." Drivers can fund ventures that offset the amount of carbon dioxide their driving adds to the atmosphere. Drivers register at a Target Neutral website, which calculates the estimated amount of carbon dioxide that may be produced by their driving over the coming year. Drivers then pay offsets based on the estimate. The typical family car, doing 10,000 miles a year, is likely to cost around £20 ($33 U.S.) to offset ("British Travel" 2006).

Congestion charging in downtown Stockholm was a controversial issue in the Swedish general election in late summer 2006. The congestion charge of up to $7 a day was narrowly approved by 52 percent of voters in a referendum that September. The Stockholm congestion charge reduced auto traffic 20 percent to 25 percent, while the use of trains, buses, and Stockholm's subway system increased. Emissions of carbon dioxide declined 10 percent to 14 percent in the inner city and 2 percent to 3 percent in Stockholm County. The project also increased the use of environmentally friendly cars (such as hybrids), which are exempt from congestion taxes. As in London, commuting by bicycle also increased. Following a trial period, the Stockholm congestion tax became permanent in August 2007.

The Car's Future

The need for a paradigm shift in how we fuel our cars—and how many of us drive them—can be illustrated by a comparison of car ownership in the United States, India, and China as played across economic trends. In 2015 in China, there were approximately 30 personal vehicles per 1,000 people of driving age; in India, there were 13. In the United States, the same figure stood at around 500. Between 2000 and 2006, sales of heavy trucks in China increased 800 percent. Sales of passenger cars increased 600 percent. Sales of new passenger vehicles tripled in India (from 500,000 a year to 1.5 million) between 1998 and 2008, a period during which the country built its first interstate highway system, the Golden Quadrilateral, which links Mumbai (Bombay), Delhi, Kolkata (Calcutta), and Bangalore.

If people in India and China drove at half the rate of those in the United States, world oil consumption, 88 million barrels a day in 2013, would balloon to more than 200 million. If drivers in India and China used cars as Americans do, world oil consumption would more than triple, with attendant impacts on both oil prices and greenhouse gas emissions (Kolbert 2007, November 5).

With China's and India's economies growing rapidly, millions of families will soon be in the market for their first private cars. Mileage for new cars in the United States has stagnated at around 20 miles per gallon—incredibly less than that of Henry Ford's Model T when it first went on the market in 1908 (Kolbert 2007, November 5). The world's climatic future, as well as the finite nature of oil supplies, are going to demand a paradigm shift in fuel technology and economy in coming years. In another 100 years, by necessity the internal combustion engine may be as antique as a horse and buggy seems today.

The "Green" Car: Future Fuel Sources

If we cannot give up our cars, could we at least downsize them? In the United States, 70 percent of car and light truck sales early in the 21st century had six- or eight-cylinder engines. In Europe, 89 percent had four cylinders or fewer. The average U.S. driver's travel mileage has increased 60 percent in 30 years. The growing numbers of sports utility vehicles roaming the roads since the 1990s recall the words of Henry Ford: "Mini-cars make mini-profits" (Commoner 1990, 80).

Henry Ford's wisdom proved to be rather obsolete (but only for a few years) once gasoline prices began rising to record levels in 2007 and the sales of U.S.-made "maxi" cars plummeted (this trend reversed in 2014 when gas prices fell by 50 percent in six months). In 2007, U.S. automakers for the first time sold fewer than half the cars purchased in the country. By 2007, some executives of the auto companies were beginning to realize this, just as the U.S. Congress sought to raise mileage standards in the United States to 35 miles a gallon by 2020, still behind most other countries and the cars they manufacture. The Toyota Prius hybrid was getting 50 miles per gallon of gasoline. However, Toyota was selling 250,000 hybrids a year in the United States by 2006.

Automobiles as Monuments to Fuel Inefficiency

Today's automobiles are monuments to fuel inefficiency. The average fuel economy of vehicles sold in the United States has remained nearly stagnant at around 20 to 25 miles a gallon for several decades. Only 13 percent of a car's energy reaches its wheels and only half of that actually propels the car. The rest is lost to idling, heat, vibration, and such accessories as air conditioning. Of the 6 percent that remains, most converts to brake heating when the car stops. As a result, less than 1 percent of the energy the average car consumes ends up propelling the driver. Amory Lovins recommends making cars much lighter as well as developing hydrogen fuel cells. He also suggests stripping the oil industry of subsidies that make gasoline cheaper by volume than bottled water (Lovins 2005).

The X Prize Foundation is sponsoring an automotive contest that is expected to carry a prize of more than $10 million, to accelerate research to lead to a car that that can travel 100 miles on a gallon of gasoline. The same foundation earlier awarded $10 million to a team that built the first private spacecraft to leave Earth's atmosphere. According to contest rules, the winning design must be commercially viable and production ready, not a "concept car" like those presented at annual auto shows. Each team must prepare a business plan for building at least 10,000 of its vehicles at a cost comparable to that of cars now available.

Hydrogen Fuel: No Free Climatic Lunch

Political correctness vis à vis global warming in the automobile industry for a time was associated with the development of hydrogen fuel cells, especially after President George W. Bush used his State of the Union address in January 2003 to propose $1.2 billion in research funding to develop hydrogen-fuel technologies. With those funds, Bush said that America could lead the world in developing clean, hydrogen-powered automobiles.

Liberal social critic and author Jeremy Rifkin published a book in September 2002 titled *The Hydrogen Economy: The Creation of the Worldwide Energy Web and the Redistribution of Power on Earth*. Rifkin believed that cheap hydrogen could make the 21st century more democratic and decentralized, a marked contrast to the way that oil transformed the 19th and 20th centuries by fueling the rise of powerful corporations and nation-states. With hydrogen, wrote Rifkin, "every human being on Earth could be 'empowered'" (Coy 2002).

Hydrogen fuel was hailed as a godsend at the time. Imagine a car fueled by the most abundant element in the atmosphere that emits nothing but a little water vapor. It is a thought worthy of high praise, but imagination ran ahead of technology's ability to deliver. Even though it has been touted as pollution free, hydrogen fuel is no free climatic lunch. Despite surface appearances, hydrogen in today's world may be no cleaner a fuel than gasoline. The hype surrounding a "hydrogen economy" has been woefully premature for one critical technological reason.

Unlike oil and coal, hydrogen does not exist in nature in a combustible form. It is usually bonded with other chemical elements, and stripping them away to produce the pure hydrogen necessary to power a fuel cell requires large amounts of energy. Unless an alternative source (such as Iceland's geothermal resource) is available, hydrogen fuel usually is produced from fossil fuels. Extraction of hydrogen from water via electrolysis and compression of the hydrogen to fit inside a tank that can be used in an automobile requires a great deal of electricity. Until electricity is routinely produced via solar, wind, and other renewable sources, the hydrogen car will require energy from conventional sources, including fossil fuels. Today, 97 percent of the hydrogen produced in the United States comes from processes that involve the burning of fossil fuels, including oil, natural gas, and coal.

Working on the Problem

Scientists have been working on this problem. In mid-2008, a team at the Massachusetts Institute of Technology (MIT) announced discovery of a cobalt–phosphorus catalyst that can split hydrogen from oxygen atoms in water to create hydrogen gas. Before that discovery by MIT chemist Daniel Nocera and colleagues, which was reported in the *Journal of the American Chemical Society*, such a reaction was possible only using chemicals that worked under toxic conditions, or platinum, which is much too expensive for industrial-scale processes. "If we are going to use solar energy in a direct conversion process, we need to cover large areas. That make a low-cost catalyst a must," said John Turner, an electrochemist at the National Renewable Energy Laboratory in Golden, Colorado (Service 2008).

Until such new technology comes into service on a commercial scale, however, basic problems with practicality of hydrogen power remain. Writing in *Nature*, Paul M. Grant provided an illustration: "Let us assume that hydrogen is obtained by 'splitting' water with electricity—electrolysis. Although this isn't the cheapest industrial approach to 'make' hydrogen, it illustrates the tremendous production scale involved—about 400 gigawatts of continuously available electric power generation [would] have to be added to the grid, nearly doubling the present U.S. national average power capacity." That, Grant calculated, would represent the power-generating capacity of 200 Hoover Dams (Grant 2003).

At $1,000 per kilowatt hour, the cost of such new infrastructure would total $400 billion. What about producing 400 gigawatts with renewable energy? Grant estimated that, "with the wind blowing hardest, and the sun shining brightest," wind-power generation would require a land area the size of New York state or a layout of state-of-the-art photovoltaic solar cells half the size of Denmark (Grant 2003, 130). Grant's preferred solution to this problem is to use energy generated by nuclear fission.

Although hydrogen is not the magic wand that some of its proponents imagine, premonitions of hydrogen-based transportation systems have been emerging on small scales in unexpected places. In Iceland, for example, 85 percent of the country's 290,000 people use geothermal energy to heat their homes (Brown 2003, 166). Working with Shell and Daimler-Chrysler, Iceland's government in 2003 began to convert Reykjavik's city buses from internal combustion to fuel-cell engines using hydroelectricity to electrolyze water and produce hydrogen. The next stage is to convert the country's automobiles, then its fishing fleet. These conversions are part of a systemic plan to divorce Iceland's economy from fossil fuels (Brown 2003, 168). An industrial-scale hydrogen-fired power plant was also being built near Venice, Italy, during 2008 by the Veneto regional government and Italian energy company ENEL. The new plant in the Porto Marghera industrial area on the Italian mainland across from the Venice lagoon and next to ENEL's coal-fired Fusina plant was designed as a "zero-emission hydrogen combustion power generation system" ("Italy to Build" 2006).

Rebirth of the Electric Car

One of Henry Ford's fantasy cars was electric, but that was before petroleum took over. Electric cars are coming back, however. In January 2007, General Motors rolled out its hybrid Chevrolet Volt, a concept car that can run 40 miles on electricity alone with a six-hour nighttime charge and gets 150 miles per gallon of gasoline running as a hybrid. For commuting trips with fewer than 20 miles each way, the car can run on a charge from a standard 110-volt garage outlet (Griscom-Little 2007). Electricity must be generated, of course, and these days that usually involves the burning of fossil fuels.

By 2017, hybrid cars had been refined and mileage had improved. A Chevrolet Volt ($34,000 to $38,000) got 53 miles per gallon (mpg) in the city and 45 on the highway. An Audi A3 Sportback e-tron ($40,0000) was rated by the U.S. Environmental Protection Agency at 86 mpg anywhere. The Ford C-Max ($25,000 to $31,000) ran at about 40 mpg, and the Honda CR-Z ($21,000 to $26,000) was rated at 31 mpg in the city and 38 on the highway.

By mid-2007, GM had committed to hiring 400 technical experts to work on fuel-saving technology. One of GM's goals was to make the Volt a production model within three to four years. Such a change from 2002, when General Motors introduced the Hummer H2, which was so fuel thirsty that the company took advantage of a loophole in federal law and refused to publish its mileage (Boudette 2007).

Even Toyota, which had sold about a million of its hybrid Prius, was losing money on the cars. "In 10 years are they [at General Motors] going to solve the technological problems with respect to the Volt? Sure," said Maryann Keller, an automotive analyst and author of a book on GM. "But are they going to be able to stake their survival, which is really more of a now to five-year proposition, on it? I'd say they can't" (Mufson 2008). Large inefficient cars will not lose their luster until gasoline prices rise—and late in 2008, they were diving to less than half the peak of $4 a gallon in the United States reached earlier that year. Since then, electric cars have gained a market niche. By 2017, gas prices had been cut in half by a steep decline in oil prices, but hybrid cars were more profitable because of an environmentally conscious clientele.

"You'd think from reading the media that we have had a burial ceremony at Arlington cemetery for the last pickup truck," said James Womack, an automotive management and writer. Womack described the time required to design a new electric vehicle, produce thousands of new parts, and adjust assembly lines. "For anything that's really new it's still about four years," he said. "To get your money back, you need to make that product for eight to 10 years with only cosmetic changes."

At the new, green GM, Lawrence Burns, vice president for research, development, and global planning, told the *Wall Street Journal:* "We have to have people think we are part of the solution, not part of the problem." The Volt, said Burns, is an effort to show consumers "we get it" on climate change (Boudette 2007).

New York City's Green Yellow Cabs

In May 2007, New York City Mayor Michael Bloomberg announced that the city's yellow taxi fleet would run entirely on gas-electric hybrids within five years. "There's an awful lot of taxicabs on the streets of New York City," Bloomberg said. "These cars just sit there in traffic sometimes, belching fumes" ("New York's Mayor" 2007). Almost 400 hybrids were tested in New York City's taxi fleet over 18 months, with models including the Toyota Prius, the Toyota Highlander Hybrid, the Lexus RX 400h, and the Ford Escape. Under Bloomberg's plan, that number was to increase to 1,000 by October 2008 and then grow by 20 percent or so each year until 2012, when every yellow cab—then numbering 13,000—would be a hybrid ("New York's Mayor" 2007). The city licenses the yellow cabs and sells licenses to individual drivers, who then purchase their own vehicles under specifications set by the Taxi and Limousine Commission. A similar hybridization of cabs was underway in San Francisco.

In 2006, the average New York taxi ran entirely on gasoline at 14 miles per gallon (mpg). By 2008, all New York taxis were required to run at 25 mpg and 30 mpg by 2009. Although hybrid vehicles cost more, Bloomberg said that increased fuel efficiency will reduce operating costs by some $10,000 a year per vehicle. Changing the New York City taxi fleet to hybrids was one piece of the former mayor's sustainability plan to help reduce carbon dioxide emissions in the city 30 percent by 2030.

Further Reading

"New York's Mayor Plans Hybrid Taxi Fleet." *The New York Times*, May 22, 2007. http://www.nytimes.com/aponline/us/AP-Green-Taxis.html (no longer available).

Automobile Efficiency: Mileage Standards

By 2007, along with many other corporations, U.S. automakers had picked up the "green" mantra, endorsing greenhouse gas limits in theory. When the debate devolved to practice, however, the same companies complained that enforcing meaningful mileage increases would deprive people of jobs and the companies of profits—this from companies that were swimming in red ink from years of slavishly manufacturing increasingly unsalable large and inefficient gas-guzzlers as fuel prices rose and more nimble companies in other countries (many of them Japanese) gobbled their market shares, even in the United States.

Automobile Efficiency: U.S. States Take Action

In late September 2004, the California Air Resources Board (CARB) defied the attitudes of the auto industry and voted unanimously to approve the world's most stringent rules reducing automobile emissions. Under the regulations, the automobile industry must cut exhaust from cars and light trucks by 25 percent and from larger trucks and sport utility vehicles by 18 percent. The industry was given until 2009 to begin introducing cleaner technology and would have until 2016 to meet the new exhaust standards. The auto industry went to court for relief from the new standards, but federal judge Anthony W. Ishii upheld the California law regulating greenhouse gas emissions from cars and trucks.

From the 2009 average for all vehicles of 25 miles per gallon, U.S. fuel economy standards rose by 2016 to 35 mpg for passenger cars and 30 mpg for trucks. Along the way, some automakers (a notable example was Volkswagen) cheated on their emissions reports, and reaped large fines.

California's plan for sharp cuts in automotive emissions of greenhouse gases could eventually lead most states on the East and West Coasts to require similar emissions cuts because the state's share of the market is so large. In turn, these requirements may provoke domestic automakers to adopt the same standards for cleaner, more fuel-efficient vehicles across their model lines. The only way to cut global warming emissions from cars is to use less fossil fuel. Because of this limitation, proposed cuts in legally allowable emissions would force automakers to increase fuel economy by roughly 35 percent to 45 percent.

During 2004, the governments of New Jersey, Rhode Island, and Connecticut said that they intended to follow California's automobile rules instead of the federal government's. New York, Massachusetts, Vermont, and Maine already had adopted the California rules. "Let's work to reduce greenhouse gases by adopting the carbon dioxide emission standards for motor vehicles which were recently proposed by the State of California," then–New York Governor George E. Pataki said in his 2003 state-of-the-state address. These seven states and California account for almost 26 percent of the U.S. auto market, according to R. L. Polk, a company that tracks automobile registrations (Hakim 2004). On January 25, 2009, President Barack Obama reversed G. W. Bush's position and directed the Environmental Protection Agency to consider the request by California and 13 other states to adopt more stringent standards for auto emissions and fuel efficiency.

In 2007, CARB voted to require automobile manufacturers to affix labels listing their vehicles' smog and greenhouse gas emissions so that California's 2 million buyers a year can compare them. "This simple tool will empower consumers to choose vehicles that help the environment," said ARB Chairman Robert Sawyer. "Most Californians recognize climate change as a very

serious problem. This label will help consumers make informed choices"
("California Air Board" 2007).

Further Reading

"California Air Board Adds Climate Labels to New Cars." *Environment News Service*,
    June 25, 2007. http://www.ens-newswire.com/ens/jun2007/2007-06-25-09.asp#
    anchor7 (no longer available).
Hakim, Danny. "Several States Likely to Follow California on Car Emissions." *The New
    York Times*, June 11, 2004, C4.

The well-paid pursuit of special interests was on display as the U.S. Congress
debated energy legislation in June 2007. On June 21, the U.S. Senate dealt domestic car manufacturers a major blow by passing a bill requiring a major increase in
mileage standards (from 25 to 35 miles a gallon by 2020), the first in two decades.
The vote, 65 to 27, sent the bill to the House. Raising the mileage standard 10 mpg
will reduce U.S. oil consumption by 1.2 million barrels a day and reduce emissions
of greenhouse gases by an amount equal to removing 30 million cars from the road.
Getting really serious about mileage, raising the standard to 55 miles per gallon by
2020, a target for which technology is now available, would cut oil demand for
transportation in the United States by half (Steinman 2007, 239).

A report in *The New York Times* noted that automakers' opposition to the mileage standard had been "ferocious." "The clashes and impasses also provided a harbinger of potentially bigger obstacles when Democrats try to pass legislation this
fall to reduce emissions of greenhouse gases tied to global warming" (Andrews
2007).

Just days after their ferocious battle against stiffer mileage standards on Capitol
Hill, Ford Motor Company and the Chrysler Group announced at a press conference in Washington, D.C., that they had joined the United States Climate Action
Partnership (USCAP), a coalition lobbying for national legislation to impose legally
binding limits on global warming emissions. At the time, the partnership included
23 of the world's largest corporations and six of the United States' best-known environmental groups. General Motors already had joined—even before the battle over
mileage standards. "We are pleased to join USCAP at a critical stage in the conversation on climate change, energy consumption and environmental protection," said
Alan Mulally, Ford's president and chief executive officer ("Automakers Join" 2007).
No explanation was forthcoming as to why the same automakers seemed extremely
opposed to practical solutions to warming.

In March 2007, just three months before their ferocious assault on higher gas-
mileage standards, the chief executives of the same automobile companies had
pledged to support mandatory caps on carbon emissions, as long as the caps covered all sectors of the economy.

Where the short-term profit-and-loss rubber meets the public relations road,
however, the U.S. automobile industry seemed by 2007 to be speaking with more

than one voice on the mileage-standard issue. On May 31, 2007, Ron Gettelfinger, president of the United Automobile Workers union, and William Clay Ford Jr., executive chairman of Ford Motor Company, told Michigan business leaders that cutting emissions and raising fuel economy standards are critical to the future of the industry and that of Michigan. Gettelfinger said that "climate change is real," and that consumers are searching for more environmentally friendly vehicles. "If the auto industry continues to be seen as dragging its feet on environmental issues," he said, "it's going to hurt our brands and vehicles in the marketplace" (Bunkley 2007).

Further Reading

Andrews, Edmund. "Senate Adopts an Energy Bill Raising Mileage for Cars." *The New York Times*, June 22, 2007. http://www.nytimes.com/2007/06/22/us/22energy.html.

"Automakers Join Call for National Greenhouse Gas Limits." Environment News Service, June 27, 2007. http://www.ens-newswire.com/ens/jun2007/2007-06-27-09.asp#anchor4 (no longer available).

"Big City Mayors Strategize to Beat Global Warming." Environment News Service, May 15, 2007. http://www.ens-newswire.com/ens/may2007/2007-05-15-01.asp (no longer available).

Boudette, Neal E. "Shifting Gears, GM Now Sees Green." *Wall Street Journal*, May 29, 2007, A8.

"British Travel Agents Launch Carbon Offset Scheme." Environment News Service, November 28, 2006. http://www.ens-newswire.com/ens/nov2006/2006-11-28-05.asp (no longer available).

Brown, Lester R. *Plan B: Rescuing a Planet Under Stress and a Civilization in Trouble.* New York: Earth Policy Institute/W.W. Norton, 2003.

Bunkley, Nick. "Detroit Finds Agreement on the Need to Be Green." *The New York Times*, June 1, 2007. http://www.nytimes.com/2007/06/01/business/01auto.html.

Cline, William R. *The Economics of Global Warming.* Washington, D.C.: Institute for International Economics, 1992.

Commoner, Barry. *Making Peace with the Planet.* New York: Pantheon, 1990.

Coy, Peter. "The Hydrogen Balm? Author Jeremy Rifkin Sees a Better, Post-Petroleum World." *Businessweek*, September 30, 2002, 83.

Grant, Paul M. "Hydrogen Lifts Off—with a Heavy Load: The Dream of Clean, Usable Energy Needs to Reflect Practical Reality." *Nature* 424 (July 10, 2003): 129–130.

Griscom-Little, Amanda. "Detroit Takes Charge." *Outside*, April 2007, 60.

"Italy to Build World's First Hydrogen-Fired Power Plant." Environment News Service, December 18, 2006. http://www.ens-newswire.com/ens/dec2006/2006-12-18-05.asp (no longer available).

Kolbert, Elizabeth. "Don't Drive, He Said." *The New Yorker*, May 7, 2007, 23–24.

Kolbert, Elizabeth. "Running on Fumes: Does the Car of the Future Have a Future?" *The New Yorker*, November 5, 2007, 87–90.

Lovins, Amory. "More Profit with Less Carbon." *Scientific American*, September 2005, 74, 76–83.

McKibben, Bill. *The End of Nature.* New York: Random House, 1989.

Mufson, Steven. "The Car of the Future—but at What Cost?" *Washington Post*, November 25, 2008, A1. http://www.washingtonpost.com/wp-dyn/content/article/2008/11/24/AR2008 112403211_pf.html.

Rifkin, Jeremy. *The Hydrogen Economy: The Creation of the Worldwide Energy Web and the Redistribution of Power on Earth.* New York: Jeremy P. Tarcher/Putnam 2002.

Rosenthal, Elisabeth. "In German Suburb, Life Goes On without Cars." *The New York Times,* May 12, 2009. http://www.nytimes.com/2009/05/12/science/earth/12suburb.html.

Schneider, Stephen H. "No Therapy for the Earth: When Personal Denial Goes Global." In Michael Aleksiuk and Thomas Nelson, eds., *Nature, Environment & Me: Explorations of Self in a Deteriorating World.* Montreal: McGill-Queens University Press, 2000.

Service, Robert F. "New Catalyst Marks Major Step in the March Toward Hydrogen Fuel." *Science* 321 (August 1, 2008): 620.

Steinman, David. *Safe Trip to Eden: 10 Steps to Save Planet Earth from Global Warming Meltdown.* New York: Thunder's Mouth Press, 2007.

Vanderkam, Laura. "Want to Save the Planet? Stay Home." *USA Today,* May 20, 2008, 9A.

White, Joseph R. "An Ecotopian View of Fuel Economy." *Wall Street Journal,* June 26, 2006, D4.

See also: Ethanol, Corn; Ethanol, Sugarcane; Global Warming, China; Global Warming, Scandinavia

COAL

Ninety percent of Earth's remaining fossil fuel reserves are in the form of coal, the most dangerous fossil fuel from a greenhouse point of view because it produces roughly 70 percent more carbon dioxide per unit of energy generated than natural gas and approximately 30 percent more than oil. Coal is also the most plentiful fossil fuel, especially in places with large populations, such as China, which controls 43 percent of remaining reserves. During the 1980s, China passed the Soviet Union as the world's largest coal producer. Coal poses environmental problems other than carbon emissions. Its mining also produces methane, and its combustion produces sulfur dioxide and nitrous oxides as well as carbon dioxide. Coal transport also usually requires more energy than any other fossil fuel.

Many scientists argue that greenhouse gas emissions worldwide will have to be reduced by at least 70 percent by 2050 to stabilize temperature rises at less-than-dangerous levels. Increasingly, a consensus has been forming that the single most compelling policy change to forestall dangerous warming will be to halt construction of coal-fired power plants pending development of adequate technology to clean and sequester (store) their emissions to a reasonable cost.

Coal is the most widely used fuel for electricity generation worldwide. Almost 40 percent of carbon dioxide emissions in the United States come from coal-fired generation of electricity. Between 2002 and 2008, worldwide consumption of coal rose by 30 percent, two-thirds of which went into power generation, the rest into industries such as steel and cement manufacturing. China was adding one to two coal-fired electricity plants per week. By 2015, however, world coal consumption seemed to have peaked and had begun to decline in China, the United States, and much of Europe.

Coal and Environmental Damage

"Clean coal" is largely an oxymoron, especially where it is mined. In Kentucky, the creeks run red with iron and more than 9,000 measurements submitted by mining companies to state regulators are largely fictitious. Most water is reported as clean. In a state where coal-country creeks run red with iron, Frasure Creek Mining has been unusually clean of late: Amid tens of thousands of measurements that it submitted to Kentucky regulators in 2013 and early 2014, fewer than 400 exceeded the state's limits for water pollution from coal-mine runoff. "Now environmental activists say they know why," *The New York Times* reported in November 2014. "Four environmental groups said many of the monthly measurement reports that Frasure [Mining] sent the state contained virtually identical data—line-for-line repeats of clean pollution reports submitted the month before" (Wines 2014). The groups sued the state for violations of the federal Clean Water Act.

"Mountaintop Removal" Coal Mining

In the Appalachian Mountains, some peaks have been blasted off to reach underlying coal deposits. The mountains are effectively transformed into what one opponent of this mining method called "strip mines on steroids" (Sorkin 2015). For years opponents have worked to shut down strip mining and have blockaded mine sites (climate scientist and former NASA executive James Hansen was arrested at one such demonstration). Opponents also rallied in Washington, D.C., in September 2010 to demand that the federal government refuse permits for such mines. In such mines, upper-elevation forests are cleared and stripped of topsoil, and explosives are used to break up rocks to access buried coal. Excess rock (mine "spoil") is discarded with bulldozers pushing it into adjacent valleys, "where it buries existing streams" (Palmer et al. 2010, 148).

"It obliterates stream ecosystems," said Emily Bernhardt, a professor of biology at Duke University and a coauthor of the study. She said 1,500 miles of streams had been destroyed so far. "They've been wiped from the landscape." At these sites, "peaks are sheared off with heavy machinery and explosives, exposing the coal seams inside. Excess rock is used to fill steep Appalachian valleys, some with streams at the bottom, to the brim" (Fahrenthold 2010).

Rainwater falling on this rock accretes harmful sulfates. "To us, it's like smoking and cancer. It's just so clear-cut that streams below mine sites are left damaged," Palmer said. "The study also linked mountaintop mining to threats to human health, citing potentially toxic dust in the air, well water contaminated with chemicals from mines, and fish tainted with toxic metals" (Fahrenthold 2010).

By 2015, several large banks were refusing to extend loans to companies engaged in mountaintop coal removal, including PNC Financial, Bank of America, Citigroup, Morgan Chase, Wells Fargo, and Credit Suisse. Pressure on financial firms to boycott mountaintop removal mining was spearheaded by the Earth Quaker Action Team and Rainforest Action Network. "In our early encounters with PNC, they didn't take us or this issue seriously," George Lakey, a retired teacher who was arrested twice during the campaign. "We showed them evidence, delivered them Appalachian water poisoned by mountaintop removal, and brought them face to face with residents hurt by this practice. We had to take direct action for them to see the light" (Sorkin 2015).

Further Reading

Fahrenthold, David A. "Scientists Say Mountaintop Mining Should Be Stopped." *Washington Post*, January 8, 2010, A3. http://www.washingtonpost.com/wp-dyn/content/article/2010/01/07/AR2010010702530_pf.html.

Palmer, M. A., et al. "Mountaintop Mining Consequences." *Science* 327 (January 8, 2010): 148–149.

Sorkin, Andrew Ross. "A New Tack in the War on Mining Mountains." *The New York Times,* March 10, 2015, B1, B4. http://www.nytimes.com/2015/03/10/business/dealbook/pnc-joins-banks-not-financing-mountaintop-coal-removal.html?ref=business.

The same legal pantomime had played out before in 2010, after which Frasure (which is owned by the India conglomerate Essar Group) and other companies immediately returned to faking their reports. Eric Chance, a water-quality specialist with the Boone, North Carolina, group Appalachian Voices, said, "Their enforcement may look good on paper, but it's a facade" (Wines 2014). That group was pursuing the matter with Kentuckians for the Commonwealth, Kentucky Riverkeeper, and Waterkeeper Alliance. All of them criticize the toothless nature of state pollution regulation. Kentucky regulators did fine Frasure $310,000 but dismissed the duplicate measurements as "inadvertent transcription errors, and absolved Frasure of any pollution violations" (Wines 2014). By 2011, however, the company's reporting improved (after a successful lawsuit by the environmental groups) as reports found "thousands of discharge measurements that exceeded legal limits by up to 13 times the permitted levels" (Wines 2014). After that, lapses resumed.

The mining companies are required by law to maintain ponds where mine runoff is allowed to stand so heavy metals like cadmium and selenium, iron, manganese, and acidic compounds are supposed to settle out. The pollutants often leak into area rivers and streams. Peter Harrison, a staff lawyer with Waterkeeper Alliance, said, "If you're a coal company in eastern Kentucky, basically you don't have to pay attention to the Clean Water Act because the primary enforcer down there is asleep at the wheel" (Wines 2014).

Europe's "Dirty Little Secret"

Along with their increasing emphasis on wind and solar power, European countries also have been burning more coal. European use of coal rose rapidly in 2012 when the price of coal declined on world markets because of increased exports from the United States because U.S. demand for coal to power generators had fallen in preference to fracked shale gas. Coal also was being used to replace nuclear energy that had been shut down. "In the United States, natural gas is now frequently less expensive than coal for power, so demand for the hard, black fuel has plummeted," the *Washington Post* reported. "Ships are steaming the coal around the world instead. U.S. coal exports to Europe were up 26 percent in the first nine months of 2012 over the same period in 2011" (Birnbaum 2013). "It's been very welcomed that U.S. greenhouse gas emissions have been going down because of the switch to gas," said David Baldock, executive director of the Institute for European Environmental Policy in London. "But if we're simply diverting the coal somewhere else, particularly to Europe, a lot of those benefits are draining away" (Birnbaum 2013).

Germany has been one of the most aggressive countries in the world in pursuing wind and solar energy, but it is also using lignite ("brown coal") despite its low energy value and high carbon dioxide output in order to replace the country's shuttered nuclear power plants. Germany has made a phaseout of nuclear power by 2022 official policy. Although solar-panel installations doubled in Germany between 2011 and 2012, several new coal-powered plants also began operations.

In addition to imports, Germany and Poland (where most electricity is generated with coal) have their own mines. "In Jaenschwalde, a stone's throw from the Polish border," wrote Michael Birnbaum in the *Washington Post* (2013), "the forested countryside quickly drops away into a 300-foot-deep pit stretching for miles. Enormous machines slowly eat away at the earth and shower soft lignite onto a conveyor belt that feeds directly into a nearby power plant. From the precipice of the mine, the 20-foot-tall trucks at the bottom look like Tonka toys."

This kind of lignite provided 25.6 percent of Germany's electricity in 2012, compared to 7.0 percent in 2010. Add that to hard black coal (19.1 percent), and Germany's ratio of power generated by coal was close to that of the United States at that time. The rise in coal generation conflicts with Germany's official policy to use renewable sources for 80 percent of its power by 2050.

Enel, Italy's largest electricity producer, raised the share of coal in its generation mix to 50 percent in 2007. As a whole, Italy was switching from expensive oil to cheaper and dirtier coal, increasing its use from 14 percent to 33 percent (Rosenthal 2008). Europe as a whole also was turning to coal with plans for 50 new coal-fired plants over the next five years, even as some countries announced ambitious plans to increase wind and solar. Scattered protests of new coal-fired power in Germany, the Czech Republic, and Great Britain have arisen over the last few years.

"In order to get over oil, which is getting more and more expensive, our plan is to convert all oil plants to coal using clean-coal technologies," said Gianfilippo Mancini, Enel's chief of generation and energy management. "This will be the cleanest coal plant in Europe. We are hoping to prove that it will be possible to make sustainable and environmentally friendly use of coal" (Rosenthal 2008). So-called clean coal only reduces visual pollution but does nothing to improve carbon dioxide emissions. The kind of carbon capture and sequestration (CCS) system that James Hansen advocates is not yet commercially available.

India: An Energy and Climate Paradox

With a population of 1.25 billion, India has been increasing its development of both fossil fuels and renewable energy. Two-thirds of Indians are so poor that cow dung is their main form of cooking fuel. Political pressure supports coal-fired power as well as solar and wind energy. By 2015, India was the world's third-highest national source of greenhouse gases after China and the United States. The country is rapidly urbanizing and has been opening new coal-fired electrical facilities to meet increasing demand. Pollution from coal-fired power as well as dung-fed cooking fires often shroud large parts of India under the world's filthiest air.

Under pressure to join an international effort to combat global warming in 2015 and 2016, India pledged to reduce the rate at which its greenhouse gas emissions are growing—but not, as many other countries had done, to lower them. As *The New York Times* reported, "Under the terms of the plan, India's economy would grow roughly sevenfold by 2030, compared with 2005 levels, while its carbon emissions would triple. Yet if India took no action, emissions would also grow seven-fold. . . . [The government maintains] that rich countries like the United States bear moral responsibility for global warming and should not deny poor countries the chance to build their economies" (Barry and Davenport 2015).

Coal Mining and Use in India

India's city air is sometimes dirtier than even China's, a mix of coal effluvia, cooking fire smoke, and vehicle exhaust. New Delhi competes with Beijing and Shanghai for the filthiest urban air in the world. The cause in India is coal-fired power generation, which has been expanding rapidly, as well as cooking fires, mainly in urban areas. Two-thirds of India's population still derive their primary cooking fuel from cow dung.

India's coal is mainly strip-mined. Its high ash content and low heat value mean that it delivers twice the pollution per unit of energy as black coal. In addition, "in a country three times more densely populated than China, India's mines and power plants directly affect millions of residents. Mercury poisoning has cursed generations of villagers in in places like Bagesati, in Uttar Pradesh, with contorted bodies, decaying teeth, and mental disorders," wrote Gardiner Harris of *The New York Times* (2014).

Harris described the city of Dhanbad, India: "Decades of strip mining have left this town in the heart of India's coal fields a fiery moonscape, with mountains of black slag, sulfurous air and sickened residents. . . . The city of Dhanbad resembles a postapocalyptic movie set" (2014). "With villages surrounded by barren slag heaps half-obscured by acrid smoke spewing from a century-old fire slowly burning through buried coal seams. Mining and fire cause subsidence that swallows homes, with inhabitants' bodies sometimes never found." The living suffer skin and respiratory problems from continually burning fires. The air is incredibly dirty. "Smog levels that would lead to highway shutdowns and near-panic in Beijing go largely unnoticed in Delhi. Pediatric respiratory clinics are overrun, but parents largely shrug when asked about the cause of their children's suffering. Face masks and air purifiers, ubiquitous among China's elite, are rare here" (Harris 2014).

Rather than reclaim these hills or rethink their exploitation, the government is digging deeper in a coal rush that could push the world into irreversible climate change and make India's cities, already among the world's most polluted, even more unlivable. "If India goes deeper and deeper into coal, we're all doomed," said Veerabhadran Ramanathan, director of the Center for Atmospheric Sciences at the Scripps Institution of Oceanography, one of the world's top climate scientists, and former chairman of the U.N. Intergovernmental Panel on Climate Change (Harris 2014). By 2015, India was the world's third largest emitter of greenhouse gases, and burgeoning consumption of cheap, dirty coal was a major reason.

Coal has a considerable domestic lobby in India. "India's development imperatives cannot be sacrificed at the altar of potential climate changes many years in the future," said India's power minister, Piyush Goyal. "The West will have to recognize we have the needs of the poor" (Harris 2014). Goyal's agency plans to superintend a doubling of India's coal consumption from 565 million tons (in 2013) to more than a billion in 2019, mainly to provide power regardless of climatic consequences. At the same time, India's federal government has pledged to expand solar power.

India is developing solar power even as it expands coal generation. In the northern central state of Madhya Pradesh, India built what was one of the world's largest solar plants on 800 acres of barren soil when it was built in 2014. However, the air at the solar plant is so dusty that panels must be continually washed.

Further Reading

Harris, Gardiner. "Coal Rush in India Could Tip Balance on Climate Change." *The New York Times*, November 17, 2014. http://www.nytimes.com/2014/11/18/world/coal-rush-in-india-could-tip-balance-on-climate-change.html.

In 2015, India pledged to produce 40 percent of its electricity from non–fossil fuel sources by 2030 and sought money and technology for low-carbon technology from other countries. Prime Minister Narendra Modi has written a book presenting a moral case for reducing global warming, but he also faces domestic pressure to alleviate poverty: "Poverty reduction is our top priority. Providing power . . . is our priority. We want faster development. My people have a right to grow. Climate change is also a priority. We have the world's largest renewable energy sector. We want to clean our air, our water, our energy, our environment" (Barry and Davenport 2015).

Delhi: "Like Living in a Gas Chamber"

Given recent publicity, many people believe that the world's worst urban air pollution is in Beijing. Actually, China's capital runs second in this noxious sweepstakes as the world capital of smog behind the Ganges River valley around Delhi, according to the World Health Organization. Delhi's highest court has called Delhi's atmosphere "like a gas chamber" (Sudarshan 2015). The Delhi airport sometimes closes when smog becomes so thick that jets cannot safely land.

Many people have persistent coughs, and some wear masks. Many days, the sun shines through a gaseous haze. Approaching Delhi from the south at 35,000 feet, the setting sun turns the Ganges River valley clouds a dull yellow, then a dun orange, then dark red before the sky goes black.

On December 16, 2015, India's Supreme Court banned registration of luxury diesel cars and sport utility vehicles with an engine capacity more than 2,000 cubic centimeters in the National Capital Region until March 31, 2016. It also imposed a one-time pollution tax on small diesel cars (Mahapatra 2015).

The same *Sunday Times of India* front page that reported the final negotiation of the Paris world climate accord (Mohan 2015) reported that pollution in Delhi and the rest of the Ganges River valley had reached the worst levels of the fall and early winter when cooler than average weather drained downhill off freshly fallen snow in the foothills of the Himalayas. Delhi was described by the newspaper as "gasping for breath" ("Winter's Officially Here" 2015). In some areas, bonfires were being lit to provide warmth for the poor and homeless.

A lively debate was taking place in the public press over what do about Delhi's pollution. Coal-fired power and dung-fueled stoves used by 600 million (for the most part rural) Indians seem mainly off-limits. People need to eat, and at least 300 million in India have no electricity. That leaves "four wheels"—private cars and trucks.

One proposal reflects Beijing's use of restrictions on when someone can drive if pollution reaches asphyxiation levels—using odd and even last numbers on license plates. This one was implemented in the Delhi area on January 1, 2016, amid howls of protest from the rich, one of whom had three drivers and three cars—all with license plates ending in odd numbers (Najar 2016).

India's *Economic Times Magazine* devoted a cover story to solutions to exploding traffic. Among other things, the story pointed that although the rate of car vehicle ownership in Delhi doubled between 2000 and 2015, the proportion lags behind many rich countries: India's car ownership in 2015 was 13 per 1,000 people, a rate that is 617 in Japan and 439 in the United States (Seetharaman 2015, 8).

Southern India is booming, and so are power production and pollution. Google recently announced plans to build its largest campus outside the

United States in Hyderabad, a city in south India with a population the size of Los Angeles. Hyderabad's traffic jams, like those of Los Angeles, have become legendary. Nearby Bangalore in the country's southwest has become India's Silicon Valley. Nearby, the southeastern state of Andhra Pradesh's government late in 2015 announced plans for two new coal-fired 800-megawatt super-critical thermal power plants to supply rapidly rising needs for energy. Traffic on the road between Hyderabad and Vijayawada 150 miles to its southeast is so thick that the trip usually takes four hours.

Further Reading

Mahapatra, Dhananjay. "Small Diesel Car Buyers to Pay One-Time Pollution Tax: SC [Supreme Court]." *Times of India*, December 17, 2015, 1.

Mohan, Vizhwa. " 'Historic' Climate Deal Done," Meets India's Key Concerns." *Sunday Times of India*, December 13, 2015, A1.

Najar, Nida. "Streets, If Not the Air, Clear Out as Delhi Tests Car Restrictions." *The New York Times*, January 5, 2016, A4.

Seetharaman, G. "Getting Past the Gridlock." *Economic Times*, December 13–19, 2015, 2–15.

Sudarshan, V. "In Delhi, Just Waiting to Exhale." *The Hindu*, December 17, 2015, 11.

"Winter's Officially Here, Hits Flights. Worsens Pollution." *Sunday Times of India*, December 13, 2015, A1.

Coal: Political Resistance

By 2015, market forces and public opinion were slowly loosening coal's hold on electricity generation worldwide as many mines and coal-fired power plants faced closure. Wind and solar power were expanding rapidly. Even so, coal was still generating almost half of the world's electric power. Before its share began to fall, coal encountered widespread political protest in Europe and the United States, along with a growing realization that much of the planet's plentiful reserves would have to remain uncombusted to preserve a habitable ecosystem and avoid climate catastrophe.

Blockading Coal at Newcastle

On October 17, 2014, people from a dozen small island nations in the Pacific Ocean who had styled themselves as "Pacific Climate Warriors" attempted to shut down the world's largest coal port in Newcastle, Australia, for the day to point up coal's role in greenhouse gas emissions, sea level rise, and the prospective drowning of their homelands. About 30 of them—from the Marshall Islands, Vanuatu Tokelau, the Solomon Islands, and Papua New Guinea—sought to shut down the port by paddling into the paths of oncoming coal ships in traditional kayaks. Several hundred Australians watched from shore. An annual flotilla blockade had been held for several years, but this was the first time that Pacific Islanders had joined.

One protester, George Nacewa from Fiji, said that effects of climate change had become "really evident back at home with coastal erosion and in terms of sea-level rise. . . . To date, there has been the relocation of two villages, they've moved further inland," he said. "That's Fiji alone, and the other Pacific islands are more affected because most of them are just atolls" (Davidson 2014).

Helen Davidson of *The Guardian* described the event:

After a procession of several canoes shipped from Vanuatu, Fiji and other islands, and a Polynesian war dance, the traditional vessels took to the water. There were laughs and cheers from the crowd and police when two men tipped and then sank their canoe almost immediately. Then the leader of an outrigger canoe with "Tai Tokelau" painted on the side, shouted for others to "bring the kayaks" and dozens went in on their plastic watercraft. (Davidson 2014)

"If we get to stop a coal ship, then all the better," Nacewa told reporters after returning to shore. "But at the end of the day we are here to highlight the impact of climate change. All these islanders in their costumes—they live the realities of climate change." Milañ Loeak, 26, of the Marshall Islands said, "Just last week we had a king tide that affected some homes in the villages." She adding that droughts and floods had become more frequent. "It's something very personal to me, especially seeing it firsthand and seeing family members and relatives and friends and their families have to leave their homes and find new places to stay" (Davidson 2014).

Within half an hour, a coal-transport ship, the 250-foot *Rhine*, emerged from a dock pulled by tugboats. The protesters formed a blockade, but police used jet skis to impede them, spraying water and provoking chaos as they formed a passageway. After half an hour, the *Rhine* continued its journey.

Further Reading

Davidson, Helen. "Pacific Islanders Blockade Newcastle Coal Port to Protest Rising Sea Levels." *The Guardian* (U.K.), October 17, 2014. http://www.theguardian.com/environment/2014/oct/17/pacific-islanders-blockade-newcastle-coal-port-to-protest-rising-sea-levels.

On July 20, 2011, then New York City Mayor Michael R. Bloomberg announced a personal donation of $50 million to the Sierra Club's Beyond Coal campaign to support its campaign to shut down coal-fired power plants. By 2011, the Beyond Coal campaign, which began in 2002, had campaigned against plans for more than 150 new coal-fired plants and played a major role in eliminating them. The Sierra Club described the donation as "a game changer" (Torres and Eilperin 2011).

A tax on coal mined on federal land was proposed in 2013. "The federal government should also take into account the economic consequences of burning coal when pricing this fuel," asserted Stanford University international law specialist David J. Hayes and Harvard economist and former presidential economic adviser James H. Stock in an editorial opinion piece in *The New York Times* (2015). "The price for taxpayer-owned coal should reflect, in some measure, the added costs associated with the impacts of greenhouse gas emissions."

Should Carbon Dioxide Capture and Sequestration Technology Be Required?

The European Commission has been weighing whether to require new power plants to include facilities that can be retrofitted to store greenhouse gas emissions via carbon capture and sequestration (CCS) technology when it is available. This is the first legal requirement of this type in the world and a large step toward making CCS a commercial reality. As written, the requirement does not contain a date by which actual CCS would be required.

Although still in its infancy, if CCS technology is widely deployed, global CO_2 emissions could be reduced by one-third by 2050 (Schiermeier 2008). Currently, Norway, Britain, China, and the United States are planning CCS pilot plants. This technology has stirred controversy because of its high cost and the fact that it does not eliminate carbon dioxide emissions but merely locks them into rock formations. In some cases, CCS is designed to pump carbon dioxide into ocean floors, even as excess CO_2 provokes ocean acidification.

Opposition to New Coal Plants Accelerates

Environmental groups tightened their focus on proposed coal-fired power plants during and after 2007. The Environmental Defense Fund and the Natural Resources Defense Council assembled "strike forces" to mobilize opposition to new plants state by state. These played a role in obtaining cancellation of plants in Florida, Texas, New Mexico, and elsewhere. They also intervened in a dispute over whether to construct a new power plant on the Navajo Nation. The plant's carbon footprint would equal that of 1.5 million average automobiles (Barringer 2007).

On June 30, 2008, Superior Court Judge Thelma Wyatt Cummings Moore for Fulton County (Georgia) blocked construction of the first coal-burning power plant proposed in Georgia in more than 20 years, ruling that the plant must limit its emissions of carbon dioxide. This was the first time a court had applied an April 2007 ruling by the U.S. Supreme Court that recognized carbon dioxide as a pollutant under the federal Clean Air Act and applied the ruling to an industrial source. The judge overturned the Georgia Environmental Protection Division's approval of an air-pollution permit for Dynegy's proposed Longleaf power plant south of Columbus, Georgia. "In a case that is being watched across the country, Judge Moore has sent a message that it is not acceptable for the state to put profits over public health," said Justine Thompson, executive director of GreenLaw, the Atlanta public interest law firm that represented the environmental groups ("Georgia Judge" 2008).

During the past decade, plans for more than 100 coal-fired power plants in the United States have been voluntarily withdrawn or denied permits by state regulators. Some utilities were also moving away from coal on their own. Xcel Energy, for example, erected 274 wind turbines in northeastern Colorado. In 2009, Xcel still generated half of its electricity from coal but was also generating almost 3,000 megawatts more than any other utility (Warner 2009).

In 2007, opposition to new coal plants was accelerating. In early August, Mayor John Engen of Missoula, Montana, a Democrat, won city council support to buy electricity from a new coal-fired plant starting in 2011 to save the city money. He then was inundated by hundreds of e-mails and phone calls from protesting constituents. Late in July, then U.S. Senate Majority Leader Harry M. Reid of Nevada told chief executives of four power companies that he would "use every means at my disposal" to stop plans to build three coal-fired plants in Nevada. "There's not a coal-fired plant in America that's clean. They're all dirty," Reid said, urging a turn to wind, solar, and geothermal power (Mufson 2007).

In June 2007, a unanimous vote of Florida's Public Service Commission rejected a Florida Power & Light proposal to build a coal-fired plant near Lake Okeechobee. Florida Governor Charlie Crist approvingly said the state's Public Service Commission had "sent a very powerful message" and that the state "should look to solar and wind and nuclear as alternatives to the way we've generated power in the Sunshine State" (Mufson 2007).

Also in 2007, Citigroup downgraded the stocks of all coal companies. "Prophesies of a new wave of coal-fired generation have vaporized, while clean coal technologies . . . remain a decade away, or more," their report said. Citigroup's analysts said that by 2008 "election politics are likely to turn progressively more bestial for coal. Candidates are already stepping up to 'ban coal.'" The Citigroup report said that coal producers' earnings would probably be hurt by "new regulatory mandates applied to a group perceived as landscape-disfiguring global warming bad guys" (Mufson 2007).

By mid-2007, several new coal-powered generating plants were being canceled or postponed across the United States. Across the country, 645 coal-fired plants were producing roughly half the nation's electricity. As recently as May 2007, more than 150 new ones had been planned to meet electricity demand that was rising at a 2.7 percent annualized rate. A private equity deal worth $32 billion involving TXU Corporation canceled eight of 11 planned coal plants, as similar plants were scuttled in Florida, North Carolina, Oregon, and other states. In the meantime, late in August 2008, Xcel Energy of Minneapolis became the first builder of coal-fired power plants to agree with Andrew Cuomo, then New York state's attorney general, to provide investors with an analysis of global warming risks posed by its business.

Climate-change concerns were often cited as coal plants were cancelled, especially in Florida, where rising sea levels from melting ice in the Arctic, Antarctic, and mountain glaciers has already eroded the state's coastlines. Florida's Public Service Commission is now legally required to give preference to alternative energy projects over new fossil fuel generation of electricity. The states of

Washington and California have been moving toward similar requirements. Xcel Energy and Public Service of Colorado were allowed to go ahead with a 750-megawatt coal-fired power plant only after they agreed to obtain 775 megawatts of wind power.

"Building new coal-fired power plants is ill conceived," said James Hansen. "Given our knowledge about what needs to be done to stabilize climate, this plan is like barging into a war without having a plan for how it should be conducted, even though information is available." "We need a moratorium on coal now," he added, "with phase-out of existing plants over the next two decades" (Rosenthal 2008).

Emission-Reducing Technologies

Many technologies have been developed to remove carbon dioxide from the emissions of coal-fired plants. One popular approach is the integrated gasification combined cycle, which creates hydrogen and CO_2 to be sequestered. Some technologies remove the carbon dioxide from the flue stream after combustion. All of these technologies (and others) require energy and cost money and, so they would probably raise electric bills 40 percent to 50 percent, barring an unforeseen technological breakthrough.

In addition, underground sequestration will succeed only where geology is suitable. Thus, space can be limited and not always in the same areas as the power plants. Errant carbon dioxide can be dangerous. In 1986, 300,000 tons of naturally occurring carbon dioxide that had been trapped in Cameroon's Lake Nyos rose to the surface, suffocating 1,700 people (Goodell 2006).

During January 2008, the U.S. Energy Department canceled the country's biggest carbon-capture demonstration project in Illinois because of its costs, which were initially budgeted in 2003 at $950 million but had spiraled to $1.5 billion and was far from complete (Rosenthal 2008).

In one experimental approach, the Warrior Run power plant in Cumberland, Maryland, captures carbon dioxide from its boiler and sells it to beverage gas distributors for use mainly in soft drinks. "If you've had a Coke today, you've probably ingested some of our product," said plant manager Larry Cantrell. The problem is that 4 megawatts of the plant's output (some 200 megawatts) goes to remove 5 percent of its carbon dioxide emissions (Kintisch 2007, 185).

Cancelations or delays of coal-power plants continued into 2009. In February, NV Energy delayed a plant in eastern Nevada until "clean coal" technology becomes available; In Montana, Southern Montana Electric Generation and Cooperative halted work on a plant near Great Falls in favor of wind turbines and another plant that will burn natural gas. Michigan's Governor Jennifer Granholm, a Democrat, told regulators not to approve any of five new coal-fired plant until "all feasible and prudent alternatives had been considered" (Watson 2009). In Kentucky, Peabody Energy dropped plans for a coal-fired energy plant in the western part of the state in favor of a plant that will convert coal to natural gas. In early March 2009, Alliant Energy dropped plans to build an enormous coal-power plant in central Iowa that would have provided enough energy to supply 500,000 homes. Several of thee actions were being challenged by coal-power advocates.

Some coal-powered electricity plants have been shut down through community organizing. In Chicago, for example, a coalition of residents with 60 organizations (including the NAACP, Greenpeace, and local groups) fought for a decade to close two 100-year-old power plants that had inundated neighborhoods with pollution. In 2012, they celebrated Chicago's renaissance as a coal-free city. The Fisk Station in the Pilsen neighborhood was closed late in 2012, and the Crawford Station in the Little Village neighborhood closed two years later. Opponents of the power plants built their case on local pollution (soot, sulfur dioxide, and nitrogen oxide) as well as the plants' emissions of greenhouse gases.

"This agreement means a cleaner, healthier environment for the communities around these coal plants," said NAACP President and CEO Benjamin Todd Jealous. "Environmental justice is a civil rights issue, and the NAACP is committed to strong regulation and monitoring of toxic coal emissions. For too long, Fisk and Crawford have been literally choking some of Chicago's most diverse neighborhoods, and some of its poorest" ("Decade-Long" 2012). "The 600,000 Chicago residents living within three miles of Fisk or Crawford have suffered long enough," said Rose Joshua, president of the NAACP South Side Chicago unit. "This is a true victory for grassroots democracy—a group of citizens who refused to be marginalized and spoke up for the health and well-being of their families and their environment" ("Decade-Long" 2012).

China, Coal, and Climate Change

By 2009, China was burning more coal than the United States, Europe, and Japan combined. The next year, China overtook the United States as the world's largest user of energy, having consumed only half as much 10 years previously. China has accelerated as a consumer of energy (mainly low-energy coal) more quickly than any other country in world history.

In 2014, however, China's coal consumption fell 2.9 percent, according to its government. Glen Peters, a scientist at the Center for International Climate and Environmental Research in Oslo, Norway, estimated that "the drop in consumption, together with slowed growth in cement production, reduced China's annual emissions of carbon dioxide, the main greenhouse gas from human activity, 0.8 percent. This was the first fall in China's emissions after more than 15 years of fast growth" (Wong and Buckley 2015).

By February 2015, China had reduced its consumption of coal by 3.7 percent below 2014 levels, according to statistics released by the Chinese government. China also leads the world in the deployment of renewable energy; in 2015, the country invested some $110 billion U.S.—twice as much as the United States, according to the World Resources Institute (Tollefson 2016). The deemphasis on coal was felt worldwide as Peabody Coal, the largest miner in the United States, skirted bankruptcy, and Union Pacific, a major coal-hauling railroad, laid off several hundred employees.

China's statistics are sometimes less than reliable, however. In 2015, for example, China admitted to having underreported its coal usage (and thus carbon dioxide emissions) by 17 percent—a billion tons of CO_2 annually. World emissions of CO_2 stabilized in 2014 at 32 billion tons. China's CO_2 emissions have risen from around 3 billion tons annually in 2000 to 8 billion in 2014. China had been underreporting its coal consumption since the year 2000, with the disparity increasing over time. By 2015, the 1 billion ton shortfall was as much coal as the United States burned in total (Buckley 2015).

While China has also become a world leader in manufacture, export, and deployment of wind and solar-power technology and equipment, the largest proportion of its industrial development (as well as rapidly growing domestic electricity supply) has been relatively dirty (and abundant) coal. China has been transforming gigatons of dirty coal with human ingenuity and labor to fill the world's retail stores with products—and its own urban skies with some of the most noxious air pollution on the planet.

China's consumption of coal increased 185 percent between 2000 and 2010, according to the International Energy Agency. During the same decade, coal use in the rest of the world rose 17.5 percent while consumption in the United States declined 1.6 percent. In 2010, coal produced 83 percent of China's electricity. The world as a whole burned 7.3 billion metric tons of coal during that year, with China burning nearly half that total (Bradsher 2011). Until 2015, when China curtailed its construction, another large coal-fired power plant opened somewhere in the country every week to 10 days that had enough capacity to serve all the households in Dallas or San Diego (Bradsher and Barboza 2006). Many of them use old, polluting technology, and they will be operating for an average of 75 years or until other sources allow authorities to shut them down.

Until 2015, China had developed a major appetite for worldwide supplies of coal where it had once been a major exporter. "At ports in Canada, Australia, Indonesia, Colombia and South Africa, ships are lining up to load coal for furnaces in China, which has evolved virtually overnight from a coal exporter to one of the world's leading purchasers," wrote Elisabeth Rosenthal in *The New York Times* (2010). Coal used to be burned mainly close to mining sites, but by 2010 that pattern had changed, with China adding the carbon emissions of transport to those of coal consumption itself.

The price of coal doubled from 2005 to 2010, spurred by Chinese demand. "This is a worst-case scenario," said David Graham-Caso, spokesman for the Sierra Club, "We don't want this coal burned here, but we don't want it burned at all. This is undermining everything we've accomplished" (Rosenthal 2010). In Australia, environmental groups have blocked coal trains en route to Newcastle as fleets of kayaking protesters on kayaks delayed cargo being loaded for Asia. Australian coal exports to China rose from $508 million in 2008 to $5.6 billion in 2009.

Vic Svec, a senior vice president at the world's largest private coal company, Peabody Energy, said it was "planning to send larger and larger amounts of coal" to China. "Coal is the fastest-growing fuel in the world and will continue to be largely driven by the enormous appetite for energy in Asia," he said (Rosenthal 2010).

China: Efforts to Reduce Emissions

By 2015, however, when China switched from coal to other sources for new power capacity, world demand crashed, along with coal's price and the market capitalization of several large companies. The coal-market crash affected the entire world. In Omaha, Nebraska, for example, several hundred people lost their jobs at Union Pacific, one of the largest coal-hauling railroads in the United States.

Even with coal consumption declining, China continues to rely on it for 75 percent of the country's energy, spewing out some 19 million tons of sulfur dioxide a year (the U.S. produces 11 million tons per year) and contributing mightily to acid rain. Many Chinese homes that once used only lightbulbs have acquired several appliances, including air conditioners, in recent years.

The personal effects of coal use in China were described in *The New York Times*:

> Wu Yiebing and his wife, Cao Waiping, used to have very little effect on their environment. But they have tasted the rising standard of living from coal-generated electricity and they are hooked, even as they suffer the vivid effects of the damage their new lifestyle creates. Years ago, the mountain village where they grew up had electricity for only several hours each evening, when water was let out of a nearby dam to turn a small turbine. They lived in a mud hut, farmed by hand from dawn to dusk on hillside terraces too small for tractors, and ate almost nothing but rice on an income of $25 a month Today, they live here in Hanjing, a small town in central China where Mr. Wu earns nearly $200 a month. He operates a large electric drill 600 feet underground in a coal mine, digging out the fuel that has powered his own family's advancement. He and his wife have a stereo, a refrigerator, a television, an electric fan, a phone and light bulbs, paying just $2.50 a month for all the electricity they can burn from a nearby coal-fired power plant. (Bradsher and Barboza 2006)

Urban China has some of the worst air quality in the world. The U.S. embassy in Beijing monitors air quality on its roof using a scale that measures particles smaller than 2.5 microns in diameter, the deadliest form of air pollution, because it easily penetrates deep into people's lungs and bloodstreams, causing permanent DNA mutations, heart attacks, lung cancer, and premature death ("Beijing Bans" 2014). On January 12, 2013, that scale reached 728. Some days, readings of small particulate matter in Beijing have reached 1,000 micrograms per cubic meter, more than three times the level that the U.S. Environmental Protection Agency regards as hazardous. During its worst "killer" smogs in the 1950s, not even industrial London reached these levels.

Air pollution is a major political issue in China, the subject of copious political pressure that has been recognized by the Communist Party on the national and local levels. The Environment News Service reported on August 6, 2014, that the Beijing Municipal Environmental Protection Bureau had "announced that the districts of Dongcheng, Xicheng, Chaoyang, Haidian, Fengtai, and Shijingshan will stop using coal and close coal-burning power plants and other coal-fired facilities over the next six-and-a-half years" ("Beijing Bans" 2014). In many instances, coal will be replaced by natural gas, a cleaner-burning fossil fuel. Beijing still must tackle pollution from cars, the source of 31 percent of the city's air pollution. Industry makes up some 18 percent and dust around 14 percent ("Beijing Bans" 2014).

Officials also curtailed burning of oil, coke, burnable trash waste, and some biomass fuels. In 2014, Beijing, China's capital and second-largest city (with 20 million people), announced plans to ban use of coal by the end of 2020.

At the end of January 2013, the Chinese national government declared a national emergency, declaring war on the thick smog that had swallowed its capital for too many days. More than 100 factories were shut down (but only for a few days), and one-third of the government's cars and trucks were removed from the roads, all to combat a midwinter spell of stagnant air from fossil fuel burning factories and vehicles that had reached hazardous levels.

"This month," reported *The New York Times*, "Beijing has writhed in the grip of the most polluted air days on recent record. The surge in pollution, which is happening across northern China, has angered residents and led the state news media to report more openly on air quality problems" (Wong 2013). Beijing's mayor, Wang Anshun, said that efforts would be undertaken to reduce the rapid rise in the number of vehicles in Beijing. The number had risen from 3.13 million in 2008 to 5.13 million in 2013. According to *The New York Times*, "Mr. Wang told the legislature on Jan. 22 that the Beijing government was aiming to cut the density of major air pollutants by 2 percent this year. To that end, officials are ordering 180,000 older vehicles off the roads, promoting the use of 'clean energy' for government vehicles and heating systems, and growing trees over 250 square miles of land in the next five years" (Wong 2013).

Beijing was facing the results of a national "growth-at-any-cost" policy that has reaped widespread pollution, other environmental damage, and soaring greenhouse gas emissions. In addition, Beijing's atmosphere collects the effluvia of coal-burning factories, power plants, and vehicles during stagnant weather, especially in winter, and seals them under an atmospheric inversion in which warm air sits on top of a colder layer near the ground. Like Mexico City and Los Angeles, Beijing is located in a valley nearly surrounded by mountains and prone to thermal inversions that hold pollution close to the ground. With 20 million people in 2007, a number that is steadily increasing, metropolitan Beijing has been undergoing on of the world's largest construction booms, adding residential and office space that requires more electricity, heating, and cooling, nearly all of it provided by coal. Automotive pollution pours into the city as well as it adds 400,000 cars and trucks a year. Although Shanghai has held new vehicle registrations to 100,000 a year with license fees as high as $7,000 per vehicle, Beijing has allowed nearly untrammeled growth.

Alternative Energy in China

Even as China continued to build coal-fired and nuclear power plants, in 2009 the country passed the United States as the world's largest market for wind energy. The country was building the world's largest wind farms (at 10,000 to 20,000 mega-watts each). China already was the world's largest producer of solar panels.

In August 2010, China announced the closure of 2,078 industrial plants that used large amounts of energy (most of it low-grade coal) as part of a drive for energy efficiency. The closures included 762 cement plants (especially heavy users of energy), 279 that produce paper, 175 steel factories, and 84 leather plants (Bradsher 2010).

In 2009, China embarked on a campaign to build 500,000 electric and hybrid cars by 2011. This is not a green strategy, however, because about 75 percent of China's electricity comes from coal, most of it low power and dirty. Indeed, with 80 percent of Chinese drivers purchasing their first cars, it is one more step up the greenhouse gas ladder (Bradsher 2009). In three years (2006–2009), China shut down more than 1,000 old coal-fired power plants whose technology resembles that still widely used in the United States.

China also has enacted tough automobile mileage standards. The booming Chinese economy and rising consumption of 1.4 billion Chinese is swamping these efforts, however. Total use of fossil fuels and production of greenhouse gases continues to rise rapidly. Apartment and office buildings continue to rise in cities, and rural people are purchasing appliances at record rates. In 2009, car sales in China surpassed those in the United States, a rise of 48 percent in one year (Bradsher 2010). Use of air conditioning is also spreading rapidly. "We really have an arduous task" even to reach China's existing energy-efficiency goals, said Gao Shixian, an energy official at the National Development and Reform Commission, in a speech at the Clean Energy Expo China in late June in Beijing (Bradsher 2010).

China has eliminated motor-vehicle fuel subsidies and set a fuel-efficiency standard for new vehicles at 36.7 miles per gallon, seven years before the United States will reach that requirement. China also has set high efficiency requirements for new coal-fired electric plants (the United States set no such standards until 2015). "Regardless of whether the United States passes its own legislation, China will take positive measures because this is a requirement for our own economy to conserve resources," said Xie Zhenhua, vice chairman of the National Development and Reform Commission (Mufson 2009). China has set a renewable-energy target 15 percent by 2020, in part by increasing its nuclear energy output 400 percent.

China's First Strategy on Global Warming

In mid-2007, China released its first strategy on global warming, which advocated improved energy efficiency and control of greenhouse gas emissions but did not include mandatory limits that could restrict the growth of the country's booming economy. This program resembled a program recommended by U.S. President George W. Bush at about the same time—as each of the world's two largest greenhouse gas sources blamed the other for taking inadequate action to address global warming. Ma Kai, head of the National Development and Reform Commission,

China's powerful economic planning agency, maintained that criticism was unfair because the Chinese produce only around one-fifth the greenhouse gases per capita as the United States (Yardley 2007).

China's plans called for a 20-percent improvement in energy efficiency. China's drive for energy efficiency is compromised, however, by its dependence on coal, much of it low in quality. By 2017, however, China was reducing the rate at which it was mining and importing coal, and had become a world leader in both the manufacture and use of solar panels. China is also in the midst of a nationwide reforestation program to help absorb greenhouse gases (Yardley 2007).

The same report, according to a description in *The New York Times*, "also painted a brief, if alarming, picture of how global warming could change China, including rising sea levels, shrinking glaciers in Tibet, rising temperatures and the likelihood of expanding deserts. By 2020, annual temperatures could rise between 1.3 [and] 2.1 [°C] from 2000. In addition, projections show that Chinese agricultural output could be reduced by 10 percent by 2030" (Yardley 2007).

China's Nuclear Construction Campaign

By 2007, China had begun construction on dozens of new nuclear power plants to address part of its growing global- arming burden from low-quality coal. China plans to spend $50 billion to build 32 nuclear plants by 2020. Some experts believe another 300 more such plants may be built by 2050 to power what will then be the largest national economy in the world. By that time, China may have half the nuclear-power capacity in the world (Eunjung Cha 2007). China also is building the world's largest repository for spent nuclear fuel in its western desert, amid the Beishan Mountains, an area that is nearly bereft of human habitation. The nuclear construction binge represents a major change for China, where only 2.3 percent of electricity was generated from nuclear power in 2006, compared with some 20 percent in the United States and almost 80 percent in France. In 2007, China had only nine nuclear power plants.

In part because of Chinese demand, the price of processed uranium ore increased from $10 to $120 a pound between 2003 and 2007. Expecting that China will be one of its major customers, Japan's Toshiba paid $5.4 billion during 2006 to acquire a U.S. company, Westinghouse Electric, that specializes in the construction of nuclear plants. The Chinese government, emphasizing safety, has been using companies such as Westinghouse to instruct its engineers who will build and operate new nuclear plants.

Further Reading

Barringer, Felicity. "Navajos and Environmentalists Split on Power Plant." *The New York Times*, July 27, 2007. http://www.nytimes.com/2007/07/27/us/27navajo.html.

Barry, Ellen, and Carol Davenport. "India Announces Plan to Lower Rate of Greenhouse Gas Emissions." *The New York Times*, October 1, 2015. http://www.nytimes.com/2015/10/02/world/asia/india-announces-plan-to-lower-rate-of-greenhouse-gas-emissions.html.

"Beijing Bans Coal Burning to Clear the Air." Environment News Service, August 6, 2014. http://ens-newswire.com/2014/08/06/beijing-bans-coal-burning-to-clear-the-air/.

Birnbaum, Michael. "Europe Consuming More Coal." *Washington Post*, February 7, 2013. http://www.washingtonpost.com/world/europe-consuming-more-coal/2013/02/07/ec21 026a-6bfe-11e2-bd36-c0fe61a205f6_print.html.

Bradsher, Keith. "China Vies to Be World's Leader in Electric Cars." *The New York Times*, April 2, 2009. http://www.nytimes.com/2009/04/02/business/global/02electric.html.

Bradsher, Keith. "China Fears Warming Effects of Consumer Wants." *The New York Times*, July 4, 2010. http://www.nytimes.com/2010/07/05/business/global/05warm.html.

Bradsher, Keith. "Cleaner China Coal May Still Feed Global Warming." *The New York Times*, June 17, 2011, B8.

Bradsher, Keith, and David Barboza. "Pollution from Chinese Coal Casts a Global Shadow." *The New York Times*, June 11, 2006. http://www.nytimes.com/2006/06/11/business/world business/11chinacoal.html.

Buckley, Chris. "China Burns Much More Coal Than Reported, Complicating Climate Talks." *The New York Times*, November 4, 2015, A1.

"Decade-Long Grassroots Campaign Shuts Two Chicago Coal Plants." Environment NEWS Service, March 2, 2012. http://ens-newswire.com/2012/03/02/decade-long-grassroots -campaign-shuts-two-chicago-coal-plants/.

Eunjung Cha, Ariana. "China Embraces Nuclear Future; Optimism Mixes with Concern as Dozens of Plants Go Up." *Washington Post*, May 29, 2007, p. D1.

"Georgia Judge Yanks Coal Power Permit on Climate Concerns." Environment News Service, June 30, 2008. http://www.ens-newswire.com/ens/jun2008/2008-06-30-091.asp (no longer available).

Goodell, Jeff. "Our Black Future." *The New York Times*, June 23, 2006. http://www.nytimes .com/2006/06/23/opinion/23goodell.html.

Hayes, David J., and James H. Stock. "The Real Cost of Coal." *The New York Times*, March 24, 2015. http://www.nytimes.com/2015/03/24/opinion/the-real-cost-of-coal.html?emc =edit_th_20150324&nl=todaysheadlines&nlid=35795487.

Kintisch, Eli. "Making Dirty Coal Plants Cleaner." *Science* 317 (July 13, 2007): 184–186.

Mufson, Steven. "Coal Rush Reverses, Power Firms Follow." *Washington Post*, September 4, 2007, D1. http://www.washingtonpost.com/wp-dyn/content/article/2007/09/03/AR2007 090301119_pf.html.

Mufson, Steven. "China Steps Up, Slowly but Surely." *Washington Post*, October 24, 2009. http://www.washingtonpost.com/wp-dyn/content/article/2009/10/23/AR200910 2304075_pf.html.

Rosenthal, Elizabeth. "Europe Turns Back to Coal, Raising Climate Fears." *The New York Times*, April 23, 2008. http://www.nytimes.com/2008/04/23/world/europe/23coal.html? pagewanted=print&_r=0.

Rosenthal, Elisabeth. "Nations That Debate Coal Use Export It to Feed China's Need." *The New York Times,* November 21, 2010. http://www.nytimes.com/2010/11/22/science/ earth/22fossil.html.

Schiermeier, Quirin. "Europe to Capture Carbon." *Nature* 451 (January 17, 2008): 232.

Tollefson, Jeff. "China's Carbon Emissions Could Peak Sooner Than Forecast." *Nature* 531 (March 24, 2016): 425. http://www.nature.com/news/china-s-carbon-emissions-could -peak-sooner-than-forecast-1.19597.

Torres, Christian, and Juliet Eilperin. "Mayor Bloomberg Gives $50 Million to Fight Coal-Fired Power Plants." *Washington Post*, July 21, 2011. http://www.washingtonpost.com/

national/health-science/mayor-bloomberg-gives-50-million-to-fight-coal-fired-power-plants/2011/07/20/gIQAEKKURI_print.html.

Warner, Melanie. "Is America Ready to Quit Coal?" *The New York Times,* February 15, 2009. http://www.nytimes.com/2009/02/15/business/15coal.html.

Watson, Traci. "Companies Rethink Coal Plants." *USA Today*, March 9, 2009, 9A.

Wines, Michael. "Clean Mining a Deception in Kentucky, Groups Say." *The New York Times*, November 18, 2014. http://www.nytimes.com/2014/11/18/us/clean-mining-a-deception-in-kentucky-groups-say.html.

Wong, Edward. "Beijing Takes Emergency Steps to Fight Smog." *The New York Times*, January 31, 2013. http://www.nytimes.com/2013/01/31/world/asia/beijing-takes-emergency-steps-to-fight-smog.html.

Wong, Edward, and Chris Buckley. "Chinese Premier Vows Tougher Regulation on Air Pollution." *The New York Times*, March 16, 2015. http://www.nytimes.com/2015/03/16/world/asia/chinese-premier-li-keqiang-vows-tougher-regulation-on-air-pollution.html.

Yardley, Jim. "China Releases Climate Change Plan." *The New York Times*, June 4, 2007. http://www.nytimes.com/2007/06/04/world/asia/04cnd-china.html.

See also: Carbon Capture; Global Warming, China; Greenhouse Gases, Effects of; Unburnable Carbon

CORPORATE SUSTAINABILITY

Concern about global warming has provoked many corporations to make genuine efforts to change the basic ways they do business. Although most corporations dismissed global warming during the 1990s, within 10 years—by 2008—rising temperatures and accumulating scientific evidence was changing the business "climate." Major companies, including utilities, manufacturing, petroleum, chemicals, and financial services—and including household names such as General Electric, DuPont, Duke Energy, Caterpillar, BP, Lehman Brothers, and Alcoa—had joined several national environmental groups in support of nationwide limits on carbon dioxide emissions. Many companies are installing solar and wind energy and reducing their energy consumption. Some are young technology companies such as Google. Others, such as Wal-Mart, may come as surprises.

In March 2014, the multinational banking firm Citigroup said the "Age of Renewables" had begun in the United States. As Giles Parkinson wrote in *Renew Economy*,

In a major new analysis released this week, Citi says the big decision makers within the U.S. power industry are focused on securing low-cost power, fuel diversity and stable cash flows, and this is drawing them increasingly to the economics of solar and wind, and how they compare with other technologies [even as] much of the mainstream media—in the US and abroad—has been swallowing the fossil fuel Kool-Aid and hailing the arrival of cheap gas, through the fracking boom, as a new energy "revolution" as if this would be a permanent state of affairs. . . . Solar costs continue to fall. (Parkinson 2014)

Corporations and Global Warming

General Electric CEO Jeffery Immelt said that regulation of greenhouse gasses is coming, and "I'd just as soon have a seat at the table than have it pushed down my throat" (Ball 2008). An account in *The New York Times* sketched the corporate agenda:

> The chief executives agreed after some discussion, to strongly discourage further construction of stationary sources that cannot easily capture carbon dioxide. This comes close to a rejection of almost all new coal-fired power plants on the drawing boards, including 11 plants recently proposed by TXU, a Texas utility. The technology that would isolate carbon dioxide emissions and bury them is still in the earliest phases of development, so this near-repudiation of existing coal technology would have a disproportionate impact on utilities that depend largely on coal, like TXU and the Southern Company. (Barringer 2007)

Shareholder Activism

Shareholders at some companies have become restive about climate change. For example, in late March 2014, ExxonMobil, the United States' largest energy company was pressed by shareholders to publish a carbon asset risk report on the company Web site. The report forces management to describe how the company is planning for a future in which controls on greenhouse gases may prohibit using some of its oil and gas reserves. The shareholders were organized by Arjuna Capital, a wealth-management company that supports sustainability. "We're gratified that ExxonMobil has agreed to drop their opposition to our proposal and address this very real risk," said Natasha Lamb, director of equity research and shareholder engagement at Arjuna Capital. "Shareholder value is at stake if companies are not prepared for a low-carbon scenario." "Companies need to acknowledge that preparing for a low-carbon future is a necessity, not a choice. Companies that prepare early for a future with reduced carbon emissions will likely perform better than those who delay," said Danielle Fugere, president of As You Sow, an environmental advocacy group that also supported the shareholder resolution ("ExxonMobil" 2014).

Shareholders also have filed global warming resolutions at the annual meetings of General Motors and Ford Motor that attempt to increase pressure on the automakers to reduce greenhouse gas emissions from their products as well as from their manufacturing processes. A coalition of shareholders—mostly members of various Catholic orders—asked that the automakers report carbon dioxide emissions from their plants and vehicles. "We believe that both General Motors and Ford face material and reputational risk in their current failure to address and reduce carbon dioxide emissions," said Sister Patricia Daly, executive director of the Tri-State Coalition for Responsible Investment (Eggert 2002).

A resolution calling for a report on how ExxonMobil will respond to the growing pressure to develop renewable energy won support from 21 percent of shareholders at the company's annual general meeting in May 2003. The 21-percent shareholder support for the renewables resolution represented $42.34 billion worth of ExxonMobil stock ("One in Five" 2003). The same day, almost a quarter of Southern Company shareholders (one of the largest U.S. utilities) voted to require the company to evaluate potential financial risks associated with its emissions of fossil fuels. Twenty-seven percent of shareholders at American Electric Power Company had previously voted in support of a similar resolution. Southern, A.E.P., Excel Energy Co., Cinergy Co., and T.X.U. Corp. are the top five emission sources of carbon dioxide in the United States. Another resolution of this type at Chevron-Texaco won 32 percent of shareholder votes in 2003, up from 9.6 percent in 2001 (Seelye 2003).

ExxonMobil pledged $10 million annually for 10 years to Stanford University for climate research, most of it to help establish a research center that will develop technologies to help generate clean energy. For many years, ExxonMobil also provided sizable donations to groups that question global warming's scientific basis, including right-wing lobbying organizations such as the Competitive Enterprise Institute, Frontiers of Freedom, George C. Marshall Institute, American Council for Capital Formation Center for Policy Research, and American Legislative Exchange Council.

Executives Discuss Climate Solutions

In late January 2014, politicians and business executives meeting at the World Economic Forum at Davos, Switzerland, devoted substantial conference time to climate change. At the same time, William D. Nordhaus, an expert and author on the economics of global warming, who was inaugurated as president of the American Economic Association, said, "There is clearly a growing recognition of this in the broader academic economic community" (Davenport 2014). Jim Yong Kim, president of the World Bank, has said that addressing climate change is central to the institution's mission, "citing global warming as the chief contributor to rising global poverty rates and falling GDPs in developing nations. In Europe, the Organization for Economic Cooperation and Development, the Paris-based club of 34 industrialized nations, has begun to warn of the steep costs of increased carbon pollution" (Davenport 2014).

Advocating that addressing climate change is critical has drawn big money that was earned building corporations. Thomas F. Steyer, whose management of hedge funds made him a billionaire, set out to spend as much as $100 million in the 2014 midterm elections to support candidates who support climate-change legislation through a hard-edged campaign of attack ads on global warming deniers. Steyer aimed to build a political campaign that rivals the conservative effort created by deniers Charles and David Koch, who have funneled millions of dollars in campaign funds to global warming skeptics. Also by 2014, Steyer had given more than $1 million to the movement seeking to stop the Keystone XL pipeline and even appeared in four 90-second ads against the pipeline (Wheaton 2014).

Another billionaire, former New York Mayor Michael R. Bloomberg, commissioned a study called *Risky Business* to assess financial risks of global warming with two former U.S. secretaries of the treasury, Henry M. Paulson Jr. and Robert E. Rubin. "This study is about one thing, the economics," Paulson said. "Business leaders are not adequately focused on the economic impact of climate change." "There are a lot of really significant, monumental issues facing the global economy, but this supersedes all else," Rubin said (Davenport 2014).

Companies Track and Reduce Carbon Emissions

Several thousand companies in the United States have been reporting their carbon emissions and energy costs through the London-based Carbon Disclosure Project, a not-for-profit group of 770 institutional investors holding $92 trillion U.S. in assets. Some are household names such as Coca-Cola and Wal-Mart, which has begun to require its suppliers to report and reduce their carbon footprints. Computer maker Dell began to neutralize the carbon impact of its operations around the world in 2008 ("Investors" 2007).

Some of the world's largest multinational companies, among them Procter & Gamble, Unilever, Tesco (the British grocery chain), and Nestle SA, were requiring their suppliers to disclose carbon dioxide emissions and global warming mitigation strategies. The companies are among several that have formed the Supply Chain Leadership Coalition that cooperates with the Carbon Disclosure Project. Eventually, products may be labeled with carbon-emission information. Cadbury Schweppes was making plans to print such information on its chocolate bars. At about the same time, Wal-Mart began a similar project by asking Oakhurst Dairy of Portland, Maine, to measure the carbon footprint of a case of milk (Spencer 2007). Sir Terry Leahy, chief executive of the Tesco supermarket chain, which sells a quarter of Britain's groceries, told Forum for the Future that the company would cut its energy consumption in half by 2010, sharply reduce the number of products it ships by air, and place a carbon label on each of the 70,000 products it sells, taking into account a product's entire manufacturing life from production to distribution to consumption.

Calculating a product's carbon footprint can be extremely complex. A year after its corporate carbon-footprint mandate, Tesco had produced only one carbon label for a house brand of Walkers crisps (potato chips). Calculating the carbon footprint of potato chips is one of the simpler food products but is still a daunting task that involves determining the amount of energy required to raise potatoes and sunflower oil, the costs of fertilizers and pesticides involved in growing them, the tractors that harvest them, the manufacture of the chips, the costs of their packaging, transportation to stores, and disposal of packaging. The company and Britain's Carbon Trust calculated that 75 grams of greenhouse gases were produced for each bag. In the meantime, Kraft Foods by 2008 was powering a plant in New York with methane produced by adding bacteria to whey, a by-product of cheese production.

Corporate Conservation: Examples

By 2012, some 25 percent of Fortune 500 companies employed a board of directors' subcommittee to oversee environmental issues (usually including climate change) up from only 10 percent 10 years before, according to Mindy Lubber, president of Ceres, a national coalition of activists, investors, and others who study environmental issues. The number of shareholder proposals related to environmental issues doubled between 2004 and 2012. Some companies (one example is American Electric Power, or AEP) had gone so far as to dock the pay of executives if their firms were cited for violation of environmental regulations more often than their board's guidelines allowed. Michael G. Morris, chief executive officer of AEP, in 2006 was docked $80,045 from a projected bonus of $2 million after the company received nine violation citations when the company's "allowance" was five (Lublin 2008).

Many companies also undertook detailed audits of their own energy use, preparing to look for ways to limit consumption and waste. Such measures were more than greening public relations. They were also saving considerable money. These companies were just beginning to adjust ways of doing business that had become especially good at turning Earth's resources into merchandise, greenhouse gases, profit, and waste heat at an accelerating rate, a world in which more (even wasteful) consumption was invariably assumed to be good for business. The production of carbon dioxide has been wedded to our ways of doing business and the ways in which we define financial success. Can profits rise without burning more fossil fuels? Many businesses were finding ways to decouple the two.

Further Reading

Lublin, Joann S. "Environmentalism Sprouts Up on Corporate Boards." *Wall Street Journal*, August 11, 2006, B6.

Corporate Sustainability Initiatives

Many companies have now appointed chief sustainability officers. "Environmental vice presidents usually spend company money, but this new breed is helping companies make money," said Eileen Claussen, president of the Pew Center on Global Climate Change. The upshot, said Geoffrey Heal, a business professor at the Columbia Business School, is that "what started out as a compliance job has evolved into one that guards the value of the brand" (Deutsch 2007).

These officers, who may go under the title "vice president for environmental affairs," negotiate with vendors and customers to create and market green products, Dow Chemical's first chief sustainability officer, David E. Kepler, met with the company's technology, manufacturing, and financial executives about strategies to encourage alternative fuels and green products. "We usually agree," Mr. Kepler said. "But if a critical environmental issue is in dispute, I'll prevail" (Deutsch 2007).

Linda J. Fisher, chief sustainability officer at DuPont, opposed purchase of a subsidiary that was not in a sustainable business. "We're building sustainability into the acquisition criteria," she said. When two business executives at General Electric (GE) opposed the cost of developing environmentally friendly products, Jeffrey R. Immelt, GE's chairman, gave Lorraine Bolsinger, vice president of GE's Ecomagination business, the research money. "I have an open door to get projects funded," she said (Deutsch 2007). Companies often do not speak with one voice, however. Although General Electric has touted its environmental initiatives, its board of directors in 2008 advised shareholders to vote against a proposal requiring the company to prepare a report on how its activities impact climate change.

In 2001, Owens-Corning named Frank O'Brien-Bernini as chief research and development and sustainability officer. He was charged with using a "lens of sustainability" to prioritize research. He then helped supervise development of a machine that makes it easier to insulate attics. Owens-Corning developed the machine after research showed that drafty attics are prime culprits in heating and air-conditioning leaks but that the "hassle factor" kept homeowners from installing insulation. "I drive innovation around products and processes," O'Brien-Bernini said. "And I make sure that our claims are backed by deep, deep science" (Deutsch 2007).

Stephen Lane, who joked that he was the "Al Gore of Citigroup," is an executive vice president whose full-time job is coaxing energy savings out of a company with 340,000 employees worldwide. He supervised an inventory of energy use in all of the company's facilities. "What you can't measure, you can't manage," he said (Carlton 2007, B1). He also wrote policies for Citigroup that govern everything from installing solar energy and timed lighting in bank branches. Lane also advised employees to switch off unused lights and climb stairs instead of using escalators, which have been stopped during nonbusiness hours. Other banks are taking similar measures. Hong King-Shanghai Bank Corp., for example, has opened a "green" prototype branch in New York state that its 400 HSBC branches in the United States soon will emulate. Lane oversees a $10-billion Citigroup plan to reduce its carbon footprint to 10 percent below 2005 levels by 2011 (Carlton 2007, B8).

Water Calls the Tune

Coca-Cola is paying closer attention to global warming because the company has found that its bottom line is affected by drought, a major factor in tropical and semitropical areas affected by expansion of Hadley cells (see "Hadley Cells and Epic Drought in the U.S. Southwest" earlier in Volume II). One of these areas is India, where Coca-Cola lost a profitable operating license because of drought in 2004. After a decade, Coca-Cola now realizes that drought often influenced by changing climate affects its income substantially. Some of Coca-Cola's executives are starting to sound a bit like climate scientists, especially Jeffrey Seabright, Coke's vice president for environment and water resources.

"Increased droughts, more unpredictable variability, 100-year floods every two years," he said, have been disrupting the company's supplies of sugar beets and sugarcane. Drought also affects the production and supply of citrus for the company's Minute Maid juices. "When we look at our most essential ingredients, we see those events as threats," Seabright said (Davenport 2014).

For example, the Coca-Cola Co. had distributed 100,000 hydrofluorocarbon-free refrigerators and vending machines worldwide by 2010. Coca-Cola's CEO Neville Isdell announced the new machines' deployment at a conference in Beijing organized by Greenpeace on May 27, 2008. The machines are cooled with compressed carbon dioxide instead of hydrofluorocarbons (HFCs), which are much more potent than CO_2 as greenhouse gases. The HFC-free machines were only a start; Coca-Cola operates 10 million vending machines and coolers worldwide ("Coca-Cola" 2008). The new machines are initial payback on Coca-Cola's investment of $40 million in researching cleaner cooling technology. The new machines also were deployed at high-profile events such as the 2004 Olympics in Athens and the 2008 games in Beijing. The new machines also are 35 percent more energy efficient than previous models.

Coca-Cola is among many companies whose executives have taken a sober look at global warming as a serious phenomenon and realized that combating it in the long run will be profitable, a bottom-line–oriented interpretation of environmentalists' belief that lack of a sustainable environment degrades all life. "Their position is at striking odds with the long-standing argument, advanced by the coal industry and others, that policies to curb carbon emissions are more economically harmful than the impact of climate change," wrote Carol Davenport in *The New York Times* (2014).

Further Reading

"Coca-Cola to Deploy 100K HFC-Free Coolers." Greenbiz.com, May 28, 2008. http://www.greenbiz.com/print/24718 (no longer available).

Davenport, Carol. "Industry Awakens to Threat of Climate Change." *The New York Times,* January 24, 2014. http://www.nytimes.com/2014/01/24/science/earth/threat-to-bottom-line-spurs-action-on-climate.html?hp&_r=0.

Corporations Act before Government

In some cases, private businesses in the United States have taken actions ahead of the federal government. "We accept that the science on global warming is overwhelming," said John W. Rowe, chairman and chief executive officer of Exelon Corp. "There should be mandatory carbon constraints" (Carey and Shapiro 2004). Exelon, the United States' largest operator of commercial nuclear power plants, probably would benefit from such constraints.

Shell Oil Company's chairman Sir Philip Watts called for global warming contrarians to accept action to limit greenhouse gas emissions "before it is too late." Watts said, "We can't wait to answer all questions [on global warming] beyond reasonable doubt," adding "there is compelling evidence that climate change is a threat." Watts made his remarks at the opening of a new Shell Center for Sustainability at Rice University in Houston. Executives at Shell have "seen and heard enough" to believe that the burning of fossil fuels poses a problem, Watts said. "We stand with those who are prepared to take action to solve that problem . . . now . . . before it is too late . . . and we believe that businesses, like Shell, can help to bridge differences that divide the U.S. and Europe on this issue" (Macalister 2003).

Shell Oil developed several alternative-energy projects through a $500 million investment in a subsidiary, Shell Renewables. Shell constructed a four-megawatt wind farm off the coast of northeastern England that was supplying 3,000 homes by 2001. BP Solar in 2000 was among the largest solar-power companies in the world. By 2000, it had sold 40 megawatts of new capacity, and sales increased in subsequent years (Schrope 2001, 516)

American Electric Power (AEP), which burns more coal than any other U.S. electric utility, once resisted the idea of combating climate change. During the late 1990s, however, then-CEO E. Linn Draper Jr. pushed for a strategy shift, preparing for limits instead of denying that global warming existed. "We felt it was inevitable that we were going to live in a carbon-constrained world," said Dale E. Heydlauff, AEP's senior vice president for environmental affairs (Carey and Shapiro 2004). AEP has invested in Chilean renewable-energy projects, retrofitted school buildings in Bulgaria for greater efficiency, and explored ways to burn coal more cleanly.

By 2000, chemical giant DuPont, once the world's largest producer of ozone-consuming chlorofluorocarbons, announced that an ambitious corporate program had eliminated half of the company's greenhouse gas emissions since 1990, nearly all of it without losing sales or profits. By 2004, DuPont had cut its greenhouse gas emissions by 65 percent since 1990, saving hundreds of millions of dollars (Carey and Shapiro 2004). By 2004, Florida Power & Light Co. had 42 wind-power facilities and had promoted energy efficiency, reducing emissions, and eliminating the need to build 10 midsized power plants, according to Randall R. LaBauve, vice president for environmental services (Carey and Shapiro 2004). By 2012, NextEra Energy (the parent company of Florida Power & Light) was the largest owner of wind generators in the United States.

Private U.S. companies in 2004 formed a Climate Group to share information about climate-related business planning, from new technologies to plugging leaks. In the meantime, BP—formerly British Petroleum but whose advertising styles more recently wanted to portray the company as "Beyond Petroleum"—reduced its greenhouse gas emissions by 10 percent and saved $650 million during three years by performing a detailed energy inventory of its operations.

Google has been buying wind power to help run its data centers, and Wal-Mart is building wind turbines for its distribution centers. Google sealed a deal with Grand River Dam Authority, an Oklahoma utility, to buy 48 megawatts of wind energy to help power its data center in that state ("A Greener" 2012). In August 2012, Wal-Mart unveiled plans for its first on-site large-scale wind-turbine pilot project at its distribution center in Red Bluff, California, providing roughly one megawatt of power, 15 percent to 20 percent of the distribution center's annual electrical use ("A Greener" 2012).

A Frito-Lay plant in Casa Grande, Arizona, took itself nearly completely off the power grid in 2010, using several strategies. One was new filters that recycled most of the water used to wash and rinse potatoes and corn destined for snack chips. The remaining sludge was used as a source of methane to power the factory's boiler. The factory also installed 50 acres of concentrating solar power for power, along with a biomass generator burning agricultural waste. The plant reduced electricity and water consumption by 90 percent, natural gas usage by 80 percent, and greenhouse gas emissions 50 percent to 75 percent. As adapted at the giant Casa Grande plant—a facility the size of two football fields—these strategies serve as prototypes for Frito Lay's 37 other processing plants in the United States and Canada (Martin 2007, A22).

Retail chains such as Safeway, Wal-Mart, Kohl's, and Whole Foods Market have installed solar arrays on the roofs of their stores to generate large amounts of electricity to save money in the long run, reducing their use of coal-fired energy. Many states also offer tax incentives, with California, New Jersey, and Connecticut in the forefront. In many urban areas, retail stores are major electricity consumers, and solar panels can supply 10 percent to 40 percent of what a store uses. "It's very clear that green energy is now front and center in the minds of the business sector," said Daniel M. Kammen, an energy expert at the University of California–Berkeley. "Not only will you see panels on the roofs of your local stores, but I suspect very soon retailers will have stickers in their windows saying, 'This is a green energy store'" (Rosenbloom 2008). Although solar power still costs 25 to 30 cents per kilowatt hour, many retailers believe that economies of scale will being the price down over time as fossil fuel prices (now as low as 6 cents per kilowatt hour for coal-fired power) will rise.

Further Reading

Ball, Jeffrey. "Economics: Creating Environmental Capital." *Wall Street Journal,* March 24, 2008, R1.

Barringer, Felicity. "A Coalition for Firm Limit on Emissions." *The New York Times,* January 19, 2007. http://www.nytimes.com/2007/01/19/business/19carbon.html.

Carey, John, and Sarah R. Shapiro. "Consensus Is Growing Among Scientists, Governments, and Business That They Must Act Fast to Combat Climate Change." *Business Week,* August 16, 2004 (LEXIS).

Carlton, Jim. "Citicorp Tries Banking on the Natural Kind of Green." *Wall Street Journal,* September 5, 2007, B1, B8.

Davenport, Carol. "Industry Awakens to Threat of Climate Change." *The New York Times,* January 24, 2014. http://www.nytimes.com/2014/01/24/science/earth/threat-to-bottom-line-spurs-action-on-climate.html?hp&_r=0.

Deutsch, Claudia. "Companies Giving Green an Office." *The New York Times,* July 3, 2007. http://www.nytimes.com/2007/07/03/business/03sustain.html.

Eggert, David. "Shareholders File Global Warming Resolutions at Ford, GM." Associated Press, December 11, 2002 (LEXIS).

"ExxonMobil Agrees to Report Carbon Stranded Asset Risk." Environment News Service, March 24, 2014. http://ens-newswire.com/2014/03/24/exxonmobil-agrees-to-report-carbon-stranded-asset-risk/.

"A Greener Bottom Line?" *Forbes*, November 7, 2012. http://www.forbes.com/sites/erika morphy/2012/10/07/green-accounting-a-greener-bottom-line/.

"Investors Sizing Up 'Carbon Footprints.'" *Omaha World-Herald*, September 30, 2007, D1.

Macalister, Terry. "Shell Chief Delivers Global Warming Warning to Bush in His Own Back Yard." *The Guardian* (U.K.), March 12, 2003, 19.

Martin, Andrew. "In Eco-Friendly Factory, Low-Guilt Potato Chips." *The New York Times*, November 15, 2007, A1, A22.

"One in Five ExxonMobil Shareholders Want Climate Action." Environment News Service. May 28, 2003. http://ens-news.com/ens/may2003/2003-05-28-09.asp#anchor3 (no longer available).

Parkinson, Giles. "Citicorp Says the 'Age of Renewables' Has Begun." *Renew Economy*, March 27, 2014. http://reneweconomy.com.au/2014/citigroup-says-the-age-of-renewa bles-has-begun-69852.

Rosenbloom, Stephanie. "Giant Retailers Look to Sun for Energy Savings." *The New York Times*, August 11, 2008. http://www.nytimes.com/2008/08/11/business/11solar.html.

Schrope, Mark. "Global Warming: A Change of Climate for Big Oil." *Nature* 411 (May 31, 2001): 516–518.

Seelye, Katharine. "Environmental Groups Gain as Companies Vote on Issues." *The New York Times*, May 29, 2003, C1.

Spencer, Jane. "Big Firms to Press Suppliers on Climate." *Wall Street Journal*, October 9, 2007, A7.

Wheaton, Sarah. "Keystone XL Pipeline Fight Lifts Environmental Movement." *The New York Times*, January 24, 2014, A9. https://www.nytimes.com/2014/01/25/us/keystone-xl -pipeline-fight-lifts-environmental-movement.html.

See also: Coal; Wind Gains Power Share; Wind Power, Denmark; Wind Power, United States; Wind Power and Solar Power in Portugal, Germany, and Japan; Unburnable Carbon

SOLUTIONS AND ACTIONS

OVERVIEW

Having surveyed evidence of worldwide warming, the science supporting it, and the political controversy attending the issue, and having examined the toll that climate change is taking on Earth and its resident species of flora and fauna, including humanity, we are left with an existential question: having created this problem, can human ingenuity solve it? What, one may ask, are the chances that humankind may dodge this bullet? Chances of forestalling seriously debilitating climate change revolve around humankind's ability to forge political and technological solutions. One will not work without the other.

This section focuses on two areas: reducing greenhouse gas production with alternative energy, including solar and wind power, and other proposals that may not be especially effective in the long run, such as corn ethanol and geoengineering, including the infusion of the upper atmosphere with sulfur. Many such efforts are underway—so much effort, on so many fronts, that an encyclopedia of solutions could be composed. What follows is merely a sampling. It does not even scratch the service of some areas, such as an entire field of sustainability oriented architecture and building design that has grown up in recent years.

By 1990, concern about global warming was rising along with carbon dioxide emissions. That year, almost 6 billion tons entered the atmosphere as a result of burning fossil fuels and deforestation. By 2010, the total was 10 billion from the same sources (Epstein and Ferber 2011, 33). The growth in emissions slowed after that as a revolution began in energy generation as concern was transmitted into action and power sources began to change.

Coal as Yesterday's Fuel

More and more, coal has become yesterday's fuel. By 2015, coal mines were closing steadily in Kentucky, Ohio, Illinois, and West Virginia, and thousands of miners had been laid off as U.S. coal production declined 15 percent over six years (Krauss 2015). The proportion of U.S. power generated by coal had fallen from 50 percent in 2000, to 40 percent in 2012, and to 30 percent in 2015, and the U.S. Department of Energy anticipated continued declines. China, once a seemingly limitless market for U.S. coal, was also curtailing its consumption just as many U.S. coal-fired electricity plants switched to cheap natural gas produced by hydraulic fracturing ("fracking"). Coal's share of U.S. electricity generation was being eroded not only by cheaper wind and solar power but also by natural gas supplied by fracking. Between 2005 and 2015, the proportion of electricity supplied in the United States by natural gas grew from 18 percent to 33 percent, according to the U. S. Energy Information Administration (Gabriel and Davenport 2016).

Between 2007 and 2013, carbon dioxide emissions from electricity generation in the United States fell 15 percent, mostly because cheap natural gas was being substituted for coal (Shellenberger 2015). In November 2015, Great Britain announced plans to close all of its coal-fired power plants by 2025. Around the same time, the government of the Canadian province of Alberta announced plans to close the last of its coal-consuming electrical plants before 2030.

Pope Francis and Climate-Change Diplomacy

During the summer of 2015, Pope Francis, although having only acceded to the throne two years earlier, was already well known for directly tackling many controversial issues. He made climate change a Vatican priority by issuing an encyclical (essentially a policy statement) detailing how the burdens of a changing climate worldwide fall disproportionally on the poor. The encyclical was part of a broader campaign by the pope to advocate protection of Earth and all creation.

Many climate contrarians, including leaders of the Republican-dominated U.S. House of Representatives, thought Francis had exceeded his authority. For his part, Francis sought to help jump-start international diplomacy in an effort to get nations to cooperate in curbing production of greenhouse gases. The encyclical was preceded by a year of meetings among prominent Vatican officials on the issue. The encyclical was issued after a 12-week campaign in the United States during which Catholic bishops raised issues related to climate change under the rubric of environmental stewardship in their sermons and homilies, as well as news media interviews and letters to editors.

"We are the first generation that can end poverty, and the last generation that can avoid the worst impacts of climate change," U.N. Secretary-General Ban Ki-moon at an international symposium on climate change at the Pontifical Academy of Sciences on April 28, 2015, one of the events leading up to the encyclical (Povoledo 2015).

Then U.S. Speaker of the House John A. Boehner, a Ohio Republican (and Catholic), had invited the pope to speak before Congress. "I think Boehner was out of his mind to invite the pope to speak to Congress," said the Rev. Thomas Reese, an analyst at the *National Catholic Reporter*. "Can you imagine what the Republicans will do when he says, 'You've got to do something about global warming'?" (Davenport and Goldstein 2015). Timothy E. Wirth, vice chairman of the United Nations Foundation, said, "We've never seen a pope do anything like this. No single individual has as much global sway as he does.

What he is doing will resonate in the government of any country that has a leading Catholic constituency." "I think this moves the needle," said Charles J. Reid Jr., a professor at the University of St. Thomas School of Law. "Benedict was an ivory-tower academic. He wrote books and hoped they would persuade by reason. But Pope Francis knows how to sell his ideas. He is engaged in the marketplace" (Davenport and Goldstein 2015).

Further Reading

Davenport, Coral, and Laurie Goldstein. "Pope Francis Steps Up Campaign on Climate Change, to Conservatives' Alarm." *The New York Times*, April 27, 2015. http:// www.nytimes.com/2015/04/28/world/europe/pope-francis-steps-up-campaign -on-climate-change-to-conservatives-alarm.html.

Povoledo, Elisabetta. "Scientists and Religious Leaders Discuss Climate Change at Vatican." *The New York Times*, April 29, 2015. http://www.nytimes.com/2015/04/29/ world/europe/scientists-and-religious-leaders-discuss-climate-change-at-vatican .html.

By 2016, several large banks, including Citicorp, JPMorgan Chase, Bank of America, and Morgan Stanley were refusing to provide financing for new coal-fired power plants and other support for the coal industry. "There are always going to be periods of boom and bust," said Chiza Vitta, a metals and mining analyst with the credit-rating firm Standard & Poor's. "But what is happening in coal is a downward shift that is permanent" (Corkery 2016).

Coal use also came under assault from new environmental regulations meant to limit greenhouse gas emissions as well as toxic gases such as mercury. Several coal producers faced bankruptcy as a share of once-mighty Peabody Energy, which had broached $780 at one point, falling during 2015 to around $2 (Krauss 2015). Peabody filed for bankruptcy in April 2016. The stock price of another producer, Arch Coal, collapsed from $360 a share in 2011 to less than $2 in 2015. A growing number of institutions were deleting coal from their stock portfolios as protests of its use (and pollution) mounted. For example, during August 2015, "1,500 activists dressed in ghostly white safety jumpsuits staged an extraordinary piece of tactical theater by invading one of the largest coal mines in Europe and shutting it down for the day" (Solnit 2015, 6).

"In the past we always knew that the demand for coal would rebound and the jobs would come back," Cecil E. Roberts Jr., the United Mine Workers of America president, said in a speech in June. "This time, there is no such certainty. Fundamental changes are underway in America and across the world that will have a lasting impact on the coal industry and our jobs" (Krauss 2015).

Wind and Solar Compete with Coal

Until recently, wind and solar power had been minor players in an energy-generation field dominated by coal, oil, and natural gas. Within roughly 15 years—since the turn of the millennium—wind and solar have become much more efficient and cost competitive. Between 2009 and 2015, the cost of generating electricity with wind declined 61 percent; solar power's cost dropped by 82 percent during the same period (Krugman 2016).

Many countries in Europe such as Germany and Denmark now generate one-third to one-half of their electricity from sun and wind. Some U.S. states (Iowa, for example) now verge on half-wind electrical generation. Iowa also lays claim to the tallest wind turbine in the United States at 554 feet, mounted on a concrete tower that raises the tip of its blade to a level equaling the top of the Washington Monument (555 feet). Customers of the Omaha Public Power District woke up one morning in 2016 to learn that their utility, which has been mainly reliant on coal, had quickly switched to one-third wind powered. In the United States as a whole, the federal Department of Energy projected that total renewable power would increase by 9.5 percent during 2016, even as oil, coal, and natural gas prices were in collapse. Utility-scale solar power was projected by the same agency to increase by 45 percent in one year (2016–2017).

By 2015, Germany as a whole was generating nearly 30 percent of its electricity by solar, wind, and other renewable sources, even as it phased out nuclear power. "Germany is the first country in the world to show they can uncouple growth from burning of fossil fuels," said Jim Yong Kim, president of the World Bank. "This is the main task of our generation" (Eddy 2015). German household consumers have borne much of the burden of this energy paradigm shift through higher rates.

Even so, given rising energy demand around the world, coal was still being used in new power plants in places such as India, even as solar and wind technologies improved and costs declined. Thus, worldwide greenhouse gas emissions (and the proportion of carbon dioxide in the atmosphere) have continued to rise "despite the undisputed worldwide technological progress and expansion of renewable technologies," according to one analysis (Edenhofer 2015, 1286). In 2013, for the first time, electric utilities worldwide installed more new capacity with renewable sources [143 gigawatts (GW), mainly wind power] than fossil fuels (141 GW), according to an analysis presented in April 2014 at the Bloomberg New Energy Finance (BNEF) annual summit in New York City. Analysts with BNEF expect that "[the] shift will continue to accelerate, and by 2030 more than four times as much renewable capacity will be added" (Randall 2015).

Generation costs for wind energy under perfect conditions (around $50 per megawatt hour in 2015) have approached that of coal-generated electricity, but supplies are likely to be interrupted by lack of wind, requiring additional capacity and fossil fuel reserve capacity. Solar generation was averaging $80 per megawatt hour in 2015 and falling at a rate that may equal that of coal power within a decade. Even so, populous nations such as China and India continue to develop new coal-generating capacity along with rapidly developing solar and wind (Edenhofer 2015, 1286).

Between 2012 and 2015, as coal prices plunged 70 percent, twice as many people were employed in the burgeoning solar-power industry than the 80,000 who were mining coal. In 2013, coal production in the United States dropped below 1 billion tons annually for the first time in 20 years—at 985 million tons, still a lot of coal, 93 percent of it being used to generate electric power (Stewart 2015).

Even as wind and solar power boomed worldwide, the International Energy Agency (IEA) warned that the change was too slow to keep temperatures from rising to levels that will seriously distort climate. In India, 25 percent of the population (300 million people) have no electricity at all (and the per capita carbon footprint is 4 percent that of the United States). Still, the country occupies an especially important position in the world energy equation as its leaders seek to increase power by all means—coal to solar. Indian leaders do recognize the growing toll of coal combustion; half the schoolchildren in New Delhi have lungs that have been scarred by the worst air pollution on Earth. Even as asthma rates rise rapidly, the production of power has taken priority as the government in 2015 announced plans to double its use of domestic coal by 2019 (Barry 2015).

In 2014, for the first time, more than half of Earth's new electricity-generating capacity came from wind and solar, according to the IEA. Even so, the IEA said, "a major course correction is still required to achieve the world's agreed climate goal" of limiting global warming to 3.6°F (2°C) above preindustrial 19th-century levels (Reed 2015). Many scientists asserted that even this goal was too high to prevent catastrophic damage in coming decades.

In 2015, the U.S. Environmental Protection Agency for the first time announced plans to limit greenhouse gas emissions from aircraft under the federal Clean Air Act of 1970 on the grounds that they "endanger human health because they significantly contribute to global warming" (Davenport 2015). Aircraft that fly within the United States represent 30 percent of world emissions from that industry, which has been anticipating such controls for many years by making its engines more fuel efficient. By 2014, jet aircraft were some 70 percent more fuel efficient per seat mile than they were in the 1960s, according to the Air Transport Action Group (Davenport 2015). In 2016, aircraft delivered after 2028 became subject to emissions controls worldwide for the first time. The standards, however, were no higher than those already attained by new aircraft being manufactured as of 2016 (Mouawad 2016).

A Wild Array of Solutions

A shift in our energy paradigm from fossil fuels to renewable and nonpolluting sources such as solar and wind power is taking place before our eyes. By the end of this century, the internal combustion engine may be as much of an antique piece as a horse and buggy are today.

When Pig Fat Flies

Will pigs fly? In the world of biomass fuel, they might. By 2007, the U.S. Department of Defense and the National Aeronautics and Space Administration (NASA) were funding exploratory projects into biofuels for jet airplanes. In 2015, United Airlines was investing in biofuels and had tested them in a jet fuel mixture on a handful of its commercial flights. The airline believes that it can produce biomass fuel from farm waste and animal fats at less than half the cost of conventional jet fuel (a large part of any airline's costs) with lower carbon dioxide emissions. On July 1, 2015, United Airlines announced a $30 million investment in Fulcrum BioEnergy. Cathay Pacific, which is based in Hong Kong, also has invested in the same company.

Fulcrum has developed technology that combines farm and household wastes with jet fuel and plans to open a refinery in Nevada by the end of 2017. Fulcrum asserts that this technology will reduce airline carbon dioxide emissions by as much as 80 percent vis à vis standard jet fuel. "There is definitely a huge interest from airlines in this market," said Angela Foster-Rice, United Airlines' managing director for environmental affairs and sustainability (Mouawad and Cardwell 2015).

Syntroleum was providing the U.S. Department of Defense with jet fuel derived from animal fats supplied by Tyson Foods, Inc. Tyson is the world's largest producer of chicken, beef, and pork, producing prodigious amounts of animal fats such as beef tallow, pork lard, chicken fat, and greases, all of which may someday be used as fuel. The two companies have planned a factory in the U.S. Southwest that produces 75 million gallons of jet fuel per year. According to Syntroleum, the U.S. Air Force has certified all of its aircraft to run on alternative fuels and wants 50 percent of its fuel to come from domestic alternative sources by 2018 ("Fueling Jets" 2007).

"There is a significant role for biofuels within the aviation sector, specifically for reducing carbon emissions," said Debbie Hammel, a senior resource specialist at the Natural Resources Defense Council who focuses on biofuel (Mouawad and Cardwell 2015).

Further Reading

"Fueling Jets with Animal Fat." Environment News Service, July 18, 2007. http://www.ens-newswire.com/ens/jul2007/2007-07-18-09.asp#anchor7 (no longer available).

Mouawad, Jad, and Diane Cardwell. "Farm Waste and Animal Fats Will Help Power a United Jet." *The New York Times*, June 30, 2015. http://www.nytimes.com/2015/06/30/business/energy-environment/farm-waste-and-animal-fats-will-help-power-a-united-jet.html.

Technological changes can simply be prosaic such as making mileage improvements on existing gasoline-burning automobiles, changing building codes, and painting building roofs white. Sometimes cutting greenhouse gas emissions can be simple and basic such as reducing methane leaks from existing oil and gas operations. The 20-year climate impact of methane escaping from oil and gas operations worldwide has the same near-term climate impact as emissions of 40 percent of total global coal combustion, according to a report issued in 2015 by the Environmental Defense Fund ("Global Oil" 2015). The 3.6 billion cubic feet of methane that escaped worldwide in just 2012 would have been worth $30 billion on the market. "Methane is both a serious climate challenge and also, in our view, a major untapped opportunity to start reversing the tide of global greenhouse gas emissions," said Mark Brownstein, chief counsel for the Environmental Defense Fund's U.S. climate and energy program (Schwartz 2015).

Writing in *Science*, S. Pacala and R. Socolow have asserted, "Humanity already possesses the fundamental scientific, technical, and industrial know-how to solve the carbon and climate problem for the next half-century" using existing technology (Pacala and Socolow 2004, 968). By "solve," the authors mean that the tools are at hand to meet global energy needs without doubling preindustrial levels of carbon dioxide. Their "stabilization strategy" involves intense attention to improved automotive fuel economy, reduced reliance on cars, more efficient building construction, improved power-plant efficiency, substitution of natural gas for coal, storage of carbon captured in power plants and well as hydrogen and synthetic fuel plants, more use of nuclear power, development of wind and photovoltaic (solar) energy sources, creation of hydrogen from renewable sources, and more intense use of biofuels such as ethanol. The strategy also advocates more intense management of natural sinks, including reductions in deforestation and aggressive management of agricultural soils through such measures as conservation tillage, or drilling seeds into soil without plowing (Pacala and Socolow 2004, 969–971).

Tokyo's Seawater Cooling Grid

In 2002, Tokyo officials announced plans to build the world's largest cooling system and fill it with seawater to reduce temperatures by 2.6°C, or the approximate amount that average temperatures in the city increased during the 20th century. General warming is being enhanced in Tokyo by an intensifying urban heat-island effect as population density increases and more air conditioning (with its waste heat) is being used to cool tall buildings.

This huge construction project cost $300 billion and requires a 250-hectare (600-acre) lattice of pipes under the city that draw in cold water from Tokyo Bay. "Heat would be forced into the network through a second set of pipes connected to air conditioning systems in the buildings above. The heat would be absorbed by the seawater, which would then flow back into the bay" (Ryall 2002). Skeptics of the project feared that dumping warmed seawater in Tokyo Bay may be causing environmental damage.

Tokyo's average temperature rise of 2.9°C during the 20th century was five times the average worldwide rate of warming. The number of days when the city experiences temperatures of 30°C or more had doubled since the 1980s (Ryall 2002). "This is a radical approach but that's what the situation demands," said Tadafumi Maejima, director of the Japan District Heating and Cooling Association, which was commissioned to draw up the plan. "Nothing like this has been attempted anywhere in the world, but we could be ready to start work in as little as two years" (Ryall 2002).

According to a report in *The Times* (London), "The pipes would initially spread beneath 123 hectares of the Marunouchi district, the city's financial heart. Later they would extend through Kasumigaseki, where the Diet (parliament), and the ministries are based. A further phase, covering an area north of Tokyo station, would use water from a sewage treatment plant" (Ryall 2002). "The amount of heat emitted by buildings needs to be cut, especially in the heart of the city where demand for air conditioners is concentrated, to break the vicious cycle of summer warming," said the association's report (Ryall 2002).

Further Reading

Ryall, Julian. "Tokyo Plans City Coolers to Beat Heat." *The Times* (London), August 11, 2002, 20.

The road to energy sustainability is strewn with bumps, however. Bear in mind that some alternative sources only sound effective. What good, for example, is a car that runs on electricity or hydrogen when its ultimate power source relies on coal, oil, or natural gas? How green is corn ethanol? Some studies suggest that, for the amount of energy generated, it is no cleaner than gasoline. Sugarcane ethanol packs more of a punch. The fact that humankind has not reduced the proportion of greenhouse gases in the atmosphere since the beginning of the fossil fueled Industrial Age almost two centuries ago has given rise to some proposals to "geoengineer" the atmosphere—that is, infuse sulfur into the stratosphere to reflect incoming solar radiation in the style of a volcano. Other proposals involve fertilizing the oceans with iron to promote the growth of phytoplankton that absorb carbon dioxide or sequestering carbon dioxide from power plants in basalt or under the ocean floor.

None of these proposals would actually reduce emissions; they would move them around. Scientists often discuss such things with a sense of desperation as last-hope solutions, but groups that deny that climate change is real such as the Heartland Institute tend to see "the sulfur solution" as a "practical, cost-effective global-warming strategy" (Hamilton 2015). Marcia McNutt, chairwoman of a committee that reported to the National Research Council in 2015 on sulfur infusion reflected the scientific consensus: people should read the report, she said, "and say 'This is downright scary.' If this is our Hail Mary, what a downright scary place we live in" (Hamilton 2014; Hamilton 2015).

This section concludes with some food for thought. How do we address the economics of addressing global warming to make solutions sustainable in a world of profit and loss? In addition, society will be forced to consider the question of "unburnable carbon"—that is, what value do we place on the reserves of fossil fuel companies that cannot be used if we are to reduce greenhouse gas emissions and dodge the climatic bullet? The answer should provoke debates in oil company boardrooms.

Further Reading

Barry, Ellen. "For Indians, Smog and Poverty Are Higher Priorities Than Talks in Paris." *The New York Times*, December 10, 2015, A19.

Corkery, Michael. "As Coal's Future Grows Murkier, Banks Pull Financing." *The New York Times*, March 21, 2016. http://www.nytimes.com/2016/03/21/business/dealbook/as-coals-future-grows-murkier-banks-pull-financing.html.

Davenport, Coral. "E.P.A. to Set New Limits on Airplane Emissions." *The New York Times*, June 2, 2015. http://www.nytimes.com/2015/06/03/business/energy-environment/epa-to-set-new-limits-on-airplane-emissions.html.

Eddy, Melissa. "Germany May Offer Model for Reining in Fossil Fuel Use." *The New York Times*, December 3, 2015. http://www.nytimes.com/2015/12/04/world/europe/germany-may-offer-model-for-reining-in-fossil-fuel-use.html.

Edenhofer, Ottmar. "King Coal and the Queen of Subsidies." *Science* 349 (September 18, 2015): 1286–1287.

Epstein, Paul R., and Dan Ferber. *Changing Planet, Changing Health: How the Climate Crisis Threatens Our Health and What We Can Do about It*. Berkeley: University of California Press, 2011.

Gabriel, Trip, and Coral Davenport. "Activists Push Fracking onto Agenda." *The New York Times*, April 5, 2016, A16.

"Global Oil and Gas Methane Reduction Opportunities." Environmental Defense Fund, April 2015. http://www.edf.org/climate/rhodium-group-report-global-oil-gas-methane-emissions?_ga=1.219423705.1787517823.1427851088.

Hamilton, Clive. *Earthmasters: The Dawn of the Age of Climate Engineering*. New Haven, CT: Yale University Press, 2014.

Hamilton, Clive. "The Risks of Climate Engineering." *The New York Times*, February 12, 2015, A27.

Krauss, Clifford. "Coal Miners Struggle to Survive in an Industry Battered by Layoffs and Bankruptcy." *The New York Times*, July 17, 2015. http://www.nytimes.com/2015/07/18/business/energy-environment/coal-miners-struggle-to-survive-in-an-industry-battered-by-layoffs-and-bankruptcy.html.

Krugman, Paul. "Wind, Sun, and Fire." *The New York Times*, February 1, 2016, A21.

Mouawad, Jad. "Deal on Aviation Emissions Sets Can't-Miss Goals." *The New York Times*, February 16, 2016. http://www.nytimes.com/2016/02/16/business/energy-environment/a-hollow-agreement-on-aviation-emissions.html?ref=business.

Pacala, S., and R. Socolow. "Stabilization Wedges: Solving the Climate Problem for the Next 50 Years with Current Technologies." *Science* 305 (August 13, 2004): 968–972.

Randall, Tom. "Fossil Fuels Just Lost the Race against Renewables: This Is the Beginning of the End." *Bloomberg Business*, April 14, 2015. http://www.bloomberg.com/news/articles/2015-04-14/fossil-fuels-just-lost-the-race-against-renewables.

Reed, Stanley. "Shift to Lower-Carbon Energy Is Too Slow, Report Warns." *The New York Times*, November 9, 2015. http://www.nytimes.com/2015/11/10/business/energy-environment/shift-to-lower-carbon-energy-is-too-slow-report-warns.html.

Schwartz, John. "Study Finds Low Cost in Reducing Methane Emissions." *The New York Times*, April 22, 2015. http://www.nytimes.com/2015/04/22/world/americas/study-finds-low-cost-in-reducing-methane-emissions.html.

Shellenberger, Michael. "The Climate War Is Over." *USA Today*, December 3, 2015, 7A.
Solnit, Rebecca. "Power in Paris." *Harper's*, December 2015, 5–7.
Stewart, James B. "Coal Industry Wobbles as Market Forces Slug Away." *The New York Times*, August 7, 2015. http://www.nytimes.com/2015/08/07/business/energy-environment/coal-industrty-wobbles-as-market-forces-slu-away.html (no longer available).

ACADEMIC PROTEST AND ACTION

A growing number of activists have been advocating divestment in fossil fuels, especially coal, by academic and charitable organizations as a way to focus attention on the dangers of global warming and the necessity of reducing dependence on energy technologies that produce carbon dioxide and other greenhouse gases.

On May 1, 2014, Harvard University police broke up a day-old blockade of the university president's office by students who were demanding that the university divest itself of any fund or investment that included fossil fuels. Student Brett Roche, class of 2015, was arrested in the demonstration and thus "became the first student arrested in the movement for fossil fuel divestment, a growing movement at dozens of universities and colleges," according to the Environment News Service ("Harvard Student" 2014). The same afternoon, Canadian novelist Margaret Atwood spoke to several hundred people at the Harvard Arts First festival. Harvard's President Drew Faust was in the audience and heard Atwood criticize the university's handling of the protests: "Any society where arrest is preferable to open dialogue is a scary place" ("Harvard Student" 2014).

Divesting Fossil Fuels

In 2016, Norway's government pension fund, the largest sovereign wealth fund in the world with $890 billion in its portfolio, divested companies that earned more than 30 percent of their profits from coal. This decision followed the Church of England's 2015 decision to divest coal and oil sands from a much smaller portfolio of $14 billion. The Norwegian decision was made by that nation's parliament. The 30-percent limit was set to restrict divestment mainly to extraction companies. It excludes power companies that burn coal as part of their electricity-generation mix. Marthe Skaar, a spokeswoman for Norges Bank Investment Management, which manages the Norwegian fund, said its divestment was "safeguarding and building financial wealth for future generations in Norway." Its reasons for divesting, she said, included "long-established climate-change risk-management expectations" (Schwartz 2015). The bank cited studies indicating that future restrictions on fossil fuel emissions probably will force many companies to leave substantial proportions of their reserves in the ground.

Within a month of Norway's vote to divest coal in 2015 (it took effect in 2016), other institutions around the world took similar stands. Edinburgh

University of Scotland, the University of Hawaii, Georgetown University (Washington, D.C.), the AXA Group (France's largest insurer), and the University of Washington all moved to divest fossil fuels from their portfolios in one way or another.

On May 6, 2014, Stanford University announced that it would sell its $18.7 billion worth of endowment funds in companies "whose principal business is coal," including past or future investment in some 100 companies worldwide that obtain more than half their revenues from coal extraction. Stanford thus became the first major university to divest a large amount of money in fossil fuel firms.

Stanford's divestment decision was the first major victory in a student-led campaign that was operating on 300 or so college and university campuses, including Fossil Free Stanford. Eleven small colleges had removed fossil fuel investments by that time (Wines 2014). Maura Cowley, executive director of Energy Action Coalition, an assemblage of groups active on climate change issues, called the decision "a huge, huge victory. Their decision, coming from such a major university and from such a huge endowment, shows that the coal industry and other fossil fuel industries are quickly becoming relics of the past," she said (Wines 2014).

By late 2015, more than 800 foundations, companies, and other groups (as well as several wealthy individuals) had pledged to divest themselves of fossil fuels. Perhaps the most surprising—and historically ironic—was the Rockefeller Brothers Fund, worth $860 million and owned by descendants of the family that built its fortune largely from Standard Oil. "This is a threshold moment," said Ellen Dorsey, executive director of the Wallace Global Fund, which has coordinated the effort to recruit foundations to the cause. "This movement has gone from a small activist band quickly into the mainstream" (Schwartz 2014).

Further Reading

Schwartz, John. "Rockefellers, Heirs to an Oil Fortune, Will Divest Charity from Fossil Fuels." *The New York Times*, September 21, 2014. http://www.nytimes.com/2014/09/22/us/heirs-to-an-oil-fortune-join-the-divestment-drive.html.

Schwartz, John. "Norway Will Divest from Coal in Push against Climate Change." *The New York Times*, June 5, 2015. http://www.nytimes.com/2015/06/06/science/norway-in-push-against-climate-change-will-divest-from-coal.html.

Wines, Michael. "Stanford to Purge $18 Billion Endowment of Coal Stock." *The New York Times*, May 7, 2014. http://www.nytimes.com/2014/05/07/education/stanford-to-purge-18-billion-endowment-of-coal-stock.html.

At noon the next day, Divest Harvard held a rally in Harvard Yard to deliver more than 60,000 petition signatures supporting divestment of $17.3 million invested by Harvard University in fossil fuels companies. Faust had earlier issued a statement opposing divestment:

Climate change represents one of the world's most consequential challenges. I very much respect the concern and commitment shown by the many members of our community who are working to confront this problem. While I share their belief in the importance of addressing climate change, I do not believe, nor do my colleagues on the Corporation, that university divestment from the fossil fuel industry is warranted or wise. ("Harvard Student" 2014)

Notable Harvard alumni such as director Darren Aronofsky and actress Natalie Portman signed a letter advocating divestment that read in part, "While we can't bankrupt the oil companies, we can start to politically bankrupt them, complicating their ability to dominate our political life" (Sorkin 2015).

Activism Spreads

Within a week after Harvard police broke up its students' sit-in and its administrators demurred on divestment. The university acted under institutional guidelines that allow trustees to weigh whether "corporate policies or practices create substantial social injury" when investing (Wines 2014). An advisory panel at Stanford of students, alumni, staff, and faculty studied the issue for five months before recommending that Stanford's trustees divest coal stocks.

In 2017, Harvard decided to "pause" its fossil-fuel investments in its $36 billion endowment, but did not divest existing holdings, after five years of student activism. Colin Butterfield, head of natural resources at the Harvard Management Company, said that climate change is a "huge problem" and that "for now, we are pausing minerals and oil and gas." He indicated that environmental concern was reinforced by, a lack of profitability, especially in coal.

Bill McKibben, co-founder of climate campaign group 350.org, said: "Harvard is divesting through the back door—testimony to the great pressure applied by students, faculty, and alumni, but also to its establishment unwillingness to simply say forthrightly: The fossil fuel age must end" (Milman, 2017).

As Harvard University "paused" its investments, the divestment movement grew to include universities such as Columbia, in New York City, as well as Cornell. By 2017, the divestment movement had spread beyond educational institutions to such institutions as the World Council of Churches, which represents half a billion Christians, which ruled out all fossil-fuel investments. The Guardian Media Group also fully divested all fossil-fuel investments, in accordance with the "Keep it in the Ground" campaign of its flagship newspaper. The city of Lyon, in France, also divested, as did the Episcopal Church of the United States, as well as several city governments (Seattle, Washington, San Francisco, California, Stockholm, Sweden, Cambridge, United Kingdom, Santa Monica, California, and many others). Several retirement funds added their names to the growing list of those selling all of their fossil-fuel investments, including the California Public Employees Retirement System and the California State Teachers' Retirement System. This is only a partial list. For a more complete, updated list, see the Cleantechnica web site at: https://cleantechnica .com/2017/05/09/divestment-fossil-fuels-may-2017-cleantechnica-update/. By October 2014, Scotland's Glasgow University and the Australian National University

(ANU) also had divested, with a university referendum showing 82 percent of ANU students supporting the idea.

Ivo Welch, professor of finance and economics at the Anderson Graduate School of Management at the University of California–Los Angeles, argued that Stanford's $18.7 billion divestment was less than a drop in the proverbial $60 trillion bucket of worldwide market capitalization. Welch and colleagues studied divestment as part of a boycott during the 1980s that had forced South Africa to abandon its policy of racial apartheid and found that it produced "no discernible effect on the valuation of companies that were being divested, either short-term or long-term. We looked hard for evidence linking boycotts and sanctions to the value of the South Africa's currency, stock market, and economy. Nothing" (Welch 2014). The divestment campaign did send an important public opinion signal, however, and apartheid fell just a decade later.

Welch acknowledged that South Africa's apartheid government was morally wrong and deserved sanction, even if he believes that divestment made no economic difference. "In the case of fossil fuels . . . the moral choice is much less clear than

Merchants of Doubt

In *Merchants of Doubt*, Naomi Oreskes, a historian of science at Harvard University, and coauthor Erik M. Conway (2010) make the case that many of the same people join several widely varied campaigns to create doubt in the public mind about the veracity of climate change as well as the dangers of acid rain, erosion of stratospheric ozone, and even tobacco smoking. The book was the basis for a documentary film of the same name, *Merchants of Doubt*, by Robert Kenner. "Her courage and persistence in communicating climate science to the wider public have made her a living legend among her colleagues," wrote two climate researchers, Benjamin D. Santer and John Abraham, in 2011 (Gillis 2015).

The debate has become highly charged and very visceral. Justin Gillis reported in *The New York Times* (2015) that "Oreskes . . . has been threatened with lawsuits and vilified on conservative websites, and routinely gets hate mail calling her a communist or worse." S. Fred Singer, one of the deniers described in *Merchants of Doubt*, asserted that before Harvard when Oreskes worked at the University of California–San Diego, she was protected by "a mostly feminist mafia." Singer also said that Oreskes had libeled him but provided no specifics (Gillis 2015).

Further Reading

Gillis, Justin. "Naomi Oreskes, a Lightning Rod in a Changing Climate." *New York Times*, June 15, 2015. http://www.nytimes.com/2015/06/16/science/naomi-oreskes-a-lightning-rod-in-a-changing-climate.html?_r=0.

Oreskes, Naomi, and Erik M. Conway. *Merchants of Doubt*. New York: Bloomsbury Press, 2010.

it was with apartheid," Welch argued (2014). "Energy is an area with no obvious solutions. Apartheid had no place in a civilized world. Fossil fuel companies are supplying a market demand, one that for the time being cannot be met by other fuel sources. Divestment won't change that calculus."

Advocates of divestment argued that similar arguments were advanced in 1855 to justify slavery in the United States when plantations were part of the established economic order supplying demand for cotton. They said that the lack-of-morality argument vis à vis fossil fuels holds only for those who do not care whether their grandchildren live in a sustainable world. For those who do, divestment has been put forth as a first (and sometimes symbolic) step. The next necessary change has been characterized as reinvestment of time, energy, and fungible assets in activities directed toward a survivable and sustainable future.

Academic Sustainability Centers

In 2007, Dow Chemical gave the University of California–Berkeley $10 million over five years to set up a sustainability center, one of a growing number of interdisciplinary study centers at academic institutions that cross many disciplinary boundaries. "We give professors a chance to step beyond their usual areas of expertise, and we give students exposure to the worlds of science and business," said Daniel C. Esty, director of the Yale Center for Business and the Environment, which combines resources from the school of management with forestry and environmental studies. The University of Tennessee consolidated environmental research programs within an Institute for a Secure and Sustainable Environment; Arizona State University initiated a degree-granting Global Institute of Sustainability, funded in part with private donations. Some "sustainability" centers (such as the Kenan-Flagler Center for Sustainable Enterprise at the University of North Carolina) study global cultural issues as well as business ethics and corporate social responsibility in the context of environmental issues (Deutsch 2007).

Some of this may be wordplay as much as a paradigm shift in academic priorities. Some old academic units have been given new, trendier "green" names. "We are seeing more centers framed as sustainability, but they may not be qualitatively different from the ethics, innovation, or globalization centers of 15 years ago," he said. "Universities realize that you can discuss sustainability with a C.E.O. and not get laughed out of the room." Others have specific environmental focuses. The Yale center hosts an "eco-services clinic" to aid corporations with environmental issues, and Duke's Corporate Sustainability Initiative (with resources from earth sciences, business, and environmental policy) has developed home-scale wind turbine, and advises local businesses to reduce their carbon footprints (Deutsch 2007).

Corporations show their supposed green credentials by funding university sustainability centers. Four multinational companies—ExxonMobil, General Electric, Schlumberger, and Toyota—provided the seed capital for the Stanford University Global Climate and Energy Project, with an emphasis on innovative energy technologies. The Shell Oil Foundation has financed Rice University's Shell Center for Sustainability since 2002. In 2007, Wal-Mart pledged funding for the Applied Sustainability Center at the University of Arkansas (Deutsch 2007).

Further Reading

Deutsch, Claudia. "A Threat So Big, Academics Try Collaboration." *The New York Times*, December 25, 2007. http://www.nytimes.com/2007/12/25/business/25sustain.html.

"Harvard Student Arrested in Fossil Fuel Divestment Protest." Environment News Service, May 2, 2014. http://ens-newswire.com/2014/05/02/ameriscan-may-2-2014/.

Milman, Oliver. "Harvard 'Pausing' Investments in Some Fossil Fuels." *The Guardian* (UK), April 27, 2017. https://www.theguardian.com/environment/2017/apr/27/harvard-university-pausing-investments-in-some-fossil-fuels.

Schwartz, John. "Rockefellers, Heirs to an Oil Fortune, Will Divest Charity from Fossil Fuels." *The New York Times*, September 21, 2014. http://www.nytimes.com/2014/09/22/us/heirs-to-an-oil-fortune-join-the-divestment-drive.html.

Sorkin, Andrew Ross. "A New Tack in the War on Mining Mountains." *The New York Times*, March 10, 2015, B1, B4. http://www.nytimes.com/2015/03/10/business/dealbook/pnc-joins-banks-not-financing-mountaintop-coal-removal.html?ref=business.

Welch, Ivo. "Why Divestment Fails." *The New York Times*, May 9, 2014. http://www.nytimes.com/2014/05/10/opinion/why-divestment-fails.html?hp&rref=opinion.

Wines, Michael. "Stanford to Purge $18 Billion Endowment of Coal Stock." *The New York Times*, May 7, 2014. http://www.nytimes.com/2014/05/07/education/stanford-to-purge-18-billion-endowment-of-coal-stock.html.

See also: Coal; Corporate Sustainability; Global Warming Economics; Unburnable Carbon; Wind Gains Power Share

ARTIFICIAL LEAF

Solar technology is undergoing a revolution that may eventually allow power to be acquired from nearly any surface on which the sun shines. One such technology is the so-called artificial leaf. At Caltech's Jorgensen Laboratory, a team of more than 190 people have been using silicon, nickel, iron, and other materials in the Joint Center for Artificial Photosynthesis (JCAP) as part of a $16-million, five-year program funded by the U.S. Department of Energy that is attempting to replicate photosynthesis as an energy source.

In 2010, a team of researchers at the Massachusetts Institute of Technology (MIT) led by Daniel Nocera created the prototype of an artificial leaf that can replicate the way plants use sunlight to knit chemical bonds. With further development, this could possibly provide a practical and inexpensive source of solar energy. The device was described by Robert F. Service in *Science* as "a silicon wafer about the size and shape of a playing card. Different catalysts coat each side of the wafer. The silicon absorbs sunlight and passes that energy to the catalysts to split water (H_2O) into molecules of hydrogen (H_2) and oxygen (O_2)" (Service 2011, April 1). The hydrogen might then be used to power a fuel cell, with the hydrogen reemerging as water. This technology makes water a cheap, clean, and available form of energy, although such a development probably remains far off in the future.

"It's spectacular," said Robert Grubbs, a chemist at the California Institute of Technology in Pasadena, after watching a demonstration of the process at a meeting of the American Chemical Society (Service 2011, April 1). The MIT team previously had developed a catalyst that used cobalt and phosphorous to break water at the molecular level and knit oxygen atoms into O_2 molecules. Hydrogen-forming catalysts also had been developed, but all of this was extremely expensive. By early 2011, Nocera had developed an inexpensive catalyst that uses three different metals. With patents pending and publication ahead, he did not name the metals. In 2011, he expressed hopes of having the technology operable on a commercial scale within two to three years with the aid of his company, Sun Catalytix. Nocera has plans to join with Ratan Tata, chair of Tata Group of India, to produce a power plant the size of a refrigerator that will manufacture electric power from sunlight and water to provide inexpensive energy to large numbers of people who currently have no access to power.

"For years the efficiency of polymer-based solar cells scraped along at a feeble 3 percent to 5 percent. But things have improved markedly over the past two years. In early April, Mitsubishi Chemical reportedly set a new efficiency record, producing organic solar cells with a 9.2 percent conversion efficiency," Robert Service reported in *Science*. "Meanwhile, three other companies—Konarka Technologies, Solarmer Energy Inc., and Heliatek—are now reporting cells with efficiencies greater than 8 percent. Many researchers in the field are confident that the figure could soon top 10 percent and possibly reach 15 percent" (Service 2011, April 15).

Jessica Marshall explained the process in *Nature* (2014, 22):

At the heart of JCAP's artificial-leaf design are two electrodes immersed in an aqueous solution. Typically, each electrode is made of a semiconductor material chosen to capture light energy from a particular part of the solar spectrum, and coated with a catalyst that will help to generate hydrogen or oxygen at useful speeds. Like many other artificial-photosynthesis devices, JCAP's system is divided by a membrane to keep the resulting gases apart and reduce the risk of an explosive reaction. Once the water has been split, the hydrogen is harvested. It can be used as a fuel by itself—perhaps in hydrogen-powered cars such as those already making their way into showrooms in California—or be reacted with carbon monoxide to make liquid-hydrocarbon fuels.

The challenge comes in finding materials that will work. Silicon, for example, works as an electrode that produces hydrogen gas, but it is stable only in an acidic solution. Photoanodes (electrodes) that produce oxygen work best only in a basic solution. Good oxygen-producing electrodes such as iridium are rare and too expensive for commercial development. These experiments may remind historians of science of the many substances that Thomas Edison tested to find the best filament for the first practical electric light bulbs, but on a much more massive scale.

"In one experiment to find the best proportions of nickel, iron, cobalt and cerium oxides to generate oxygen from water," *Nature* reported (Marshall 2014), "the team screened nearly 5,500 combinations for stability and function using a miniaturized chemical lab that glided over the glass plates tirelessly." The best-performing combination is not the most effective catalyst ever found for this reaction, but it is transparent, allowing light to pass through to the photo-absorber, and it has good chemical compatibility with that material.

Thus far, the best combination of economy and stability (with 10 to 20 percent efficiency) has eluded the researchers (Reece et al. 2011, 645). A coating of titanium dioxide on silicon may be the answer. "That's basically the last piece of the puzzle to create the first-generation prototype," said a spokesman for JCAP (Marshall 2014, 24). A less expensive fabrication method for silicon must be found for single-crystal silicon, however. The United States effort is being spurred by similar work in Japan, so a race of sorts has developed.

By 2016, as *Science* reported, scientists were studying a form of artificial photosynthesis that "when combined with solar photovoltaic cells, solar-to-chemical conversion rates should become nearly an order of magnitude more efficient than natural photosynthesis" (Liu et al. 2016, 1210). Liu and colleagues wrote that "on a previous artificial photosynthesis design, [we] combined the hydrogen-oxidizing bacterium *Raistonia eutropha* with a cobalt-phosphorus water-splitting catalyst. This biocompatible self-healing electrode circumvented the toxicity challenges of previous designs and allowed it to operate aerobically" (Liu et al. 2016, 1210).

As reported in *Science*, by 2016 scientists were studying a form of artificial photosynthesis. "When combined with solar photovoltaic cells, solar-to-chemical conversion rates should become nearly an order of magnitude more efficient than natural photosynthesis" (Liu et al. 2016, 1210). As Liu and colleagues wrote, "on a previous artificial photosynthesis design, [we] combined the hydrogen-oxidizing bacterium *Raistonia eutropha* with a cobalt–phosphorus water-splitting catalyst. This biocompatible self-healing electrode circumvented the toxicity challenges of previous designs and allowed it to operate aerobically" (Liu et al. 2016, 1210).

Richard Martin reported in the *MIT Technology Review* (2016) that Daniel G. Nocera and colleague Pamela Silver

> have devised a system that completes the process of making liquid fuel from sunlight, carbon dioxide, and water. And they've done it at an efficiency of 10 percent, using pure carbon dioxide—in other words, one-tenth of the energy in sunlight is captured and turned into fuel. That is much higher than natural photosynthesis, which converts about 1 percent of solar energy into the carbohydrates used by plants, and it could be a milestone in the shift away from fossil fuels. . . . Bill Gates has said that to solve our energy problems, someday we need to do what photosynthesis does, and that someday we might be able to do it even more efficiently than plants," says Nocera. "That someday has arrived. . . . [In other words,] solar energy, water, and carbon dioxide [have now been used] to produce energy-dense liquid fuels. Nocera and Silver's system uses a pair of catalysts to split water into oxygen and hydrogen, and feeds the hydrogen to bacteria along with carbon dioxide. The bacteria, a microorganism that has been bioengineered to specific characteristics, converts the carbon dioxide and hydrogen into liquid fuels. (Martin 2016)

This new work "is really quite amazing," said Peidong Yang of the University of California, Berkeley, who had developed a similar system with much lower efficiency. "The high performance of this system is unparalleled in any other artificial photosynthesis system reported to date" (Martin 2016).

Further Reading

Liu, Chong, et al. "Water Splitting–Biosynthetic System with CO2 Reduction Efficiencies Exceeding Photosynthesis." *Science* 352 (June 3, 2016): 1210–1213.

Marshall, Jessica. "Solar Energy: Springtime for the Artificial Leaf." *Nature* 510 (June 4, 2014): 22–24. http://www.nature.com/news/solar-energy-springtime-for-the-artificial-leaf-1.15341.

Martin, Richard. "A Big Leap for an Artificial Leaf: A New System for Making Liquid Fuel from Sunlight, Water, and Air Is a Promising Step for Solar Fuels." *MIT Technology Review*, June 7, 2016. https://www.technologyreview.com/s/601641/a-big-leap-for-an-artificial-leaf/.

Reece, Steven Y., et al. "Wireless Solar Water Splitting Using Silicon-Based Semiconductors and Earth-Abundant Catalysts." *Science* 334 (November 4, 2011): 645–648.

Service, Robert F. "Artificial Leaf Turns Sunlight into a Cheap Energy Source." *Science* 332 (April 1, 2011): 25.

Service, Robert F. "Outlook Brightens for Plastic Solar Cells." *Science* 332 (April 15, 2011): 293.

CARBON CAPTURE

To pose the issue of whether carbon capture and sequestration (CCS) may be useful as a weapon against global warming, consider Ben McGrath's comment in *The New Yorker*:

> Consider carbon capture and storage, sometimes known as "sequestration": the prospect of burying our energy emissions in reservoirs of rock, so that we might continue to provide heat and air conditioning for our cars without hastening the demise of island nations. It's almost like a get-out-of-jail-free card for gratuitous trips to the mall. "It's amoral," the geophysicist Dennis Kent said recently. "It's enabling," his colleague David Goldberg said. (McGrath 2013, 24)

The science journal *Nature* commented on CCS in a 2014 editorial:

> Over the past decade or so, carbon capture and storage . . . has been the fairy godmother of climate change, or at least of the politicians who have pledged in ever more ambitious terms to tackle the problem. . . .
>
> Dig into most political promises to slash greenhouse gas emissions by headline amounts—80 percent by 2050, that kind of thing—and there she is. A significant proportion of the promised cuts are the result not of declines in carbon dioxide production, but of attempts to trap damaging emissions at source and divert them under the ground rather than into the atmosphere. Clean coal,

CCS technology, capture-ready: the idea has spawned its own terminology. Regulations on carbon pollution permitted from new fossil fuel-fired power plants are also being drawn up, on the assumption that CCS is feasible, and that it can be implemented on a massive scale. ("No Magic" 2014)

Problems with CCS

CCS supporters assert that the storage process will be necessary despite its costs and environmental debits because coal will remain the preferred source of electricity for decades to come, especially in large and growing markets such as India and China. By 2016, energy markets had devastated this assumption, as coal consumption declined worldwide. Even so, the costs of experiments in CCS have been considerable. In 2012, the U.S. Congressional Budget Office said that Congress had authorized $6.9 billion for research and development of technology that remains unproved, whereas wind and solar have produced results that are without attendant pollution. If CCS fails, it will not be for lack of government largesse. "That's a lot of money for a technology whose adoption faces three potentially insurmountable hurdles," wrote Robert Bryce, Manhattan Institute senior fellow, and author of *Power Hungry: The Myths of 'Green' Energy and the Real Fuels of the Future*: "It greatly reduces the output of power plants; pipeline capacity to move the newly captured carbon dioxide is woefully insufficient; and the volume of waste material is staggering. Lawmakers should stop perpetuating the hope that the technology can help make huge cuts in the United States' carbon dioxide emissions" (Bryce 2010).

In addition, the Pacific Northwest National Laboratory estimates that as many as 23,000 miles of pipelines will be required to conduct captured carbon from plants to storage sites, an expensive investment for a waste product with little market value. (A tiny fraction of the CO_2 could be used to force oil from old oil wells, so it still has some value.) Also consider that all of this investment may be going into a fossil fuel that is rapidly losing favor in the United States, where dependence on coal for electricity generation has fallen from half to less than one-third in less than a decade as of 2015.

The third problem involves the size of storage space required for CCS:

In 2009, carbon dioxide emissions in the United States totaled 5.4 billion tons. Let's assume that policy makers want to use carbon capture to get rid of half of those emissions—say, 3 billion tons per year. That works out to about 8.2 million tons of carbon dioxide per day, which would have to be collected and compressed to about 1,000 pounds per square inch (that compressed volume of carbon dioxide would be roughly equivalent to the volume of daily global oil production). In other words, we would need to find an underground location (or locations) able to swallow a volume equal to the contents of 41 oil supertankers each day, 365 days a year. (Bryce 2010)

U.S. Government Generosity for CCS

The United States has been the world leader in CCS, and the federal government has been extremely generous. By 2013, the U.S. Department of Energy had spent some $6.5 billion over 30 years to research CCS (Nijhuis 2014, 35). In 2013, the U.S. Environmental Protection Agency and the Obama administration proposed that all new coal-fired power plants include CCS technology, even as Norway's government was abandoning its $1-billion Mongstad project that had tested CCS on a large scale at an oil refinery. Meanwhile, Howard Herzog of the Massachusetts Institute of Technology compiled a list of more than 25 CCS projects around the world that had been cancelled or suspended because they did not perform as planned.

"Power plants that can capture and store their carbon are initially expected to cost about 75 percent more than regular coal plants," according to one analysis. "And those costs won't come down unless there's either a huge technological breakthrough or utilities invest a lot more of their own money in building new plants. Neither appears imminent" (Plumer 2012). Even so, the boondoggle juggernaut rolls on. Its sponsors say that socialized CCS on a large "demonstration" scale will push private power companies into the market, which has not happened. Many of these companies are instead building wind turbines and distribution networks. Witness Mid-American Energy of Iowa, which is building wind power that met almost 50 percent of its customers' needs by 2016 (and aims for 89 percent by 2020). The Omaha Public Power District, heretofore two-thirds coal generated (the rest was nuclear) by 2017 generated 33 percent of its electricity from wind.

According to the Global CCS Institute's 2013 report, "while CCS projects are progressing, the pace is well below the level required for CCS to make a substantial contribution to climate change mitigation." *The New York Times* reported that "Gareth Lloyd, the general manager of corporate affairs at the CCS Institute (the corporate lobbying group), said . . . that despite the growth in renewable energy, about 60 percent of energy in 2060 will still come from fossil fuels, so CCS is not an optional technology if we're to address climate change" (Wald 2013).

Coal-fired power using CCS will cost at least 40 percent more than conventional coal-generated electricity. Such costs threatened to price it out of a market in which natural gas could be substituted as its price fell (Kintisch 2013, 1439). The costs of wind and solar power also have been declining steadily over time. A major problem with CCS is its own immense appetite for energy. Capturing carbon dioxide at a coal-fired electric plant requires as much as 28 percent of its output, which means it is not an energy-conservation measure. How expensive will full-scale CCS become? No one knows for certain, but an experimental facility at the Mountaineer coal-fired power plant in West Virginia required a 10-story building covering 14 acres to sequester 0.25 percent of the plant's effluent (Nijhuis 2014, 37).

Geophysicists Mark Zoback and Steven Gorlick of Stanford University have expressed concern that many sequestration sites are composed of shale that is brittle enough to crack when infused with liquid carbon dioxide and thus provoke small earthquakes and leaks. For these reasons, they regard CCS as "an extremely expensive and risky strategy" (Nijhuis 2014, 36). Sequestration of CO_2 deep underground also could cause leaks into aquifers near the surface, increasing drinking water contamination as much as 10 times background levels, according to a study by scientists at Duke University. "We found the potential for contamination is real, but there are ways to avoid or reduce the risk," said Robert B. Jackson, Nicholas Professor of Global Environmental Change and professor of biology at Duke ("Leaking" 2010). The study was published in the online edition of *Environmental Science & Technology*. This study showed that "there are a number of potential sites where CO_2 leaks drive contaminants up tenfold or more, in some cases to levels above the maximum contaminant loads set by the EPA for potable water," Jackson says (Little and Jackson 2010).

Problems with Disposal of CO_2 in the Oceans

Disposal of carbon dioxide in the deep oceans has been proposed as one method of mitigating global warming. However, many proposals to inject human-generated carbon dioxide into the oceans ignore the possible effects of such sequestration on life at these levels. Brad A. Seibel and Patrick J. Walsh examined these effects, finding that increased deepwater carbon dioxide levels result in decreases of seawater pH; such increased acidity can be harmful to sea creatures "as has been demonstrated for the effects of acid rain on freshwater fish" (Seibel and Walsh 2001, 319). They find that "a drop in arterial pH by just 0.2 would reduce bound oxygen in the deep-sea crustacean *Glyphocrangon vicaria* by 25 percent" (Seibel and Walsh, 2001, 320). The same drop in arterial pH would reduce bound oxygen in the midwater shrimp species *Gnathophausia ingens* by 50 percent.

According to Seibel and Walsh, "Deep-sea fish hemoglobins are even more sensitive to pH" (Seibel and Walsh 2001, 320). Small increases in carbon dioxide levels and resulting decreases of pH levels "may trigger metabolic suppression in a variety of organisms" because "low pH has been shown to inhibit protein synthesis in trout living in lakes rendered acidic through anthropogenic effects" (Seibel and Walsh 2001, 320). Seibel and Walsh cited research by R. L. Haedrich that "any change that takes place too quickly to allow for a compensating adaptive change within the genetic potential of finely adapted deepwater organisms is likely to be harmful" (Seibel and Walsh 2001, 320). According to the researchers, "Available data indicate that deep-sea organisms are highly sensitive to even modest pH changes. . . . Small perturbations in CO_2 or pH may thus have important consequences for the ecology of the deep sea . . ." (Seibel and Walsh 2001, 320).

"Through various feedback mechanisms, the ocean circulation could change and affect the retention time of carbon dioxide injected into the deep ocean, thereby indirectly altering oceanic carbon storage and atmospheric carbon dioxide concentration," said Atul Jain, a professor of atmospheric sciences at the University of Illinois at Urbana–Champaign. "Where you inject the carbon dioxide turns out to be a very important issue" ("Global Warming" 2002). To help them investigate the possible effects of feedbacks between global climate change, the ocean carbon cycle, and oceanic carbon sequestration, Jain and Long Cao, a graduate student, developed a model addressing atmosphere–ocean and climate–carbon cycle relationships. The researchers then studied whether oceanic carbon sequestration was effective by directly injecting carbon dioxide in different locations and at different ocean depths ("Global Warming" 2002).

By 2014, the Sleipner gas field in the North Sea had stored several million tons of carbon dioxide over 17 years in brine-saturated sandstone a half-mile below the sea floor by Statoil. The Norwegian oil company asserts that the stored carbon is safe, but independent reports express fears that unexpected fractures could cause the carbon to leak into ocean waters already overloaded with CO_2 above the world's biggest storage site. "At the Sleipner field in the North Sea, waste carbon dioxide is separated from natural gas and injected into a rock formation 800 meters below the seabed, busting the fantasy that "billions of tons of carbon dioxide can be stored below the sea floor for centuries, keeping it from warming the planet" (Monastersky 2013, 339).

By 2013, European partners had spent the equivalent of $13.8 million learning that stuffing and trying to contain carbon dioxide under the ocean floor is no sure bet. Data from an autonomous underwater vehicle and other tools deployed by the European Commission's €10-million ($13.8 million U.S.) ECO_2 research project suggest that a scarred, fractured seabed provides possible escape routes for carbon dioxide from reservoirs. "We are saying it is very likely something will come out in the end," said Klaus Wallmann, ECO_2 coordinator and a marine geochemist at the GEOMAR Helmholtz Centre for Ocean Research in Kiel, Germany (Monastersky 2013, 339). However, oil-industry sponsors of carbon-storage efforts in the North and Barents Seas admit to no possible leakage.

Why Dump CO_2 in an Acidifying Ocean?

Some of the worst problems from rising carbon dioxide levels in the atmosphere have nothing to do with the higher temperatures we associate with global warming. Most are nearly invisible. Take, for example, the intensifying acidity of the oceans. We are killing the oceans by overloading them with carbon dioxide, and acidity—not temperature—is the problem.

Already, without further human intervention, some 30 percent of the carbon dioxide that humans inject into the atmosphere is being sequestered into the oceans. Ocean pH is now lower than it has been in 55 million years and continues to decline, according to marine chemist Richard Feely of the National Oceanic and Atmospheric Administration's Marine Environmental Laboratory in Seattle. Models by Feely and colleagues anticipate that ocean pH will decline from 8.2 before the Industrial Revolution to 7.8 by 2100, increasing acidity by 150 percent. Shells of microscopic oceanic animals already have thinned by as much as one-third because of rising CO_2 levels. These animals are at the base of the oceanic food chain. Deployment of CCS under the oceans risks leaks that will accelerate damage from human-cause increases in acidity.

A Moratorium on Coal-Fired Electricity Generation

The development of carbon capture and sequestration technology is running into another problem. As pervasive as coal-generated electricity has been in the past, it is now obsolete and in decline. The Earth system simply cannot afford it. James E. Hansen, director of NASA's Goddard Institute for Space Studies, has proposed a moratorium on construction of new coal-fired power plants until technology for carbon dioxide capture and sequestration is available. A quarter or so of carbon dioxide emissions from power plants will remain in the air "forever"—that is, more than 500 years, long after new technology is refined and deployed. As a result, Hansen expects that all power plants without adequate sequestration will be obsolete and slated for closure (or at least retrofitting) before midcentury (Hansen 2007, March 19).

According to Hansen,

> Coal will determine whether we continue to increase climate change or slow the human impact. Increased fossil fuel CO_2 in the air today, compared to the preindustrial atmosphere, is due 50 percent to coal, 35 percent to oil and 15 percent to [natural] gas. As oil resources peak, coal will determine future CO_2 levels. Recently, after giving a high school commencement talk in my hometown, Denison, Iowa, I drove from Denison to Dunlap, where my parents are buried. For most of 20 miles there were trains parked, engine to caboose, half of the cars being filled with coal. If we cannot stop the building of more coal-fired power plants, those coal trains will be death trains—no less gruesome than if they were boxcars headed to crematoria, loaded with uncountable irreplaceable species. (Hansen 2007, July 23).

Fusing Carbon Dioxide with Basalt

Some of the problems with CCS, such as potential for leakage, were being solved by 2016 at a project in Iceland that fuses carbon dioxide with volcanic basalt, triggering a reaction that rapidly forms a mineral called *calcite* that may lock the gas into the rock forever. The process involves dissolving the gas into water and pumping the mixture, which resembles soda water, into basalts that contain iron, calcium, and magnesium that react with carbon dioxide.

"This study demonstrates for the first time the permanent disposal of CO_2 as environmentally benign carbonate minerals in basaltic rocks," Matter et al. wrote in *Science*.

> We find that over 95 percent of the CO_2 injected into the CarbFix site in Iceland was mineralized to carbonate minerals in less than two years. This result contrasts with the common view that the immobilization of CO_2 as carbonate minerals within geologic reservoirs takes several hundreds to thousands of years. Our results, therefore, demonstrate that the safe long-term storage

of anthropogenic CO_2 emissions through mineralization can be far faster than previously postulated. (Matter et al. 2016, 1312)

Most other tests of carbon capture and sequestration have taken place in sandstone formations. Sandstone, however, "is too chemically inert to foster CO_2-trapping reactions, and scientists worry that gas injected into it could leak back into the atmosphere. Scientists say the new results could help solve some of the technical problems that have kept CCS projects from being commercially successful. But they say the main obstacle—high cost—is one that only changes in policy can overcome" (Kintisch 2016, 1262).

This is no automatic fix. As of 2016, the process had not been tried on a commercial scale. It uses 25 tons of water for every ton of carbon dioxide processed, which may be prohibitive in dry areas. The process also could not be used in areas that contain little or no basalt, which is common but hardly universal.

Further Reading

Bryce, Robert. "A Bad Bet on Carbon." *The New York Times*, May 12, 2010. http://www.nytimes.com/2010/05/13/opinion/13bryce.html.

"Global Warming Could Hamper Ocean Sequestration." Environment News Service, December 4, 2002. http://ens-news.com/ens/dec2002/2002-12-04-09.asp (no longer available).

Hansen, James E. "Political Interference with Government Climate Change Science." Testimony of James E. Hansen, 4273 Durham Road, Kintnersville, PA, to Committee on Oversight and Government Reform United States House of Representatives, March 19, 2007.

Hansen, James E. "Coal Trains of Death." James Hansen's E-mail List, July 23, 2007.

Kintisch, Eli. "U.S. Carbon Plan Relies on Uncertain Capture Technology." *Science* 341 (September 27, 2013): 1438–1439.

Kintisch, Eli. "New Solution to Carbon Pollution?" *Science* 352 (June 10, 2016): 1262–1263.

"Leaking Underground CO_2 Storage Could Contaminate Drinking Water." EurekaAlert! November 11, 2010. http://www.eurekalert.org/pub_releases/2010-11/du-luc111110.php.

Little, Mark G., and Robert B. Jackson. "Potential Impacts of Leakage from Deep CO_2 Geo-Sequestration on Overlying Freshwater Aquifers." *Environmental Science and Technology* 44 (2010): 9225–9232. http://pubs.acs.org/doi/abs/10.1021/es102235w.

Matter, Juerg M., et al. "Rapid Carbon Mineralization for Permanent Disposal of Anthropogenic Carbon Dioxide Emissions." *Science* 352 (June 10, 2016): 1312–1314.

McGrath, Ben. "Drill, Baby, Drill." *The New Yorker*, December 2, 2013, 24.

Monastersky, Richard. "Seabed Scars Raise Questions over Carbon-Storage Plan." *Nature* 504 (December 19, 2013): 339–340. http://www.nature.com/news/seabed-scars-raise-questions-over-carbon-storage-plan-1.14386.

Nijhuis, Michelle. "Can Coal Ever Be Clean?" *National Geographic*, April 2014, 28–61.

"No Magic Fix for Carbon." *Nature* 509 (May 1, 2014): 7.

Plumer, Brad. "CBO: Carbon Capture Efforts Aren't Going So Well." *Washington Post*, July 2, 2012. https://www.washingtonpost.com/pb/news/wonk/wp/2012/07/02/cbo-carbon-capture-efforts-arent-going-so-well/?outputType=accessibility&nid=menu_nav_accessibilityforscreenreader.

Seibel, Brad A., and Patrick J. Walsh. "Potential Impacts of CO_2 Injection on Deep-Sea Biota." *Science* 294 (October 12, 2001): 319–320.
Wald, Matthew L. "Despite Climate Concern, Global Study Finds Fewer Carbon Capture Projects." *The New York Times,* October 10, 2013. http://www.nytimes.com/2013/10/11/science/earth/study-finds-setbacks-in-carbon-capture-projects.html.

See also: Coal; Global Warming, China

CONCENTRATING SOLAR POWER

Concentrating (or concentrated) solar power (CSP) is much more powerful than photovoltaic cells. A rooftop photovoltaic complex might power a small office building, but a CSP complex can produce much more. A CSP near Seville, Spain, for example, generates 11 megawatts, enough electricity for a small town (Abboud 2006). A development of similar size in California's Mojave desert serves 400,000 homes. CSP mirrors track the sun and concentrate its power on single points, generating steam that runs a turbine. Some of the heat also is stored in oil or molten salt to run the turbine after sunset or when clouds obscure the sun.

According to the consulting firm Emerging Energy Research, more than 60 CSP projects were in planning stages around the world by 2014, including some in the United States. Spain is currently subsidizing CSP development and requiring utility companies to buy their power at above-market rates. Abengoa plans to eventually build enough CSP capacity to supply all of Seville, some 180,000 homes.

Solar arrays are spreading across the northern Sahara. In Morocco, which has 3,000 hours of sunshine a year, Noor 1 was being developed at the "door of the desert" in the south-central Moroccan town of Ouarzazate in 2016. When it becomes operational in 2020, this will be the largest CSP plant in the world at 580 megawatts created by 500,000 12-meter-tall parabolic mirrors that focus energy onto a fluid-filled pipeline. The resulting hot fluid—393°C (739°F)—will heat water and create steam. By that time, according to NASA Earth Observatory, "almost half of Morocco's energy is expected to come from renewables, about one-third of which will be from solar" ("Solar" 2016).

Several enormous arrays of solar thermal power plants across the Sahara desert could supply most if not all of the power needs for Europe, the Middle East, and North Africa, where plans were being made by 2007 to establish a renewable energy supergrid. Solar energy is stored in a heat-retaining fluid and used to drive turbines. One plan calls for roughly 1,000 100-megawatt power plants worth of solar and other renewable energies (Feresin 2007). The technology for this system is now available, says Gerhard Knies, a retired physicist based in Hamburg, Germany. The grid "could offer unlimited, cheap, and carbon dioxide-free energy to Europe" (Feresin 2007). Knies has joined with other advocates of the idea. Although the energy will be cheap once infrastructure has been built, getting to that point could cost at least 400 billion euros (around $480 billion U.S.). Europeans also may be unwilling to place their energy fates in the hands of North African states. The idea of the Sahara as a solar-energy bank actually goes back as far as Frank

Shuman, an inventor who was working from Philadelphia in 1913 when he built the first prototype thermal solar plant in Egypt. Today's advocates propose solar as the major component of an energy system that also will use wind and biomass fuels.

In 2007 in the desert north of Tucson, Arizona Public Service (APS) was experimenting with CSP, using an array of mirrors that focuses sunlight and heats mineral oil upto 550°F; the heat then evaporates a liquid hydrocarbon that runs a generator to make electricity. The array includes six rows of mirrors, each nearly a quarter-mile long, or some 100,000 square feet of reflective space This demonstration project produces one megawatt of power, enough to power a large shopping center. Acciona Solar Power, the company that installed the APS experimental array was also planning a 350-acre plant near Boulder City, Nevada, with a capacity of 64 megawatts, also using CSP. Arizona Public Service and several other utilities also were considering a joint project to build a 250-megawatt CSP plant (Wald 2008).

Arizona Public Service has built a 1,900-acre, 290-million watt CSP facility, Starwood Solar I west of Phoenix, that has provided power to 73,000 homes since 2013. The facility uses 3,500 parabolic mirrors to focus solar thermal energy in a fluid that, once heated, converts water into steam that turns the plant's turbines to create electricity. Using molten salt, the plant produces and stores to solar energy six hours after sunset during peak usage hours. Starwood is APS's second large solar facility. The first, Abengoa Solar's 280-megawatt Solana solar plant 70 miles southwest of Phoenix (near Gila Bend, Arizona), began operation in 2012 ("Second" 2009).

Further Reading

Abboud, Leila. "Sun Reigns on Spain's Plains: Madrid Leads a Global Push to Capitalize on New Solar-Power Technologies." *Wall Street Journal*, December 5, 2006, A4.

Feresin, Emiliano. "Europe Looks to Draw Power from Africa." *Nature* 450 (November 29, 2007): 595.

"Second Giant Solar Plant Planned for Arizona Electricity Users." Environment News Service, May 25, 2009. http://www.ens-newswire.com/ens/may2009/2009-05-25-091.html.

"Solar in the Sahara." NASA Earth Observatory, January 10, 2016. http://earthobservatory .nasa.gov/IOTD/view.php?id=87293&src=eoa-iotd.

Wald, Matthew. "Turning Glare into Watts." *The New York Times*, March 6, 2008. http://www .nytimes.com/2008/03/06/business/06solar.html.

See also: Corporate Sustainability; Solar Power; Unburnable Carbon; Wind Gains Power Share

ETHANOL, CORN

In 2004, with pledges from President George W. Bush that corn ethanol would help make the United States energy independent, the "blender's credit"—also known as the Volumetric Ethanol Excise Tax Credit (VEETC)—was created, providing oil refiners with an incentive to mix corn-based fuel with gasoline. In 2009, the 51-cent-a-gallon subsidy on pure ethanol was reduced to 45 cents, and on

December 31, 2012, it expired as doubts intensified regarding corn's suitability as a green fuel. In the meantime, corn ethanol had experienced a classic bubble.

Despite doubts about ethanol as a green fuel, in 2007 U.S. corn acreage hit 93.6 million acres, a record, and up 20 percent from 2006. One-fourth of U.S. corn went into ethanol by that time. Ethanol production capacity in the United States grew by 400 percent between 2000 and 2006 to 5.6 billion gallons a year. Ethanol increasingly was being produced at large plants such as VeraSun Energy near Charles City, Iowa, the largest single producer in the United States in 2008. Steven Mufson of the *Washington Post* described VeraSun:

> The plant is hard to miss. Its two massive concrete silos reach 150 feet into the air; each one holds half a million bushels of corn, delivered by an average of 110 brimming trucks every day. The silos are connected to a distillery with giant shiny steel vats for milling the corn, then fermenting and distilling it into 200-proof, fuel-grade ethanol. The ethanol is shipped out by train, 84 black tanker cars at a time. (Mufson 2008)

Corn—A Poor Source of Combustion

Scientists did the math and learned that corn is an extremely poor source of combustible fuel. An ear of corn contains a piddling amount of energy when used to propel a vehicle. Filling the gas tank of a sport utility vehicle with ethanol requires 450 pounds of processed corn, enough to feed several people for a year. Nevertheless, corn ethanol is now an established interest in Washington, D.C., and, according to the *Omaha World-Herald*, "an economic engine in the Hawkeye State [Iowa]." With his liberal colleagues, Representative Steven King, a right-wing Republican who opposes most other government subsidies, was working "to blunt gathering momentum for reductions in the renewable fuel standard, the federal mandate that oil companies blend a certain amount of biofuel every year into the gasoline supply" (Morton 2013).

As Robert Bryce, a senior fellow at the Manhattan Institute and the author of a report from the institute, *The Hidden Corn-Ethanol Tax* explained in the *New York Times*:

> Ethanol contains about 76,000 [British thermal units] per gallon. Gasoline contains about 114,000 BTUs per gallon. Therefore, to get the same amount of energy contained in a gallon of gasoline, a motorist must buy about 1.5 gallons of ethanol. As Fueleconomy.gov, a site run by the federal government, advises that vehicles running on the most common form of ethanol-blended fuel, E10 (which contains 10 percent ethanol and 90 percent gasoline), will typically get "3 percent to 4 percent fewer miles per gallon" than they would if they were running on pure gasoline. That mileage penalty—in essence, a tax—must be paid at the pump through the purchase of additional fuel.
>
> Since 1982, officials in Nebraska . . . have been monitoring . . . wholesale, or "rack," prices for ethanol and gasoline at fuel depots in Omaha. In

December 2014, the rack price of a gallon of ethanol was $2.40, while a gallon of unleaded gasoline was $1.73. [At] 1.5 gallons of ethanol to match the energy in a gallon of gasoline, . . . you would need to pay about $3.60 to get the same amount of energy as from a gallon of gasoline, making ethanol about twice as expensive. . . . Since 1982, the price of an energy-equivalent amount of ethanol has, on average, been about 2.4 times the price of gasoline. (Bryce 2015)

In the Corn Belt, opposing corn ethanol is almost as uncommon as coming out against Husker football, apple pie, and the Republican Party. Or, as the Omaha *World-Herald* (which loves all four effusively) put it in an editorial: "One need only look at how ethanol has helped the economies of Iowa and Nebraska, the top-producing states. It's not just farm jobs; it's implement dealers, bankers, major industrial operations, home builders, and vehicle dealers. . . . Corn prices more than doubled after the renewable fuels standards Congress passed in 2007" ("Sound Reasons" 2013). The price of corn shot up from $1.90 to $3.75 a bushel between 2006 and 2007 partially because of the heavy buzz over ethanol, and a large number of farmers were loving the fact that the era of cheap corn had ended.

Researchers at the University of Minnesota in the meantime estimated that converting the entire U.S. corn crop to ethanol would replace only one-eighth of U.S. gasoline consumption, wrote *New York Times* columnist Paul Krugman (2007). In addition, corn must be grown and transported, after which ethanol must be manufactured. Replacing a gallon of gasoline with a gallon of ethanol does not save a gallon of gasoline because most of the energy that goes into corn comes from fossil fuels. The real savings, Krugman noted, is more like a quarter of a gallon—so make that a 3-percent savings in gasoline consumption for the entire U.S. corn crop.

With all of our corn in our gas tanks, what would we *eat*? At the peak of the ethanol boom, the rising price of corn had already raised the price or tortillas out of the reach of many poor people in Mexico and elsewhere in Latin America. Twenty years ago, the corn market caught Mexico in a trap as trade restrictions were eased during the early 1990s, allowing corn imports from the United States to rise from 8 percent in 1994 to 34 percent of the market there in 2012.

"The higher your import dependence, the higher your dependence to global price spikes," said Timothy A. Wise of the Global Development and Environment Institute at Tufts University in Medford, Massachusetts (Nowakowski 2014, 18). At the same time, the ethanol boom was one (not the only) factor in wild swings in corn's price—between $3 and $8 a bushel. Spurred by financial incentives from the U.S. government, the production of biofuels soared 500 percent during the first decade of the 21st century. Tortilla prices in Mexico soared 150 percent between 2000 and 2011, severely pinching the budgets of the poor, provoking riots and government price controls. As early as January 31, 2007, rising prices for corn were igniting demonstrations by 75,000 people in Mexico City, where the price of tortillas hit record highs as President George W. Bush touted corn ethanol in his State of the Union message. Prices later fell as the United States phased out ethanol price incentives, and the price of corn, two-thirds of the cost of a Mexican tortillas, fell as well.

How "Green" Is It?

Corn ethanol is a carbon-based fossil fuel when it is burned to propel cars. Depending on its source (and whose statistics are on offer), ethanol emits only 10 percent to 34 percent less greenhouse gases than plain old garden-variety gasoline. Even that savings is not what it seems. Corn must be grown on factory farms, a highly energy-intensive business. Under some circumstances (if a biomass field replaced a forest, for example) this type of fuel might actually produce a net increase in emissions of greenhouse gases, considering the entire production process.

While it was subsidizing corn ethanol, the United States' national commitment elevated to nearly religious status by George W. Bush, become monumentally expensive. If they had been extended 15 years from 2008 at the rates then in effect (they were not), tax credits, grants, and loan guarantees would have cost the federal treasury $140 billion. New proposals under consideration in Congress at the time could have raised the cost to $205 billion. The biggest single prospective expense was an extension of ethanol tax credits, which would have cost an estimated $131 billion through 2022.

The costs did not end there. The United States is a major corn producer, and its emphasis on corn as a fuel helped feed the inflation of corn prices worldwide. The use of corn has become surprisingly widespread in our food supply, aside from the obvious products such as corn flakes. Meat animals eat it, so it affects the price of chicken (including eggs), beef, and pork. The manufacture of soda pop consumes huge amounts of corn syrup, so its price also rises. A night at the movies even got pricier as the price of popcorn rose 40 percent between 2006 and 2007.

Some farmers have expressed concerns about ethanol's impact, especially its appetite for scarce water that may deplete aquifers. The rising number of ethanol plants in the Great Plains region drained billions of gallons of water each year from the Ogallala Aquifer, which is already depleted. The aquifer is one of the world's largest, a vast, shallow underground pool beneath portions of Colorado, Kansas, Nebraska, New Mexico, Nebraska, Oklahoma, South Dakota, Texas, and Wyoming.

A report written by Martha Roberts and Theodore Toombs for the Environmental Defense Fund described how the Ogallala Aquifer's water table is declining as rates of groundwater pumping regularly exceed replacement. A gallon of ethanol requires four gallons of water to produce. "This dramatic expansion of ethanol production has substantial implications for already strained water and grassland resources in the Ogallala Aquifer region," the authors said ("Ethanol Production" 2007). Residents in Webster County, Missouri, sued to stop construction of an ethanol plant on grounds that it would use more water than all of the county's 33,000 residents combined. Other concerns included truck traffic in rural areas and air pollution with a sticky-sweet smell that resembles that of a barroom floor after a busy Saturday night (Barrett 2007, A1).

Yet another environmental debit of corn ethanol is its contribution to the expansion of the Gulf of Mexico's seasonal "dead zone," an area of extremely low oxygen that kills most sea life, according to a study by Simon Donner of the University of British Columbia and Chris Kucharik of the University of Wisconsin–Madison, who modeled the effects of its production. Putting it briefly: more corn ethanol requires more fertilizer; when fertilizer runs off, it is washed into the gulf as nitrogen overloads and enlarges the dead zone. By 2007, the dead zone by summer already covered 7,700 square miles, the size of New Jersey. Nitrogen and phosphorus from agricultural fertilizers provoke the growth of algae that suck up nearly all available oxygen. Animals in the water leave the affected area or die. The dead zone in the Gulf of Mexico was the largest on record in 2008, growing in large part because of increasing U.S. corn production. The dead zone had grown to some 8,800 square miles, nearly the size of New Hampshire.

A Critique of Corn as Fuel from a Climatic Point of View

When the full emissions costs of producing biofuels are calculated, most of them are environmentally more expensive, producing more greenhouse gases than fossil fuels, according to studies published in *Science* early in 2008. Growth of feedstock for many biofuels, from corn to sugarcane to palm oil, destroys natural ecosystems (most notably rain forests in the tropics and South American grasslands). Destruction of these older natural ecosystems also removes carbon sinks. In addition to the greenhouse gases caused by growing biofuels, refining and transporting them also add their own emissions.

"When you take this into account, most of the biofuel that people are using or planning to use would probably increase greenhouse gasses substantially," said Timothy Searchinger, a researcher in environment and economics at Princeton University. "Previously there's been an accounting error: land-use change has been left out of prior analysis" (Rosenthal 2008). Joseph Fargione and colleagues agreed:

> Biofuels are a potential low-carbon energy source, but whether biofuels offer carbon savings depends on how they are produced. Converting rain forests, peat lands, savannas, or grasslands to produce food crop-based biofuels in Brazil, Southeast Asia, and the United States creates a "biofuel carbon debt" by releasing 17 to 420 times more CO_2 than the annual greenhouse gas (GHG) reductions that these biofuels would provide by displacing fossil fuels. In contrast, biofuels made from waste biomass or from biomass grown on degraded and abandoned agricultural lands planted with perennials incur little or no carbon debt and can offer immediate and sustained GHG advantages. (Fargione et al. 2008, 1235)

Searchinger and colleagues wrote:

> Most prior studies have found that substituting biofuels for gasoline will reduce greenhouse gases because biofuels sequester carbon through the growth of the feedstock. These analyses have failed to count the carbon emissions that occur as farmers worldwide respond to higher prices and convert forest and grassland to new cropland to replace the grain (or cropland) diverted to biofuels. By using a worldwide agricultural model to estimate emissions from

land-use change, we found that corn-based ethanol, instead of producing a 20 percent savings, nearly doubles greenhouse emissions over 30 years and increases greenhouse gases for 167 years. Biofuels from switchgrass, if grown on U.S. corn lands, increase emissions by 50 percent. This result raises concerns about large biofuel mandates and highlights the value of using waste products. (Searchinger et al. 2008, 1238)

Environmentally, ethanol is a nonstarter, according to James E. Hansen, former director of the NASA Goddard Institute for Space Studies, and one of the leading climate scientists in the United States:

A proposed national plan for 20 percent ethanol in vehicle fuels, envisaged to be derived in large part from corn, does more harm to the planet than good. It would do little to reduce CO_2 emissions, it would degrade retention of carbon in soils and forests, and it would strike hard at the world's poor through increased food prices. There are a variety of ways that renewable or other CO_2-free energies may eventually power vehicles. Governments should not dictate the nature of those solutions. Biofuels are likely to play a major part in our energy future. As a native Iowan, I like to imagine that the Midwest will come to the rescue of compatriots threatened by rising seas. Native grasses appropriately cultivated, perhaps with improved varieties, can draw down atmospheric CO_2. The prairies from Texas to North Dakota may contribute, if we get on with solving the climate problem before super-drought spreads from the west to the prairies. If we act soon, we can keep the prairies as productive land. (Hansen 2007)

Further Reading

Barrett, Joe. "Ethanol Reaps a Backlash in Small Midwestern Towns." *Wall Street Journal*, March 23, 2007, A1, A8.

Bryce, Robert. "End the Ethanol Rip-Off." *The New York Times*, March 10, 2015. http://www.nytimes.com/2015/03/10/opinion/end-the-ethanol-rip-off.html.

"Ethanol Production Threatens Plains States with Water Scarcity." Environment News Service, September 21, 2007. http://www.ens-newswire.com/ens/sep2007/2007-09-21-091.asp (no longer available).

Fargione, Joseph, et al. "Land Clearing and the Biofuel Carbon Debt." *Science* 319 (February 29, 2008): 1235–1238.

Hansen, James E. Personal communication, April 12, 2007.

Krugman, Paul. "The Sum of All Ears." *The New York Times*, January 29, 2007, A23.

Morton, Joseph. "Ethanol Rises above Partisanship in Iowa." *Omaha World-Herald*, November 11, 2013, 1A, 3A.

Mufson, Steven. "Siphoning Off Corn to Fuel Our Cars." *Washington Post*, April 30, 2008, A1. http://www.washingtonpost.com/wp-dyn/content/article/2008/04/29/AR2008042903092_pf.html.

Nowakowski, Kelsey. "When Food Is Fuel." *National Geographic*, October 2014, 18–19.

Rosenthal, Elisabeth. "Studies Deem Biofuels a Greenhouse Threat." *The New York Times*, February 8, 2008. http://www.nytimes.com/2008/02/08/science/earth/08wbiofuels.html.

Searchinger, Timothy, et al. "Use of U.S. Croplands for Biofuels Increases Greenhouse Gases through Emissions from Land-Use Change." *Science* 319 (February 29, 2008): 1238–1240.
"Sound Reasons for Ethanol." Omaha *World-Herald*, November 14, 2014, 4B.

See also: Agriculture, Future of; Ethanol, Sugarcane; Greenhouse Gas Emissions

ETHANOL, SUGARCANE

Ethanol can be produced through fermentation of many natural substances, but sugarcane offers advantages over most others, including corn. According to scientists at the Center for Sugarcane Technology and other Brazilian research institutes, for each unit of energy expended to turn sugarcane into ethanol, 8.3 times as much energy is created—compare that to a maximum of 1.3 times for corn. "There's no reason why we shouldn't be able to improve that ratio to 10 to 1," said Suani Teixeira Coelho, director of the National Center for Biomass at the University of São Paulo. "It's no miracle. Our energy balance is so favorable not just because we have high yields, but also because we don't use any fossil fuels to process the cane, which is not the case with corn" (Rohter 2006). Sugarcane is generally more economical than oil with the per-barrel price under $30.

Use of ethanol in Brazil accelerated after 2003, following introduction of "flex fuel" engines, designed to run on ethanol, gasoline or any mixture of the two. Gasoline sold in Brazil contains about 25 percent alcohol. By 2013, more than half of Brazil's transportation fuel was supplied by sugarcane ethanol. Three-quarters of new cars sold in Brazil were flex-fuel. By 2013, more than 80 percent of the 1.3 million automobiles sold in Brazil had flex-fuel engines, compared to 20 percent in the United States), and most fueling stations offered E85 fuel, compared to 1 percent in the United States.

Brazil, using ethanol from sugarcane, became energy self-sufficient in 2006, even as demand for fuel grew. Brazil's full-court press on ethanol was three decades old by that time. During the 1970s, Brazil's government began developing the ethanol industry by subsidizing the sugarcane industry and requiring its use in government vehicles, By the late 1990s, however, the subsidies were phased out as the cost of producing ethanol dropped to 80 cents a gallon, less than $1.50, the worldwide average for producing gasoline (Luhnow and Samor 2006, A-1, A-8). Many gas stations in Brazil have two sets of pumps, marked "A" for alcohol and "G" for gasoline. "Renewable fuel has been a fantastic solution for us," Brazil's minister of agriculture, Roberto Rodrigues, said. "And it offers a way out of the fossil fuel trap for others as well" (Rohter 2006).

In the past, the material remaining when sugarcane stalks were compressed to squeeze out juice was discarded. Today, Brazilian sugar mills use that residue to generate electricity to process cane into ethanol, and they use other by-products to fertilize the fields where cane is planted. Some mills are now producing so much electricity that they sell their excess to the national grid. Brazil's government taxes ethanol at 9 cents per gallon compared to 42 cents for gasoline, and all gasoline is legally required to contain at least 10 percent ethanol. Researchers in Brazil have decoded the genetics of sugarcane and used the knowledge to breed varieties with even higher sugar content. Brazil has increased the per-acre productivity of sugarcane threefold since 1975 (Luhnow and Samor 2006). Politicians from corn-growing states in the United States have obtained a stiff protective tariff of 45 cents a gallon on Brazilian sugar-derived alcohol. Brazil exported $600 million worth of ethanol in 2005, but nearly none of it went to the United States. Combined with a corn-ethanol subsidy of 45 cents a gallon, corn-ethanol manufacturers during George W. Bush's second term enjoyed almost a dollar advantage over sugarcane ethanol imports (a subsidy that has since been removed).

Sugarcane ethanol may be an energy winner, but local Native people and sugarcane field workers in Brazil have often become victims of the boom. *Birdwatchers*, an Italian film, describes the Brazilian Guarani-Kaiowá Indians' struggle against factory farming biofuels that is crowding them off their land. *Birdwatchers* parses the issues (land invasion, suicides, and rebellion) against the background of a love story involving the daughter of a wealthy landowner and a young Guarani who has become a shaman apprentice in the Brazilian state of Mato Grosso do Sul. The Guarani in Brazil have lost much of their land to sugarcane cultivation.

"Mato Grosso" means "thick forest" in Portuguese. Today, however, most of the trees in the area have been felled. Since the 1950s, the Guarani and neighboring peoples have been evicted from their land by cattle ranchers and sugarcane and soya planters. Many now work on the ranches farms for subsistence wages in perpetual peonage. Over 20 years, more than 500 Native people, some as young as nine years of age, have killed themselves. Others have been shot to death while trying to reoccupy alienated land.

If sugarcane is such a good source of biofuel, then why not use sugar beets for ethanol? The processing equipment for sugar beets and sugarcane is somewhat similar. The problem with sugar beets lies in their harvest cycle in temperate regions. Sugarcane is grown nearly year-round (and thus could supply a processing plant almost all the time), but sugar beets are grown on an annual cycle and harvested in the fall. Thus, according to Kenneth P. Vogel, a professor in the University of Nebraska–Lincoln's agronomy department, a processing plant that costs hundreds of millions of dollars would run only for a few months a year. This hurdle could perhaps be overcome with technology (not yet designed) to switch from sugar beets to other types of ethanol sources. Instead, ethanol interests on the Great Plains are looking at switchgrass (which is handled like hay), corn stover, and other plant biomass. The manufacturing technology for these needs to be developed as well ("No Sugar-Beet" 2007).

Further Reading

Luhnow, David, and Geraldo Samor. "As Brazil Fills Up on Ethanol, It Weans Off Energy Imports." *Wall Street Journal*, January 9, 2006, A1, A8.
"No Sugar-Beet Answer." *Omaha World-Herald*, April 7, 2007, 6B.

Rohter, Larry. "With Big Boost from Sugar Cane, Brazil Is Satisfying Its Fuel Needs." *The New York Times*, April 10, 2006. http://www.nytimes.com/2006/04/10/world/americas/10brazil.html.

See also: Agriculture, Future of; Ethanol, Corn; Greenhouse Gas Emissions

GEOENGINEERING

Geoengineering involves a gallery of grand plans that assume the human race is an addict that will never learn to live without its carbon dioxide fix. From bombing the atmosphere with sulfur (to cool it) to building gigantic space mirrors to deflect sunlight, all of these ideas are expensive, short term, and would do little—at great cost—and not postpone the eventual reckoning.

Benjamin Franklin may have first noticed the idea that sulfur cools the atmosphere. Franklin was on diplomatic assignment for the aborning United States of America in France in 1783 when several Icelandic volcanoes erupted over eight months. Temperatures in France and eastern North America dropped. "There existed a constant fog all over Europe and [a] great part of North America," Franklin wrote (Funk 2014, 266).

A few decades after Franklin made his observations, nature illustrated just how abruptly a quick dose of sulfur could cool Earth. On April 5, 1815, the largest volcanic eruption in recorded history (before or since) exploded from Mount Tambora on the island of Sumbawa, now part of Indonesia. A witness said that the entire mountain "appeared like a body of liquid fire" (Kintisch 2010, 61). The huge eruption ejected a plume of sulfur into the stratosphere worldwide, and 1816 became known as "the year without a summer." Snow fell in Maine in June, and farmers in upstate New York lost their corn crops as the period between frosts shrank from the usual 130 days to just 70. Mohawks at Akwesasne in far northern New York state reported frosts into June.

Sulfur's Effects

Sulfur certainly does cool the atmosphere. Suspended particulates caused by emissions of sulfur dioxide and some other urban air pollutants (aerosols) increase Earth's net albedo, thus usually exerting a cooling influence on planetary temperatures. James E. Hansen has estimated that aerosols cool the climate by approximately one watt per square meter, "which has substantially offset greenhouse warming" (Hansen et al. 1997, 231).

When Mount Pinatubo erupted in the Philippines in 2001, ejecting 10 million tons of sulfur into the atmosphere, it was enough to cool Earth's near-surface atmosphere by 0.5°C or so for a year or two, or roughly the increase in temperature attributable to global warming during the previous century. Had an El Niño not occurred at the same time, the temperature drop may have been on the order of 0.7°C (Crutzen 2006, 212; Morton 2007, 132).

Sulfur's effects are temporary—unlike carbon dioxide, which often remains for several hundred years. Sulfur dioxide washes out within two weeks. Because of its brief residency in the atmosphere, any attempt to cool Earth using it would have to be persistent. The physical requirement to lift that much sulfur into the atmosphere year after year would be quite a challenge, especially as increases in greenhouse gas levels over time require more of it. The sulfur dioxide would have to be refreshed at least twice a month.

Acid rain would not be aggravated by injecting sulfur into the stratosphere. As Eli Kintisch wrote in *Hack the Planet* (2010, 65), "sulfur dioxide is a pollutant that comes out of smokestacks and forms acid rain in the lower atmosphere. . . . [G]as . . . released in the upper atmosphere, even in aggressive doses . . . would add only slightly to the global atmospheric sulfur load." However, a continuous sulfurous haze would inhibit solar power from concentrated solar power facilities, which gather direct sunlight. Solar panels, which use diffuse sunlight, would not be affected.

A Serious Debate

Elizabeth Kolbert writes frequently about climate change in *The New Yorker*. In 2009, she provided a witty send-up of the idea that geoengineers can solve global warming by pumping the stratosphere full of sulfur dioxide. In her review of Steven D. Levitt and Stephen J. Dubner's book *SuperFreakonomics* (2009), she described the book's use of the "parable of horseshit"—the fact that urban planners during the late 19th century feared that the rising volume of horse excrement would inundate large cities such as New York until a technological fix (the automobile) remedied the problem. The authors of *SuperFreakonomics* believe that extending an 18-mile-long hose into the stratosphere and pumping it full of sulfur dioxide would mimic an ongoing volcanic eruption and cool the Earth. The authors ignore climate science completely, not to mention the technological problems of keeping a hose aloft. Kolbert concluded, "All of which goes to show that while some forms of horseshit are no longer a problem, others will always be with us" (Kolbert 2009, 77). This project, fetchingly named "Stratoshield" by Nathan Myhrvold's Intellectual Ventures, proposed to hold the hose above Earth with helium balloons, adding a Jules Verne touch (Klein 2014, 262, 264).

Humor aside, for a period shortly after the turn of the millennium, the idea of counteracting human warming with stratospheric sulfur enjoyed a spasm of serious scientific review. A meeting of top scientists was convened at Harvard University to consider it. The British Royal Society devoted an entire issue of its journal *Philosophical Transactions* to the idea, and the U.S. National Academy of Sciences devoted a workshop to it. The American Meteorological Society called for serious study to "offer strategies of last resort if abrupt, catastrophic, or otherwise unacceptable climate-change impacts become unavoidable" (Kintisch 2010, 12).

Tim Kruger, who manages the Oxford University Geoengineering Programme, defended robust research into the subject because the threat of global warming, if left unaddressed, may require its use, even with problems, as the lesser of evils. Critics, he argues, compare the downsides with "the climate of today, rather than that of a climate-changed future. This is the equivalent of condemning a drug for having side-effects in healthy people before even considering whether the benefits would outweigh any side-effects on the ill. . . . Humanity may yet find itself in the position of having to decide which option is the least worse" (Kruger 2014, 457). Paul Crutzen of the Max Planck Institute for Chemistry in Germany, who earned his bona fides describing the dangers of stratospheric ozone depletion, prepared a paper advancing the sulfur solution to *Climatic Change* that was widely circulated before its publication (2006, 211–219). Crutzen acknowledged that bombing the atmosphere with sulfur should be a strategy of last resort but argued that humankind was not mounting an adequate response to global warming. He said that 5.3 million tons of sulfur per year delivered by airplane (or otherwise) at an annual cost of $50 billion would compensate for the greenhouse gas warming expected during the 21st century.

The idea is not new (Russian scientist Mikhail Budyko proposed it in the 1970s), but this was the first time a Nobel Prize–winning scientist with environmental credentials had made such a serious proposal. Crutzen also has coined the word *Anthropocene* to describe as the geologic epoch in which human beings are the primary force shaping the future of Earth. The term was meant to supersede *Holocene*, the epoch in which most geology textbooks said we live until Crutzen's idea was widely adopted. He was not the first with this idea (the late 19th-century geologist Antonio Stoppani spoke of the "anthropozoic era"), but his nomenclature has coincided with humankind's obvious control of Earth's climatic future via global warming—as well as the planetwide scope of proposed solutions such as impregnating the atmosphere with sulfur.

A Tone of Paradox and Desperation

Among advocates of the sulfur solution there exists a paradoxical mixture of fear that humanity is not up to the task of dealing with global warming. There is also a large dose of techno-fixit hubris that assumes that squirting prodigious amounts of sulfur into the atmosphere will fix the problem. A fascinating array of people have stepped into this debate. Bill Gates, for example, has funded geoengineering research "with his near-mystical quest for energy miracles' [tapping] into what may be our culture's most intoxicating narrative: the belief that technology is going to save us from the effects of our actions . . . our . . . most powerful form of magical thinking" (Klein 2014, 255). In *This Changes Everything: Capitalism and the Climate* (2014, 268), writer Naomi Klein called Gates geoengineering's "sugar daddy." Lowell Wood, a coinventor of the sky-hose idea, also advances the idea of terraforming Mars, with its thin atmosphere that is 95 percent carbon dioxide. There exists, according to Wood, "a 50/50 chance that young children now alive will walk on Martian meadows . . . [and] will swim in Martian lakes" (Klein 2014, 288).

A tone of desperation is palpable in climate-change science when well-known people seriously propose that filling the stratosphere with sulfur dioxide may be the only way to stop runaway greenhouse warming. Do we really want to pump the stratosphere full of sulfur to shroud the surface from warmth and then live in a perpetual acid mist?

When Crutzen advanced the sulfur shield idea in 2006, he cited a "grossly dis-appointing international political response" to increasing evidence of global warm-ing (Kerr 2006, 401). Atmospheric scientist Ken Caldeira said that countries need to "undertake studies on what we might do" in a climate crisis, given the current trajectory of carbon concentrations in the atmosphere. "Nobody likes the idea of engineering Earth's climate. . . . Unfortunately, at some point, our other options may be even more unpleasant," said Caldeira (Eilperin 2010).

David Keith's Advocacy of Geoengineering

Canadian David Keith became a professor at Harvard University and recipient (with others) of $6 million in grant money from the Bill Gates Foundation. Since the 1990s, he has become perhaps the world's most prominent single advocate of geo-engineering (Keith 2013). Keith, who has an affinity for huge-scale, expensive techno-fixes, also promoted carbon capture and sequestration in a report with four Canadian energy executives. By 2014, Keith was sponsoring a "field experiment aimed at understanding another geoengineering technology that would use a bal-loon to release sun-blocking particles of sulfuric acid in the stratosphere" (Kintisch 2013, 307). Keith's support for the "sulfur solution" is unequivocal: "It is possible to cool the planet by injecting reflective particles of sulfuric acid into the upper atmosphere. . . . To say that it's 'possible' understates the case: it is cheap and tech-nically easy . . . for the price of a Hollywood blockbuster" (Keith 2013, ix). Keith writes with occasional humor and nuance and factors critics of his ideas into the discourse. For example, he proposes using "solar radiation management" to counter half of global warming's effects to mitigate effects on precipitation distribution, most notably the Asian monsoon (Keith 2013, 13–14).

Keith argues that the idea must be tested. He presents himself as a friend of envi-ronmentalists who also subject their assertions to intellectual rigor. He finds James Hansen to be something of an extremist, a "climate scientist turned activist" (Keith 2013, 23). Keith realizes global warming is serious business, however, even as he ridicules forecasts of "imminent doom" (2013, 23). Instead, Keith believes that alarmism incites "disaster fatigue," a response that numbs people and prevents action (2013, 170).

Distortion of rainfall patterns from sulfur dousing of the atmosphere is a real threat, as Mike Hulme commented in *Can Science Fix Climate Change?* (2014, 51):

And these are not isolated results. Nearly all the modeling studies that have simulated the regional effects of stratospheric aerosol injection show that the existing mosaic of regional and local climates ends up being reconfigured. Sta-bilizing global temperature to avoid the danger zone of more than two degrees [Celsius] of warming only ends up *destabilising* regional climates around the world.

Even so, Hulme describes a "slippery slope" down which humankind may slide that combines "technological and sociological 'lock-in'" with a bias toward innovation and "the vested interests of fortune, fear, fame, and fanaticism pushing the technology onwards" and compelled by intensifying climate catastrophe (which he anticipates in acute detail) until "full-scale deployment [of stratospheric aerosol injection] eventually becomes unstoppable" (Hulme 2014, 69, 76–81).

More Problems Arising

An increase in atmospheric particulate matter does not always exert a cooling influence, according to some reports. Researchers working with the National Oceanic and Atmospheric Administration (NOAA) have assembled data indicating that periodic increases in atmospheric dust concentrations during glacial periods of the last 100,000 years may have resulted in significant regional warming, and that this warming may have triggered some of the abrupt climatic changes observed in paleoclimate records.

Jonathan T. Overpeck, working with the Paleoclimatology Program at NOAA's National Geophysical Data Center in Boulder, Colorado, led a team of scientists who conducted global climate model simulations in 1996 to examine the potential role of tropospheric dust in glacial climates. Comparing "modern dust" with "glacial dust" conditions, they found patterns of regional warming that increased at progressively higher latitudes. The warming was greatest (up to 4.4°C) in regions with dust over areas that were covered with snow and ice ("Abrupt Climate" 1996). Under some circumstances, "aerosols can reduce cloud cover and thus significantly offset aerosol-induced radiative cooling at the top of the atmosphere on a regional scale" (Ackerman et al. 2000, 1042). Simply put, soot and other such pollutants may not always mitigate other warming influences on climate as much as many proponents assert.

Filling the upper atmosphere with sulfur also may deplete stratospheric ozone and reduce overall precipitation, most notably during the African and Indian monsoons, which are crucial for hundreds of millions of subsistence farmers (Robock 2008, 1166). Globally, one-third of Earth's people depend on Asian and African monsoons rains for their staple foods. Reducing the intensity of sunlight also could reduce evaporation and rainfall.

Some scientists object to such a scheme on grounds that it would remove pressure to deal with the problem at its source—that is, to use energy sources other than fossil fuels. "I refuse to go down that road," said biochemist Meinrat Andreae of the Max Planck Institute for Chemistry in Mainz, Germany. "You're papering over the problem so people can keep inflicting damage on the climate system" (Kerr 2006, 403). "The biggest risk of geoengineering is that it eliminates pressure to decrease greenhouse gases," said Ken Caldeira (Kerr 2006).

Further Reading

"Abrupt Climate Change during Last Glacial Period Could Be Tied to Dust-Induced Global Warming." Press release NOAA 96-78, December 4, 1996. http://www.publicaffairs.noaa.gov/pr96/dec96/noaa96-78.html.

Ackerman, A. S., et al. "Reduction of Tropical Cloudiness by Soot." *Science* 288 (May 12, 2000): 1042–1047.

Crutzen, Paul. "Albedo Enhancement by Stratospheric Sulfur Injections: A Contribution to Resolving Policy Dilemma?" *Climatic Change* 77 (2006): 211–220.

Eilperin, Juliet. "Geo-Engineering Sparks International Ban, First-Ever Congressional Report." *Washington Post,* October 29, 2010. http://www.washingtonpost.com/wp-dyn/content/article/2010/10/29/AR2010102906365_pf.html.

Funk, McKenzie. *Windfall: The Booming Business of Global Warming.* New York: Penguin Press, 2014.

Hansen, James E., Makiko Sato, and R. Ruedy. "The Missing Climate Forcing." *Philosophical Transactions of the Royal Society of London B* 352 (1997): 231–240.

Hulme, Mike. *Can Science Fix Climate Change? A Case against Climate Engineering.* Cambridge, UK: Polity Press, 2014.

Keith, David. *A Case for Climate Engineering.* Cambridge, MA: MIT/Boston Review, 2013.

Kerr, Richard A. "Pollute the Planet for Climate's Sake?" *Science* 314 (October 20, 2006): 401–403.

Kintisch, Eli. *Hack the Planet: Science's Best Hope—or Worst Nightmare—for Averting Climate Catastrophe.* New York: Wiley, 2010.

Kintisch, Eli. "Dr. Cool." *Science* 342 (October 18, 2013): 307–309.

Klein, Naomi. *This Changes Everything: Capitalism and the Climate.* New York: Simon & Schuster, 2014.

Kolbert, Elizabeth. "Hosed: Is There a Quick Fix for the Climate?" *The New Yorker,* November 16, 2009, 75–77.

Kruger, Tim. "Stratospheric Folly." *Nature* 508 (April 24, 2014): 457.

Levitt, Steven D., and Stephen J. Dubner. *SuperFreakonomics: Global Cooling, Patriotic Prostitutes, and Why Suicide Bombers Should Buy Life Insurance.* New York: William Morrow, 2009.

Morton, Oliver. "Is This What It Takes to Save the World?" *Nature* 447 (2007): 132–136.

Robock, Alan. "Whither Geo-Engineering?" *Science* 320 (May 30, 2008): 1166–1167.

See also: Atmospheric Circulation; Temperatures, Global; Temperatures, Greenhouse Gas Levels and

GLOBAL WARMING ECONOMICS

More than 1 million homes and places of businesses along the East, West, and Gulf Coasts of the United States may become subject to serial flooding because of rising seas provoked by melting ice and thermal expansion of seawater. Inland, farming could become unsustainable in areas of the U.S. Midwest and South, where heat and humidity may restrict outdoor activity. Agricultural production will shift northward, with North Dakota and Minnesota becoming more important. And this is only one country, 5 percent of Earth's land area, providing examples of costs incurred by one species (humans) in a rapidly warming world. Oceans, two-thirds of the planet's surface, could become too warm and too acidic to sustain many forms of life.

Risky Business: The Risks of Inaction

When the costs of challenging global warming are discussed, many people do not summarize the many debits of failing to act. Such costs were examined in a report titled *Risky Business* issued during 2014 by senior officials in the United States, including three former U.S. treasury department secretaries (George P. Schultz, secretary under Richard Nixon and secretary of state under Ronald Reagan; Henry M. Paulson Jr., Republican secretary for George W. Bush; and Robert E. Rubin, a Democrat, who held the same post in the Clinton administration). The report was funded mainly by three wealthy businesspeople: Paulson; Thomas F. Steyer, a retired hedge-fund executive and Democratic Party activist; and Michael R. Bloomberg, the former nominally Republican mayor of New York City.

Schultz said that he had become concerned enough about climate change to have solar panels installed on the roof of his home in California and to buy an electric car. "I say we should take out an insurance policy," he said (Gillis 2014). "I actually do believe that we're at a tipping point with the planet," Paulson said. "A lot of things are going to happen that none of us are going to like to see" (Gillis 2014). Rubin said that the U.S. Securities and Exchange Commission should require publicly held corporations to disclose climate-related risks in their businesses and disclose them to investors. "Here we have this existential threat that I really do think has the possibility of being catastrophic, and I don't think people have any sense of that," said Rubin (Gillis 2014).

The report said that the 40 percent of U.S. population who live near the coasts will face some of the greatest risks. Rising seas churned by more violent storms pushing surges "could swallow more than $370 billion worth of property in Florida and Louisiana alone by the end of the century" (Gillis 2014). The report's models say that sea levels may rise one foot on average by 2050 or so—and two to three feet by century's end, although a more rapid rise (six to eight feet) is possible shortly after that. The risk is compounded by the fact that land is subsiding along much of the East and Gulf Coasts as sea level rises.

If global warming is not addressed, the world economy's production could fall significantly, perhaps by as much as $20 trillion by the end of the 21st century, according to figures compiled by some economists. Such a figure would represent 6 percent to 8 percent of global economic output in 2100. These figures do not include the costs of lost biodiversity or unpredictable events such as extreme weather.

"The climate system has enormous momentum, as does the economic system that emits so much carbon dioxide," said coauthor Frank Ackerman, an economist with the Global Development and the Environment Institute at Tufts University. "We have to start turning off greenhouse gas emissions now in order to avoid catastrophe in decades to come." Limiting temperature rise to 2°C could avoid $12 trillion in damage annually and cost a quarter of that amount, according to this study. The report "Climate Change: the Costs of Inaction" was compiled by economists at the Global Development and Environment Institute at Tufts University ("Failure" 2006).

"An Economic Catastrophe"

Costs of neglecting global warming would include many costly environmental impacts such as decreased crop yields in the developing world, increasingly intense droughts and water shortages, nearly complete loss of coral reefs, the spread of tropical diseases such as malaria, and the extinction of several Arctic species, including polar bears.

"This report demonstrates that climate change will not only be an environmental and social disaster [but also] an economic catastrophe, especially if global temperatures are allowed to increase by more than two degrees centigrade," said Elizabeth Blast of Friends of the Earth, which commissioned the report ("Failure" 2006). A 3°C temperature rise would compound damage to agriculture around the world, according to the report, including the collapse of the Amazon River valley's ecosystem and the loss of boreal and alpine ecosystems.

"Setting a real price on carbon emissions is the single most important policy step to take," said Robert N. Stavins, director of the environmental economics program at Harvard University, "Pricing is the way you get both the short-term gains through efficiency and the longer-term gains from investments in research and switching to cleaner fuels" (Lohr 2006).

How much money will be required to avert catastrophic climate change? A consensus estimate averages around 1 percent of world economic activity annually for 50 years. As with all forecasts so far into the future, these can only be broad estimates. In 2006, 1 percent of the United States economic activity was more than $120 billion a year, or $400 per person, roughly the money being spent at that time on the Iraq and Afghanistan wars (Lohr 2006). "There's no easy way around the fact that if global warming is a serious risk, there will be serious costs," said W. David Montgomery, an economist at Charles River Associates (Lohr 2006).

The Stern Review

The Stern Review on the Economics of Climate Change, which has been called "the most comprehensive review ever carried out on the economics of climate change," warns that "global warming could inflict worldwide disruption as great as that caused by the two World Wars and the Great Depression" ("Failure" 2006). *The Stern Review*, named after its principal author, Sir Nicolas Stern, head of the British Government Economic Service and a former chief economist at the World Bank, estimates that ignoring warming's impacts could cost $9 trillion U.S. "Hundreds of millions of people could suffer hunger, water shortages, and coastal flooding as the world warms," the report said ("Failure" 2006).

"The task is urgent," Stern warned. "Delaying action, even by a decade or two, will take us into dangerous territory. We must not let this window of opportunity close" ("Failure" 2006). Stern's report calculates that "tackling climate change would cost 20 times less than doing nothing." The report estimates the cost of business as usual (or doing nothing) at 5 percent to 20 percent of global gross domestic product (GDP) by 2100 (Giles 2006, 6). Although the coal and oil industries have concentrated on the cost of change, the *Stern Review* says that *not* confronting the problem will be much worse for the world economy in the long run. Its own calculations peg the cost of stabilizing greenhouse gas levels at twice preindustrial levels would cost only 1 percent of global GDP. Thus, the report supports restrictions on carbon emissions and development of alternative-energy sources.

"The Greatest Market Failure"

The second half of the *Stern Review* describes the policy challenges involved in moving to a low-carbon economy worldwide. This study indicates that lack of action to counter climate change is "the greatest market failure the world has seen" ("Failure" 2006). Stern believes that three elements of policy are required for an effective response. "First, we must establish a carbon price via tax, trade, and regulation. Without this price there is no incentive to decarbonize," he said ("Failure" 2006).

Stern continued:

Second, we must promote technology: through research and development. Further, private sector investors need confidence that there will be markets for their products: that is why deployment policy also makes sense. . . . And third we must deal with market failure—for example problems in property and capital markets inhibit investments for energy efficiency. Further, the sticks and carrots of incentives, rightly emphasized by we economists, need to be supported by information. And still further, greater understanding of the issues can itself change the behavior of individuals and firms. ("Failure" 2006).

Climate-change contrarians complain that dealing with global warming will draw money away from other economic activities, but the *Stern Review* anticipates that markets for low-carbon technologies will be worth at least $500 billion, and perhaps much more, by 2050. *Stern* sees tackling climate change as a progrowth strategy, believing that ignoring it will ultimately undermine economic growth ("Failure" 2006).

Lack of action is not an option, according to *Stern*. "The consequences are stark, for our planet and for the people who live on it, threatening the basic elements of life—access to water, food production, health and our environment" ("Failure" 2006).

Blair said that a temperature rise of 2 to 3°C would portend the following:

Disappearing glaciers that will significantly reduce water supply to more than 1 billion people.
Rising sea levels could lead to 200 million people being displaced.
Declining crops yields will lead to famine and death, particularly in Africa.
Diseases such as malaria will become more widespread.
As much as 40 percent of Earth's animal species may become extinct ("Failure" 2006).

Right-wing critics attacked the *Stern Review's* models, especially its assumption that world population will reach 15 billion, a figure that critics regarded as unlikely. Richard Tol, an economist at Princeton University, asserted that *Stern's* use of existing data was unduly pessimistic. Tol said that *Stern* had underplayed efforts at adaptation (Giles 2006, 6). Roger Pielke Jr., a climate change policy expert at the University of Colorado, said that *Stern* had "cherry-picked" the literature (Giles 2006, 6).

Further Reading

"Failure to Manage Global Warming Would Cripple World Economy." Environment News Service, October 13 and October 30, 2006. http://www.ens-newswire.com/ens/oct2006/2006-10-30-06.asp (no longer available).

Giles, Jim. "How Much Will It Cost to Save the World?" *Nature* 444 (November 2, 2006): 6–7.

Gillis, Justin. "Bipartisan Report Tallies High Toll on Economy from Global Warming." *The New York Times*, June 24, 2014. http://www.nytimes.com/2014/06/24/science/report-tallies-toll-on-economy-from-global-warming.html.

Lohr, Steve. "Global Warming Is a Challenge for Economic Policy, Too - Business - International Herald Tribune" *New York Times*, November 12, 2006. http://www.nytimes.com/2006/12/12/business/worldbusiness/12iht-climate.3877222.html.

See also: Biodiversity; Carbon Cycle Feedbacks; Extinctions; Extreme Weather; Greenhouse Gases, Effects of; Land Use

IRON FERTILIZATION, OCEAN

Nearly half of Earth's photosynthesis is performed by phytoplankton in the world's seas and oceans. This fact has led to proposals to geoengineer a carbon dioxide sink using the same "biological pump" that is believed to have driven at least some of Earth's past climate cycles. The chemistry of the oceans varies widely with regard to the amount of iron necessary to prime this pump and substantially increase carbon dioxide sequestration. In the equatorial Pacific and Southern Oceans, wrote Sallie W. Chisholm, a marine biologist at the Massachusetts Institute of Technology, it "is possible to stimulate the productivity of hundreds of square kilometers of ocean with a few barrels of fertilizer" (Chisholm 2000, 686).

Seed the Oceans with Iron?

Should the oceans be seeded with large amounts of iron ore that will stimulate the growth of phytoplankton that consume carbon dioxide? The idea has attracted support among corporations and foundations looking for ways to minimize the effects of carbon dioxide without changing the world's basic energy generation mix. The idea is simple on its face: iron stimulates the growth of phytoplanktonic algae that are believed to be responsible for approximately half the world's biological absorption of carbon dioxide. By 2009, ocean fertilization had been largely rejected as a solution, in large part because to succeed it would disrupt large-scale ocean ecosystems in unpredictable ways while holding little promise of significant reductions in atmospheric carbon dioxide (Strong et al. 2009, 347).

Ulf Riebesell, a marine biologist at the Alfred Wegener Institute for Polar and Marine Research in Bremerhaven, Germany, believes that ambitious iron seeding of the oceans could remove 3 billion to 5 billion tons of carbon dioxide per year, or some 10 percent to 20 percent of human-generated emissions at a factor of 10 times less expensive than planting forests to achieve the same end (Schiermeier 2003, 110). Patents have been issued for ocean fertilization, and demonstration projects have been undertaken. One such project has been described and evaluated in *Nature* (Watson et al. 2000).

In an experiment conducted between Tasmania and Antarctica, researchers confirmed that vast stretches of the world's southern oceans are primed to explode with photosynthesis but lack only iron. The researchers, who described their work in *Nature*, said it is too soon to start large-scale iron seeding because the new experiment raised as many questions as it answered. At best, they said, iron seeding would absorb only a small amount of the carbon dioxide in the atmosphere. They also said that their experimental bloom of plankton was not tracked long enough to determine whether the carbon harvested from the air sank into the deep sea or was again released into the environment as carbon dioxide gas. Atsushi Tsuda and colleagues have studied iron fertilization and found that, under some circumstances, iron fertilization can dramatically increase phytoplankton mass (Tsuda et al. 2003).

"There are still fundamental scientific questions that need to be addressed before anyone can responsibly promote iron fertilization as a climate control tactic," said Kenneth H. Coale, an oceanographer who has helped design studies of iron's effects in the tropical Pacific (Revkin 2000). This seemingly simple proposition has some potential problems, however. First, there exists no way to measure the amount of carbon taken up by phytoplankton. In addition, the algae produce dimethylsulphide, which plays a role in cloud formation.

Problems with Iron Fertilization

Phytoplankton also increase the amount of sunlight and heat energy absorbed by ocean water. They produce compounds such as methyl halides, which play a role in stratospheric ozone depletion. The iron also could promote the growth of toxic algae that may kill other marine life and change the chemistry of ocean water by removing oxygen. "The oceans are a tightly linked system, one part of which cannot be changed without resonating through the whole system," wrote Sallie W. Chisholm. "There is no free lunch" (Schiermeier 2003, 110; Chisholm, 2000, 685).

So much iron may be required to produce the desired effect that fertilization of this type will never be practical or even possible. "The experiments enabled us to make an initial determination about the amount of iron that would be required and the size of the area to be fertilized," said Ken O. Buesseler of the Woods Hole Oceanographic Institution, who coauthored a study of the idea. "Based on the studies to date, the amount of iron needed and [the] area of ocean that would be impacted is too large to support the commercial application of iron to the ocean as a solution to our greenhouse gas problem," he explained ("Iron Link" 2003). "It may not be an inexpensive or practical option if what we have seen to date is true in further experiments on larger scales over longer time spans," Buesseler said. "The oceans are already naturally taking up human-produced carbon dioxide, so the changes to the system are already underway," he said. "We need to first ask will it work and then what are the environmental consequences?" ("Iron Link" 2003).

As one study, "To assess whether iron fertilization has potential as an effective sequestration strategy, we need to measure the ratio of iron added to the amount of carbon sequestered in the form of particulate organic carbon to the deep ocean in field studies" (Buesseler and Boyd 2003, 67). To date, wrote Buesseler and Boyd, experiments of this type have "produced notable increases in biomass and associated decreases in dissolved inorganic carbon and macronutrients. However, evidence of sinking carbon particle carrying [particulate organic carbon] to the deep ocean was limited" (Buesseler and Boyd 2003, 67). Small-scale experiments with iron fertilization indicate that it can do more than produce oxygen—it also can increase production of nitrous oxide, a greenhouse gas that is more potent than carbon dioxide.

Given the limits of current technology, this study estimates that an area a magnitude larger than the Southern Ocean (waters southerly of 50° south) would have to be fertilized to remove 30 percent of the carbon that human activity currently injects into the atmosphere. Thus, according to this study, "ocean iron fertilization may not be a cheap and attractive option if impacts on carbon export and sequestration are as low as observed to date" (Buesseler and Boyd 2003, 68).

Trial Runs for Iron Fertilization

Despite prospective problems, iron fertilization is considered potentially viable in some quarters. Scientists who fed tons of iron into the Southern Ocean reported evidence during 2004 that stimulating the growth of phytoplankton in this way may strengthens the oceans' viability as a carbon sink. Writing in *Science*, ocean biologists and chemists from more than 20 research centers said they triggered two huge blooms of phytoplankton that turned the ocean green for weeks and consumed hundreds and perhaps thousands of tons of carbon dioxide.

"These findings would be encouraging to those considering iron fertilization as a global geoengineering strategy," said Ken Coale, a chief scientist at the Moss Landing Marine Laboratories north of Monterey, California. "But the scientists involved in this experiment realize that this looked only skin deep at the functioning of ocean ecosystems and much more needs to be understood before we recommend such a strategy on a global scale" (Hoffman 2004). "From my work, I don't think this could solve a significant fraction of our greenhouse gas problem while causing unknown ecological consequences," said Buesseler (Hoffman 2004).

A scientific team led by Stéphane Blain reported observations of a phytoplankton bloom induced by natural iron fertilization,

an approach that offers the opportunity to overcome some of the limitations of short-term experiments. We found that a large phytoplankton bloom over the Kerguelen plateau in the Southern Ocean was sustained by the supply of iron and major nutrients to surface waters from iron-rich deep water below. The efficiency of fertilization, defined as the ratio of the carbon export to the amount of iron supplied, was at least ten times higher than previous estimates from short-term blooms induced by iron-addition experiments. This result sheds new light on the effect of long-term fertilization by iron and macronutrients on carbon sequestration, suggesting that changes in iron supply from below—as invoked in some palaeoclimatic and future climate change scenarios—may have a more significant effect on atmospheric carbon dioxide concentrations than previously thought. (Blain 2007, 1070)

Iron Fertization Banned

Iron fertilization faces another important problem—the ocean food chain. In 2008 and early 2009, tiny shrimplike crustaceans ate 159 square miles (about 300 square kilometers) of algae that were part of an iron-fertilization experiment in the South Atlantic, according to the Alfred Wegener Institute of Germany. *National Geographic News* reported that:

> Working aboard the German research vessel *Polarstern,* German and Indian scientists in recent weeks mixed ten tons of ferrous sulfate with seawater. The team then pumped the artificially enhanced water back into the Atlantic outside Argentina's coastal waters. As expected, the experiment created a massive, CO_2-eating algae bloom. But it was the wrong algae. The blooms were mostly tiny haptophytes, not the larger diatom algae the team had expected. The smaller algae variety is typically found only in coastal waters, and it's a favorite food of tiny shrimplike crustaceans called copepods. The copepods wolfed down the algae shortly after the new South Atlantic bloom appeared. (Hearn 2009)

Meeting in Berlin in late May 2008, the U.N. Convention on Biological Diversity issued a moratorium signed by representatives of 191 nations on large-scale commercial projects to fertilize the oceans in pursuit of climate-change mitigation. The ban will hold until scientists better understand the effects of such mitigation on the ocean food chain. Iron fertilization, for example, could increase ocean acidity and reduce oxygen levels.

Delegates from the 85 nations that make up the London Convention Treaty (created in 1972 by the U.N. International Maritime Organization) on October 31, 2008, placed limits on iron fertilization, limiting it to "legitimate scientific research," a move to inhibit premature commercial application that exploits awards of carbon credits under diplomatic agreements such as the Kyoto Protocol. The body is charged with regulating pollution in international waters through compacts such as the London Protocol. The group acted after announcements of several plans to commercialize iron fertilization. Any research must consider carbon flux (such as emissions of nitrous oxides), impacts on the food web and oxygen levels, and the possibility that toxic species may grow (Kintisch 2008).

Further Reading

Blain, Stéphane, et al. "Effect of Natural Iron Fertilization on Carbon Sequestration in the Southern Ocean." *Nature* 446 (April 26, 2007): 1070–1074.

Buesseler, Ken O., and Philip W. Boyd. "Will Ocean Fertilization Work?" *Science* 300 (April 4, 2003): 67–68.

Chisholm, Sallie W. "Stirring Times in the Southern Ocean." *Nature* 407 (October 12, 2000): 685–686.

Hearn, Kelly. "Huge Man-Made Algae Swarm Devoured—Bad for Climate?" *National Geographic News*, March 27, 2009. http://news.nationalgeographic.com/news/2009/03/090327-iron-seeding.html.

Hoffman, Ian. "Iron Curtain over Global Warming; Ocean Experiment Suggests Phytoplankton May Cool Climate." *Daily Review* (Hayward, CA), April 17, 2004, n.p. (LEXIS).

"Iron Link to CO_2 Reductions Weakened." Environment News Service, April 10, 2003. http://ens-news.com/ens/apr2003/2003-04-10-09.asp#anchor8.

Kintisch, Eli. "Rules for Ocean Fertilization Could Repel Companies." *Science* 322 (November 7, 2008): 835.

Revkin, Andrew C. "Antarctic Test Raises Hope on a Global-Warming Gas." *The New York Times*, October 12, 2000, A18.

Schiermeier, Quirin. "The Oresmen." *Nature* 421 (January 9, 2003): 109–110.

Strong, Aaron, et al. "Ocean Fertilization: Time to Move On." *Nature* 461 (September 17, 2009): 347–348.

Tsuda, Atsushi, et al. "A Mesoscale Iron Enrichment in the Western Subarctic Pacific Induces a Large Centric Diatom Bloom." *Science* 300 (May 9, 2003): 958–961.

Watson, A. J., et al. "Effect of Iron Supply on Southern Ocean CO_2 Uptake and Implications for Glacial Atmospheric CO_2." *Nature* 407 (October 12, 2000): 730–733.

See also: Coral Reefs; Fisheries; Monsoons; Oceans' Absorption of Heat

SOLAR POWER

Solar power has advanced significantly since the days of inefficient photovoltaics. In California, solar power is being built into roof tiles, and nanotechnology may someday make any surface touched by the sun a source of power—windows, for example. Experiments have been undertaken with concentrating solar power (CSP), a mirror in the shape of a parabola to focus light onto a black pipe with a heat-transfer fluid inside. The fluid is used to boil water into steam, which turns a generator. Another form of solar thermal technology involves a "power tower," a tall structure flanked by thousands of mirrors, each of which pivots to focus light on the tower to heat fluid. That design can work even in places with weaker sunlight than a desert (Wald 2008). Many private homes feed solar power into the electrical grid, their meters running backward, paying householders for contributed power.

The cost of solar power has been declining sharply. Economies of scale, as well as improvements in efficiency and less expensive construction materials, brought solar energy down to a cost that competes with fossil fuel generation under some conditions by 2015 (Service 2008, 718). In 2014, 50 gigawatts of new solar energy was installed worldwide, equaling all of the solar energy on the planet just 10 years previously.

Solar: "A Postcard from the Future"

"We are in the middle of a low-carbon-energy revolution," Gernot Wagner and colleagues wrote in *Nature* in 2015. The cost of crystalline silicon photovoltaic (PV) modules fell by 99 percent between 1978 and 2014. Eighty percent of that decline occurred between 2008 and 2014 (Wagner et al. 2015). China's climate, energy, and industrial policies have boosted the manufacturing scale of renewable technologies, expanding solar PV production more than 100-fold between 1997 and 2005. By 2015, in a few areas of Germany, solar energy met more than 50 percent of electricity demand on a sunny Sunday afternoon, under optimal, low-demand conditions.

By 2015, with residential solar power capacity growing at more than 30 percent a year in some places (including Hawaii, California, Arizona, and Japan), so much power was being fed back into the grid that it was wreaking havoc with the budgets of some utility companies because it was cutting into their revenues. In Hawaii, for example, some 12 percent of private homes by 2015 were producing enough solar power to feed power back into the grid and receive payment for it.

"Hawaii is a postcard from the future," said Adam Browning, executive director of Vote Solar, a California-based advocacy organization. Some utilities have been adding fees for use of solar on their grids. "Hawaii's case is not isolated," said Massoud Amin, University of Minnesota electrical and computer engineering professor. "When we push year-on-year 30 to 40 percent growth in this market, with the number of installations doubling, quickly—every two years or so—there's going to be problems" (Cardwell 2015).

A 2014 study from investment banking firm Lazard said that utility-scale solar was selling as low as 5.6 cents per kilowatt-hour, a rate comparable to coal and natural gas. "It is really quite notable, when compared to where we were just five years ago, to see the decline in the cost of these technologies," said Jonathan Mir, managing director at Lazard. Wind and solar often were used because of their intermittent nature in combination with other fuels. By 2014, the cost of solar power had dropped 70 percent in the U.S. Southwest (Cardwell 2014).

Further Reading

Cardwell, Diane. "Solar and Wind Energy Start to Win on Price vs. Conventional Fuels." *The New York Times*, November 23, 2014. http://www.nytimes.com /2014/11/24/business/energy-environment/solar-and-wind-energy-start-to-win-on-price-vs-conventional-fuels.html.

Cardwell, Diane. "Solar Power Battle Puts Hawaii at Forefront of Worldwide Changes." *The New York Times*, April 18, 2015. http://www.nytimes.com/2015/04/19/business/energy-environment/solar-power-battle-puts-hawaii-at-forefront-of-worldwide-changes.html.

Wagner, Gernot, et al. "Energy Policy: Push Renewables to Spur Carbon Pricing." *Nature* 525 (September 3, 2015): 27–29. http://www.nature.com/news/energy-policy-push-renewables-to-spur-carbon-pricing-1.18260.

In the meantime, a study from the investment banking firm Lazard said that by 2014 utility-scale solar sold for as low as 5.6 cents a kilowatt-hour, a price comparable to coal and natural gas. "It is really quite notable, when compared to where we were just five years ago, to see the decline in the cost of these technologies," said Jonathan Mir, managing director at Lazard. Wind and solar often were used because of their intermittent nature in combination with other fuels. By 2014, the cost of solar power had dropped 70 percent in the U.S. Southwest (Cardwell 2014).

Breakthroughs in solar power have been sought since the days of Thomas Edison. In a conversation with Henry Ford and tire tycoon Harvey Firestone in 1931, shortly before Edison died, he said, "I'd put my money on the sun and solar energy. What a source of power! I hope we don't have to wait until oil and coal run out before we tackle that" (Revkin 2006).

More is on the way. Combining existing fossil fuel power plants with solar technology may increase electricity production while reducing emissions. Vishwanath Haily Dalvi, working in Mumbai's Institute of Chemical Technology, has been examining solar thermal technology that converts solar input to heat. That heat then drives turbines in plants built to run on coal, natural gas, or oil. This hybrid technology may reduce fossil fuel emissions for an equivalent amount of energy by as much as half and at minimal cost ("Sun's Heat" 2015).

Electrifying India with Household Solar Power

Solar systems financed by small loans may eventually help millions of people in India acquire electric power for the first time and skip the fossil fuel age entirely. By 2015, several companies in India were financing home systems in Indian villages that cost almost $200 (more than 13,000 Indian rupees), but doing it in monthly payments of $3.50 to $5.00 a month that equaled what families who support themselves on $2 to $3 a day had been paying for kerosene. One company, Selco (Solar Electric Light Company) has announced that such a business model will allow access to solar-powered electricity to 300 million people in India (and 1.2 billion in the world as a whole) who currently have no power, skipping the fossil fuel age entirely.

India's Prime Minister Narendra Modi has promised total electrification in India by the end of 2022, mainly by adding coal-fired generation even when smog has reached the worst levels on Earth in the Delhi area. Solar, which provided only 1 percent of India's power in 2015, may reduce reliance on coal but will not replace it.

As *The New York Times* reported early in 2016,

> Without financing, decentralized renewable energy could never compete in India with kerosene, which is cheap because the government subsidizes its sale at a cost of more than $5 billion a year. Use of kerosene contributes to carbon emissions, but also to more personal and immediate hazards like skin irritation, respiratory problems and a significant fire risk. Ultimately, it provides only dim, flickering lighting. (Bearak 2016)

Further Reading

Bearak, Max. "Electrifying India, with the Sun and Small Loans." *The New York Times,* January 2, 2015. http://www.nytimes.com/2016/01/03/business/energy-environment/electrifying-india-with-the-sun-and-small-loans.html.

As Robert Service reported in *Science* in 2014,

> Researchers from three research groups reported . . . that they beat the previous efficiency record for converting the energy in sunlight to electricity in solar cells called perovskites. Though perovskite solar cells are only five years old, their efficiency has skyrocketed from 3.8 percent to 19.3 percent. Ultimately, engineers may be able to layer perovskite solar cells atop conventional photovoltaics made from crystalline silicon, creating cells that are up to 32 percent efficient. That could make solar electricity as cheap as power produced from fossil fuels. (Service 2014)

The number of jobs in the U.S. solar-power industry rose to 335,000 in 2017, compared with 94,000 in 2010. In 2014, the United States installed as much solar-power capacity in three weeks as it did in all of 2008 (Hughes 2016).

Organolead halide perovskite solar cells were first developed in 2009. By 2015, they had taken the lead among emergent photovoltaic technologies, demonstrating power-conversion efficiencies of as much as 20 percent. The perovskite precursor compounds are abundant and inexpensive and can easily be converted into thin films, anticipating a time when solar energy will be harvested from nearly any surface on which the sun shines. So, in principle, perovskite photovoltaics can generate electricity at an extremely low cost (Sessolo and Bolink 2015, 917).

Solar Power on Federal Land

In 2010, the U.S. Department of Interior began approving large solar-power arrays on government land, some of which could service as many as one-half million homes. Two plants were approved in the California deserts. The larger plant was built by Tessera Solar on 6,360 acres in California's Imperial Valley. Its "suncatchers"—"reflectors in the shape of radar dishes—concentrate solar energy and activate a four-cylinder engine to generate" 709 megawatts (Barringer 2010). A similar, smaller system generates 45 megawatts on 422 acres in the Lucerne Valley. Both generate energy for 566,000 homes.

"Solar power has captured the public imagination," wrote Andrew C. Revkin and Matthew L. Wald in *The New York Times*. "Panels that convert sunlight to electricity are winning supporters around the world—from Europe, where gleaming arrays cloak skyscrapers and farmers' fields, to Wall Street, where stock offerings for panel makers have had a great ride, to California, where Gov. Arnold Schwarzenegger's 'Million Solar Roofs' initiative is promoted as building a homegrown industry and fighting global warming" (Revkin and Wald 2007). For all the excitement, however, solar power in 2006 contributed only 0.01 percent of the United States' electricity supply. "Most of the environmental stuff out there now is toys compared to the scale we need to really solve the planet's problems," said Vinod Khosla, a prominent Silicon Valley entrepreneur who focuses on energy (Revkin and Wald 2007).

Solar Arrays around the World

In 2015, NASA satellite photos showed large solar arrays spreading across China's Gobi desert and occupying three times the area covered in the same area three years earlier. In 2006, satellites showed the area near Dunhuang in Gansu Province of northwestern China bare of solar panels. *China Daily* reported that Gansu Province's solar capacity had reached 5.2 gigawatts by 2014. That same year, China's total capacity reached 28 gigawatts, a 200-percent increase in one year ("Growth of Solar" 2015).

By 2011 in the United States, major builders were beginning to sell midrange tract houses with installed solar power and utility bills close to zero. In some sunny areas, including Tucson, Arizona, and Las Vegas, Nevada, houses were being sold by Arizona-based Meritage Homes, the ninth-largest residential construction company in the United States, for $140,000 to $170,000, that produce as much energy as they consume. The homes come with a nine-panel solar array that reduces electric bills by one-third. For $10,000 more, 24 more panels reduce power bills to zero, more or less, depending on consumption. The homes have smart meters that feed surplus energy into the electrical grid, running backward when kilowatts are being credited.

In 2010, solar power was three to five times as expensive as coal per kilowatt hour, but the cost was plummeting as inventors worked on new technologies and mass production. As a measure of the interest in solar in California (spurred by state law requiring that 20 percent of the state's electricity come from renewable sources by 2010), the U.S. Bureau of Land Management by mid-2008 had received applications for projects that could cover 78,490 acres adjacent to Joshua Tree National Park; if built, 125 projects could replace 70 industrial-sized coal-fired power plants (Maloney 2008).

In 2010, Dow Jones installed a 4.1-megawatt solar power system at its 200-acre campus in New Brunswick, New Jersey, which was then one of the largest commercial solar sites in the United States. The solar array provides 15 percent of the energy used by 2,000 people on the campus and as much as 50 percent at peak use or during morning "high sun" ("Dow Jones" 2010).

Solar Power's Scale and Efficiency Grow

In 2013, photovoltaic solar power installations had been constructed in San Luis Obispo County, California, that produced 800 megawatts of power under bright sun, enough to equal the output of a large coal-fired or small nuclear plant.. The solar array spanned two locations and covered 12.5 square miles, many times the scale of any previous solar array. The largest previous photovoltaic solar array in the United States was at Nellis Air Force Base in Nevada and produced 14 megawatts. A concentrated solar power array in Nevada (NevadaOne) produced 64 megawatts in 2008.

Until recently, photovoltaic cells were only able to convert 20 percent of the solar energy they received into energy. By 2007, however, a team of researchers at the University of Delaware had converted 42.8 percent in an experimental project (a prototype has yet to be built) and are aiming at 50 percent by 2018. The group uses a novel light-splitting technique that increases efficiency markedly (Kintisch 2007).

The silicon solar panels that dominate the industry today may be replaced by new technologies that combine several light-absorbing materials that capture different portions of the solar spectrum or solar cells manufactured in rolls of thin copper indium gallium selenide in a film atop a metal foil. Nanotechnology plays a role in some designs for future solar-generating technology that has been theorized but not yet commercialized. Although today's silicon cells convert from 15 percent to 20 percent of sunlight to electricity in the field (up to 24 percent under perfect laboratory conditions), new technologies that have broached the realm of theory (and some in design but not commercialization) raise that figure to 40 percent, 60 percent, and even 80 percent (Service 2008). Photovoltaics made of plastic may dramatically reduce manufacturing costs.

Sometimes solar power leaps over the fossil fuel age. Consider Tecnosol, a small company in Nicaragua, where more than half of rural citizens have no access to electricity. In some areas, most notably in the remote eastern provinces, firewood is still the main source of fuel for cooking. Many people suffer respiratory diseases from wood smoke or spend scarce money on kerosene. Tecnosol in 2007 was installing 25,000 solar units and cutting carbon dioxide emissions by 150,000 tons over the life of the equipment (Mallaby 2007).

"Zero Energy" Homes and Retail Stores

Some new housing developments already have been designed to stop the electric meter completely by supplying them with enough solar power from their roofs.

Premier Gardens in the Sacramento area touted "zero energy homes" in a subdivision that produced 300 megawatts of electricity a year using photovoltaic roof tiles with integrated solar technology developed and sold by General Electric (Rogers 2006, 2RE).

In 2007 and 2008, corporations and retail stores seized on solar power as a money-saving strategy. Japanese TV maker Sharp has equipped its Kameyama plant with solar panels and window film that are capable of producing 5.2 megawatts. In Mountain View, California, Google has built a rooftop solar-powered generation system at its headquarters that generates 1.6 megawatts, which is enough to power some 1,000 California homes.

By 2008, Google was using solar power for almost a third of the electricity consumed by office workers at its headquarters, excluding power consumed by data centers that power many of Google's Web services worldwide. Data centers usually consume about 10 times more electricity than buildings that house office workers. "We are going to be producing roughly 30 percent of the power that we use," David Radcliffe, vice president of real estate at Google, told Reuters in an interview. "This is for our corporate-office people center," he said (Auchard and Anderson 2006).

In 2008, the Fresno Yosemite International Airport began installing the largest solar electric project at any airport in the United States, according to Fresno Mayor Alan Autry. At about the same time, the Airports Division or the Hawaii Department of Transportation solicited proposals from private companies to develop photovoltaic solar power systems within two years

In England, Susan Roaf's solar roof fuels her electric car. An architect at England's Oxford Brookes University, Roaf has designed a solar house fitted with photovoltaic cells that harvest the sun's energy and convert it into electricity. According to a report in *The Times* (London), "Her system is so efficient that she uses it to charge her electric car's batteries and makes a profit by selling 57 percent back to the national network" (O'Connell 2002). Using similar technology, cities could someday become self-sufficient power generators without using fossil fuels.

In some northern Chinese farming villages, homes that only a few years ago relied solely on low-grade coal for heat and cooking now use solar-heated rooftop water tanks (Landauer 2002). Some 85 percent of Israeli households have solar water heaters, which the government estimates lighten the country's overall energy burden by 3 percent. Solar heaters have been used in Israel since the 1950s. The global energy crisis of 1979 reminded Israelis of their reliance on foreign sources for oil and coal, and by law since 1980 solar water heaters have been required in new homes. Twenty companies employing 4,000 people now manufacture the systems and sell them for $300 to $1,000 each. A midrange system can pay for itself in energy savings in three or four years; if well maintained, it can last more than 10 years (Kaplow 2001).

Eric Doub, owner of Ecofutures Building of Boulder, Colorado, has built a five-bedroom family home with his wife, Catherine Childs, and their two children that produces more energy than it consumes. It even has a hot tub. He spent $1.38 million (the house itself cost $987,000) and cut his firm's employment by 10 people, 40 percent of the company's workforce. By 2008, the four-year-old project had become a showplace with 2,600 visitors. The house now has a name—Solar Harvest—and was built mainly with salvaged materials. It includes 12 used solar panels that heat water in a 6,000-gallon underground hot-water tank purchased from a dairy farm (Feder 2008).

Retail chains such as Safeway, Wal-Mart, Kohl's, and Whole Foods Market have installed solar arrays on the roofs of their stores to generate large amounts of electricity to reduce their use of coal-fired energy and save money in the long run. In many urban areas, retail stores are major electricity consumers, and solar panels can supply from 10 percent to 40 percent of what a store uses.

"It's very clear that green energy is now front and center in the minds of the business sector," said Daniel M. Kammen, an energy expert at the University of California–Berkeley. "Not only will you see panels on the roofs of your local stores, but I suspect very soon retailers will have stickers in their windows saying, 'This is a green energy store'" (Rosenbloom 2008).

Some solar proposals verge on the fantastic. One would build a solar power station on the moon, which, says its proponent,

in 10 years in which concentrated solar thermal energy obtained from parabolic concentrators fabricated from lunar materials is converted to electricity. . . . This power station would beam 21 trillion kilowatt hours/year of totally clean electric energy to the earth, using the proven technology of microwave power beaming, which passes through cloud cover. This energy would represent only 10 percent of the yearly world energy use in 2020. (Falbel 2015)

By 2016, Ukraine was involved in a large-scale shift to solar power to relieve dependence on natural gas and oil imports from Russia. As part of this shift, the wasteland surrounding the former Chernobyl nuclear plant was being replaced by a $1.1-billion U.S. solar panel array and a biogas facility. With Russia using supply lines for political leverage, "energy independence has become a matter of national security for Ukraine," said Sergiy Savchuk, head of the state agency on energy efficiency and energy saving in Kiev. "That's why renewable-energy development is now a priority issue for the Ukrainian government" (Schiermeier 2016).

Further Reading

Auchard, Eric, and Leonard Anderson. "Google Plans Largest U.S. Solar-Powered Office." *Washington Post*, October 16, 2006. http://www.washingtonpost.com/wp-dyn/content/article/2006/10/16/AR2006101601100_pf.html (no longer available).

Barringer, Felicity. "Solar Power Plants to Rise on U.S. Land." *The New York Times*, October 6, 2010. http://www.nytimes.com/2010/10/06/science/earth/06solar.html.

Cardwell, Diane. "Copenhagen Lighting the Way to Greener, More Efficient Cities." *The New York Times*, December 8, 2014. http://www.nytimes.com/2014/12/09/business/energy-environment/copenhagen-lighting-the-way-to-greener-more-effecient-cities.html.

"Dow Jones Plans Large Solar Installation on N.J. Campus." *USA Today*, April 12, 2010, 2B.

Falbel, Gerald. "Proposal for Reduction in Global Warming and Climate Change." July 25, 2015. http://www.freelunchclimatechangesolution.com.

Feder, Barnaby J. "The Showhouse That Sustainability Built." *The New York Times*, March 26, 2008. http://www.nytimes.com/2008/03/26/business/businessspecial2/26boulder.html.

"Growth of Solar in the Gobi Desert." NASA Earth Observatory, June 17, 2015. http://earthobservatory.nasa.gov/IOTD/view.php?id=86060&src=eoa-iotd.

Hughes, Trevor. "It's a New Dawn for Solar Industry—Mainly." *USA Today*, January 14, 2016, 6B.

Kaplow, Larry. "Solar Water Heaters: Israel Sets Standard for Energy; Cutting Dependence: Jerusalem's Alternative Energy Use a Lesson for United States." *Atlanta Journal Constitution*, August 5, 2001, 1P.

Kintisch, Eli. "Light-Splitting Trick Squeezes More Electricity Out of Sun's Rays." *Science* 317 (August 3, 2007): 583–584.

Landauer, Robert. "Big Changes in Our China Suburb." *Sunday Oregonian*, October 20, 2002, F4.

Mallaby, Sebastian. "Carbon Policy That Works; Avoiding the Pitfalls of Kyoto Cap-and-Trade." *Washington Post*, July 23, 2007, A-17. http://www.washingtonpost.com/wpdyn/content/article/2007/07/22/AR2007072200884_pf.html.

Maloney, Peter. "Solar Projects Draw New Opposition." *The New York Times*, September 24, 2008, H2.

O'Connell, Sanjida. "Power to the People." *The Times* (London), May 20, 2002, n.p. (LEXIS).

Revkin, Andrew C. "Budgets Falling in Race to Fight Global Warming." *The New York Times*, October 30, 2006. http://www.nytimes.com/2006/10/30/business/worldbusiness/30energy.html.

Revkin, Andrew C., and Matthew L. Wald. "Solar Power Wins Enthusiasts but Not Money." *The New York Times*, July 16, 2007. http://www.nytimes.com/2007/07/16/business/16solar.html.

Rogers, Paul. "Solar Energy Heats Up." *Omaha World-Herald*, October 15, 2006, 1RE, 2RE.

Rosenbloom, Stephanie. "Giant Retailers Look to Sun for Energy Savings." *The New York Times*, August 11, 2008. http://www.nytimes.com/2008/08/11/business/11solar.html.

Schiermeier, Quirin. "Solar on the Steppe: Ukraine Embraces Renewables Revolution: Former Soviet Nation Bids for Independence from Russian Fossil Fuels." *Nature* 537 (September 29, 2016): 598. doi: 10.1038/537598a.

Service, Robert F. "Solar Power: Can the Upstarts Top Silicon?" *Science* 319 (February 8, 2008): 718–720.

Service, Robert. "Perovskite Solar Cells Keep On Surging." *Science* 344 (May 2, 2014): 458.

Sessolo, M., and H. Bolink. "Perovskite Solar Cells Join the Major League." *Science* 350 (November 20, 2015): 917.

"Sun's Heat Could Cut Fossil-Fuel Use." *Nature* 523 (July 23, 2015): 384.

Wald, Matthew. "Two Large Solar Plants Planned in California." *The New York Times*, August 15, 2008. http://www.nytimes.com/2008/08/15/business/15solar.html.

See also: Concentrating Solar Power; Corporate Sustainability; Unburnable Carbon; Wind Gains Power Share

UNBURNABLE CARBON

Given the general consensus among world climate scientists and diplomats that global temperatures cannot rise more than 2°C without doing irreparable harm to humans and the rest of the planet's flora and fauna, what becomes of trillions of dollars worth of recoverable fossil fuels that will have to remain locked away to preserve a habitable Earth? When the threat of a warming climate is truly taken seriously by the fossil fuel industry, it will have to deal with the question of how to lock away reserves that have been defined as corporate assets but that could become worthless.

The value of these assets already have been estimated by scientists. Some of their calculations appeared in the British journal *Nature* early in 2015. According to Michael Jakob and Jerome Hilaire, who work with the Potsdam Institute for Climate Impact, "Cumulative carbon dioxide emissions must be less than 870 to 1,240 gigatons between 2011 and 2050 if we are to have a reasonable chance of limiting global warming to 2°C above average global temperature of preindustrial times" (Jakob and Hilaire 2015, 150). In this case, "reasonable" is defined as 50 percent as calculated with a model.

This target represents one-fourth to one-third of the oil, natural gas, and coal held on company balance sheets as provable reserves. What is provable and what companies own in the future tense is a liquid figure that can be subject to upward or downward revision, given market prices and development of mining and drilling technology.

Peak Oil: A Moving Target

Oil is a "finite" resource because Earth contains only so much of it. A common misconception endures that it may soon run out, that we have or soon will experience "peak oil." Changing technology, however, opens access to more oil over time. Production of oil in the United States rose some 80 percent between 2006 and 2014, much of it from sources that had not earlier been tapped. Canada's tar sands oil increased production as well.

In 1950, world oil production was around 10 million barrels per day (mbd). By 1970, production passed 50 mbd. In 2010, it doubled that to almost 100 mbd. Even in the infancy of the automotive age, fears were expressed that oil supplies would be exhausted. Each time, reserves rose and more oil was drilled. All of this adds to the atmosphere's load of carbon dioxide. Renewables, reduced coal use, more investment in energy inefficiency, cuts in fossil fuel subsidies, and caps in methane missions will keep world emission of greenhouse gases from rising further. The International Energy Agency's latest report did not recommend that world oil production be limited. "The Age of Oil Endures," wrote Robert J. Samuelson in the *Washington Post* (2015).

Further Reading

Samuelson, Robert J. "The Retreat of 'Peak Oil.'" *Washington Post*, June 15, 2015. http:// www.washingtonpost.com/opinions/the-retreat-of-peak-oil/2015/06/14/76a24ae4 -1124-11e5-9726-49d6fa26a8c6_story.html.

Oil Reserves That Cannot Be Burned

A few decades ago, oil from shale and tar sands was known to exist, but the exact size of the reserves was largely unknown. By 2010, oil booms had developed in Alberta, North Dakota, and (on a smaller scale) other parts of North America using methods such as hydraulic fracturing ("fracking") and oil (tar) sands mining, sending oil prices up to $100 a barrel. As prices fell by more than 70 percent in 2015 and 2016, however, some of this production shut down because it became unprofitable. The United States and Canada are not the only countries with oil-shale potential, but merely the first to apply the technology to production. "Encouraged by the recent shale-gas production boom in the United States, several world regions, including China, India, Africa, and the Middle East, are seeking to unlock their large endowments or increase existing production," wrote Michael Jakob and Jerome Hilaire (2014, 151).

The question of unburnable carbon raises profound social, political, and economic questions a world in which the fossil fuel industry is superlatively equipped with payrolls, equipment, and political influence to continue mining and drilling without limit on a 19th-century model even as levels of greenhouse gases and temperatures continue to rise. At what point will climatic changes force a landmark change in how carbon-based fuel reserves are valued? Will it then be too late, given thermal inertia, to forestall planetary disaster?

Writing in *Nature* (2014, 187) Christophe McGlade and Paul Ekins calculated the amount of reserves that cannot be burned to avoid such a disaster: "Our results suggest that, globally, a third of oil reserves, half of gas reserves and over 80 percent of current coal reserves should remain unused from 2010 to 2050 in order to meet the target of 2°C," they wrote. "We show that development of resources in the Arctic and any increase in unconventional oil production are incommensurate with efforts to limit average global warming to 2°C."

Unburnable Carbon by Region

In the Middle East, which is home to much of Earth's most easily and least costly oil that can be recovered, some 40 percent of reserves would have to remain unexploited if temperature rises are to be kept within the 2°C limit, according to calculations by McGlade and Ekins (2014, 188). Coal reserves are much larger than oil in terms both of energy potential and carbon dioxide emissions, but would also have to remain mainly in the ground. China and India would face 66 percent sequester rates and Africa 85 percent. The United States, Australia, and the territory of the former Soviet Union would be required to leave some 90 percent of coal reserves locked away to provide climate stability. Nearly all (at least 90 percent) of oil and gas reserves attributed to unconventional sources (fracking and tar sands) would have to be shut down or remain in the ground to maintain any reasonable chance to stabilize global climate over the long term, according to McGlade and Ekins (2015, 188–189).

Furthermore, McGlade and Ekins wrote, "Our results show that policy makers' instincts to exploit rapidly and completely their territorial fossil fuels are, in aggregate, inconsistent with their commitments to this temperature limit. Implementation of this policy commitment would also render unnecessary continued substantial expenditure on fossil fuel exploration, because any new discoveries could not lead to increased aggregate production" (McGlade and Ekins 2014, 187).

Temperature Rises with Complete Combustion

Another calculation of unburnable carbon was provided in 2015 by Michael Greenstone, Milton Friedman professor of economics at the University of Chicago and director of that university's Energy Policy Institute (and chief economist of President Obama's Council of Economic Advisers in 2009 and 2010). He wrote in *The New York Times* that recoverable reserves and resources of coal, oil, and natural gas, once combusted, would raise average global temperatures 16.2°F. ("Reserves" are known deposits recoverable at or near market prices; "resources" are known deposits that could be exploited with today's technology above that level.) "If we use all of the fossil fuels in the ground, the planet will warm in a way that is difficult to imagine. Unless the economics of energy markets change, we are poised to use them," Greenstone commented. And there may be more: "Indeed, it is well known that there are ample supplies of coal deeper beneath the Earth's surface that do not yet qualify as resources, and there is increasing evidence that energy from methane hydrates may become relevant commercially" (Greenstone 2015).

Given that fossil fuel reserves had an estimated value of $27 trillion U.S. at 2014 prices, any effort to sequester substantial amounts of them would produce a financial earthquake in the fossil fuel industry and, according to Jakob and Hilaire, force the companies to "ask themselves whether they should continue to invest in exploration for, and processing of, oil, gas, and coal, or risk losing billions of dollars of stranded assets" (Jakob and Hilaire 2015, 151).

Further Reading

Greenstone, Michael. "If We Dig Out All Our Fossil Fuels, Here's How Hot We Can Expect It to Get." *The New York Times*, April 9, 2015. http://www.nytimes.com/2015/04/09/upshot/if-we-dig-out-all-our-fossil-fuels-heres-how-hot-we-can-expect-it-to-get.html.

Jakob, Michael, and Jerome Hilaire. "Climate Science: Unburnable Fossil-Fuel Reserves." *Nature* 517 (January 8, 2015): 150–152. doi: 10.1038/517150a.

McGlade, Christophe, and Paul Ekins. "The Geographical Distribution of Fossil Fuels Unused When Limiting Global Warming to 2°C." *Nature* 517 (January 8, 2015): 187–190. doi: 10.1038/nature14016.

See also: Carbon Capture; Coal; Corporate Sustainability; Global Warming, China

WIND GAINS POWER SHARE

Global installations of wind power increased by 35,467 and 51,447 megawatts (MW) in 2013 and 2014, respectively. As of the end of 2014, total worldwide cumulative installed capacity from wind power totaled 369,553 MW, an increase of 16 percent in one year. In 2014, China was adding half of the world's new capacity and was the world's leader at 114,763 MW, 31.7 percent of the total. The United States was second at 65,879 (17.8 percent) followed by Germany (39,165 MW, 10.6%), Spain (22,987 MW, 6.2%), India (22,465 MW, 6.1 percent), United Kingdom (12,440 MW, 3.4 percent), and Canada (9,694 MW, 2.6 percent) (Global Wind Energy Council 2015).

Wind-power advocates now watch the share of electricity generated by turbines day by day. "A glorious wind blows on the prairie," remarked one such observer in Omaha in mid-April 2015 as the share of wind power in the Southwest Power Pool (the regional transmission organization that balances demand and supply) passed 25 percent on April 11 at 1:10 p.m. during a warm, breezy afternoon in a region from North Dakota southward through northern Texas. Wind has been gaining power share so quickly that we can watch it happen. "It passed 28 percent later the same day," the same observer remarked (Atkeison 2015).

Wind power has become more adaptable from huge turbines with blades the length of football fields to household-scale units. Turbines may be mounted on land or offshore; engineering is being developed to install floating wind farms in depths of 200 meters (660 feet) or more. Even oil companies are exploring wind power. In 2009, Shell Oil joined with Dallas-based Luminant Co. to build a large (3,000-megawatt) wind farm in the Texas panhandle.

Even in Wyoming, where coal has heretofore been king, wind is catching up. In the largest coal-producing state, jobs in that industry are declining, with construction of North America's largest wind farm. Ironies abound in Carbon County, 200 miles southwest of Gillette, where the state's first coal mine opened a century ago (and the last one closed a decade ago). The major owner of the new 2,000-acre wind farm is Colorado billionaire Philip F. Anschutz, a major Republican donor, and his Anschutz Corporation. The wind farm will produce enough power to supply a million homes (far more than Wyoming's 584,000 people ever will need), and will be built with transmission lines for export. "We can produce wind power here that's competitive with anything: coal, natural gas," said Bill Miller, the president and chief executive of the Anschutz Corporation (Davenport 2016).

Wind Power Becomes Cost Competitive

By the early 21st century, wind power was becoming competitive in cost with electricity generated by fossil fuels as its use surged. After 2004, wind became less expensive to generate electricity than fossil fuels in many areas as its use exploded by 20 to 25 percent annually. The unit price falls as wind use grows (Flannery 2005, 268–269).

A Chic Wind Turbine Out Back

Wind turbines were popping up in unexpected places. Parts of the American "rust belt" along the south shores of the Great Lakes are retooling as wind-energy centers. Witness Lackawanna, New York, a suburb of Buffalo, where, according to a report in *The New York Times*,

> on the 2.2-mile shoreline above a labyrinth of pipes, blackened buildings and crumbling coke ovens that was once home to a behemoth Bethlehem Steel plant: eight gleaming white windmills with 153-foot blades slowly turning in the wind off Lake Erie, on a former Superfund site where iron and steel slag and other industrial waste were dumped during 80 years of production. (Staba 2007)

This wind farm is part of New York state's "brownfields" program, which turns former low-level toxic sites into productive uses. To local boosters, the "rust belt" has become the "wind belt." The turbines (called "steel winds") cost $4.5 million each to build. Power lines once used by the steel plant now carry electricity from the turbines, while paved roads, rail lines, and an industrial port built by Bethlehem are now wind-turbine infrastructure.

A wind turbine out back has become a corporate chic status symbol, a brandishing of "green" credentials. They have been popping up all over the country, many of them ad hoc corporate efforts. *Outside Magazine* (a subsidiary of *Rolling Stone*) switched 90 percent of its Santa Fe, New Mexico, office's electricity consumption to wind power. Burgerville, a chain in Washington and Oregon, was running all its restaurants on wind power.

Further Reading

Staba, David. "An Old Steel Mill Retools to Produce Clean Energy." *The New York Times*, May 22, 2007. http://www.nytimes.com/2007/05/22/nyregion/22wind.html.

Some areas of Europe (Denmark and parts of Germany and Spain) were using wind as a major energy source by the turn of the 21st century. Germany's northernmost state of Schleswig-Holstein has been using wind for 28 percent of its electricity; by 2003, 18 percent of Denmark's electricity was being derived from wind (Brown 2003, 201). All of these areas had increased capability substantially by 2015. In some U.S. states, wind was gaining power share rapidly. Iowa, for example, reached half wind in 2017. Mid-America Energy, which supplies about half of Iowa's electric power, plans to be 85 percent wind and solar by 2020.

Advances in wind-turbine technology adapted from the aerospace industry have reduced the cost of wind power from 38 cents per kilowatt-hour (during the early 1980s) to 3 to 6 cents. This rate is competitive with costs of power generation from fossil fuels, but costs vary according to site. Major corporations, including Shell International and BP, have been moving into wind power.

Capacity Increases Rapidly

Wind and solar power are now fastest-growing sources of energy in the United States and the world. Wind power has become so popular that a shortage of parts causing installations to fall behind demand. A new phrase in the English language, "wind-rich," was coined to describe an area with relatively steady, unimpeded access to turbine-ready breezes. Some surprising pitchmen for wind (one of them being President George W. Bush) believe that the United States may derive as much as 20 percent of its electrical power from wind by roughly 2020. Electricity generation accounts for some 40 percent of U.S. greenhouse gases.

By 2008, wind-power capacity in the United States was increasing rapidly, but the electrical grid had not kept up with the additional supply in many areas. For example, a $320-million, 200-turbine array at Maple Ridge Wind farm in upstate New York was forced to shut down for lack of transmission capacity even when the wind blew. In places, the grid is almost a century old with 200,000 miles of lines and 500 owners. "We need an interstate transmission superhighway system," said Suedeen G. Kelly, a member of the Federal Energy Regulatory Commission. "The windiest sites have not been built, because there is no way to move that electricity from there to the load centers," he said (Wald 2008). Many of the best wind-power production areas are sparsely populated. According to Bill Richardson, former governor of New Mexico and secretary of the Department of Energy under President Bill Clinton, "We still have a third-world grid. . . . With the federal government not investing, not setting good regulatory mechanisms, and basically taking a back seat on everything except drilling and fossil fuels, the grid has not been modernized, especially for wind energy" (Wald 2008).

Despite a recession and tight credit, wind-power capacity in the United States grew 39 percent in 2009 and rose to some 2 percent of electrical capacity, enough to supply the equivalent of 9.7 million homes. More than one-quarter of that power was being generated in Texas. In 2009, 80 percent of new electrical generating capacity came from wind. Federal tax credits helped. The American Wind Energy Association, however, cautioned that many deliveries were "in the pipeline" from before the recession and that growth in 2010 might decline (Mouawad 2010, B1). In 2009, U.S. wind capacity was around 35,000 megawatts, up from 25,000 in 2008 and 17,000 in 2007. In 2009, wind-energy manufacturing employed 85,000 people in the United States. By 2009, 29 states also had passed laws requiring that a certain percentage of power come from renewable sources by a certain date. By 2008, wind turbines in the United States generated around 48 billion kilowatt hours of electricity, enough to power 4.5 million homes. That is a significant rise from previous years but still only 1.2 percent of the country's electrical demand (Tomsho 2008). Wind power in the United States in 2007 grew by 45 percent, adding 5,244 megawatts of capacity, a third of all the new electrical capacity in the country. In one year, wind-energy employment in the United States doubled to about 20,000 (Smith 2008). Capacity and employment more than doubled by 2015.

Problems with Wind Turbines: Visual Pollution, Bird Kills, and Fickleness

Wind turbines have not been universally welcomed. Some neighbors complain that they make for "visual pollution." In their book *Cape Wind* (2007), Wendy Williams and Robert Whitcomb detailed the battle over this issue on Cape Cod in Massachusetts that included some high-octane names such as the late Ted Kennedy. The Cape Wind farm at Nantucket Sound is no garden-variety project. Perhaps a coal strip mine would be a more salubrious sight, but no one receiving coal-fired energy on Cape Cod has to watch coal being mined. The Cape Wind project is highly visible and has plans for 130 turbines to produce 3.6 megawatts each with a 175-foot blade. On a windy day and at maximum capacity, all 130 machines would produce as much power as a modestly sized electricity plant burning coal or natural gas.

The power capacity of wind turbines may be restricted only by the size of a blade that can be hauled to a site, mounted, and maintained with some degree of structural integrity. A turbine at 250 feet takes advantage of winds that are 20 percent stronger at that elevation than at 150 feet, said Dr. Mark Z. Jacobson, an associate professor at Stanford University's department of civil and environmental engineering (Wald 2007). One problem with bigger and higher turbines is that they have a greater chance of killing birds; some 40,000 of them were being killed each year in the United States by 2006—one bird per 30 turbines per year, a small fraction of the millions killed each year by domestic cats (Marris and Fairless 2007, 126).

One major problem for wind energy is that hot days—when power demand is highest—are usually the least windy. "As a result, wind turns out to be a good way to save fuel, but not a good way to avoid building plants that burn coal," wrote Matthew Wald in *The New York Times* (2007).

A wind machine is a bit like a bicycle that a commuter keeps in the garage for sunny days. It saves gasoline, but the commuter has to own a car anyway. Without major advances in ways to store large quantities of electricity or big changes in the way regional power grids are organized, wind may run up against its practical limits sooner than expected. In many places, wind tends to blow best on winter nights, when demand is low.

According to Robert E. Gramlich, a policy director with the American Wind Energy Association, wind energy could be integrated into an electrical grid on a large enough area that ebbing wind in one part could be balanced by continuing breezes in another (Wald 2006).

Some people have compared the level of noise and vibrations from large wind turbines to a jet taking off, and some neighbors of wind turbines have complained of insomnia and headaches. Nina Pierpont, a pediatrician who lives in Malone, New York, has given the condition a name—wind turbine syndrome—and plans to write a book about it. Physician Amanda Harry of Great Britain says the turbines also can cause anxiety, vertigo, and depression. Acoustical engineer Mariana Alves-Pereira of Portugal asserts that wind-turbine noise and vibration may contribute to strokes and epilepsy. Despite such anecdotal testimonies, in 2008 the European Union exonerated wind turbines of culpability in any medical condition except occasional loss of sleep (Keen 2008).

Curing the syndrome is usually simple: a matter of distance. In real-estate legal language, this means providing proper setbacks, usually 1,000 to 2,500 feet from the nearest neighbors. Some neighbors, however, are insisting on setbacks of a mile or more to keep their home values from declining.

Household Wind Turbines

Herschel Carter is one of a growing number of people in the United States who have been installing household scale wind-power generators, which are available for $10,000 to $15,000. The 62-year-old retired insurance agent watched the amount of electricity he consumes from the grid fall nearly to half (1,752 to 993 kilowatt hours) in March 2007, the first month his wind turbine operated. His electric bill for that month fell $68. Payback will take several years, but Carter regards the turbine as a form of energy security because his wife is on oxygen and needs electricity for it. The unit has a battery that stores energy from windy days for calm ones (McNulty 2007).

Household-scale wind turbines were invented in the United States around 1920, and the country has since led the world in developing their technology, with more than 90 percent of all household wind turbines installed in the United States. Growth in the worldwide number of household wind turbines has been averaging 14 percent to 25 percent a year between 2000 and 2006, according to Randall Swisher, executive director of the American Wind Energy Association. At least one-half acre of land is required for a residential wind turbine, and regulators must permit a tower of at least 35 feet in height (the higher the tower, the more efficient the turbine).

Smaller wind turbines may be available. A miniature wind turbine developed by the Scottish company Windsave can be mounted on an urban roof. Its four-foot blades can provide part of an average home's electricity (company publicity says one-third) at a cost of around $2,000 (McGuire 2005, 175).

Further Reading

Atkeison, John. Personal correspondence, April 11, 2015.

Brown, Lester R. *Plan B: Rescuing a Planet Under Stress and a Civilization in Trouble.* New York: Earth Policy Institute/W.W. Norton, 2003.

Davenport, Carol. "As Wind Power Lifts Wyoming's Fortunes, Coal Miners Are Left in the Dust." *The New York Times,* June 19, 2016.

Flannery, Tim. *The Weather Makers: How Man Is Changing the Climate and What It Means for Life on Earth.* New York: Atlantic Monthly Press, 2005.

Global Wind Energy Council. *Global Wind Statistics, 2014*. February 10, 2015. http://www. gwec.net/wp-content/uploads/2015/02/GWEC_GlobalWindStats2014_FINAL_10.2 .2015.pdf. Accessed April 18, 2015.

Keen, Judy. "Neighbors at Odds Over Noise from Wind Turbines." *USA Today*, November 4, 2008, A3.

Marris, Emma, and Daemon Fairless. "Wind Farms' Deadly Reputation Hard to Shift." *Nature* 447 (May 10, 2007): 126.

McGuire, Bill. *Surviving Armageddon: Solutions for a Threatened Planet*. New York: Oxford University Press, 2005.

McNulty, Sheila. "U.S. Power Generation Answer Is Blowing in the Wind." *Financial Times* (London), April 24, 2007, 7.

Mouawad, Jad. "Wind Power Grows 39 Percent for the Year." *The New York Times*, January 25, 2010, B1, B6.

Smith, Rebecca. "Wind, Solar Power Gain Users." *Wall Street Journal*, January 18, 2008, A6.

Tomsho, Robert. "Wind Shift in Energy Debate." *Wall Street Journal*, June 19, 2008, A11.

Wald, Matthew. "It's Free, Plentiful and Fickle." *The New York Times*, December 28, 2006. http://www.nytimes.com/2006/12/28/business/28wind.html.

Wald, Matthew. "What's So Bad about Big?" *The New York Times*, March 7, 2007. http:// www.nytimes.com/2007/03/07/business/businessspecial2/07big.html.

Wald, Matthew. "Wind Energy Bumps into Power Grid's Limits." *The New York Times*, August 27, 2008. http://www.nytimes.com/2008/08/27/business/27grid.html.

Williams, Wendy, and Robert Whitcomb. *Cape Wind: Money, Celebrity, Class, Politics, and the Battle for Our Energy Future on Nantucket Sound*. New York: Public Affairs, 2007.

See also: Solar Power; Wind Power, Denmark; Wind Power, United States; Wind Power and Solar Power in Portugal, Germany, and Japan

WIND POWER, DENMARK

In the 1970s, Denmark was dependent on imported oil and saw its economy devastated by the oil embargoes of that decade. The nation has since made an enduring commitment to achieve energy independence with wind power as an important part of this strategy. Denmark has become a world leader in wind-turbine technology, and Danish work on wind turbines is one major reason why the technology now generates electricity is price competitive with oil, coal, and nuclear power. Denmark has also built a wind-turbine infrastructure that provides several thousand jobs. Wind power may well help Denmark in its goal of becoming the world's first carbon-neutral nation by 2050.

In matters of advanced technology, Denmark dominated the worldwide wind-power industry until about 2010 when the Chinese surpassed them. Until that time, Danish companies supplied more than half the wind turbines in use worldwide, making wind-energy technology one of the country's largest exports. Some Danish wind turbines have blades almost 300 feet long—the length of a football field. During January 2007, an extremely stormy month, Denmark harvested 36 percent of its electricity from wind, almost double the usual. By 2016, Danish wind power routinely supplied half of the nation's electricity.

In 2007, a new center-right coalition government reduced subsidies for wind-power development, and its growth slowed. Suddenly, the tax environment for new wind power development was better in Texas than in Denmark. The Danes, however, had an extremely long head start.

Danish wind energy has also experienced technical setbacks. For example, Danish wind operators hoped to bypass local objections and take advantage of stronger, steadier air currents by building giant turbines at sea. In one case in 2004, turbines at Horns Reef, some 10 miles off the Danish coast, broke down when their equipment was damaged by storms and saltwater. Vestas, a Danish manufacturer, fixed the problem by replacing the equipment at a cost of 38 million euros. But Peter Kruse, the head of investor relations for Vestas, warned that the lesson from Horns Reef was that wind farms at sea would remain far more expensive than those on land. "Offshore wind farms don't destroy your landscape," Mr. Kruse said, but the added installation and maintenance costs are "going to be very disappointing for many politicians across the world" (Kanter 2007).

The Danish wind-power industry also provides 150,000 families with shares of profits from the national electricity grid. Thousands of rural Danish residents have joined wind-power cooperatives, buying turbines and leasing sites to build them, often on members' land (Woodard 2001). At the same time, power companies in Denmark are now taxed for each ton of carbon dioxide emitted above a low limit that will gradually become even lower.

By 2014, Denmark's 5.6 million people were drawing more than 40 percent of their electric power from renewables, mainly wind, and had plans to be completely off of all forms of fossil fuels—including those used for transportation—by 2060. Denmark's renewable energy commitment has been the most ambitious on Earth, but it has not been accomplished without technical and political friction.

One problem might be thought of as a benefit for consumers: renewable energy was causing power prices to crash during high-demand periods. However, this is when conventional plants make their money, and Denmark needs to keep them running in case the sun does not shine or the wind does not blow. "We are really worried about this situation," said Anders Stouge, the deputy director general of the Danish Energy Association. "If we don't do something, we will in the future face higher and higher risks of blackouts" (Gillis 2014). The Danes have been experimenting with attaching a premium value to standby capacity, which would reward old-source plants when their power is needed most.

The biggest problem with going completely off fossil fuels, however, is transportation. The Danes are already more reliant on bicycles than anyone else for short-range transport, but they still need cars for longer trips. Electric cars have been tried but have not done well with consumers because of their high cost, limited power, and short range. The technology requires considerable work.

Wind Power on Samsø Island, Denmark

The nearly 4,000 residents of Denmark's largely agricultural Samsø Island accepted a challenge from the Danish government during the 1990s to convert to a carbon-neutral style of life. By 2007, these Danes had largely accomplished the task with no notable sacrifice of comfort. Farmers grow rapeseed and use the oil to power their machinery, homegrown straw is used to power centralized home-heating plants, solar panels heat water and store it for use on cloudy days (of which the 40 square mile island has many), and wind power provides electricity from turbines in which most families on the island have a share of ownership. The island is now building more wind turbines to export power.

During the late 1990s, most of the homes of the Samsingers were heated with oil imported on tankers, and their electricity was generated from coal and imported from the Danish mainland on cables. The average resident of the island produced 11 tons of carbon dioxide a year (Kolbert 2008). In 1997, the Danish Ministry of Environment and Energy sponsored a renewable-energy contest that the city government entered after an engineer who did not live on the island filed a proposal after asking the mayor. To everyone's surprise, Samsø was picked as Denmark's "renewable-energy island." The designation carried no prize money, so nothing happened for a few months. After that, residents established energy cooperatives and began to tutor themselves in such things as insulation and wind power.

According to Elizabeth Kolbert, writing in *The New Yorker*,

> They removed their furnaces and replaced them with heat pumps. By 2001, fossil fuel use on Samsø had been cut in half. By 2003, instead of importing electricity, the island was exporting it, and by 2005 it was producing from renewable sources more energy than it was using. "People on Samsø started thinking about energy," Ingvar Jørgensen, a farmer who heats his house with solar hot water and a straw-burning furnace, told me. "It became a kind of sport." (Kolbert 2008)

Ten years later, Samsø found itself a subject of study by energy tourists as researchers traveled great distances, burning copious fossil fuels, to observe Samsø's experiment with its 11 wind turbines spinning in a relatively constant wind that makes an ideal wind-power site.

Samsø also built district heating plants that run on biomass, mainly bales of locally grown straw, burned to warm water that circulates among local homes, where it provides both heat (the houses are tightly insulated, and the climate usually mild) and hot water. One district plant in Nordby burns wood chips, and solar panels provide hot water when the sun is shining. Kolbert (2008) explained that "Burning straw or wood . . . produces CO_2 but while fossil fuels release carbon that would have remained sequestered, biomass releases carbon that would have entered the atmosphere through decomposition."

A few Samsø farmers also have converted their cars and tractors to operate on canola oil. Some of them grow their own canola seeds, which also can be used in cooking.

Further Reading

Gillis, Justin, "A Tricky Transition from Fossil Fuel." *The New York Times,* November 10, 2014. http://www.nytimes.com/2014/11/11/science/earth/denmark-aims-for-100-percent -renewable-energy.html.

Kanter, James. "Across the Atlantic, Slowing Breezes." *The New York Times*, March 7, 2007. http://www.nytimes.com/2007/03/07/business/businessspecial2/07europe.html.

Kolbert, Elizabeth. "The Island in the Wind: A Danish Community's Victory over Carbon Emissions." *The New Yorker*, July 7, 2008. http://www.newyorker.com/maga zine/2008/07/07/the-island-in-the-wind.

Woodard, Colin. "Wind Turbines Sprout from Europe to U.S." *Christian Science Monitor*, March 14, 2001, 7.

See also: Solar Power; Wind Gains Power Share; Wind Power, United States; Wind Power and Solar Power in Portugal, Germany, and Japan

WIND POWER, UNITED STATES

The costs of wind power have been declining rapidly so that it now competes favorably with fossil fuels under some conditions. This is an important reason why wind has suddenly become popular as a source of electrical generation in many American states, especially in the Midwest.

The Omaha Public Power District, for example, used to rely entirely on coal and nuclear power. By 2018, however, it will generate one-third of its capacity with wind. Construction began in mid-2015 on a 400-megawatt wind farm in Holt County, Nebraska, some 180 miles northwest of Omaha under the aegis of BHE Renewables, a wholly owned subsidiary of Berkshire Hathaway, which in 2018 will move Omaha's wind power share from barely zero to 33 percent. Since 2012, BHE Renewables has invested in a growing array of solar, wind, solar, geothermal, and hydroelectric projects. By 2016, one-half of Iowa's electricity was generated by wind power. The cost of wind power is now being quoted at one to two cents per kilowatt hour, which is often lower than prices for electricity from coal, oil, or natural gas. In 2014, California's government had a target of 33 percent renewable power by 2020.

Wind capacity in the Pacific Northwest, where it is often used with hydroelectric, soared from only 25 megawatts in 1998 to 3,800 megawatts by 2009. During 2006 alone, Washington state added 428 megawatts of wind power, trailing only Texas in new installations. The electrical grid in the Northwest is especially inviting to wind-power developers because of its hydroelectric distribution network, relatively reliable wind, progressive utility companies, and state laws that establish preferences for renewable energy. Washington state law requires utilities to work toward generating 15 percent to 25 percent of their electricity from sources that do not generate carbon dioxide. Hydroelectric dams also have been built in river gorges and other valleys that are natural conduits for wind. The transmission line connecting the dams with urban areas have been built with considerable surplus capacity—a built-in network for wind power. Farmers in the area have been earning $2,000 to $4,000 per year by allowing per site for wind turbines (Harden 2007).

Wind Power in Texas Oil Country

Although Texas has a reputation as a conservative state run by oil barons, it has become one of the most progressive environments for wind energy in the United States. George W. Bush, himself once an oilman, helped start the "wind rush" when he was governor by signing a bill requiring the state's power infrastructure to include 3 percent from alternative-energy sources by 2009. By December 2004, state planners recommended that 10 percent of Texas' power should come from renewable sources by 2025.

"That's just money you're hearing," said wind farmer Louis Brooks, who receives $500 a month for each of 78 wind turbines on his land near Sweetwater, Texas, as the spinning turbines emit a low hum (Krauss 2008). By 2008, Texas was obtaining more than 3 percent of its electricity from wind turbines, enough to supply power to 1 million homes. Some of the turbines are twice as high as the Statue of Liberty, with blades that span as wide as the wingspan of a jumbo jet. "Texas has been looking at oil and gas rigs for 100 years, and frankly, wind turbines look a little nicer," said Jerry Patterson, the Texas land commissioner, whose responsibilities include leasing state lands for wind-energy development. "We're No. 1 in wind in the United States, and that will never change" (Krauss 2008).

Cielo Wind Power of Austin has been buying 60-year "wind rights" from ranchers in Texas in its plans to build the Noelke Hill Wind Ranch for $130 million. The company hopes to generate 240 megawatts of power, enough to provide electricity to 80,000 homes through TXU, a utility based in Dallas. Texas law now requires that 3 percent of the state's electricity come from renewable sources. Several companies, among them American Electric Power and General Electric, have plans to spend as much as $1 billion on wind energy in the midst of the Permian basin, which has heretofore been known as oil country. Wind speed in the area averages 16 miles per hour and is consistent enough to bend juniper trees.

By late 2002, Cielo had built $300 million worth of wind-power turbines in Texas and sold them to various power companies. The turbines jut "more than 200 feet into the air with blades 200 feet in diameter to capture high-velocity wind patterns" (Herrick 2002, B3). Texas by late 2002 was generating enough wind power to supply 300,000 homes (Herrick 2002, B3). A standard line used by wind developers with the ranchers in the area is, "You've been getting your hat blown off your head your whole life. It's time to stop cussin' and make some money" (Herrick 2002, B3). Wind turbines placed roughly one every 25 acres can earn the average landowner around $3,200 each (Herrick 2002, B3).

Wind has become truly big business in Texas. By 2007, Royal Dutch Shell and a wind-development company owned by Goldman Sachs Group Inc. were building wind-power infrastructure in the Texas Hill Country around Silverton. The companies were building some of the largest wind farms in the world and planning hundreds of turbines costing about $2 million each (Ball 2007). The area, once home to oil rigs, has a relentless wind that blows so steadily that sagebrush and mesquite grow at an angle. The area's canyons tend to funnel the wind.

In the same area, Shell, a company diversifying from its oil roots, installed a 120-square-mile wind farm, which was touted in 2007 as "the largest in the world" by occupying a land area five times the size of Manhattan Island. Shell's new wind farm will produce electricity equal to a coal-fired power plant (Ball 2007, A11).

Shell and other companies are betting that wind power's time has come. Their lobbying contacts in Austin, the state capital, are poised to make a reality of this hunch. Large companies such as General Electric, which manufactures wind turbines, also have joined the wind lobby. Texas law also requires utilities to buy wind power when and where it is available as a proportion of generating capacity. In 2007, Shell was selling wind power from its 30-square-mile Brazos wind farm to TXU Corporation of Dallas, which is under pressure by environmentalists to reduce its reliance on coal-generated power. The Brazos wind farm contains 160 turbines that have a capacity of 160 megawatts when the wind blows. Local ranchers can collect as much as $80,000 a year from energy companies for a 640-acre section of land under a wind farm (Ball 2007, A11).

Meanwhile, in mid-July 2008, Texas regulators approved a $4.93-billion wind-power transmission project, including transmission to carry enough power to supply 18,500 megawatts, or 3.7 million homes, from wind-rich western parts of the state to major population centers such as Dallas, Houston, Austin, and San Antonio. This trend accelerated in 2014 as many companies signed power-purchase agreements for solar or wind, especially in the Great Plains and Southwest. The time was coming when wind power would compete without subsidies. "In Texas," reported Diane Cardwell of the *New York Times* (2014):

> Austin Energy signed a deal . . . for 20 years of output from a solar farm at less than 5 cents a kilowatt-hour. In September, the Grand River Dam Authority in Oklahoma announced its approval of a new agreement to buy power from a new wind farm expected to be completed next year. Grand River estimated the deal would save its customers roughly $50 million from the project. Also in Oklahoma, American Electric Power ended up tripling the amount of wind power it had originally sought after seeing how low the bids came in last year. "Wind was on sale—it was a Blue Light Special," said Jay Godfrey, managing director of renewable energy for the company. He noted that Oklahoma, unlike many states, did not require utilities to buy power from renewable sources. "We were doing it because it made sense for our ratepayers," he said.

Wind and Solar at the Ballpark

The Philadelphia Eagles' football stadium began going green in a major way in 2010, installing 2,500 solar panels, 80 20-foot-high wind turbines, and a generator that runs on natural gas and biodiesel. The aim was to make the sports stadium the first to generate all of its own electricity. Some of the fat used to fry hot dogs is recycled to generate power (Belson 2010, B18).

Following nine years of review and copious controversy, the federal government on April 28, 2010, approved the first American offshore wind farm, the highly contested Cape Wind Associates project in Nantucket Sound, a 130-turbine array near

Cape Cod, that could allow the United States to install wind energy on a scale already used in Europe and China. "I am approving the Cape Wind project," said Interior Secretary Ken Salazar in Boston. "This will be the first of many projects up and down the Atlantic coast" (Seelye 2010). Opponents went to court, however, and stalled Cape Wind for several years. Supporters believe that the $1-billion project could provide energy for as much as 75 percent of the power needs on Cape Cod, Martha's Vineyard, and Nantucket, along with several hundred construction jobs.

In Seattle by 2009, 90 percent of electricity was from nonfossil sources— traditional hydropower plus an aggressive emphasis on wind by the Bonneville Power Administration. Under the best wind conditions, 18 percent of the power generated by the BPA was coming from wind by 2009, a proportion that increased thereafter.

Further Reading

Ball, Jeffrey. "The Texas Wind Powers a Big Energy Gamble." *Wall Street Journal*, March 12, 2007, A1, A11.

Belson, Ken. "Just Wind, Baby." *The New York Times*, November 18, 2010, B18, B20.

Cardwell, Diane. "Copenhagen Lighting the Way to Greener, More Efficient Cities." *The New York Times*, December 8, 2014. http://www.nytimes.com/2014/12/09/business/energy-environment/copenhagen-lighting-the-way-to-greener-more-effecient-cities.html.

Harden, Blaine. "Air, Water Powerful Partners in Northwest." *Washington Post*, March 21, 2007, A3. http://www.washingtonpost.com/wp-dyn/content/article/2007/03/20/AR200 7032001634_pf.html.

Herrick, Thaddeus. "The New Texas Wind Rush: Oil Patch Turns to Turbines, as Ranchers Sell Wind Rights; A New Type of Prospector." *Wall Street Journal*, September 23, 2002, B1, B3.

Krauss, Clifford. "Move Over, Oil, There's Money in Texas Wind." *The New York Times*, February 23, 2008. http://www.nytimes.com/2008/02/23/business/23wind.html.

Seelye, Katharine. "Regulators Approve First Offshore Wind Farm in U.S." *The New York Times*, April 28, 2010. http://www.nytimes.com/2010/04/29/us/29wind.html (no longer available).

See also: Solar Power; Wind Gains Power Share; Wind Power, Denmark; Wind Power and Solar Power in Portugal, Germany, and Japan

WIND POWER AND SOLAR POWER IN PORTUGAL, GERMANY, AND JAPAN

Even Germany, land of fog and weeping skies, was finding a place for solar cells that are so sensitive that a hot shower on a cloudy day is no longer a problem. In 2006, half the world's new solar capacity was installed in Germany, despite the fact that the country has only half as many sunny days as Portugal, a more obvious solar success story (Whitlock 2007). The German solar panels work on drizzly days, although they generate only a quarter to half as much electricity as a sunny day.

Coal mining was Espenhain's largest employer under the East German regime, providing 8,000 jobs, until the country collapsed in 1989. After German reunification, the mining jobs vanished. "This region was known as the dirtiest in all of

Europe," said Juergen Frisch, mayor of Espenhain. "The solar plant came at a very good time for Espenhain. It's helped to change our image. Unlike the coal mines, the solar plant makes almost no noise, save for the low thrum of a few outdoor air-conditioning units that cool the electrical transformers. The plant, with 33,500 solar panels, sits on a 37-acre site in a field off a rural road and requires scarcely any maintenance" (Whitlock 2007).

A German law enacted in 2000 requires the country's utility companies to subsidize new solar installations by buying their electricity and using it on their grids at marked-up rates that allow small solar enterprises to earn profits. Wind and biofuels also have preferred status under German law. As the world's sixth-biggest producer of carbon dioxide emissions, Germany is trying to slash its output of greenhouse gases and wants renewable sources to supply a quarter of its energy needs by 2020 (Whitlock 2007). Germany also has decided to phase out all nuclear power plants by 2020

In northwestern Germany, Freiburg has been working to become the world's first "solar city" with a "solar-powered train station, energy-efficient row houses, innumerable rooftop photovoltaic systems and, high on a hill overlooking the vineyards, the world-famous Heliotrop, a high-tech cylindrical house that rotates to follow the sun" (Roberts 2004, 192). Freiberg is home to the Fraunhofer Institute for Solar Energy Systems, where scientists have been seeking breakthrough research that will reduce the cost of photovoltaic solar energy to competitive levels with fossil fuels and nuclear power, including a "multilayer" cell with an efficiency of 40 percent, which would be twice current levels.

Wind, Solar, and Water in Portugal

By 2010, nearly 45 percent of Portugal's electricity was coming from renewable sources, up from 17 percent in 2005. The transition also has sparked gripes about rising power rates, which also went up 15 percent during the same period (Rosenthal 2010, A1). Roughly half of Portugal's renewable energy comes from wind and the other half from hydropower, sometimes operated together, as wind turbines pump water uphill at night, which flows downhill by day, creating electricity to meet customer demand. Portugal's two-way grid also uses rooftop solar panels in a minor role. The grid uses fossil fuels for backup (efficient plants fueled by natural gas are the biggest players), and different components of load are constantly balanced to match weather conditions with demand. The power grid has been overhauled to transfer power from remote but wind-rich areas to cities.

The energy transition has placed little stress on Portugal's national budget because construction costs have been largely absorbed by declining costs of imported fossil fuels, especially oil, which doubled in cost while the grid was changing. Imported fuel had accounted for half of Portugal's trade deficit. Portugal plans to use renewables for 60 percent of its electrical power and 31 percent of its total energy by 2020 (Rosenthal 2010, A8).

By 2007, a new solar array spread across 150 hilly acres near Serpa, some 120 miles southeast of Lisbon, the Portuguese capital. This project was an old-fashioned photovoltaic constructed by General Electric Energy Financial Services and Power Light Corporation of the United States in partnership with the Portuguese company Catavento. "This is the most productive [largest capacity] solar plant in the world," its backers asserted. "It will produce 40 percent more energy than the second-largest one, Gut Erlasse in Germany," said Howard Wenger, speaking for Power Light ("Portugal Celebrates" 2007).

Unlike frequently misty Germany, southern Portugal is among the sunniest places in Europe, with as many as 3,300 hours of sunlight annually, nine hours during an average day. This solar array, when finished, produced enough power to supply 8,000 homes. It replaced 30,000 tons worth of annual greenhouse gas emissions ("Portugal Celebrates" 2007). This plant's photovoltaic system uses silicon solar-cell technology to convert sunlight directly into electricity, producing 20 gigawatts per year. "This project is successful because Portugal's sunshine is plentiful, the solar power technology is proven, government policies are supportive, and we are investing . . . to help our customers meet their environmental challenges," said Kevin Walsh, managing director and leader of renewable energy at GE Energy Financial Services ("Portugal Celebrates" 2007).

Germany: Energy Transition as National Policy

The small North Sea island of Heligoland off the coast of Germany has become a wind-power colony with "wind turbines, standing as far as 60 miles from the mainland, stretching as high as 60-story buildings and costing up to $30 million apiece. On some of these giant machines, a single blade roughly equals the wingspan of the largest airliner" (Gillis 2014). Long-range plans call for wind farms as far as 125 miles offshore at a cost of as much as $340 million per turbine—machine, power cables, installation, and other costs.

In 2015 also in the North Sea, the Netherlands was building a large array of wind turbines 50 miles offshore at a cost of $3 billion that will provide power for 1.5 million homes beginning in 2017. Wind power from sea-based platforms costs as much as three times that of land-based wind farms because of relatively high constriction, maintenance, and transmission costs, but government subsidies have been used to support the Dutch program.

The Germans have a term for "the energy transition"—*energiewende*—being undertaken by the country in pursuit of renewable energy (Gillis 2014). By 2014, 30 percent of Germany's electricity was coming from wind and solar, the highest ratio of any large nation. Solar panels were providing 7 percent of power and wind turbines around 10 percent in 2014. Because renewable power is intermittent, however, German utilities retain their conventional power plants and have engaged in a complex dance in which they are quickly used when the sun does not shine and the wind does not blow.

Germany's demand for wind turbines and solar panels has stimulated the manufacturing of both (a large proportion of which is done in China) and driven down costs. Wind power in Germany provided an early template for several American states that were generating 25 percent to 50 percent of their electricity from wind by 2016. "It's pretty amazing what's happening, really," said Gerard Reid, an Irish financier who worked in Berlin on energy projects. "The Germans call it a transformation, but to me it's a revolution" (Gillis 2014).

By 2014, Germany's price tag for its energiewende was more than $140 billion, which includes:

> guaranteed returns for farmers, homeowners, businesses and local cooperatives willing to install solar panels, wind turbines, biogas plants and other sources of renewable energy. The plan is paid for through surcharges on electricity bills that cost the typical German family roughly $280 a year, though some of that has been offset as renewables have pushed down wholesale electricity prices. The program has expanded the renewables market and created huge economies of scale, with worldwide sales of solar panels doubling about every 21 months over the past decade, and prices falling roughly 20 percent with each doubling. (Gillis 2014)

Reacting to the 2011 Fukushima earthquake and nuclear plant disaster in Japan, Germany also has pledged to shut down all of its nuclear power plants within a decade.

Japan's Wind Power Platforms

Japan's wind-energy industry got a boost after 2011. As Hiroko Tabuchi wrote in *The New York Times*, this was symbolized by "a giant floating wind turbine [that] signals the start of Japan's most ambitious bet yet on clean energy . . . from the severely damaged and leaking nuclear reactors at Fukushima" (Tabuchi 2013). A single turbine, 2,350 feet tall, rose first, supplying enough power for 1,700 homes.

"Unremarkable, perhaps," Tabuchi wrote, "but consider the goal of this offshore project: to generate over 1 gigawatt of electricity from 140 wind turbines by 2020. That is equivalent to the power generated by a nuclear reactor" (Tabuchi 2013). As Japan turned away from nuclear power, wind turbines became a symbol of a new way of supplying energy to a nation that has next to no fossil fuels of its own. With coastlines of several islands offering thousands of turbine sites, the potential seems limitless.

"It's Japan's biggest hope," said Hideo Imamura, a spokesman for Shimizu, a Japanese engineering and construction firm, during a recent trip to the turbine ahead of its test run. "It's an all-Japan effort, almost 100 percent Japan-made" (Tabuchi 2013). The Japanese have designed platforms for substation and electrical transformer equipment that expand potential locations for wind farms offshore. By one estimate, Japan can use such platforms, according to Tabuchi, "to generate as much as 1,570 gigawatts of electricity, roughly eight times the current capacity of all of Japan's power companies combined, according to computer simulations based on historical weather data by researchers at Tokyo University, one of the project's main participants" (2013).

By one estimate, if the Japanese technology were to be applied worldwide, it could supply as much as 7.5 terawatts of electricity. According to a study by researchers at the University of Delaware and Stanford University in 2012, that would meet half the world's projected power demand in 2030 (Tabuchi 2013). "We're opening a new page in the history of offshore wind power," said Takeshi Ishihara, a civil engineer at Tokyo University who leads the project.

However, Paul J. Scalise, a research fellow at the University of Tokyo Institute of Social Science, said these estimates were too grandiose. "We shouldn't forget the obvious reality check. The farther from the coast they place these floating wind farms, the more expensive it becomes to build them and transmit the power back to Japan," he said. "It becomes a cost-plus benefit analysis in which you weigh the benefits of the electricity versus the cost to build and maintain the infrastructure" (Tabuchi 2013).

Other Japanese citizens have challenged the wind farm, especially local fishermen whose range has been restricted by the platforms. Cost is also a possible obstacle. All told, the platforms will require around eight times as much money per kilowatt hour as land-based turbines. Sponsors of the project hope that design improvements will narrow that gap.

Further Reading

Gillis, Justin. "Sun and Wind Alter Global Landscape, Leaving Utilities Behind." *The New York Times*, September 13, 2014. http://www.nytimes.com/2014/09/14/science/earth/sun-and-wind-alter-german-landscape-leaving-utilities-behind.html.

"Portugal Celebrates Massive Solar Plant" *The New York Times*, March 28, 2007. http://www.nytimes.com/aponline/technology/AP-Portugal-Solar-Power-Plant.html (no longer available).

Roberts, Paul. *The End of Oil: The Edge of a Perilous New World.* Boston: Houghton-Mifflin, 2004.

Rosenthal. Elisabeth. "Portugal Gives Itself a Clean-Energy Makeover." *The New York Times*, August 10, 2010, A1, A8.

Tabuchi, Hiroko. "To Expand Offshore Power, Japan Builds Floating Windmills." *The New York Times,* October 24, 2013. http://www.nytimes.com/2013/10/25/business/international/to-expand-offshore-power-japan-builds-floating-windmills.html.

Whitlock, Craig. "Cloudy Germany a Powerhouse in Solar Energy." *Washington Post,* May 5, 2007, A1. http://www.washingtonpost.com/wp-dyn/content/article/2007/05/04/AR2007050402466_pf.html.

See also: Solar Power; Wind Gains Power Share; Wind Power, Denmark; Wind Power, United States

Primary Documents

INTRODUCTION

Our swiftly changing climate has been extremely well documented in scientific journals and governmental reports worldwide for more than three decades. Reports from many of the major scientific journals are listed in the Selected Bibliography for this set. This section includes a highly selective sample of excerpts from some of the many reports on recent climate change from governmental organizations, including the Intergovernmental Panel on Climate Change (IPCC) and World Meteorological Organization (WMO), both part of the United Nations; the U.S. government's National Climate Assessment (2014); and the Inuit Tapiriit and Inuit Circumpolar Conference, both representing Arctic indigenous peoples. The geographic range of these excerpts ranges from the Canadian Arctic, through the United States, and Africa, with more generalized applications around the world. Subject matter includes current evidence of climate change and advice on what should be done to confront the problem.

UNITED NATIONS INTERGOVERNMENTAL PANEL ON CLIMATE CHANGE (2014)
Accelerating Changes: Evidence and Impact

Human influence on the climate system is clear, and recent anthropogenic emissions of greenhouse gases are the highest in history. Recent climate changes have had widespread impacts on human and natural systems. Warming of the climate system is unequivocal, and since the 1950s, many of the observed changes are unprecedented over decades to millennia. The atmosphere and ocean have warmed, the amounts of snow and ice have diminished, and sea level has risen

Each of the last three decades has been successively warmer at the Earth's surface than any preceding decade since 1850. The period from 1983 to 2012 was likely the warmest 30-year period of the last 1400 years in the Northern Hemisphere, where such assessment is possible.

Ocean warming dominates the increase in energy stored in the climate system, accounting for more than 90 percent of the energy accumulated between 1971 and 2010, with only about 1 percent stored in the atmosphere. On a global scale, the ocean warming is largest near the surface, and the upper 75 m warmed by 0.11 [0.09 to 0.13]°C per decade over the period 1971 to 2010. It is virtually certain that the upper ocean (0–700 meters) warmed from 1971 to 2010, and it likely warmed between the 1870s and 1971.

Glaciers have continued to shrink almost worldwide. Northern Hemisphere spring snow cover has continued to decrease in extent. There is high confidence that permafrost temperatures have increased in most regions since the early 1980s in response to increased surface temperature and changing snow cover. The annual mean Arctic sea ice extent decreased over the period 1979 to 2012, with a rate that was very likely in the range of 3.5 percent to 4.1 percent per decade. Arctic sea ice extent has decreased in every season and in every successive decade since 1979, with the most rapid decrease in decadal mean extent in summer.

It is very likely that the annual mean Antarctic sea ice extent increased in the range of 1.2 to 1.8 percent per decade between 1979 and 2012. However, there is high confidence that there are strong regional differences in Antarctica, with extent increasing in some regions and decreasing in others.

Over the period 1901 to 2010, global mean sea level rose by 0.19 [0.17 to 0.21] meters. The rate of sea-level rise since the mid-19th century has been larger than the mean rate during the previous two millennia.

Anthropogenic greenhouse gas emissions have increased since the pre-industrial era, driven largely by economic and population growth, and are now higher than ever. This has led to atmospheric concentrations of carbon dioxide, methane and nitrous oxide that are unprecedented in at least the last 800,000 years. Their effects, together with those of other anthropogenic drivers, have been detected throughout the climate system and are extremely likely to have been the dominant cause of the observed warming since the mid-20th century. Anthropogenic greenhouse gas (GHG) emissions since the pre-industrial era have driven large increases in the atmospheric concentrations of CO_2, CH_4 [methane], and N_2O [Nitrous oxides].

Between 1750 and 2011, cumulative anthropogenic CO_2 emissions to the atmosphere were 2040±310 Gt CO_2. About 40 percent of these emissions have remained in the atmosphere (880±35 Gt CO_2); the rest was removed from the atmosphere and stored on land (in plants and soils) and in the ocean. The ocean has absorbed about 30 percent of the emitted anthropogenic CO_2, causing ocean acidification. About half of the anthropogenic CO_2 emissions between 1750 and 2011 have occurred in the last 40 years. Total anthropogenic greenhouse gas emissions have continued to increase over 1970 to 2010 with larger absolute increases between 2000 and 2010, despite a growing number of climate change mitigation policies. Emissions of CO_2 from fossil fuel combustion and industrial processes contributed about 78 percent of the total greenhouse gas emissions increase from 1970 to 2010, with a similar percentage contribution for the increase during the period 2000 to 2010. Globally, economic and population growth continued to be the most important drivers of increases in CO_2 emissions from fossil fuel combustion. The contribution of population growth between 2000 and 2010 remained roughly identical to the previous three decades, while the contribution of economic growth has risen sharply. Increased use of coal has reversed the long-standing trend of gradual decarbonization (i.e., reducing the carbon intensity of energy) of the world's energy supply.

The evidence for human influence on the climate system has grown since the Fourth Assessment Report. It is extremely likely that more than half of the observed increase in global average surface temperature from 1951 to 2010 was caused by the anthropogenic increase in greenhouse gas concentrations and other anthropogenic forcings together. The best estimate of the human-induced contribution to warming is similar to the observed warming over this period. Anthropogenic forcings have likely made a substantial contribution to surface temperature increases since the mid-20th century over every continental region except Antarctica.

Anthropogenic influences have likely affected the global water cycle since 1960 and contributed to the retreat of glaciers since the 1960s and to the increased surface melting of the Greenland ice sheet since 1993. Anthropogenic influences have very likely contributed to Arctic sea-ice loss since 1979 and have very likely made a substantial contribution to increases in global upper ocean heat content (surface to a depth of 700 meters) and to global mean sea-level rise observed since the 1970s. For Antarctica, large observational uncertainties result in low confidence that anthropogenic forcings have contributed to the observed warming averaged over available stations.

In recent decades, changes in climate have caused impacts on natural and human systems on all continents and across the oceans. Impacts are due to observed climate change, irrespective of its cause, indicating the sensitivity of natural and human systems to changing climate. Evidence of observed climate-change impacts is strongest and most comprehensive for natural systems. In many regions, changing precipitation or melting snow and ice are altering hydrological systems, affecting water resources in terms of quantity and quality. Many terrestrial, freshwater, and marine species have shifted their geographic ranges, seasonal activities, migration patterns, abundances, and species interactions in response to ongoing climate change. Some impacts on human systems have also been attributed to climate change, with a major or minor contribution of climate change distinguishable from other influences. Assessment of many studies covering a wide range of regions and crops shows that negative impacts of climate change on crop yields have been more common than positive impacts. Some impacts of ocean acidification on marine organisms have been attributed to human influence.

Based on the available scientific literature since the Fourth Assessment, there are substantially more impacts in recent decades now attributed to climate change. Attribution requires defined scientific evidence on the role of climate change. Absence from the map of additional impacts attributed to climate change does not imply that such impacts have not occurred. The publications supporting attributed impacts reflect a growing knowledge base, but publications are still limited for many regions, systems and processes, highlighting gaps in data and studies.

Changes in many extreme weather and climate events have been observed since about 1950. Some of these changes have been linked to human influences, including a decrease in cold temperature extremes, an increase in warm temperature extremes, an increase in extreme high sea levels and an increase in the number of heavy precipitation events in a number of regions. It is very likely that the number of cold days and nights has decreased and the number of warm days and nights has increased on the global scale. It is likely that the frequency of heat waves has increased in large parts of Europe, Asia and Australia.

It is very likely that human influence has contributed to the observed frequency of heat waves for each region, including the potential for risk reduction through adaptation and mitigation, as well as limits to adaptation. Each key risk is assessed as very low, low, medium, high, or very high.

Climate change is projected to undermine food security. Due to projected climate change by the mid-21st century and beyond, global marine species redistribution and marine biodiversity reduction in sensitive regions will challenge the sustained provision of fisheries productivity and other ecosystem services. For wheat, rice, and maize in tropical and temperate regions, climate change without adaptation is projected to negatively impact production for local temperature increases of 2°C or more above late-20th century levels, although individual locations may benefit.

Global temperature increases of about 4°C or more above late-20th century levels, combined with increasing food demand, would pose large risks to food security globally. Climate change is projected to reduce renewable surface water and groundwater resources in most dry subtropical regions, intensifying competition for water among sectors.

These risks are amplified for those lacking essential infrastructure and services or living in exposed areas. Rural areas are expected to experience major impacts on water availability and supply, food security, infrastructure, and agricultural incomes, including shifts in the production areas of food and nonfood crops around the world. Aggregate economic losses accelerate with increasing temperature, but global economic impacts from climate change are currently difficult to estimate. From a poverty perspective, climate-change impacts are projected to slow down economic growth, make poverty reduction more difficult, further erode food security, and prolong existing and create new poverty traps, the latter particularly in urban areas and emerging hotspots of hunger.

International dimensions such as trade and relations among states are also important for understanding the risks of climate change at regional scales. Climate change is projected to increase displacement of people. Populations that lack the resources for planned migration experience higher exposure to extreme weather events, particularly in developing countries with low income. Climate change can indirectly increase risks of violent conflicts by amplifying well-documented drivers of these conflicts such as poverty and economic shocks.

Many aspects of climate change and associated impacts will continue for centuries, even if anthropogenic emissions of greenhouse gases are stopped. The risks of abrupt or irreversible changes increase as the magnitude of the warming increases. A large fraction of anthropogenic climate change resulting from CO_2 emissions is irreversible on a multicentury to millennial time scale, except in the case of a large net removal of CO_2 from the atmosphere over a sustained period.

Source: Intergovernmental Panel on Climate Change. United Nations. *Climate Change 2014: Synthesis Report*. Rajendra K. Pachauri and Leo Meyer, eds. https://www.ipcc.ch/pdf /assessment-report/ar5/syr/SYR_AR5_FINAL_full.pdf.

UNITED NATIONS INTERGOVERNMENTAL PANEL ON CLIMATE CHANGE (2014)

Mitigation Strategies

Stabilization of global average surface temperature does not imply stabilization for all aspects of the climate system. Shifting biomes, soil carbon, ice sheets, ocean temperatures and associated sea-level rise all have their own intrinsic long timescales which will result in changes lasting hundreds to thousands of years after global surface temperature is stabilized. There is high confidence that ocean acidification will increase for centuries if CO_2 emissions continue, and will strongly affect marine ecosystems. It is virtually certain that global mean sea-level rise will continue for many centuries beyond 2100, with the amount of rise dependent on future emissions. Abrupt and irreversible ice loss from the [Greenland and] Antarctic ice sheet[s] are possible, but current evidence and understanding is insufficient to make a quantitative assessment.

Magnitudes and rates of climate change associated with medium-to high-emission scenarios pose an increased risk of abrupt and irreversible regional-scale change in the composition, structure, and function of marine, terrestrial and freshwater ecosystems, including wetlands. A reduction in permafrost extent is virtually certain with continued rise in global temperatures.

Adaptation and mitigation are complementary strategies for reducing and managing the risks of climate change. Substantial emissions reductions over the next few decades can reduce climate risks in the 21st century and beyond, increase prospects for effective adaptation, reduce the costs and challenges of mitigation in the longer term, and contribute to climate-resilient pathways for sustainable development

Limiting the effects of climate change is necessary to achieve sustainable development and equity, including poverty eradication. Countries' past and future contributions to the accumulation of greenhouse gases in the atmosphere are different, and countries also face varying challenges and circumstances and have different capacities to address mitigation and adaptation. Mitigation and adaptation raise issues of equity, justice, and fairness. Many of those most vulnerable to climate change have contributed and contribute little to GHG emissions.

Delaying mitigation shifts burdens from the present to the future, and insufficient adaptation responses to emerging impacts are already eroding the basis for sustainable development. Comprehensive strategies in response to climate change that are consistent with sustainable development take into account the co-benefits, adverse side-effects and risks that may arise from both adaptation and mitigation options. The design of climate policy is influenced by how individuals and organizations perceive risks and uncertainties and take them into account. Methods of valuation from economic, social and ethical analysis are available to assist decision-making. These methods can take account of a wide range of possible impacts, including low-probability outcomes with large consequences. But they cannot identify a single best balance between mitigation, adaptation, and residual climate impacts.

Climate change has the characteristics of a collective action problem at the global scale, because most greenhouse gases accumulate over time and mix globally, and emissions by any agent (e.g., individual, community, company, country) affect other agents. Effective mitigation will not be achieved if individual agents advance their own interests independently. Cooperative responses, including international cooperation, are therefore required to effectively mitigate GHG emissions and address other climate change issues. The effectiveness of adaptation can be enhanced through complementary actions across levels, including international cooperation. The evidence suggests that outcomes seen as equitable can lead to more effective cooperation.

Without additional mitigation efforts beyond those in place today, and even with adaptation, warming by the end of the 21st century will lead to very high risk of severe, widespread, and irreversible impacts globally. Mitigation involves some level of co-benefits and of risks due to adverse side-effects, but these risks do not involve the same possibility of severe, widespread, and irreversible impacts as risks from climate change, increasing the benefits from near-term mitigation efforts.

Mitigation and adaptation are complementary approaches for reducing risks of climate change impacts over different time scales. Mitigation, in the near-term and through the century, can substantially reduce climate change impacts in the latter decades of the 21st century and beyond. Benefits from adaptation can already be realized in addressing current risks, and can be realized in the future for addressing emerging risks.

Five "Reasons for Concern"

The Five Reasons For Concern are associated with:

1. Unique and threatened systems,
2. Extreme weather events,
3. Distribution of impacts,
4. Global aggregate impacts, and
5. Large-scale singular events.

Without additional mitigation efforts beyond those in place today, and even with adaptation, warming by the end of the 21st century will lead to high to very high risk of severe, widespread, and irreversible impacts globally. In most scenarios without additional mitigation efforts (those with 2100 atmospheric concentrations more than 1000ppm CO_2-equivalent), warming is more likely than not to exceed 4°C above pre-industrial levels by 2100. The risks associated with temperatures at or above 4°C include substantial species extinction, global and regional food insecurity, consequential constraints on common human activities, and limited potential for adaptation in some cases. Some risks of climate change, such as risks to unique and threatened systems and risks associated with extreme weather events, are moderate to high at temperatures 1°C to 2°C above pre-industrial levels.

Substantial cuts in greenhouse gas emissions over the next few decades can substantially reduce risks of climate change by limiting warming in the second half of the 21st century and beyond. Cumulative emissions of CO_2 largely determine global mean surface warming by the late 21st century and beyond. Limiting risks across reasons for concern would imply a limit for cumulative emissions of CO_2. *Such a limit would require that global net emissions of CO_2 eventually decrease to zero and would constrain annual emissions over the next few decades* [emphasis added].

But some risks from climate damages are unavoidable, even with mitigation and adaptation. Mitigation involves some level of co-benefits and risks, but these risks do not involve the same possibility of severe, widespread, and irreversible impacts as risks from climate change. Inertia in the economic and climate system and the possibility of irreversible impacts from climate change increase the benefits from near-term mitigation efforts. Delays in additional mitigation or constraints on technological options increase the longer-term mitigation costs to hold climate change risks at a given level.

Adaptation can reduce the risks of climate change impacts, but there are limits to its effectiveness, especially with greater magnitudes and rates of climate change. Taking a longer-term perspective, in the context of sustainable development, increases the likelihood that more immediate adaptation actions will also enhance future options and preparedness. Adaptation can contribute to the well-being of populations, the security of assets, and the maintenance of ecosystem goods, functions and services now and in the future. Adaptation is place-and context-specific.

A first step towards adaptation to future climate change is reducing vulnerability and exposure to present climate variability. Integration of adaptation into planning, including policy design, and decision making can promote synergies with development and disaster risk reduction. Building adaptive capacity is crucial for effective selection and implementation of adaptation options.

Adaptation planning and implementation can be enhanced through complementary actions across levels, from individuals to governments. National governments can coordinate adaptation efforts of local and subnational governments, for example, by protecting vulnerable groups, by supporting economic diversification, and by providing information, policy and legal frameworks, and financial support. Local government and the private sector are increasingly recognized as critical to progress in adaptation, given their roles in scaling up adaptation of communities, households, and civil society and in managing risk information and financing medium evidence.

Adaptation planning and implementation at all levels of governance are contingent on societal values, objectives, and risk perceptions. Recognition of diverse interests, circumstances, social-cultural contexts, and expectations can benefit decision-making processes. Indigenous, local, and traditional knowledge systems and practices, including indigenous peoples' holistic view of community and environment, are a major resource for adapting to climate change, but these have not been used consistently in existing adaptation efforts. Integrating such forms of knowledge with existing practices increases the effectiveness of adaptation. Constraints can interact to impede adaptation planning and implementation.

Common constraints on implementation arise from the following: limited financial and human resources; limited integration or coordination of governance; uncertainties about projected impacts; different perceptions of risks; competing values; absence of key adaptation leaders and advocates; and limited tools to monitor adaptation effectiveness. Another constraint includes insufficient research, monitoring, and observation and the finance to maintain them. Greater rates and magnitude of climate change increase the likelihood of exceeding adaptation limits.

Limits to adaptation emerge from the interaction among climate change and biophysical and/or socioeconomic constraints. Further, poor planning or implementation, overemphasizing short-term outcomes, or failing to sufficiently anticipate consequences, can result in maladaptation, increasing the vulnerability or exposure of the target group in the future or the vulnerability of other people, places, or sectors. Underestimating the complexity of adaptation as a social process can create unrealistic expectations about achieving intended adaptation outcomes.

Significant co-benefits, synergies, and tradeoffs exist between mitigation and adaptation and among different adaptation responses; interactions occur both within and across regions. Increasing efforts to mitigate and adapt to climate change imply an increasing complexity of interactions, particularly at the intersections among water, energy, land use, and biodiversity, but tools to understand and manage these interactions remain limited. Examples of actions with co-benefits include

improved energy efficiency and cleaner energy sources, leading to reduced emissions of health-damaging climate-altering air pollutants;
reduced energy and water consumption in urban areas through greening cities and recycling water;
sustainable agriculture and forestry; and
protection of ecosystems for carbon storage and other ecosystem services.

Transformations in economic, social, technological, and political decisions and actions can enhance adaptation and promote sustainable development.

At the national level, transformation is considered most effective when it reflects a country's own visions and approaches to achieving sustainable development in accordance with its national circumstances and priorities. Restricting adaptation responses to incremental changes to existing systems and structures, without considering transformational change, may increase costs and losses, and miss opportunities.

Planning and implementation of transformational adaptation could reflect strengthened, altered or aligned paradigms and may place new and increased demands on governance structures to reconcile different goals and visions for the future and to address possible equity and ethical implications. Adaptation pathways are enhanced by iterative learning, deliberative processes, and innovation.

There are multiple mitigation pathways that are likely to limit warming to below 2°C relative to pre-industrial levels. These pathways would require substantial emissions reductions over the next few decades and near zero emissions of CO_2 and other long-lived GHGs by the end of the century. Implementing such reductions poses substantial technological, economic, social, and institutional challenges, which increase with delays in additional mitigation and if key technologies are not available. Limiting warming to lower or higher levels involves similar challenges, but on different timescales.

Without additional efforts to reduce GHG emissions beyond those in place today, global emissions growth is expected to persist, driven by growth in global population and economic activities. Global mean surface temperature increases in 2100 in baseline scenarios—those without additional mitigation—range from 3.7 to 4.8°C above the average for 1850 to 1900 for a median climate response. They range from 2.5°C to 7.8°C when including climate uncertainty (5th to 95th percentile range).

Emissions scenarios leading to GHG concentrations in 2100 of about 450 ppm CO_2-equivalent or lower are likely to maintain warming below 2°C over the 21st century relative to pre-industrial levels. These scenarios are characterized by 40 percent to 70 percent global anthropogenic greenhouse-gas emissions reductions by 2050 compared to 2010 and emissions levels near zero or below in 2100. Mitigation scenarios reaching concentration levels of about 500 ppm CO_2-equivalent by 2100 are more likely than not to limit temperature change to less than 2 degrees C, unless they temporarily overshoot concentration levels of roughly 530 ppm CO_2-eq [equivalent] before 2100, in which case they are about as likely as not to achieve that goal. In these 500 ppm CO_2-eq scenarios, global 2050 emissions levels are 25 to 55 lower than in 2010.

Source: Intergovernmental Panel on Climate Change. United Nations. *Climate Change 2014: Synthesis Report*. Rajendra K. Pachauri and Leo Meyer, eds. https://www.ipcc.ch/pdf/assessment-report/ar5/syr/SYR_AR5_FINAL_full.pdf.

WORLD METEOROLOGICAL ORGANIZATION (2015)
The Climate in Africa

The World Meteorological Organization has issued a report on the *Climate in Africa, 2013*, the first in what is intended to be an annual series. It examines temperatures, rainfall and extreme events on a continent which is especially vulnerable to natural climate variability and long-term climate change due to greenhouse gas emissions.

The year 2013, was one of the warmest years on the continent since at least 1950, with temperatures above average in most regions. Precipitation at the continental scale was near average, according to the report.

Several extreme events hit the region. The floods that hit Mozambique in January 2013 were among the 10 most severe in the world that year, based on the number of deaths. In contrast, the rains in Namibia and neighboring countries fell well below normal, leading to a severe drought.

"This regional statement provides further evidence that weather and climate services are vital for protecting life and property in Africa," said WMO Secretary-General Michel Jarraud. "The need to inform decisions about disaster risk, agriculture, water resources, public health and other climate-sensitive sectors is the driving force behind the WMO-led and Norway-funded Global Framework for Climate Services (GFCS) Adaptation Programme in Africa," he said.

The Climate in Africa: 2013 is WMO's first peer-reviewed statement on regional climate conditions. It is modeled on the global statements issued by WMO every year. It was produced by an African Task Team composed of experts representing every sub-region of the Continent, in consultation with the African Centre of Meteorological Applications for Development (ACMAD).

The publication highlighted other long term environmental issues that need high attention, including the risk of a total depletion of Kilimanjaro glacier by mid of this century, the only existing glacier in Africa and the urgent need for improving the adequacy and sustainability of GHG observations over the continent.

WMO's Regional Association for Africa, which is currently meeting in Cabo Verde, approved a resolution that the Africa Climate statement should be compiled and issued on an annual basis.

Weather and Climate Services Underpin Sustainable Development

3 February 2014, Geneva/Cabo Verde (WMO)—The urgent need to boost Africa's resilience to natural hazards like the devastating floods which recently hit southern Africa is high on the agenda of meetings of ministers responsible for meteorological services and their directors.

The African Ministerial Conference on Meteorology (AMCOMET) on 13–14 February will focus on how to improve the provision and use of weather and climate services which will be vital to help the continent cope with shocks caused by extreme weather and climate change. It will be preceded by a meeting from 3–9 February of Permanent Representatives of African countries with WMO, who are directors of National Meteorological and Hydrological Services and by Technical preparation meeting for AMCOMET on 10–12 February. All meetings are being hosted by the Government of Cabo Verde.

"The global warming trend continued in 2014, resulting in heat-waves, flooding and droughts. The first weeks of this year have been marked by flooding in Malawi and Mozambique and a deadly tropical cyclone in Madagascar," said World Meteorological Organization (WMO) Secretary-General Michel Jarraud.

"Africa is extremely vulnerable to the increasing impacts of climate change, which will worsen food insecurity and water stress for the continent's growing population," said Mr. Jarraud.

"National Meteorological and Hydrological Services in Africa are vital to public safety and well-being. They are critical enablers of sustainable development and indispensable partners of economic sectors like agriculture health, water management, energy and transport. They should receive the resources and recognition necessary to fulfill their mandates. This would save the continent millions of dollars and many, many lives," said Mr. Jarraud.

Institutional frameworks differ from country to country, resulting in insufficient cooperation and operational policies. In order to remedy this situation, the AMCOMET session in Cabo Verde will discuss implementation of the Integrated

African Strategy for Meteorology (Weather and Climate). It is the third session of AMCOMET, which was set up in 2010 under the auspices of WMO and the African Union Commission.

The strategy identifies five key pillars for action:

1. Increase political support and recognition of National Meteorological and Hydrological Services and related Regional Climate Centres;
2. Enhance weather and climate service delivery for sustainable development;
3. Improve access to meteorological services for in particular marine and aviation sectors;
4. Support the provision of weather and climate services for climate change adaptation and mitigation;
5. Strengthen partnerships with relevant institutions and funding mechanisms.

The AMCOMET session is also due to decide on the establishment of a Regional Climate Centre for Central Africa to consolidate research and forecasting capabilities. Ministers will also consider input from the meteorological community to a pan-African Space Policy and Strategy.

Ahead of the AMCOMET session, NMHS directors from WMO's Regional Association for Africa will hold their quadrennial meeting to discuss challenges and priorities to be reported to the 17th World Meteorological Congress in May. These include the provision of quality-assured aeronautical, marine and other meteorological services; disaster risk reduction and the expansion of severe weather forecasting demonstration projects; and the improvement of Africa's observing and communication networks. Africa covers one fifth of the world's total land area but has a totally inadequate observation network required for accurate forecasts.

The meetings in Cabo Verde will also discuss progress in rolling out the WMO-spearheaded Global Framework for Climate Services (GFCS). This aims to improve climate services such as seasonal outlooks and make them more widely available, especially in the priority sectors of food security, water management, health and disaster risk reduction. Flagship GFCS projects, involving many different partners, are currently being implemented in several African countries. Increasing numbers of African countries are developing national frameworks for climate services.

Outcomes of the meetings in Cabo Verde will inform discussions on Sustainable Development Goals and feed into the 3rd World Conference on Disaster Risk Reduction in Japan in March 2015 and the U.N. Climate Change negotiations in Paris in December 2015.

Source: "WMO Issues: The Climate in Africa in 2013." World Meteorological Organization, February 3, 2015. https://public.wmo.int/en/media/news/wmo-issues-climate-africa-2013.

WORLD METEOROLOGICAL ORGANIZATION (2014)
Warming Trend Continues in 2014
14 of 15 Hottest Years Have Been in 21st Century

Geneva, 2 February 2015 (WMO) The World Meteorological Organization (WMO) has ranked 2014 as the hottest year on record, as part of a continuing trend. After consolidating leading international datasets, WMO noted that the difference in temperature between the warmest years is only a few hundredths of a degree—less than the margin of uncertainty. [Note: The years 2015 and 2016 were substantially warmer worldwide than 2014.]

Average global air temperatures over land and sea surface in 2014 were 0.57°C (1.03°F) above the long-term average of 14.00°C (57.2°F) for the 1961–1990 reference period. By comparison, temperatures were 0.55°C (1.00°F) above average in 2010 and 0.54°C (0.98°F) above average in 2005, according to WMO calculations. The estimated margin of uncertainty was 0.10°C (0.18°F).

"The overall warming trend is more important than the ranking of an individual year," said WMO Secretary-General Michel Jarraud. "Analysis of the datasets indicates that 2014 was nominally the warmest on record, although there is very little difference between the three hottest years," said Mr. Jarraud. "Fourteen of the fifteen hottest years have all been this century. We expect global warming to continue, given that rising levels of greenhouse gases in the atmosphere and the increasing heat content of the oceans are committing us to a warmer future," he said.

Around 93% of the excess energy trapped in the atmosphere by greenhouse gases from fossil fuels and other human activities ends up in the oceans. Therefore, the heat content of the oceans is key to understanding the climate system. Global sea-surface temperatures reached record levels in 2014.

It is notable that the high 2014 temperatures occurred in the absence of a fully developed El Niño. El Niño occurs when warmer than average sea-surface temperatures in the eastern tropical Pacific combine, in a self-reinforcing loop, with atmospheric pressure systems. This has an overall warming impact on the climate. High temperatures in 1998—the hottest year before the 21st century—occurred during a strong El-Niño year.

"In 2014, record-breaking heat combined with torrential rainfall and floods in many countries and drought in some others—consistent with the expectation of a changing climate," said Mr. Jarraud.

Strong weather and climate services are now more necessary than ever before to increase resilience to disasters and help countries and communities adapt to a fast changing and, in many places, less hospitable climate, said Mr. Jarraud.

WMO released the global temperature analysis in advance of climate change negotiations to be held in Geneva from 8 to 13 February. These talks will help to pave the way for an agreement on action to be adopted by the Parties to the UN Framework Convention on Climate Change next December in Paris.

The WMO analysis is based, amongst others, on three complementary datasets maintained by the Hadley Centre of the UK's Met Office and the Climatic Research Unit, University of East Anglia, United Kingdom (combined); the U.S. National Oceanic and Atmospheric Administration (NOAA) National Climatic Data Centre; and the Goddard Institute of Space Studies (GISS) operated by the National Aeronautics and Space Administration (NASA).

Global average temperatures are also estimated using reanalysis systems, which use the most advanced weather forecasting systems to combine many sources of data to provide a complementary analysis approaches. WMO in particular uses data from the reanalysis produced by the European Centre for Medium-Range Weather Forecasts, which also ranks 2014 as among the four warmest.

Source: "Warming Trend Continues in 2014." World Meteorological Organization. February 2, 2015. https://www.wmo.int/media/content/warming-trend-continues-2014.

UNITED STATES NATIONAL CLIMATE ASSESSMENT, EXTRACTS (2014)

The National Climate Assessment summarizes the impacts of climate change on the United States, now and in the future. A team of more than 300 experts guided by a 60-member Federal Advisory Committee produced the report, which was extensively reviewed by the public and experts, including federal agencies and a panel of the National Academy of Sciences.

Global climate is changing and this is apparent across the United States in a wide range of observations. The global warming of the past 50 years is primarily due to human activities, predominantly the burning of fossil fuels.

Evidence for changes in Earth's climate can be found from the top of the atmosphere to the depths of the oceans. Researchers from around the world have compiled this evidence using satellites, weather balloons, thermometers at surface stations, and many other types of observing systems that monitor the Earth's weather and climate. The sum total of this evidence tells an unambiguous story: the planet is warming.

Temperatures at Earth's surface, in the troposphere (the active weather layer extending up to about 5 to 10 miles above the ground), and in the oceans have all increased over recent decades. The largest increases in temperature are occurring closer to the poles, especially in the Arctic. This warming has triggered many other changes to the Earth's climate. Snow and ice cover have decreased in most areas. Atmospheric water vapor is increasing in the lower atmosphere because a warmer atmosphere can hold more water. Sea level is increasing because water expands as it warms and because melting ice on land adds water to the oceans. Changes in other climate-relevant indicators such as growing season length have been observed in many areas. Worldwide, the observed changes in average conditions have been accompanied by increasing trends in extremes of heat and heavy precipitation events, and decreases in extreme cold. It is the sum total of these indicators that leads to the conclusion that warming of our planet is unequivocal.

Global annual average temperature (as measured over both land and oceans) has increased by more than 1.5°F (0.8°C) since 1880 (through 2012). . . . While there is a clear long-term global warming trend, some years do not show a temperature increase relative to the previous year, and some years show greater changes than others. These year-to-year fluctuations in temperature are due to natural processes, such as the effects of El Niños, La Niñas, and volcanic eruptions.

Sea ice in the Arctic has decreased dramatically since the satellite record began in 1979. Minimum Arctic sea ice extent (which occurs in early to mid-September) has decreased by more than 40 percent. This decline is unprecedented in the historical record, and the reduction of ice volume and thickness is even greater. Ice thickness decreased by more than 50 percent from 1958–1976 to 2003–2008. The percentage of the March ice cover made up of thicker ice (ice that has survived a summer melt season) decreased from 75 percent in the mid-1980s to 45 percent in 2011.

The retreat of sea ice has occurred faster than climate models had predicted. . . . Arctic minimum sea ice extent in 1984, was about 2.59 million square miles, the average minimum extent for 1979–2000. . . . The extent of sea ice dropped to 1.32 million square miles at the end of summer 2012. The dramatic loss of Arctic sea ice increases warming and has many other impacts on the region. Marine mammals including polar bears and many seal species depend on sea ice for nearly all aspects of their existence. Alaska Native coastal communities rely on sea ice for many reasons, including its role as a buffer against coastal erosion from storms and as a platform for hunting.

Ice loss increases Arctic warming by replacing white, reflective ice with dark water that absorbs more energy from the sun. More open water can also increase snowfall over northern land areas and increase the north-south meanders of the jet stream, consistent with the occurrence of unusually cold and snowy winters at mid-latitudes in several recent years., Significant uncertainties remain in interpreting the effect of Arctic ice changes on mid-latitude weather patterns.

In addition to the rapid decline of Arctic sea ice, rising temperatures are reducing the volume and surface extent of ice on land and lakes. Snow cover on land has also decreased over the past several decades, especially in late spring.

U.S. average temperature has increased by 1.3°F to 1.9°F since record keeping began in 1895; most of this increase has occurred since about 1970. The most recent decade was the nation's warmest on record. Because human-induced warming is superimposed on a naturally varying climate, the temperature rise has not been, and will not be, uniform or smooth across the country or over time.

While surface air temperature is the most widely cited measure of climate change, other aspects of climate that are affected by temperature are often more directly relevant to both human society and the natural environment. Examples include shorter duration of ice on lakes and rivers, reduced glacier extent, earlier melting of snowpack, reduced lake levels due to increased evaporation, lengthening of the growing season, changes in plant hardiness zones, increased humidity, rising ocean temperatures, rising sea level, and changes in some types of extreme weather.

Taken as a whole, these changes provide compelling evidence that increasing temperatures are affecting both ecosystems and human society.

A longer growing season provides a longer period for plant growth and productivity and can slow the increase in atmospheric CO_2 concentrations through increased CO_2 uptake by living things and their environment. The longer growing season can increase the growth of beneficial plants (such as crops and forests) as well as undesirable ones (such as ragweed). In some cases where moisture is limited, the greater evaporation and loss of moisture through plant transpiration (release of water from plant leaves) associated with a longer growing season can mean less productivity because of increased drying and earlier and longer fire seasons.

Increased frost-free season length, especially in already hot and moisture-stressed regions like the Southwest, can lead to further heat stress on plants and increased water demands for crops. Higher temperatures and fewer frost-free days during winter can lead to early bud burst or bloom of some perennial plants, resulting in frost damage when cold conditions occur in late spring. In addition, with higher winter temperatures, some agricultural pests can persist year-round, and new pests and diseases may become established.

The lengthening of the frost-free season has been somewhat greater in the western U.S. than the eastern U.S., increasing by 2 to 3 weeks in the Northwest and Southwest, 1 to 2 weeks in the Midwest, Great Plains, and Northeast, and slightly less than 1 week in the Southeast. These differences mirror the overall trend of more warming in the north and west and less warming in the Southeast.

Average annual precipitation over the U.S. has increased in recent decades, although there are important regional differences. For example, precipitation since 1991 (relative to 1901–1960) increased the most in the Northeast (8%), Midwest (9%), and southern Great Plains (8%), while much of the Southeast and Southwest had a mix of areas of increases and decreases.

Climate has changed naturally throughout Earth's history. However, natural factors cannot explain the recent observed warming.

In the past, climate change was driven exclusively by natural factors: explosive volcanic eruptions that injected reflective particles into the upper atmosphere, changes in energy from the sun, periodic variations in the Earth's orbit, natural cycles that transfer heat between the ocean and the atmosphere, and slowly changing natural variations in heat-trapping gases in the atmosphere.

All of these natural factors, and their interactions with each other, have altered global average temperature over periods ranging from months to thousands of years. For example, past glacial periods were initiated by shifts in the Earth's orbit, and then amplified by resulting decreases in atmospheric levels of carbon dioxide and subsequently by greater reflection of the sun's energy by ice and snow as the Earth's climate system responded to a cooler climate.

Natural factors are still affecting the planet's climate today. The difference is that, since the beginning of the Industrial Revolution, humans have been increasingly affecting global climate, to the point where we are now the primary cause of recent and projected future change.

The majority of the warming at the global scale over the past 50 years can only be explained by the effects of human influences, especially the emissions from burning fossil fuels (coal, oil, and natural gas) and from deforestation.

The emissions from human influences affecting climate include heat-trapping gases such as carbon dioxide (CO_2), methane, and nitrous oxide, and particles such as black carbon (soot), which has a warming influence, and sulfates, which have an overall cooling influence. In addition to human-induced global climate change, local climate can also be affected by other human factors (such as crop irrigation) and natural variability.

Oil used for transportation and coal used for electricity generation are the largest contributors to the rise in carbon dioxide that is the primary driver of recent climate change.

Carbon dioxide has been building up in the atmosphere since the beginning of the industrial era in the mid-1700s, primarily due to burning coal, oil, and gas, and secondarily due to clearing of forests. Atmospheric levels have increased by about 40% relative to pre-industrial levels.

Methane levels in the atmosphere have increased due to human activities including agriculture (with livestock producing methane in their digestive tracts and rice farming producing it via bacteria that live in the flooded fields); mining coal, extraction and transport of natural gas, and other fossil fuel-related activities; and waste disposal including sewage and decomposing garbage in landfills. Since pre-industrial times, methane levels have increased by 250%.

Other heat-trapping gases produced by human activities include nitrous oxide, halocarbons, and ozone. Nitrous oxide levels are increasing, primarily as a result of fertilizer use and fossil fuel burning. The concentration of nitrous oxide has increased by about 20% relative to pre-industrial times.

The conclusion that human influences are the primary driver of recent climate change is based on multiple lines of independent evidence. The first line of evidence is our fundamental understanding of how certain gases trap heat, how the climate system responds to increases in these gases, and how other human and natural factors influence climate. The second line of evidence is from reconstructions of past climates using evidence such as tree rings, ice cores, and corals. These show that global surface temperatures over the last several decades are clearly unusual, with the last decade (2000–2009) warmer than any time in at least the last 1,300 years and perhaps much longer.

The third line of evidence comes from using climate models to simulate the climate of the past century, separating the human and natural factors that influence climate. When the human factors are removed, these models show that solar and volcanic activity would have tended to slightly cool the earth, and other natural variations are too small to explain the amount of warming. Only when the human influences are included do the models reproduce the warming observed over the past 50 years.

Another line of evidence involves so-called "fingerprint" studies that are able to attribute observed climate changes to particular causes. For example, the fact that the stratosphere (the layer above the troposphere) is cooling while the Earth's surface and lower atmosphere are warming is a fingerprint that the warming is due to increases in heat-trapping gases. In contrast, if the observed warming had been due to increases in solar output, Earth's atmosphere would have warmed throughout its entire extent, including the stratosphere. In addition to such temperature analyses, scientific attribution of observed changes to human influence extends to many other aspects of climate, such as changing patterns in precipitation, increasing humidity, changes in pressure, and increasing ocean heat content.

Source: "Our Changing Climate." National Climate Assessment. United States. 2014. http://nca2014.globalchange.gov/highlights/report-findings/our-changing-climate.

IMPACT OF CLIMATE CHANGE IN THE ARCTIC: OBSERVATIONS DRAWN FROM SEVERAL GOVERNMENT REPORTS (2002)

This is a world where, as Jose A. Kusugak, president of the Inuit Tapiriit Kanatami described it, "a Storm is a Symphony and Land and Ice are One. . . . Here was our own symphony of wind instruments" (Krupnik and Jolly, 2002, v). Kusugak said that "For Inuit, during most of the year, sea ice is really a large extension of land. In winter, it was rare to find igloos and camps built on land. The land was colder than building igloos on sea or lake ice. The radiant heat of the water made that much difference."

Kusugak said:

At least for now global warming is not all bad. Many northerners, who love to boat, actually are enjoying longer boating seasons. Many Inuit fish with fish nets under ice and are happy that the ice is not as thick as it once was. Fish are fat and plentiful at the moment. Berry picking in the fall has never been better. In biblical terms, it would be the seven years of plenty. But life as we knew it and know it now, is changing fast. *Sirmik* (permanent ice) in many areas is melting causing lakes to drain. *Aniuvat* (permanent snow patches) are disappearing. *Aniuvat* produces our favorite tea water, and caribou frequent *aniuvat* patches to get away from mosquitoes and flies. What I fear is that lives will be lost, because of thinning of ice, and because after lake ice melts and snow on the land is gone in the late spring people are still travelling on sea ice to the beginning of July. Will the ice still be safe for them?

Arctic Native peoples have navigated many changes in the past, and many see climate change as one more peril tossed to them by nature.

The Arctic is by nature highly variable. The availability of many resources is cyclical or unpredictable. For example, there are always uncertainties in caribou migrations. The lesser snow geese may arrive early or late; the nesting success varies considerably from year to year. Weather is changeable and inconsistent. There are large natural variations in the extent of sea-ice cover from year to year, and in freeze-up and break-up dates. Erosion, uplift, thermokarst, and plant succession change landscapes and coastlines. The peoples of the Arctic are familiar with these characteristics of their homeland, and recognize that surprises are inherent in their ecosystems and ways of life. That is why indigenous peoples tend to be flexible in their ways and to have cultural adaptations, such as mobile hunting groups and strong sharing ethics, that help deal with environmental variability and uncertainty.

However, Arctic Native life is changing fundamentally:

At the same time, the settled way of life has reduced both the flexibility of indigenous peoples to move with the seasons to obtain their livelihoods and the extent of their day-to-day contact with their environment, and thus the depth of their knowledge of precise environmental conditions. Instead, they have become dependent on mechanized transportation and fossil fuels to carry out their seasonal rounds while based in one central location. Together, these dependencies have increased the vulnerability of Arctic communities to the impacts of climate change.

Just how dramatic is *silaup aulaninga* (these days, roughly, "climate change")? According to the Arctic Climate Impact Assessment:

Change has been so dramatic that during the coldest month of the year . . . [in December, 2001], torrential rains have fallen in the Thule region so much that there appeared a thick layer of solid ice on top of the sea ice and the surface of the land. . . . The snow that usually covers the sea ice became *nilak* (freshwater ice), and the lower level became *pukak* (crystalized ice), which was very bad for the paws of our sled dogs.

The Arctic Climate Impact Assessment, (2004) demonstrates that the Arctic is currently experiencing some of the most rapid and severe climate change on Earth. Two of its key findings are (1) that marine species dependent upon sea ice, including polar bears, seals, walrus, and various species of birds, are declining and could face extinction, and (2) that the Inuit culture, which is heavily dependent upon sea ice and these species, will experience severe disruption. Sheila Watt-Cloutier, Chairwoman of the ICC, said that the petition had been filed in commitment to cultural survival: "Inuit are an ancient people. Our way of life is dependent upon the natural environment and the animals. Climate change is destroying our environment and eroding our culture. But we refuse to disappear. We will not become a footnote to globalization."

Native Alaskans who live off the land see "good" weather very differently from urban non-Natives, a contrast illustrated in the ACIA by a description of television weathercasters:

This disparity in how weather is perceived by the rural Alaskan communities versus the urban mainstream is apparent from watching weather forecasters across the country, including urban centers in Alaska, such as Anchorage. The premise of these forecasters is based on the urban view that "good" weather should be sunny and warm. For the Qikiktagrugmiut, however, weather is "good" if it is favorable to the country's productivity and the ability of people to access the land's resources. Thus, "good" weather may include rain in July to produce a bountiful berry crop and extremely low temperatures during early autumn so that Kotzebue Sound and the surrounding rivers and lakes freeze quickly and reliably for safe travel. These two conditions, rain and extreme low temperatures, are almost universally portrayed as "bad" weather in urban settings. In addition, the Qikiktagrugmiut's ability to cope with extreme weather events differs from that of most urban communities across the nation. Blizzards that would shut down entire cities and be portrayed as mini-disasters by the urban media are looked upon favorably by rural Alaskan communities.

Increasing variations in weather across the Arctic have become a constant challenge year round. Not only are winters often shorter and warmer, with shorter periods of intense cold, but summers can be cooler. A case study of the Native Village of Kotzebue (home of about 3,000 Iñupiaq, also called *Qikiktagrugmiut*), on the Baldwin Peninsula 30 miles above the Arctic Circle in northwestern Alaska provided the Arctic Climate Impact Assessment (ACIA) an account of a brief, very unusual snowstorm in July, 2000 that covered the ground for a day, long enough to devastate the year's berry harvest. This out-of-season snow arrived in the general content of rising temperatures, especially in autumn and winter.

Arviat, Nunavut: Weather: Winters warmer; in 1940s and 1950s frostbite only took seconds; it is not that cold anymore.

Baker Lake, Nunavut: Caribou not as healthy in recent years. Meat is tough and skin around the neck area tears too easily. More liquid in joints and more white pustules on meat. Caribou less fat and undernourished due to heat and dryness in summer. Caribou skins are weak and tear easily during field dressing. More diseased caribou, e.g., sores in mouth and on tongue. More grizzly bear sightings and encounters around Baker Lake area. Changes in fish (mainly char and trout); trout darker in color, little fat observed between meat layers when boiled, less fish in usual fishing spots, fish eating things they are not supposed to eat, fish too skinny, smell different—"like earth," mushy meat. At least ten new kinds of insects in the area, all winged insects, some recognized from tree line area. Strange kinds of flying insects being observed. Warmer temperatures may be responsible for arrival of flying insects from the south and for insects being active longer in the year.

Bering Strait, USA: Seals: Spotted seals declined from late 1960s/early 1970s to present. 1996 and 1997 spring break-up came early resulting in more strandings of baby ringed seals on the beach. Fewer seals in Nome area these days, perhaps as result of less ice for ringed seal dens. Walrus: Physical condition of walrus generally poor 1996–1998—animals skinny and productivity low. One cause was reduced sea ice, which forces walrus to swim farther between feeding areas. Walrus in good condition in spring 1999 after cold winter and good ice in Bering Sea.

> **St. Lawrence Island, Alaska:** More extreme weather conditions in the last 10 to 20 years (e.g., winds and storms that cause dangerous ice conditions that impede hunting). More warm weather. More intense storms which last longer than in the past. Winds constantly change from one direction to another and with more intensity. Delays in ice packing and freeze-up due to changing winds. Increased frequency of windy and warm conditions creating dangerous ice conditions such as thin ice, snow-covered small open leads, and very rough ice.
>
> **Iqaluit, Nunavut:** Ice conditions becoming more unpredictable through the 1990s with several accidents occurring in late 1990s. In mid/late 1990s, sea ice breaks up more quickly near Iqaluit and earlier than usual. Less sea ice near shorelines.

Reference

Krupnik, I., and D. Jolly, eds. 2002. *The Earth is Faster Now: Indigenous Observations of Arctic Environmental Change.* Fairbanks, AK: Arctic Research Consortium of the United States. http://www.arcus.org/publications/eifn.

WHO IS LIABLE FOR RUINING A CULTURE? THE INUIT SUE THE UNITED STATES OF AMERICA (2005)

Petition to the Inter-American Commission on Human Rights Seeking Relief From Violations Resulting From Global Warming Caused by Acts and Omissions of the United States. Submitted by Sheila Watt-Cloutier With the Support of the Inuit Circumpolar Conference, on Behalf of All Inuit in the Arctic. December 7, 2005. Iqualuit, Nunavut.

"The Climate Change Petition by the Inuit Circumpolar Conference to the Inter-American Commission on Human Rights: Presentation by Sheila Watt-Cloutier, Chair, Inuit Circumpolar Conference [at the] Eleventh Conference of Parties to the UN Framework Convention on Climate Change, December 7, 2005. Montreal. [For the text of the entire petition, See Ansari, Anna. "Climate Change and Human Rights: Case Law." December 8, 2015. Updated by Rosemary Campagna. http://guides.brooklaw.edu/c.php?g=330929&p=2223232.]

The debate over global warming now turns more frequently to legal liability—which individuals, corporations, and governments are responsible, and how should they be held to account? The Inuit Circumpolar Conference, which filed a petition and obtained a hearing March 1, 2007 before the Organization of American States, seeks to establish a legal basis for legal responsibility in world forums for violating their human rights because of its contributions to rapid warming in the Arctic.

The ICC represents the interests of roughly 150,000 Inuit spread around the North Pole from Nunavut (which means "our home" in the Inuktitut language) to

Alaska and Russia. Nunavut itself, a territory four times the size of France, has a population of roughly 27,000, 85 percent of whom are Inuit. From the top of the world, having been exposed to atmospheric perils from toxic chemicals and global warming, the Inuit find themselves in an unwilling but necessary position of international arbiters in an emerging "law of the air" that will eventually govern our shared atmospheric commons.

The OAS responded affirmatively early in 2007 to a request by Nobel Peace Prize nominee Sheila Watt-Cloutier, president until 2006 of the ICC, and two environmental law organizations, Earthjustice and the Center for International Environmental Law. The Petition sought a declaration from the Commission that emissions of greenhouse gases from the United States—source of more than 25 percent of the world's greenhouse gases during the last century—is violating Inuit human rights as outlined in the 1948 American Declaration on the Rights and Duties of Man.

The Petition asserts that practice of the Inuits' rights to culture, life, health, physical integrity and security, property, and subsistence have been imperiled by global warming. With accelerating loss of ice and snow, hunting, travel and other subsistence activities have become more dangerous, and in some cases impossible, the Inuit assert. In addition, drinking-water sources have been threatened. Some coastal communities may be forced to move to escape rising waters provoked by rising seas and increasing storminess.

The petition does not seek monetary damages. Instead, it seeks cessation of United States actions that violate Inuit rights to live in a cold environment—not a small task, since such action will involve major restructuring of economic base to sharply curtail emissions of greenhouse gases. The Petition anticipates the types of actions that will have to take place on a worldwide scale, including the rapidly expanding economies of China and India, to preserve the Inuit way of life, as well as a sustainable worldwide biosphere. The Inuit in this case are very conscious of their pivotal role in a natural world that will not survive climatic business as usual. The Inuit ask United States leadership in an international effort involving sharp reductions in greenhouse-gas emissions to avoid destruction of the order of nature that has sustained them for many thousands of years.

The Inter-American Commission on Human Rights, based in Washington D.C., was created in 1959 by the Organization of American States to safeguard personal freedom and security among the citizens of member states. The ICC's 167-page Petition alleges violations of fundamental human rights among Arctic peoples who ring the Arctic Ocean from Nunavut (a semi-sovereign Inuit province of Canada), to Greenland, Russia, and Alaska. The Petition, which was compiled in defense of Inuit rights as a people within evolving international human-rights law, asks that members of the Commission visit the Arctic to learn first-hand the effects of climate change on the environment and Inuit people. The Petition also seeks a declaration from the commission that emissions of greenhouse gases from the United States—source of more than 25 percent of the world's greenhouse gases during the last century—is violating Inuit human rights as outlined in the 1948 American

Declaration on the Rights and Duties of Man. As of this writing, such a declaration has not been issued.

The Petition asserts that practice of the Inuit's rights to culture, life, health, physical integrity and security, property, and subsistence has been imperiled by global warming. With accelerating loss of ice and snow, hunting, travel and other subsistence activities have become more dangerous, and in some cases impossible, the Inuit assert. In addition, their drinking-water sources have been threatened. Some coastal communities may be forced to move to escape rising waters provoked by rising seas and increasing storminess.

The Petition begins by establishing that global warming is harming "every aspect of Inuit life and culture." It next associates warming with human-provoked emissions of greenhouse gases. Detail then follows regarding the specific perils faced by Inuit people, including dangers to hunters and others who need to travel and obtain food from the land, dangers of melting permafrost, harm to animals in the Arctic, coastal erosion and storm surges, and heat-related health problems. Next, the Petition establishes that the United States of America historically has been the world's greatest national source of greenhouse gases. Because carbon dioxide, the major greenhouse gas, may reside in the atmosphere for a century or longer, the burden is cumulative, so while China during 2007 passed the United States in present-day emissions, the United States remains by far the largest source of the historical burden, at about 27.5 percent of the worldwide total.

The Petition then documents violations of international law, especially the American Declaration. Combining traditional knowledge of hunters and elders with wide-ranging peer-reviewed science, the Petition alleges that Inuit human rights are being violated in several ways, including:

1. The right to life and physical security;
2. The right to personal property;
3. The right to health;
4. The right to practice indigenous culture;
5. The right to use traditionally occupied land; and
6. The right to traditional means of subsistence.

According to Watt-Cloutier, The ICC Petition illustrates three basic messages that arise from Inuit experience in the Arctic.

1. "Dangerous" climate change already is occurring;
2. Climate change in the Arctic is quickly going to become worse; and
3. Climate change in the Arctic is important globally.

The Petition alleges that the United States has "consistently denied, distorted, and suppressed scientific evidence of the causes, rate, and magnitude of global warming" as it reiterates that many of the dangers currently facing the Inuit, including retreat of protective sea ice, impaired access to vital resources, loss of homes, and other infrastructure, are a direct result of human-rights violations committed by the United States, which should be held to account. The Petition then concludes by documenting a lack of remedies within the United States legal system, as well as that country's lack of cooperation under international law in this venue under the administration of George W. Bush.

The ICC and its co-petitioners acknowledge that the declaration they seek from the Commission will not be legally enforceable, but that it would have great moral value. The petition is intended to educate and encourage the United States to join the community of nations in a global effort to combat climate change. The focus has been placed on the United States because of its refusal to join in the Kyoto Protocol and other diplomatic efforts to reduce greenhouse-gas emissions.

Summarizing, the Petition asks the Commission to recommend that the United States:

1. Adopt mandatory measures to limit its emissions of greenhouse gases in co-operation with the community of nations;
2. Take into account the impact of US greenhouse gas emissions on the Arctic and Inuit before approving all major government actions;
3. In consultation with the Inuit, develop a plan to protect Inuit culture and the Arctic environment and to mitigate any harm caused by US greenhouse gas emissions; and
4. In co-ordination with Inuit, develop a plan to help Inuit adapt to unavoidable climate change.

The Petition also asks the Commission to declare that the United States of America has an obligation to take into account the impact of its emissions on the Arctic and Inuit people in all major government actions.

In addition to detailed documentation of the perils faced by Inuit people as a result of global warming, the ICC Petition also contains an extensive legal justification making a case that international law is a proper forum for this dispute, given that one of the fundamental norms of customary international law establishes every state's obligation not to knowingly allow its territory to be used for acts contrary to the rights of other states. Because the emission of greenhouse gases in one state causes harm to others, this norm provides context for assessing states' human-rights obligations with respect to global warming. Under international law, a strong presumption exists compelling nations are to work within the international framework to correct human-rights violations.

UNIKKAAQATIGIIT: PUTTING THE HUMAN FACE ON CLIMATE CHANGE—PERSPECTIVES FROM INUIT IN CANADA (2005)

For more detail, see Intergovernmental Panel on Climate Change. *Climate Change 2007: Synthesis Report*. IPCC Fourth Assessment Report (AR4). 2007. https://www.ipcc.ch/publications_and_data/publications_ipcc_fourth_assessment_report_synthesis_report.htm.

Sometimes it is tough to explain to people from "the south" just how swiftly and fundamentally life in the Arctic has been altered by changing climate. Unikkaaqatigiit: Putting the Human Face on Climate Change: Perspectives from Inuit in Canada *puts these changes into words describing daily life. The information presented in* Unikkaaqatigiit—Putting the Human Face on Climate Change *was collected from the workshops held from 2002 to 2005, involving 17 Inuit communities.*

For example, one such chain was presented by a hunter in the workshops. It involved changes in the environment that affected the diet of caribou, which in turn affected the traditional food security of Inuit. As explained by the hunter, ice layers formed by midwinter thawing or rare winter. Caribou, unable to reach lichen sealed under ice, had been raiding town garbage dumps—not unlike polar bears who found themselves stranded ln shore when ice receded prematurely. One report even had it that hungry bears were hitching rides on garbage trucks on their way to a dump. Verification for this was scanty, however.

Page 9

Some communities . . . noted that individuals are dressing more appropriately for the new conditions brought on by changes in the insect population. They are also beginning to install mosquito netting on windows in their homes in response to the increased incidence of biting flies. Warmer water temperatures are reported to be causing fish caught in nets to rot more quickly. In response, fishers are checking nets and collecting fish more often. They are also harvesting fish earlier in the year. In Nunatsiavut, geese are harvested earlier in the year, and in Nunavut, seals are increasingly harvested in the summer when they are fatter and less likely to sink when shot. The ability to shift the time and place of harvesting activities by a community was reported to support their ability to adapt to the changes taking place.

Page 63

Hunters, elders, and community residents alike have noticed a change in the quality and type of snow that falls in their regions. Residents in Nunatsiavut and Nunavut today more frequently see snow that has a top layer of ice—commonly referred to as "glitter" by Nunatsiavumiut—as a result of freeze-thaw events in the winter. Other communities have started to notice that snow has become drier, more grainy and of a different consistency. While snow seems to be getting drier and less sticky in Nunavik and Nunatsiavut, it is getting harder around Repulse Bay and heavier in Inuvik.

Later fall freeze-ups are also causing difficulties with travel. For example, in the Nunavik region, the communities of Kangiqsujuaq and Ivujivik have found that the timing of ice formation has moved from November to December. Residents in Puvirnituq, in this same region, expressed concern that freeze-up is sometimes not completed until early January. In the ISR community of Aklavik, residents reported that they used to gather at the river's edge as a community during the height of the spring ice break-up to watch and listen to this remarkable spectacle. For several years now, because the break-up has been less dramatic, community members no longer observe this ritual.

Page 74

Seals are sinking faster after being shot (as a result of having less fat), so hunters need to drive their boats faster, which increases fuel consumption in order to catch the seals before they sink. Also hunters have shifted the times they go seal hunting in an attempt to avoid periods when seals are thinnest. Similar changes to hunting practices have been made in Holman Island in the ISR, where residents reported that thinner sea ice has led to a shorter nursing period for seal pups, which is contributing to a decrease in the amount of fat seen on seals. Also in this region, Inuit state that the priming of the fur of most land mammals has been negatively affected by warming temperatures. Fur quality has decreased because of the lack of extreme cold temperatures, resulting in a drop in fur prices and a decrease in income generated by hunting and trapping activities.

Page 76

Overall, the combination of changes—such as decreased amounts of snow around Arctic Bay, stronger winds and rougher ice conditions near Paulatuk, and earlier break-up and later freeze-up in all regions—has meant that travel has become more dangerous, less predictable and consequently less frequent. Spring travel, in particular, has become much more risky than ever before. Despite these reports, some individuals reported positive changes with regard to travel in their region.

Page 78

All regions expressed concerns about the consumption of caribou. The diet of caribou has been affected by changes in the quality of snow cover and increased icing events that restrict winter access to lichen, the primary food for this species. This has led to more unhealthy and thin animals being taken by Inuit hunters. In addition to changes in the diet of caribou, other factors—the increased number of abnormalities reported in the meat, the increased number of flies around the herds and the concern that some caribou may be infected by parasites—contribute to an increased vigilance and concern about consumption of this country food item. As a result of changes in the condition of animals and meat, more care is being taken when selecting animals to consume. All regions also reported that much more meat is being thrown out.

Page 85

In addition, the need to purchase new equipment not previously needed was expressed in some regions. With the prevalence of new and different types of insects in the Arctic, residents . . . reported that protective clothing and insect repellents are now required more than ever before. Residents are installing screens in house windows and doors to increase ventilation during the hotter summer months while keeping out these biting insects. Inuit across the North discussed the need to buy and wear protective sunscreen lotion more often because of the increased incidence of more frequent and harsh sunburns, blisters and rashes. Many more people are buying sunblock.

Page 86

The impacts to food availability, accessibility and quality related to increasing accounts of wildlife abnormalities, changes in migration and distribution patterns of animals and weather and other challenges to access were reported as resulting in an increase in consumption of some store-bought foods as replacements. This shift is economically significant, as a resident from Nunatsiavut stated that one caribou is worth the equivalent of $600–$800 in store-bought foods. Communities identified the need for maintenance and better upkeep of community freezer facilities and programs to ensure a more secure and consistent supply of country foods throughout the year.

However, the operation and maintenance of community freezers is quite costly, being reported in Inuvik as between $10,000 and $12,000 per month.

The seal pelt industry has also been affected by environmental changes, according to residents of Holman Island. In this community, fewer seal pelts are harvested now because the animals tend to be thinner. Also, less fat on these animals means they are less healthy, and therefore yield poorer quality pelts. In Repulse Bay, similar reports emerged, as did those concerning changes in ice conditions affecting the opportunities for seal hunting, further decreasing the potential revenue gained from the sale of seal pelts. However, an increase of harp seals reported in Nunatsiavut represents a potentially positive economic change, as the market for these pelts has returned despite the fact that there is a general decrease in the number of seal products worn in the community.

Page 102

Discussions occurred in Nunavik, Nunatsiavut . . . about the need to protect against insect bites as a result of the increasing numbers of insects and the appearance of new insect species in the region. In Ivujivik, people have found that mosquito repellants do not always work well, and they are now buying bug nets to better protect themselves. Residents mentioned the need for all houses in the communities to have insect screens installed in windows (not previously existing) to keep insects out when it is warm enough to open windows, and for mosquito repellants to be used more often. Previously, when average summer temperatures were not as warm as they are today, this was not an issue, as many residents did not feel the need to open their windows. However, today, with warmer summer temperatures and new insects appearing in the community, the need for screens in all windows is crucial. Many residents suggested that the municipal councils should furnish houses with them.

Reference

Nickels, S., et al. *Unikkaaqatigiit: Putting the Human Face on Climate Change: Perspectives from Inuit in Canada.* Quebec City, Quebec: Inuit Tapiriit Kanatami, Nasivvik Centre for Inuit Health and Changing Environments at Université Laval and the Ajunnginiq Centre at the National Aboriginal Health Organization, Ottawa, 2005. https://www.itk.ca/sites /default/files/unikkaaqatigiit01_0.pdf.

Selected Bibliography

Abboud, Leila. "Sun Reigns on Spain's Plains: Madrid Leads a Global Push to Capitalize on New Solar-Power Technologies." *Wall Street Journal*, December 5, 2006, A4.

"About Senator Inhofe." U.S. Senate Official Biography. No date. http://www.inhofe.senate .gov/biography. Accessed January 10, 2015.

Abrahamson, Dean Edwin. "Global Warming: The Issue, Impacts, Responses." Pp. 3–34 in Dean Edwin Abrahamson, ed., *The Challenge of Global Warming*. Washington, DC: Island Press, 1989.

Abram, N. J., et al. "Coral Reef Death during the 1997 Indian Ocean Dipole Linked to Indonesian Wildfires." *Science* 301 (August 15, 2003): 952–955.

"Abrupt Climate Change during Last Glacial Period Could Be Tied to Dust-Induced Global Warming." Press release NOAA 96-78, December 4, 1996. http://www.publicaffairs.noaa .gov/pr96/dec96/noaa96-78.html.

"Abrupt Climate Change: Should We Be Worried?" Woods Hole Oceanographic Institution, February 10, 2003. https://www.whoi.edu/page.do?pid=83339&tid=3622&cid=9986.

Achenbach, Joel. "Global Warming Did It! Well, Maybe Not." *Washington Post*, August 3, 2008, B1. http://www.washingtonpost.com/wp-dyn/content/article/2008/08/01/AR2008 080103014_pf.html.

"Acidic Oceans Shrink Plankton." *Nature* 510 (June 12, 2014): 190.

"Acidic Waters Do Not Toughen Corals." *Nature* 498 (June 20, 2013): 274–275.

Ackerman, A. S., et al. "Reduction of Tropical Cloudiness by Soot." *Science* 288 (May 12, 2000): 1042–1047.

Adam, David. "Hatchooooh! Record Numbers of People Are Complaining of Hay Fever." *The Guardian* (U.K.), June 18, 2003, 4.

Adams, Jonathan, Mark Maslin, and Ellen Thomas. "Sudden Climate Transitions during the Quaternary." *Physical Geography* 23 (1999): 23:1–36.

Adams, Marilyn. "Strapped Insurers Flee Coastal Areas." *USA Today*, April 26, 2006, B1.

"Afghanistan's Kabul Basin Faces Dry and Thirsty Future." Environment News Service, June 18, 2010. http://www.ens-newswire.com/ens/jun2010/2010-06-18-02.html.

AghaKouchak, Amir, et al. "Water and Climate: Recognizing Anthropogenic Drought." *Nature* 524 (August 27, 2015): 409–411. http://www.nature.com/news/water-and-climate -recognize-anthropogenic-drought-1.18220.

Aguillera, Mario. "Antarctic Ice Shelves Rapidly Thinning; New Study Reveals Accelerating Losses over Two Decades." UC San Diego News Center, March 26, 2015. http://ucsdnews .ucsd.edu/pressrelease/antarctic_ice_shelves_rapidly_thinning.

Ainsworth, Tracy D., et al. "Climate Change Disables Coral Bleaching Protection on the Great Barrier Reef." *Science* 532 (April 15, 2016): 338–342.

Aitken, A. R. A., et al. "Repeated Large-Scale Retreat and Advance of Totten Glacier Indicated by Inland Bed Erosion." *Nature* 533 (May 19, 2016): 385–389. http://www.nature .com/nature/journal/v533/n7603/full/nature17447.html.

Aitken, Mike. "St. Andrews Stymied by Natural Hazard." *The Scotsman* (Edinburgh), April 18, 2001, 20.

"Alaskan Glaciers Retreating." Environment News Service, December 11, 2001. http://ens-news.com/ens/dec2001/2001L-12-11-09.html.

"Alaska's Biggest (Ice) Losers Are Inland." NASA Earth Observatory, July 7, 2015. http://earthobservatory.nasa.gov/IOTD/view.php?id=86168&src=eoa-iotd.

"Alaska Sinks as Climate Change Thaws Permafrost." *USA Today*, October 9, 2013. http://www.usatoday.com/search/Alaska%20Sinks%20as%20Climate%20Change%20Thaws%20Permafrost%20October%202013/.

"Alaska: Walrus Again Crowd onto Shore." *The New York Times*, September 11, 2015, A15.

"Alberta Wildfires Prompt Oil Firms to Suspend Production and Evacuate Staff." *The Guardian* (U.K.), May 26, 2015. http://www.theguardian.com/world/2015/may/26/alberta-wildfires-oil-production-suspended-evacuations.

Albright, Rebecca, et al. "Reversal of Ocean Acidification Enhances Net Coral Reef Calcification." *Nature* 530 (February 24, 2016). http://www.nature.com/nature/journal/vaop/ncurrent/full/nature17155.html.

Alford, Ross A. "Bleak Future for Amphibians." *Science* 480 (December 22 and 29, 2011): 461–462.

Alford, Ross A., Kay S. Bradfield, and Stephen J. Richards. "Ecology: Global Warming and Amphibian Losses." *Nature* 447 (May 31, 2007): E3, E4.

Allan, Richard P. "Climate Change: Human Influence on Rainfall." *Nature* 470 (February 17, 2011): 344–445.

Allan, Richard P., and Brian J. Soden. "Atmospheric Warming and the Amplification of Precipitation Extremes." *Science* 321 (September 12, 2008): 1481–1484.

Allen, Lee. "Southwest Tribes Struggle with Climate Change Fallout." Indian Country Today Media Network, June 14, 2012. http://indiancountrytodaymedianetwork.com/article/southwest-tribes-struggle-with-climate-change-fallout-118386.

Alley, Richard B. "Ice-Core Evidence of Abrupt Climate Changes." *Proceedings of the National Academy of Sciences of the United States of America* 97(4) (February 15, 2000): 1331–1334.

Alley, Richard B. "A Heated Mirror for Future Climate." *Science* 352 (April 8, 2016): 151–152.

Alley, Richard B., ed. *Abrupt Climate Change: Inevitable Surprises*. Washington, DC: National Academy Press, 2002.

Alley, Richard B., et al. "Abrupt Climate Change." *Science* 299 (March 28, 2003): 2005–2010.

Alley, Richard B., et al. "Ice-Sheet and Sea-Level Changes." *Science* 310 (October 21, 2005): 456–460.

Allison, Melissa. "The Effect of Rising Sea Levels on Coastal Homes." Zillow Porchlight, August 2, 2016. http://www.zillow.com/blog/rising-sea-levels-coastal-homes-202268/.

Almendares, J., and M. Sierra. "Critical Conditions: A Profile of Honduras." *Lancet* 342 (December 4, 1993): 1400–1403.

"The Alutiiq Pride Shellfish Hatchery." Web page. http://alutiiqpridehatchery.com. Accessed January 31, 2015.

"Amazon Carbon Sink Threatened by Drought." NASA Earth Observatory. March 5, 2009. http://earthobservatory.nasa.gov/Newsroom/view.php?id=37493&src=eoa (no longer available).

"Amazon Destruction More Rapid Than Expected." April 10, 1999. http://www.gsreport.com/articles/art000099.html (no longer available).

"American Indians Feel the Effects of Climate Change at Higher Rate Than Other Groups." Indian Country Today Media Network. August 9, 2011. http://indiancountry

todaymedianetwork.com/article/american-indians-feel-the-effects-of-climate-change-at -higher-rate-than-other-groups-46365.

Amstrup, S. C., et al. "Recent Observations of Intraspecific Predation and Cannibalism among Polar Bears in the Southern Beaufort Sea." *Polar Biology*, 2006. doi: 10.1007/s00300-006-0142-5.

"Analysis: Climate Warming at Steep Rate." *Omaha World-Herald*, February 23, 2000, 12.

Ananthaswamy, Anil. "Fatal Cloudburst Devastates Himalayan Desert Town." *New Scientist*, August 2010. http://www.NewScientist.com/article/dn19323-fatal-cloudburst-devastates -himalayan-desert-town.html?DCMP=OTC-rss&nsref=environment.

"Anatomy of an Ice Shelf's Demise." *Nature* 503 (November 28, 2013): 441.

"Ancient Fossils Show Arctic Now Near Climate Tipping Point." Environment News Service, June 29, 2010. http://www.ens-newswire.com/ens/jun2010/2010-06-29-01.html.

Anderegg, William R. L., et al. "Tree Mortality Predicted from Drought-Induced Vascular Damage." *Nature GeoScience*, March 30, 2015. doi: 10.1038/ngeo2400.

Anderson, David M., Jonathan T. Overpeck, and Anil K. Gupta. "Increase in the Asian Southwest Monsoon during the Past Four Centuries." *Science* 297 (July 26, 2002): 596–599.

Anderson, James G., et al. "UV Dosage Levels in Summer: Increased Risk of Ozone Loss from Convectively Injected Water Vapor." *Science* 337 (August 17, 2012): 835–839.

Anderson, J. G., W. H. Brune, and M. H. Proffitt. "Ozone Destruction by Chlorine Radicals within the Antarctic Vortex: The Spatial and Temporal Evolution of ClO/O_3, Anticorrelation Based on In Situ ER-2 Data." *Journal of Geophysical Research* 94 (1989): 11465–11479.

Anderson, John Ward. "Europe's Summer of Wild, Wild Weather; Fires, Droughts and Floods Leave Wake of Destruction." *Washington Post*, August 2, 2007, A11. http://www.washingtonpost.com/wp-dyn/content/article/2007/08/01/AR2007080102347_pf.html.

Anderson, Rebecca K., et al. "A Millennial Perspective on Arctic Warming from 14-C in Quartz and Plants Emerging from Beneath Ice Caps." *Geophysical Research Letters* 35 (2008): L01502. doi: 10.1029/2007/GL032057.

Andrews, Edmund. "Senate Adopts an Energy Bill Raising Mileage for Cars." *The New York Times*, June 22, 2007. http://www.nytimes.com/2007/06/22/us/22energy.html.

Angier, Natalie. "Ice Dwellers Are Finding Less Ice to Dwell On." *The New York Times*, May 20, 2008. http://www.nytimes.com/2008/05/20/science/20count.html.

Annan, Kofi. "As Climate Changes, Can We?" *Washington Post*, November 8, 2006, A27. http://www.washingtonpost.com/wp-dyn/content/article/2006/11/07/AR2006110701 229_pf.html.

Annan, Kofi. "Global Warming an All-Encompassing Threat." Address to United Nations Conference on Climate Change, Nairobi, Kenya. Environment News Service, November 15, 2006. http://www.ens-newswire.com/ens/nov2006/2006-11-15-insann.asp (no longer available).

"Annual Peak of Arctic Sea Ice Hit a Record Low." NASA Earth Observatory, April 8, 2016. http://earthobservatory.nasa.gov/IOTD/view.php?id=87831&src=eoa-iotd.

"Annual Peak of Arctic Sea Ice Is Far Below the Norm." NASA Earth Observatory, March 21, 2015. http://earthobservatory.nasa.gov/IOTD/view.php?id=85562&src=eoa-iotd.

Ansari, Anna. "Climate Change and Human Rights: Case Law." December 8, 2015. Updated by Rosemary Campagna. http://guides.brooklaw.edu/c.php?g=330929&p=2223232.

"Antarctic Ice Loss Speeds Up, Nearly Matches Greenland Loss." NASA Earth Observatory. January 23, 2008. Accessed March 12, 2008. http://www.jpl.nasa.gov/news/news .cfm?release=2008-010.

"Antarctic Ice Shelf Collapse Tied to Global Warming." Environment News Service, October 16, 2006. [http://www.ens-newswire.com/.

"Antarctic Ice Shelf Disappears, Arctic Melting Rapidly." Environment News Service, April 3, 2009. http://www.ens-newswire.com/ens/apr2009/2009-04-03-01.asp (no longer available).

"Antarctic Sea Ice Reaches New Maximum Extent." NASA Earth Observatory, October 11, 2012. http://earthobservatory.nasa.gov/IOTD/view.php?id=79369&src=eoa-iotd.

Antonowicz, Anton. "Baking Alaska: As World Leaders Bicker, Global Warming Is Killing a Way of Life." *Daily Mirror* (U.K.), November 28, 2000, 8–9.

Appenzeller, Tim. "The Big Thaw." *National Geographic*, June 2007, 56–71.

Archambeault, William G. "Louisiana Indians: Survivors in a Post-Katrina Environment." *Indigenous Policy Journal*, 2006. Indigenous Studies Network (ISN). http://indigenouspolicy .org/Articles/VolXVIINo3/LouisianaIndians/tabid/181/Default.aspx. Accessed May 16, 2012.

Archer, David. *The Long Thaw: How Humans Are Changing the Next 100,000 Years of Earth's Climate*. Princeton, NJ: Princeton University Press, 2009.

Arctic Climate Impact Assessment. *Impacts of a Warming Arctic*. Cambridge, UK: Cambridge University Press, 2004.

"Arctic Ice Melting Rapidly, Study Says." *The New York Times*, September 14, 2006. http:// www.nytimes.com/aponline/us/AP-Warming-Sea-Ice.html (no longer available).

"Arctic Ice Retreating 30 Years Ahead of Projections." Environment News Service, April 30, 2007. http://www.ens-newswire.com/ens/apr2007/2007-04-30-04.asp (no longer available).

"Arctic Ocean Ice Thinner by Half in Six Years." Environment News Service, September 14, 2007. http://www.ens-newswire.com/ens/sep2007/2007-09-14-03.asp (no longer available).

"Arctic Oscillation Chills North America, Warms Arctic." NASA Earth Observatory, January 26, 2011. http://earthobservatory.nasa.gov/IOTD/view.php?id=48882&src=eoa-iotd.

"Arctic Oscillation Chills U.S. and Europe." NASA Earth Observatory, December 17, 2010. http://earthobservatory.nasa.gov/IOTD/view.php?id=47880&src=eoa-iotd.

"Arctic Ozone Loss." NASA's Earth Observatory March 30, 2011. http://earthobservatory .nasa.gov/IOTD/view.php?id=49874&src=eoa-iotd.

"Arctic Sea Ice." NASA Earth Observatory. May 3, 2011. http://earthobservatory.nasa.gov/ IOTD/view.php?id=50365&src=eoa-iotd.

"Arctic Sea Ice Extent Hits Record Low." Environment News Service, August 20, 2007. http:// www.ens-newswire.com/ens/aug2007/2007-08-20-01.asp.

"Arctic Sea Ice Melt Accelerating." Environment News Service, October 4, 2006. http://www .ens-newswire.com/ens/oct2006/2006-10-04-02.asp (no longer available).

"Arctic Sea Ice Melt May Set Off Climate Change Cascade." Environment News Service, March 19. 2006. http://www.ens-newswire.com/ens/mar2007/2007-03-19-06.asp (no longer available).

"Arctic Sea Ice Reaches Annual Low." NASA Earth Observatory, September 16, 2015. http:// earthobservatory.nasa.gov/IOTD/view.php?id=86607&src=eoa-iotd.

"Arctic Snowpack Thins." *Nature* 512 (August 14, 2014): 115.

"The Arena Profile: Senator James Inhofe." Politico.com. No date. Accessed January 11, 2015. http://www.politico.com/arena/bio/sen_james_inhofe.html.

Arendt, Anthony A., et al. "Rapid Wastage of Alaska Glaciers and Their Contribution to Rising Sea Level." *Science* 297 (July 19, 2002): 382–386.

"Are White Christmases Just a Memory?" Environment News Service, December 21, 2001. http://ens-news.com/ens/dec2001/2001L-12-21-09.html (no longer available).

Arnold, David. "Global Warming Lends Power to a Jellyfish in Narragansett Bay and Long Island Sound; Non-Native Species Are Taking Over." *Boston Globe*, July 2, 2002, C1.

Aronson, Richard B., et al. "Coral Bleach-Out in Belize." *Nature* 405 (May 4, 2000): 36.

Arrhenius, Svante. "On the Influence of Carbonic Acid in the Air upon the Temperature of the Ground." *The London, Edinburgh, and Dublin Philosophical Magazine and Journal of Science*, 5th ser. (April 1896), 237–276.

Arrigo, Kevin R., et al. "Massive Phytoplankton Blooms under Arctic Sea Ice." *Science* 336 (June 15, 2012): 1408.

Arthur, Charles. "Snows of Kilimanjaro Will Disappear by 2020, Threatening World-Wide Drought." *The Independent* (London), October 18, 2002, 7.

Arthur, Charles. "Temperature Rise Kills 90 Per Cent of Ocean's Surface Coral." *The Independent* (London), September 18, 2003, n.p. (LEXIS).

"As Arctic Sea Ice Melts, Experts Expect New Low." *The New York Times*, August 28, 2008. http://www.nytimes.com/2008/08/28/science/earth/28seaice.html.

Asner, Gregory P., et al. "Selective Logging in the Brazilian Amazon." *Science* 310 (October 21, 2005): 480–481.

Asokan, Shyamantha. "Indian States Fight over River Usage." *Washington Post*, April 1, 2012. http://www.washingtonpost.com/world/asia_pacific/indian-states-fight-over-river-usage /2013/04/01/73026ae0-9895-11e2-b68f-dc5c4b47e519_print.html.

Assel, Raymond A., Frank H. Quinn, and Cynthia E. Sellinger. "Hydroclimatic Factors of the Recent Record Drop in Laurentian Great Lakes Water Levels." *Bulletin of the American Meteorological Society* 85(8) (August 2004): 1143–1150.

"Athabasca Oil Sands." NASA Earth Observatory. November 30, 2011. http://earthobservatory .nasa.gov/IOTD/view.php?id=76559&src=eoa-iotd.

Atlas, James. "Is This the End?" *The New York Times*, November 24, 2012. http://www .nytimes.com/2012/11/25/opinion/sunday/is-this-the-end.html.

Auchard, Eric, and Leonard Anderson. "Google Plans Largest U.S. Solar-Powered Office." *Washington Post*, October 16, 2006. http://www.washingtonpost.com/wp-dyn/content/ article/2006/10/16/AR2006101601100_pf.html (no longer available).

Augenbraun, Harvey, Elaine Matthews, and David Sarma. "The Greenhouse Effect, Greenhouse Gases, and Global Warming." No date. http://icp.giss.nasa.gov/research/methane/ greenhouse.html (no longer available).

Austen, Ian. "Canada: Storm Floods Toronto." *The New York Times*, July 8, 2013. http://www .nytimes.com/2013/07/09/world/americas/canada-storm-floods-toronto.html.

"Australia Assesses Fire Damage in Capital." *Omaha World-Herald*, January 20, 2003, A4.

"Australian Army Started Bushfire Blazes, Climate Also Blamed." Environment News Service, October 23, 2013. http://ens-newswire.com/2013/10/23/australian-army-started -bushfire-blazes-climate-also-blamed/.

"Automakers Join Call for National Greenhouse Gas Limits." Environment News Service, June 27, 2007. http://www.ens-newswire.com/ens/jun2007/2007-06-27-09.asp#anchor4 (no longer available).

Ayres, Ed. *God's Last Offer: Negotiating a Sustainable Future*. New York: Four Walls Eight Windows, 1999.

Azzioni, Tales. "Drought Has Amazon Tributary at Record Low Levels." Associated Press in *Washington Post*, October 25, 2010. http://www.washingtonpost.com/wp-dyn/content/ article/2010/10/25/AR2010102502661_pf.html (no longer available).

Baar, Hein J. W. de, and Michel H. C. Stoll. "Storage of Carbon Dioxide in the Oceans." Pp. 143–177 in F. Stuart Chapin III et al., eds., *Arctic Ecosystems in a Changing Climate: An Ecophysiological Perspective.* San Diego: Academic Press, 1992.

Bacon, John. "1,000-year Storm Slams S.C." *USA Today*, October 5, 2015, 1A.

"Baffin Island Ice Caps Shrink by Half in 50 Years." Environment News Service, January 28, 2008. http://www.ens-newswire.com/ens/jan2008/2008-01-28-02.asp.

Bagla, Pallava. "No Sign Yet of Himalayan Meltdown, Indian Report Finds." *Science* 326 (November 13, 2009): 924–925.

Bajaj, Vikas. "Crops in India Wilt in a Weak Monsoon Season." *The New York Times*, September 3, 2012. http://www.nytimes.com/2012/09/04/business/global/drought-in-india-devastates-crops-and-farmers.html.

"Baked Alaska." NASA Earth Observatory, May 27, 2015. http://earthobservatory.nasa.gov/IOTD/view.php?id=85932&src=eoa-iotd.

Baker, Andrew C. "Reef Corals Bleach to Survive Change." *Nature* 411 (June 14, 2001): 765–766.

Baker, Andrew C., et al. "Corals' Adaptive Response to Climate Change." *Nature* 430 (August 12, 2004): 741.

Balch, Jennifer. "Atmospheric Science: Drought and Fire Change Sink to Source." *Nature* 506 (February 6, 2014): 41–42.

Baldwin, Mark P., et al. "Weather from the Stratosphere?" *Science* 301 (July 18, 2003): 317–318.

Baldocchi, D., and E. Waller. "Winter Fog Is Decreasing in the Fruit Growing Region of the Central Valley of California." *Geophysical Research Letters* 41(9) (May 15, 2014): 3251–3256. http://onlinelibrary.wiley.com/doi/10.1002/2014GL060018/abstract.

Ball, Jeffrey. "The Texas Wind Powers a Big Energy Gamble." *Wall Street Journal*, March 12, 2007, A1, A11.

Ball, Jeffrey. "Economics: Creating Environmental Capital." *Wall Street Journal*, March 24, 2008, R1.

Ball, Philip. "Climate Change Set to Poke Holes in Ozone." *Nature* March 3, 2004. http://info.nature.com/cgi-bin24/DM/y/eOCB0BfHSK0Ch0JVV0AY.

Balmaseda, Magdalena A., Kevin E. Trenberth, and Erland Källén. "Distinctive Climate Signals in Reanalysis of Global Ocean Heat Content." *Geophysical Research Letters* 40(9) (May 16, 2013): 1754–1759.

Bamber, Jonathan L., David G. Vaughan, and Ian Joughin. "Widespread Complex Flow in the Interior of the Antarctic Ice Sheet." *Science* 287 (February 18, 2000): 1248–1250.

Bamber, Jonathan L., et al. "Reassessment of the Potential Sea-Level Rise from a Collapse of the West Antarctic Ice Sheet." *Science* 324 (May 15, 2009): 901–903.

"BAMS State of the Climate: 2011." *Bulletin of the American Meteorological Society*, Special Supplement, 93(7) (July 2012).

Barbeliuk, Anne. "Warmer Globe Choking Ocean." *The Mercury* (Hobart, Australia), March 16, 2002, n.p. (LEXIS).

Barkham, Patrick. "Going Down: Tuvalu, a Nation of Nine Islands" *The Guardian* (U.K.), February 16, 2002, 24.

Barley, Shanta. "Major Antarctic Glacier Is 'Past Its Tipping Point.'" *New Scientist*, January 13, 2010. http://www.newscientist.com/article/dn18383-major-antarctic-glacier-is-past-its-tipping-point.html?DCMP=OTC-rss&nsref=environment.

Barnett, Tim P., et al. "Penetration of Human-Induced Warming into the World's Oceans." *Science* 309 (July 8, 2005): 284–287.

Baron, Ethan. "Beetles Could Chew Up 80 Percent of B.C. Pine: Report: "Worst-Case Scenario' by 2020 Blamed on Global Warming." *Ottawa Citizen*, September 12, 2004, A3.

Barrett, Joe. "Ethanol Reaps a Backlash in Small Midwestern Towns." *Wall Street Journal*, March 23, 2007, A1, A8.

Barringer, Felicity. "A Coalition for Firm Limit on Emissions." *The New York Times*, January 19, 2007. http://www.nytimes.com/2007/01/19/business/19carbon.html.

Barringer, Felicity. "Navajos and Environmentalists Split on Power Plant." *The New York Times*, July 27, 2007. http://www.nytimes.com/2007/07/27/us/27navajo.html.

Barringer, Felicity. "Precipitation across U.S. Intensifies over 50 Years." *The New York Times*, December 5, 2007. http://www.nytimes.com/2007/12/05/us/05storms.html.

Barringer, Felicity. "Lake Mead Could Be within a Few Years of Going Dry, Study Finds." *The New York Times*, February 13, 2008. http://www.nytimes.com/2008/02/13/us/13mead.html.

Barringer, Felicity. "Polar Bear Is Made a Protected Species." *The New York Times*, May 15, 2008. http://www.nytimes.com/2008/05/15/us/15polar.html.

Barringer, Felicity. "With Push toward Renewable Energy, California Sets Pace for Solar Power." *The New York Times*, July 16, 2009, A17.

Barringer, Felicity. "Solar Power Plants to Rise on U.S. Land." *The New York Times*, October 6, 2010. http://www.nytimes.com/2010/10/06/science/earth/06solar.html.

Barringer, Felicity. "A Significant Ozone Hole Is Reported over the Arctic." *The New York Times*, October 3, 2011. http://www.nytimes.com/2011/10/04/science/earth/04ozone.html.

Barringer, Felicity, and Kenneth Chang. "Experts See New Normal as a Hotter, Drier West Faces More Huge Fires." *The New York Times*, July 1, 2013 http://www.nytimes.com/2013/07/02/us/experts-see-a-hotter-drier-west-with-more-huge-fires.htm (no longer available).

Barriopedro, David, et al. "The Hot Summer of 2010: Redrawing the Temperature Record Map of Europe." *Science* 332 (April 8, 2011): 220–224.

Barry, Ellen. "For Indians, Smog and Poverty Are Higher Priorities Than Talks in Paris." *The New York Times*, December 10, 2015, A19.

Barry, Ellen, and Carol Davenport. "India Announces Plan to Lower Rate of Greenhouse Gas Emissions." *The New York Times*, October 1, 2015. http://www.nytimes.com/2015/10/02/world/asia/india-announces-plan-to-lower-rate-of-greenhouse-gas-emissions.html.

Barry, R. G. "Cryospheric Responses to a Global Temperature Increase." Pp. 169–180 in Jill Williams, ed., *Carbon Dioxide, Climate, and Society*. Proceedings of IIASA workshop cosponsored by WMO, UNEP, and SCOPE, February 21–24, 1978. Oxford, UK: Pergamon Press, 1978.

Barta, Patrick. "Parched Outback: In Australia, a Drought Spurs a Radical Remedy." *Wall Street Journal*, July 11, 2007, A1, A12.

Basu, Janet. "Ecologists' Statement on the Consequences of Rapid Climatic Change—May 20, 1997." http://www.dieoff.com/page104.htm.

Bates, Albert K., and Project Plenty. *Climate in Crisis: The Greenhouse Effect and What We Can Do.* Summertown, TN: The Book Publishing Co., 1990.

Battisti, David S., and Rosamond L. Naylor. "Historical Warnings of Future Food Insecurity with Unprecedented Seasonal Heat." *Science* 323 (January 9, 2009): 240–244.

Bazzaz, Fakhri A., and Eric D. Fajer. "Plant Life in a CO_2-Rich World." *Scientific American*, January 1992, 68–74.

Bearak, Max. "Electrifying India, with the Sun and Small Loans." *The New York Times*, January 2, 2015. http://www.nytimes.com/2016/01/03/business/energy-environment/electrifying-india-with-the-sun-and-small-loans.html.

"Bears Face Days-Long Swims." *Omaha World-Herald*, April 23, 2016, 3A.

Beaugrand, Gregory, et al. "Reorganization of North Atlantic Marine Copepod Biodiversity and Climate." *Science* 296 (May 31, 2002): 1692–1694.

Beaugrand, Gregory, et al. "Plankton Effect on Cod Recruitment in the North Sea." *Nature* 426 (December 11, 2003): 661–664.

Behrenfeld, Michael J., et al. "Climate-Driven Trends in Contemporary Ocean Productivity." *Nature* 444 (December 7, 2006): 752–755.

"Beijing Bans Coal Burning to Clear the Air." Environment News Service, August 6, 2014. http://ens-newswire.com/2014/08/06/beijing-bans-coal-burning-to-clear-the-air/.

Beinecke, Frances. "A Climate for Change: Next Steps in Solving Global Warming." Pp. 175–185 in Steven Kazlowski, *The Last Polar Bear: Facing the Truth of a Warming World*. Seattle: Braided River, 2008.

Belford, Aubrey, and Meraiah Foley. "Australian Floods Rage through Brisbane." *The New York Times*, January 13, 2011. http://www.nytimes.com/2011/01/14/world/asia/14australia.html.

Belkin, Douglas. "Northern Vintage: Canada's Wines Rise with Mercury." *Wall Street Journal*, October 15, 2007, A1, A20.

Belmecheri, Soumaya, et al. "Multi-Century Evaluation of Sierra Nevada Snowpack." *Nature Climate Change*, September 2015. http://www.nature.com/nclimate/journal/vaop/ncurrent/full/nclimate2809.html. doi: 10.1038/nclimate2809.

Belson, Ken. "Just Wind, Baby." *The New York Times*, November 18, 2010, B18, B20.

Belt, Don. "The Coming Storm." *National Geographic*, May 2011, 58–83.

Bender, Morris A., et al. "Modeled Impact of Anthropogenic Warming on the Frequency of Intense Atlantic Hurricanes." *Science* 327 (January 22, 2010): 454–458.

Bengtsson, L., M. Botzet, and M. Esch. "Will Greenhouse Gas-Induced Warming over the Next 50 Years Lead to a Higher Frequency and Greater Intensity of Hurricanes?" *Tellus* 48A (1996): 57–73.

Benjamin, Craig. "The Machu Picchu Model: Climate Change and Agricultural Diversity." *Native Americas* 16(3–4) (Summer–Fall 1999): 76–81.

Bennartz, R., et al. "July 2012 Greenland Melt Extent Enhanced by Low-Level Liquid Clouds." *Nature* 496 (April 4, 2013): 83–86.

Benson, Simon. "Giant Squid 'Taking over the World.'" *Daily Telegraph* (Sydney), July 31, 2002, 4.

Benton, Michael J. *When Life Nearly Died: The Greatest Mass Extinction of All Time*. London: Thames & Hudson, 2003.

Bernard, Harold W., Jr. *Global Warming: Signs to Watch For*. Bloomington: Indiana University Press, 1993.

Bernstein, Lenny. "World's Fish Have Been Moving to Cooler Waters for Decades, Study Finds." *Washington Post*, May 15, 2013. http://www.washingtonpost.com/national/health-science/worlds-fish-have-been-moving-to-cooler-waters-for-decades-study-finds/2013/05/15/730292e8-bcd7-11e2-9b09-1638acc3942e_print.html.

Berry, Carol. "Alberta Oil Sands Up Close: Gunshot Sounds, Dead Birds, a Moonscape." Indian Country Today Media Network, February 2, 2012. http://indiancountrytodaymedianetwork.com/article/alberta-oil-sands-up-close%3A-gunshot-sounds%2C-dead-birds%2C-a-moonscape-95444 (no longer available).

Berry, Carol. "Tribal Members Sign Treaty Calling for an End to Alberta Oil Sands Development and Keystone XL." Indian Country Today Media Network. January 31, 2013. http://indiancountrytodaymedianetwork.com/2013/01/31/tribal-members-sign-treaty-calling-end-alberta-oil-sands-development-and-keystone-xl.

Bethel, M. B., et al. "Blending Geospatial Technology and Traditional Ecological Knowledge to Enhance Restoration Decision-Support Processes in Coastal Louisiana." *Journal of Coastal Research*, 27(3) (2011): 555–571.

Betsill, Michelle M. "Impacts of Stratospheric Ozone Depletion." Pp. 913–923 in Thomas D. Potter and Bradley R. Colman, eds., *Handbook of Weather, Climate, and Water: Atmospheric Chemistry, Hydrology, and Societal Impacts*. Hoboken, NJ: Wiley InterScience, 2003.

Bhanoo, Sindya N. "Seas Grow Less Effective at Absorbing Emissions." *The New York Times*, November 19, 2009. http://www.nytimes.com/2009/11/19/science/earth/19oceans.html.

Bhanoo, Sindya N. "The Ozone Hole Is Mending. Now for the 'But.'" *The New York Times*, January 26, 2010. http://www.nytimes.com/2010/01/26/science/earth/26ozone.html.

Bhanoo, Sindya N. "Penguins Harmed by Tracking Bands, Study Finds." *The New York Times*, January 18, 2011, D3.

Bhattacharya, N. C. "Prospects of Agriculture in a Carbon-Dioxide-Enriched Environment." Pp. 487–505 in Richard A. Geyer, ed., *A Global Warming Forum: Scientific, Economic, and Legal Overview*. Boca Raton, FL: CRC Press, 1993.

"Big Chunk of Ice Breaks Off of Greenland Glacier." LiveScience, July 12, 2010. http://www.ouramazingplanet.com/big-chunk-of-ice-breaks-off-of-greenland-glacier-0334/.

"Big City Mayors Strategize to Beat Global Warming." Environment News Service, May 15, 2007. http://www.ens-newswire.com/ens/may2007/2007-05-15-01.asp (no longer available).

"Big Freeze Plunged Europe into Ice Age in Months." NASA Earth Observatory, November 29, 2009. http://earthobservatory.nasa.gov/Newsroom/view.php?id=41518&src=eoa-manews (no longer available).

"Big Swings in Weather to Come." *Nature* 517 (January 29, 2015): 530–531.

Bilger, Burkhard. "The Great Oasis." *The New Yorker*, December 10 and 26, 2011, 110–121.

Bingham, Robert G., et al. "Inland Thinning of West Antarctic Ice Sheet Steered along Subglacial Rifts." *Nature* 487 (July 26, 2012): 468–471.

"Biomass Burning Boosts Stratospheric Moisture." Environment News Service, February 20, 2002. http://ens-news.com/ens/feb2002/2002L-02-20-09.html (no longer available).

Birnbaum, Michael. "Europe Consuming More Coal." *Washington Post*, February 7, 2013. http://www.washingtonpost.com/world/europe-consuming-more-coal/2013/02/07/ec21026a-6bfe-11e2-bd36-c0fe61a205f6_print.html.

Bishop, Greg. "In Wilting Heat, Sharapova Escapes 2nd Round." *The New York Times*, January 16, 2014. http://www.nytimes.com/2014/01/17/sports/tennis/heat-wilts-players-and-crowd-at-australian-open.html.

Black, David E. "The Rains May Be A-Comin'." *Science* 297 (July 26, 2002): 528–529.

Blain, Stéphane, et al. "Effect of Natural Iron Fertilization on Carbon Sequestration in the Southern Ocean." *Nature* 446 (April 26, 2007): 1070–1074.

Blanchon, Paul, et al. "Rapid Sea-Level Rise and Reef Back-Stepping at the Close of the Last Interglacial Highstand." *Nature* 458 (April 16, 2009): 881–884.

Bloom, Arnold J., et al. "Nitrate Assimilation Is Inhibited by Elevated CO_2 in Field-Grown Wheat." *Nature Climate Change*. April 6, 2014. http://www.nature.com/nclimate/journal/vaop/ncurrent/full/nclimate2183.html.

Blunier, Thomas. "'Frozen' Methane Escapes from the Sea Floor." *Science* 288 (April 7, 2000): 68–69.

Boero, Fernando. *Review of Jellyfish Blooms in the Mediterranean and Black Sea.* Rome: Food and Agriculture Organization of the United Nations, 2013. http://www.fao.org/docrep/017/i3169e/i3169e.pdf.

Bohannon, John. "The Nile Delta's Sinking Future." *Science* 327 (March 19, 2010): 1444–1447.

Bollasina, Massimo A., Yi Ming, and V. Ramaswamy. "Anthropogenic Aerosols and the Weakening of the South Asian Summer Monsoon." *Science* 334 (October 28, 2011): 502–505.

Bond, Nicholas A., et al. "Causes and Impacts of the 2014 Warm Anomaly in the NE Pacific." *Geophysical Research Letters*, 42 (April 6, 2015). http://onlinelibrary.wiley.com/doi/10.1002/2015GL063306/abstract?campaign=wlytk-41855.5282060185.

Booth, Ben B., et al. "Aerosols Implicated as a Prime Driver of 20th-Century North Atlantic Climate Variability." *Nature* 484 (April 12, 2012): 228–232.

Borenstein, Seth. "Arctic Lost 60 Percent of Ozone Layer; Global Warming Suspected." Knight-Ritter News Service, April 6, 2000 (LEXIS).

Borenstein, Seth. "Scientists Worry about Evidence of Melting Arctic Ice." *Seattle Times*, February 18, 2005, A6.

Borenstein, Seth. "Scientists Find New Global Warming 'Time Bomb' Methane Bubbling Up from Permafrost." Associated Press, September 6, 2006 (LEXIS).

Borenstein, Seth. "Future Forecast: Extreme Weather; Study Outlines a Climate Shift Caused by Global Warming." Associated Press, October 21, 2006, A-2. http://www.washingtonpost.com/wp-dyn/content/article/2006/10/20/AR2006102001454_pf.html.

Borenstein, Seth. "January Weather Hottest by Far." Associated Press, February 15, 2007 (LEXIS).

Borenstein, Seth. "Earth's Tropics Belt Expands, May Mean Drier Weather for U.S. Southwest, Mediterranean." Associated Press, December 2, 2007 (LEXIS).

Borenstein, Seth. "Jump in Atlantic Hurricanes Not Global Warming, Says Study that Predicts Fewer Future Storms." Associated Press, May 19, 2008 (LEXIS).

Borenstein, Seth. "Scientists: Americans Are Becoming Weather Wimps." *Omaha World-Herald*, January 10, 2014, 3A. http://www.omaha.com/article/20140109/AP01/301099812 (no longer available).

Borenstein, Seth. "UN Panel: Warming Worsens Food, Hunger Problems." Associated Press, March 31, 2014. http://bigstory.ap.org/article/un-panel-warming-worsens-food-hunger-problems.

Borenstein, Seth. "Study: Polar Bears Disappearing from Key Region." Associated Press, November 17, 2014. http://bigstory.ap.org/article/28aa5cf3d43f44e1aa1f37b836a0e82b/study-polar-bears-disappearing-key-region.

Borenstein, Seth. "Report: Arctic Loses Snow, Ice; Absorbs More Heat." *Houston Chronicle*, December 20, 2014. http://www.houstonchronicle.com/news/politics/article/Report-Arctic-loses-snow-ice-absorbs-more-heat-5971210.php#/0.

Borenstein, Seth. "Syria's Civil War Linked Partly to Drought, Global Warming." Associated Press, March 3, 2015. http://federalnewsradio.com/government-news/2015/03/syrias-civil-war-linked-partly-to-drought-global-warming/.

Borenstein, Seth. "Top Doctors' Prescription for Feverish Planet: Cut Out Coal." Associated Press, June 22, 2015. http://bigstory.ap.org/article/7645057cf40e444196b269af02fec021/top-doctors-prescription-feverish-planet-cut-out-coal.

Borenstein, Seth. "Beyond Record Hot, February [2016] Was 'Astronomical' and 'Strange.'" Associated Press, March 17, 2016. http://phys.org/news/2016-03-hot-february-astron omical-strange.html.

Borenstein, Seth. "Study: Warming to Trigger 3 Times as Many Downpours in U.S." Associated Press, December 5, 2016. http://bigstory.ap.org/article/9aacfd5603cb475eb46196 0f4e05b948/study-warming-trigger-3-times-many-downpours-us.

Bornemann, Andre, et al. "Isotopic Evidence for Glaciation during the Cretaceous Supergreenhouse." *Science* 319 (January 11, 2008): 189–192.

Both, Christiaan, et al. "Climate Change and Population Declines in a Long-Distance Migratory Bird." *Nature* 441 (May 4, 2006): 81–83.

Boudette, Neal E. "Shifting Gears, GM Now Sees Green." *Wall Street Journal*, May 29, 2007, A8.

Bowen, Gabriel J., et al. "A Humid Climate State during the Palaeocene/Eocene Thermal Maximum." *Nature* 432 (November 25, 2004): 495–499.

Bowen, Mark. *Thin Ice: Unlocking the Secrets of Climate in the World's Highest Mountains.* New York: Henry Holt, 2006.

Bowman, David M. J. S., et al. "Fire in the Earth System." *Science* (April 24, 2009): 481–484.

Boxall, Bettina. "Epic Droughts Possible, Study Says." *Los Angeles Times*, October 8, 2004, A-17.

Boyce, Daniel G., Marlon R. Lewis, and Boris Worm. "Global Phytoplankton Decline over the Past Century." *Nature* 466 (July 29, 2010): 591–596.

Boyd, Robert S. "Rising Tides Raises Questions; Satellites Will Provide Exact Measurements." Pittsburgh *Post-Gazette*, December 9, 2001, A3.

Boyd, Robert S. "Arctic Temperatures Hit Record High." Knight Ridder Washington Bureau, October 16, 2008 (LEXIS).

Braasch, Gary. *Earth under Fire: How Global Warming Is Changing the World.* Berkeley: University of California Press, 2007.

Bradbury, Roger. "A World without Coral Reefs." *The New York Times*, July 13, 2012. http://www.nytimes.com/2012/07/14/opinion/a-world-without-coral-reefs.html.

Bradley, Raymond S., et al. "Threats to Water Supplies in the Tropical Andes." *Science* 312 (June 23, 2006): 1755–1756.

Bradsher, Keith. "China Prospering but Polluting; Dirty Fuels Power Economic Growth." *International Herald-Tribune*, October 22, 2003, 1.

Bradsher, Keith. "A Drought in Australia, a Global Shortage of Rice." *The New York Times*, April 17, 2008. http://www.nytimes.com/2008/04/17/business/worldbusiness/17warm .html.

Bradsher, Keith. "With First Car, a New Life in China." *The New York Times*, April 24, 2008. http://www.nytimes.com/2008/04/24/business/worldbusiness/24firstcar.html.

Bradsher, Keith. "China Vies to Be World's Leader in Electric Cars." *The New York Times*, April 2, 2009. http://www.nytimes.com/2009/04/02/business/global/02electric.html.

Bradsher, Keith. "Recession Elsewhere, but It's Booming in China." *The New York Times*, December 10, 2009. http://www.nytimes.com/2009/12/10/business/economy/10con sume.html.

Bradsher, Keith. "China Fears Warming Effects of Consumer Wants." *The New York Times*, July 4, 2010. http://www.nytimes.com/2010/07/05/business/global/05warm.html.

Bradsher, Keith. "U.N. Food Agency Issues Warning on China Drought." *The New York Times*, February 8, 2011. http://www.nytimes.com/2011/02/09/business/global/09food.html.

Bradsher, Keith. "Cleaner China Coal May Still Feed Global Warming." *The New York Times*, June 17, 2011, B8.

Bradsher, Keith. "Struggle for Survival in Philippine City Shattered by Typhoon." *The New York Times*, November 11, 2013. http://www.nytimes.com/2013/11/12/world/asia/philippines-storm-surge-leaves-scenes-of-devastation.html.

Bradsher, Keith, and David Barboza. "Pollution from Chinese Coal Casts a Global Shadow." *The New York Times*, June 11, 2006. http://www.nytimes.com/2006/06/11/business/worldbusiness/11chinacoal.html.

Braem, Nicole M. "Greenpeace Activists Visit Yup'ik and Inupiat Villages in Alaska to Gather Information about Global Warming." *Nunatsiaq News*, August 8, 1997. http://www.nunanet.com/~nunat/week/70808.html (no longer available).

Brahic, Catherine. "Parts of Amazon Close to Tipping Point." *New Scientist*, March 5, 2009. http://www.newscientist.com/article/dn16708-parts-of-amazon-close-to-tipping-point.html.

Braine, Theresa. "Native Alaska Village of Point Lay Hailed for Stewardship of 35,000 Walruses." Indian Country Today Media Network, October 3, 2014. http://indiancountrytodaymedianetwork.com/2014/10/03/native-alaska-village-point-lay-hailed-stewardship-35000-walruses-157175.

Brandenburg, John E., and Monica Rix Paxson. *Dead Mars, Dying Earth*. Freedom, CA: The Crossing Press, 1999.

Brasseur, Guy P., et al. "Natural and Human-Induced Perturbations in the Middle Atmosphere: A Short Tutorial." In David E. Siskind et al., eds., *Atmospheric Science across the Stratopause*. Washington, DC: American Geophysical Union, 2000.

Brewer, Peter G., et al. "Direct Experiments on the Ocean Disposal of Fossil Fuel CO_2." *Science* 284 (May 7, 1999): 943–945.

Brienen, R. J. W., et al. "Long-Term Decline of the Amazon Carbon Sink." *Nature* 519 (March 19, 2015): 344–348.

Brierley, Chris M., et al. "Greatly Expanded Tropical Warm Pool and Weakened Hadley Circulation in the Early Pliocene." *Science* 323 (March 27, 2009): 1714–1718.

Brigham-Grette, Julie, et al. "Pliocene Warmth, Polar Amplification, and Stepped Pleistocene Cooling Recorded in NE Arctic Russia." *Science* 345 (June 21, 2013): 1421–1427.

Brikowski, Tom H., Yair Lotan, and Margaret S. Pearle. "Climate-Related Increase in the Prevalence of Urolithiasis in the United States." *Proceedings of the National Academy of Sciences* 105(28) (July 15, 2008): 9841–9846.

"Britain's Sea Levels 'Will Rise 5 Feet This Century.'" *Daily Mail* (London), December 17, 2007 (LEXIS).

"British Travel Agents Launch Carbon Offset Scheme." Environment News Service, November 28, 2006. http://www.ens-newswire.com/ens/nov2006/2006-11-28-05.asp (no longer available).

Britt, Robert Roy. "Antarctic Ice Shelves Falling Apart." Explorezone.com, April 9, 1999. http://www.explorezone.com/archives/99_04/09_antarctic_ice.htm (no longer available).

Broad, William J. "How to Cool a Planet (Maybe)." *New York Times*, June 2, 2006. http://www.nytimes.com/2006/06/27/science/earth/27cool.html.

Broad, William J. "From a Rapt Audience, a Call to Cool the Hype." *The New York Times*, March 13, 2007. http://www.nytimes.com/2007/03/13/science/13gore.html.

Broccoli, A. J., and S. Manabe. "Can Existing Climate Models Be Used to Study Anthropogenic Changes in Tropical Cyclone Intensity?" *Geophysical Research Letters* 17 (1990): 1917–1920.

Broder, John M. "Climate Change Report Outlines Perils for U.S. Military." *The New York Times*, November 9, 2012. http://www.nytimes.com/2012/11/10/science/earth/climate-change-report-outlines-perils-for-us-military.html.

Broecker, Wallace S. "Are We Headed for a Thermohaline Catastrophe?" Pp. 83–95 in Lee C. Gerhard, William E. Harrison, and Bernold M. Hanson, eds., *Geological Perspectives of Global Climate Change*. American Association of Petroleum Geologists Studies in Geology No. 17. Tulsa, OK: AAPG, 2001.

Broecker, Wallace S., et al. "A Possible 20th-Century Slowdown of Southern Ocean Deep Water Formation." *Science* 286 (November 5, 1999): 1132–1135.

Broeke, Michiel van den, et al. "Partitioning Recent Greenland Mass Loss." *Science* 326 (November 13, 2006): 984–986.

Bromaghin, Jeffrey F., et al. "Polar Bear Population Dynamics in the Southern Beaufort Sea during a Period of Sea Ice Decline." *Ecological Applications*, 2014. http://www.esajournals .org/doi/abs/10.1890/14-1129.1.

Bromwich, David H., et al. "Central West Antarctica among the Most Rapidly Warming Regions on Earth." *Nature GeoScience*, December 23, 2012. doi: 10.1038/ngeo1671.

Brooks, H. "Severe Thunderstorms and Climate Change." *Atmospheric Research* 123 (April 1, 2013): 129–138.

Brown, DeNeen L. "Waking the Dead, Rousing Taboo; In Northwest Canada, Thawing Permafrost Is Unearthing Ancestral Graves." *Washington Post*, October 17, 2001, A27.

Brown, DeNeen L. "Greenland's Glaciers Crumble." *Washington Post*, October 13, 2002, A30. https://www.washingtonpost.com/archive/politics/2002/10/13/greenlands-glaciers -crumble/21482aa2-9160-4338-82cc-21cd7e33c2c1/?utm_term=.8c37041f460b.

Brown, Lester R. *Plan B: Rescuing a Planet under Stress and a Civilization in Trouble*. New York: Earth Policy Institute/W.W. Norton, 2003.

Brown, Lester R. "The Earth Is Shrinking." Environment News Service, November 20, 2006. http://www.ens-newswire.com (no longer available).

Brown, Paul. "Overfishing and Global Warming Land Cod on Endangered List." *The Guardian* (U.K.), July 20, 2000, 3.

Brown, Paul. "Melting Permafrost Threatens Alps; Communities Face Devastating Landslides from Unstable Mountain Ranges." *The Guardian* (U.K.), January 4, 2001, 3.

Brown, Paul. "Geographers' Conference: Ice Field Loss Puts Alpine Rivers at Risk: Global Warming Warning to Europe." *The Guardian* (U.K.), January 5, 2002, 9.

Brown, Paul. "Scientists Warn of Himalayan Floods: Global Warming Melts Glaciers and Produces Many Unstable Lakes." *The Guardian* (U.K.), April 17, 2002, 13.

Brown, Paul. "Inuit Blame Bush for Impending Extinction." *The Guardian* (U.K.), December 13, 2003. http://www.smh.com.au/articles/2003/12/12/1071125653832.html.

Brown, Paul, and Tony Sutton. "Global Warming Brings New Cash Crop to West Country as Rising Water Temperatures Allow Valuable Shellfish to Thrive." *The Guardian* (U.K.), December 10, 2002, 8.

Brown, Robbie, and Katherine Q. Seelye. "A Chaotic Flurry of Twisters That Spread the Devastation Fast and Wide." *The New York Times*, April 29, 2011, A19.

Browne, Anthony. "How Climate Change Is Killing Off Rare Animals." *The Observer* (London), February 10, 2002, 15.

Browne, Anthony. "Canute Was Right! Time to Give up the Coast." *The Times* (London), October 11, 2002, 8.

Browne, Anthony, and Paul Simons. "Euro-Spiders Invade as Temperature Creeps up." *The Times* (London), December 24, 2002, 8.

Browne, Malcolm W. "Will Humans Overwhelm the Earth? The Debate Goes On." *New York Times*, December 8, 1998, n.p.

"Brush Fires Collapsing Bear Dens." *Calgary Sun*, November 2, 2002, 18.

Bryant, D., et al. *Reefs at Risk: A Map-Based Indicator of Threats to the World's Coral Reefs.* Washington, DC: World Resources Institute, 1998.

Bryce, Robert. "A Bad Bet on Carbon." *The New York Times*, May 12, 2010. http://www.nytimes.com/2010/05/13/opinion/13bryce.html.

Bryce, Robert. "End the Ethanol Rip-Off." *The New York Times*, March 10, 2015. http://www.nytimes.com/2015/03/10/opinion/end-the-ethanol-rip-off.html.

Bryden, Harry L., Hannah R. Longworth, and Stuart A. Cunningham. "Slowing of the Atlantic Meridional Overturning Circulation at 25° North." *Nature* 438 (December 1, 2005): 655–657.

Buckley, Chris. "China Burns Much More Coal Than Reported, Complicating Climate Talks." *The New York Times*, November 4, 2015, A1.

Buesseler, Ken O., and Philip W. Boyd. "Will Ocean Fertilization Work?" *Science* 300 (April 4, 2003): 67–68.

Buncombe, Andrew, and Severin Carrell. "Melting Planet: Species Are Dying Out Faster Than We Have Dared Recognize, Scientists Will Warn This Week," *The Independent* (London), October 2, 2005.

Bunkley, Nick. "Detroit Finds Agreement on the Need to Be Green." *The New York Times*, June 1, 2007. http://www.nytimes.com/2007/06/01/business/01auto.html.

Büntgen, Ulf, et al. "Drought-Induced Decline in Mediterranean Truffle Harvest." *Nature Climate Change* 2 (December 2012): 827–829.

Bunting, Madeleine. "Confronting the Perils of Global Warming in a Vanishing Landscape: As Vital Talks Begin at the Hague, Millions Are Already Suffering the Consequences of Climate Change." *The Guardian* (U.K.), November 14, 2000, 1.

Burke, Lauretta, et al. *Reefs at Risk: Revisited.* Washington, DC: World Resources Institute, 2011. http://www.wri.org/sites/default/files/pdf/reefs_at_risk_revisited.pdf.

Burnett, Adam W., et al. "Increasing Great Lake-Effect Snowfall during the Twentieth Century: A Regional Response to Global Warming?" *Journal of Climate* 16(21) (November 1, 2003): 3535–3542.

Burns, Judith. "Methane Seeps from Arctic Sea-Bed." BBC News. August 18, 2009. http://news.bbc.co.uk/go/pr/fr/-/2/hi/science/nature/8205864.stm.

"Bushfire Devastates Australian Town." NASA Earth Observatory, January 12, 2016. http://earthobservatory.nasa.gov/IOTD/view.php?id=87302&src=eoa-iotd.

"Bushfires Menace Towns in Western Australia." NASA Earth Observatory, February 6, 2015. http://earthobservatory.nasa.gov/IOTD/view.php?id=85225&src=eoa-iotd.

Buxton, James. "Suspects in the Mystery of Scotland's Vanishing Salmon: Fish Farms, Seals, and Global Warming Are All Blamed for What Some See as a Crisis." *Financial Times* (London), June 13, 2000, 11.

Byers, Stephen, and Olympia Snowe. *Meeting the Climate Challenge: Recommendations of the International Climate Change Task Force.* London: Institute for Public Policy Research, January 2005.

"Cabinet to Meet on Mt. Everest." Associated Press in *The New York Times*, November 2, 2009. http://www.nytimes.com/2009/11/03/world/africa/03melt.html (no longer available).

Cai, Wenju, et al. "More Extreme Swings of the South Pacific Convergence Zone Due to Greenhouse Warming." *Nature* 488 (August 16, 2012): 365–369.

Cai, Wenju, et al. "Increasing Frequency of Extreme El Niño Events Due to Greenhouse Warming." *Nature Climate Change* 4 (February 2014): 111–116.

Cai, Wenju, et al. "Increased Frequency of Extreme Indian Ocean Dipole Events Due to Greenhouse Warming." *Nature* 510 (June 12, 2014): 254–258.

Calamai, Peter. "Atlantic Water Changing: Scientists." *Toronto Star*, June 21, 2001, A18.

Calamai, Peter. "Alert over Shrinking Ozone Layer." *Toronto Star*, March 18, 2002, A8.

Calamai, Peter. "Global Warming Threatens Reindeer." *Toronto Star*, December 23, 2002, A23.

Caldeira, Ken, and Philip B. Duffy. "The Role of the Southern Ocean in the Uptake and Storage of Anthropogenic Carbon Dioxide." *Science* 287 (January 28, 2000): 620–622.

Caldeira, Ken, and Michael E. Wickett. "Oceanography: Anthropogenic Carbon and Ocean pH." *Nature* 425 (September 25, 2003): 365.

"California Air Board Adds Climate Labels to New Cars." Environment News Service, June 25, 2007. http://www.ens-newswire.com/ens/jun2007/2007-06-25-09.asp#anchor7 (no longer available).

"California's Temperature Rising." Environment News Service, April 9, 2007. http://www.ens-newswire.com/ens/apr2007/2007-04-09-09.asp#anchor6 (no longer available).

Callaway, Ewen. "Melting Glaciers Release Toxic Chemical Cocktail." *New Scientist*, May 7, 2008. http://www.newscientist.com/article/dn13848.

Campbell, Bruce. "Climate Change: Call for UN to Act on Food Security." *Nature* 509 (May 15, 2014): 288.

"Canadian Tundra Turning Green." Environment News Service, March 3, 2007. http://www.ens-newswire.com/ens/mar2007/2007-03-06-02.asp (no longer available).

"Can Coral Cope with Climate Change?" *Nature* 484 (April 19, 2012): 290.

Canfeld, Clarke. "Warming Drives off Cape Cod's Namesake, Other Fish." Associated Press in ABC News. November 17, 2009. http://abcnews.go.com/print?id=9062135 (no longer available).

Cantin, Neal E., et al. "Ocean Warming Slows Coral Growth in the Central Red Sea." *Science*, July 16, 2010: 322–325.

"Cape Hatteras, N.C. Lighthouse Lights Up Sky from New Perch." *Omaha World-Herald*, November 14, 1999, p. A16.

Capella, Peter. "Disasters Will Outstrip Aid Effort as World Heats Up." *The Guardian* (U.K.), June 29, 2001, 15.

Capella, Peter. "Europe's Alps Crumbling; Glaciers Melting in Heat Wave." Agence France Presse, August 7, 2003 (LEXIS).

"Carbon Neutral Chic." *Wall Street Journal*, July 9, 2007, A14.

Cardwell, Diane. "Solar and Wind Energy Start to Win on Price vs. Conventional Fuels." *The New York Times*, November 23, 2014. http://www.nytimes.com/2014/11/24/business/energy-environment/solar-and-wind-energy-start-to-win-on-price-vs-conventional-fuels.html.

Cardwell, Diane. "Copenhagen Lighting the Way to Greener, More Efficient Cities." *The New York Times*, December 8, 2014. http://www.nytimes.com/2014/12/09/business/energy-environment/copenhagen-lighting-the-way-to-greener-more-effecient-cities.html.

Cardwell, Diane. "Solar Power Battle Puts Hawaii at Forefront of Worldwide Changes." *The New York Times*, April 18, 2015. http://www.nytimes.com/2015/04/19/business/energy-environment/solar-power-battle-puts-hawaii-at-forefront-of-worldwide-changes.html.

Carey, John, and Sarah R. Shapiro. "Consensus Is Growing among Scientists, Governments, and Business That They Must Act Fast to Combat Climate Change." *Businessweek*, August 16, 2004 (LEXIS).

Carlton, Jim. "Some in Santa Fe Pine for Lost Symbol, but Others Move On." *Wall Street Journal*, July 31, 2006, A1, A8.

Carlton, Jim. "Citicorp Tries Banking on the Natural Kind of Green." *Wall Street Journal*, September 5, 2007, B1, B8.

Carlton, Jim. "A Winter without Walrus: Harvesting of Food Staple for Remote Eskimo Villages Plummets; Disaster Declared." *Wall Street Journal*, October 4, 2013, A4.

Carpenter, Betsy. "Feeling the Sting: Warming Oceans, Depleted Fish Stocks, Dirty Water—They Set the Stage for a Jellyfish Invasion." *U.S. News and World Report*, August 16, 2004, 68–69.

Carpenter, Kent E., et al. "One-Third of Reef-Building Corals Face Elevated Extinction Risk from Climate Change and Local Impacts." *Science* 321 (July 25, 2008): 560–563.

Cars and Climate Change. Paris, France: International Energy Agency, 1993.

Carswell, Cally. "Bumblebees Aren't Keeping Up with a Warming Planet." *Science* 349 (July 10, 2015): 126–127.

Chan, Johnny. "Comment on 'Changes in Tropical Cyclone Number, Duration, and Intensity in a Warming Environment.'" *Science* 311 (March 24, 2006): 1713.

Chang, Kenneth. "Ozone Hole Is Now Seen as a Cause for Antarctic Cooling." *The New York Times*, May 3, 2002, A16.

Chang, Kenneth. "Warming Is Blamed for Antarctica's Weight Gain." *The New York Times*, May 20, 2005, A22.

Chang, Kenneth. "Strongest Storms Grow Stronger Yet, Study Says." *The New York Times*, October 4, 2008, A18.

Chang, Kenneth. "Study Finds New Evidence of Warming in Antarctica." *The New York Times*, January 22, 2009. http://www.nytimes.com/2009/01/22/science/earth/22climate.html.

Chang, Kenneth. "Research Cites Role of Warming in Extremes." *The New York Times*, September 5, 2013. http://www.nytimes.com/2013/09/06/science/earth/research-cites-role-of-warming-in-extremes.html.

Chang, Kenneth. "Snow Is Down and Heat Is Up in the Arctic, Report Says." *The New York Times*, December 18, 2014. http://www.nytimes.com/2014/12/18/science/snow-is-down-and-heat-is-up-in-the-arctic-report-says.html.

"Changes in Climate Bring Hops Northward." *The Herald* (Glasgow), September 29, 2000, 13.

Chapman, James. "Early Spring Misery for 12 Million Hay Fever Sufferers." *The Daily Mail* (London), February 4, 2003, 23.

Charmantier, Anne, et al. "Adaptive Phenotypic Plasticity in Response to Climate Change in a Wild Bird Population." *Science* 320 (May 9, 2008): 800–803.

Chase, Marilyn. "As Virus Spreads, Views of West Nile Grow Even Darker." *Wall Street Journal*, October 14, 2004, A1, A10.

Chea, Terence. "Environmental Groups Call for Planes to Reduce Emissions." *USA Today*, December 6, 2007, 6B.

Chen, J., et al. "Accelerated Antarctic Ice Loss from Satellite Gravity Measurements." *Nature GeoScience* 2 (November 22, 2009): 859–862.

Chen, Xianyao, and Ka-Kit Tung. "Varying Planetary Heat Sink Led to Global-warming Slowdown and Acceleration." *Science* 345 (August 22, 2014): 897–903.

Chen, Yang, et al. "Forecasting Fire Season Severity in South America Using Sea Surface Temperature Anomalies." *Science* 334 (November 11, 2011): 787–791.

Chengappa, Raj. "The Monsoon: What's Wrong with the Weather?" *India Today*, August 12, 2002, 40.

Cheung, William W. L., Reg Watson, and Daniel Pauly. "Signature of Ocean Warming in Global Fisheries Catch." *Nature* 497 (May 16, 2013): 365–368.

Chin, Gilbert. "No Deepwater Slowdown?" *Science* 293 (July 27, 2001): 575, citing Alejandro Orsi et al., *Geophysical Research Letters* 28 (2001): 2923.

Chisholm, Sallie W. "Stirring Times in the Southern Ocean." *Nature* 407 (October 12, 2000): 685–686.

Christensen, Jens H., and Ole B. Christensen. "Severe Summertime Flooding in Europe." *Nature* 421 (February 20, 2003): 805.

Christensen, Torben R., et al. "Thawing Sub-Arctic Permafrost: Effects on Vegetation and Methane Emissions." *Geophysical Research Letters* 31(4) (February 20, 2004): L04501. doi: 10.1029/2003GL018680.

Christianson, Gale E. *Greenhouse: The 200-Year Story of Global Warming.* New York: Walker & Co., 1999.

Christie, Maureen. *The Ozone Layer: A Philosophy of Science Perspective.* Cambridge, UK: Cambridge University Press, 2001.

Ciborowski, Peter. "Sources, Sinks, Trends, and Opportunities." Pp. 213–230 in Edwin Abrahamson, ed., *The Challenge of Global Warming.* Washington, DC: Island Press, 1989.

Cifuentes, Luis, et al. "Hidden Health Benefits of Greenhouse Gas Mitigation." *Science* 252 (August 17, 2001): 1257–1259.

Clark, Jayne. "Tours Bear Witness to Earth's Sentinel Species." *USA Today*, November 2, 2007, 1D, 2D.

Clark, Jayne. "Lake Powell on the Rise." *USA Today*, May 9, 2008, 1D, 2D.

Clark, Peter U., et al. "The Role of the Thermohaline Circulation in Abrupt Climate Change." *Nature* 415 (February 21, 2002): 863–868.

Clark, Peter U., et al. "Sea-Level Fingerprinting as a Direct Test for the Source of Global Meltwater Pulse." *Science* 295 (March 29, 2002): 2438–2441.

Clark, Peter U., et al. "Rapid Rise of Sea Level 19,000 Years Ago and Its Global Implications." *Science* 304 (May 21, 2004): 1141–1144.

Clark, Peter U., et al. "Consequences of Twenty-First-Century Policy for Multi-Millennial Climate and Sea-Level Change." *Nature Climate Change*, February 2016. http://www.nature.com/nclimate/journal/v6/n4/abs/nclimate2923.html.

Clark, Peter U., and Peter Huybers. "Interglacial and Future Sea Level." *Nature* 462 (December 17, 2009): 856–857.

Clarke, Tom. "Boiling Seas Linked to Mass Extinction." *Nature*, August 22, 2003, online. http://info.nature.com/cgi-bin24/DM/y/eLod0BfHSK0Ch0DYy0AL.

Clarkson, M. O., et al. "Ocean Acidification and the Permo-Triassic Mass Extinction." *Science* 348 (April 10, 2015): 229–232.

Clavel, Guy. "Global Warming Makes Polar Bears Sweat." Agence France Presse, November 3, 2002.

"Clear to Take Off." Review of Inhofe, *The Greatest Hoax*, Amazon.com. September 4, 2014. http://www.amazon.com/The-Greatest-Hoax-Conspiracy-Threatens/dp/1936488493.

Clift, Peter, and Karen Bice. "Earth Science: Baked Alaska." *Nature* 419 (September 12, 2002): 129–130.

"Climate Blamed for Blah Foliage." Associated Press in *Omaha World-Herald*, October 29. 2007, 3A.

"Climate Change Dislocates Migratory Animals, Birds." Environment News Service, November 17, 2006. http://www.ens-newswire.com (no longer available).

"Climate-Change Ecology: Extinctions Underestimated." *Nature* 481 (January 12, 2012): 116–117.

"Climate Change Likely to Make Everest Even Riskier." *Newsday*, April 23, 2014. http://www.newsday.com/travel/climate-change-likely-to-make-everest-even-riskier-1.7800984 (no longer available).

"Climate Change Linked to Doubling of Atlantic Hurricanes." Environment News Service, July 30, 2007. http://www.ens-newswire.com/ens/jul2007/2007-07-30-01.asp (no longer available).

"Climate Change Puts Health of Arctic Villagers on Thin Ice." Indian Country Today Media Network, March 07, 2011. http://indiancountrytodaymedianetwork.com/article/climate-change-puts-health-of-arctic-villagers-on-thin-ice-21391.

"Climate Change Threatens Loggerhead Turtles." Environment News Service, February 22, 2007. http://www.ens-newswire.com/ens/feb2007/2007-02-22-02.asp (no longer available).

"Climate Change Pushing Bird Species to Oblivion." Environment News Service, November 14, 2006. http://www.ens-newswire.com/ens/nov2006/2006-11-14-01.asp (no longer available).

"Climate Deal Must Cover Acid Oceans, World Scientists Urge." Environment News Service, June 1, 2009. http://www.ens-newswire.com/ens/jun2009/2009-06-01-01.asp (no longer available).

"Climate Scoreboard." Climate Interactive. 2015. https://www.climateinteractive.org/tools/scoreboard/.

"Climate Warms Twice as Fast." British Broadcasting Company Monitoring Asia-Pacific, January 6, 2003 (LEXIS).

Cline, William R. *The Economics of Global Warming*. Washington, DC: Institute for International Economics, 1992.

Clover, Charles. "Air Travel Is a Threat to Climate." *Daily Telegraph* (London), June 5, 1999.

Clover, Charles. "Thousands of Species 'Threatened by Warming.'" *Daily Telegraph* (London), August 31, 2000, 9.

Clover, Charles. "Geographers' Conference: Alps May Crumble as Permafrost Melts." *Daily Telegraph* (London), January 4, 2001, 12.

Clover, Charles. "Global Warming Is Driving Fish North." *Daily Telegraph* (London), May 31, 2002, 14.

Clover, Charles, and David Millward. "Future of Cheap Flights in Doubt; Ban New Runways and Raise Fares, Say Pollution Experts." *Daily Telegraph* (London), November 30, 2002, 1, 4.

"Coca-Cola to Deploy 100K HFC-Free Coolers." Greenbiz.com, May 28, 2008. http://www.greenbiz.com/print/24718 (no longer available).

Cochrane, Joe. "Southeast Asia, Choking on Haze, Struggles for a Solution." *The New York Times*, October 8, 2015. http://www.nytimes.com/2015/10/09/world/asia/indonesia-forest-fires-haze-singapore-malaysia.html.

Cochrane, Joe. "Rain in Indonesia Dampens Fires That Spread Toxic Haze." *The New York Times*, October 29, 2015, A4.

Cockerham, Sean. "Report Links Extreme Weather, Melting of Sea Ice." *Seattle Times*, December 13, 2013, A13.

Cody, Edward. "Mild Weather Takes Edge Off Chinese Ice Festival; Residents of Tourist City Blame Global Warming." *Washington Post*, February 25, 2007, A19. http://www.washingtonpost.com/wp-dyn/content/article/2007/02/24/AR2007022401421.html.

Cohn, Nate. "How El Niño Might Alter the Political Climate." *The New York Times*, May 20, 2014. http://www.nytimes.com/2014/05/20/upshot/how-el-nino-might-alter-the-political-climate.html.

Collins, James P., and Martha L. Crump. *Extinction in Our Time: Amphibian Decline*. New York: Oxford University Press, 2009.

Collins, Simon. "Birds Starve in Warmer Seas." *New Zealand Herald*, November 14, 2002, n.p. (LEXIS).

"Coming Surge in Storm Surges." *Nature* 483 (March 22, 2012): 377.

Comiso, Josefino C. "Large Decadal Decline of the Arctic Multiyear Ice Cover." *Journal of Climate* 25 (2012): 1176–1193. doi: http://dx.doi.org/10.1175/JCLI-D-11-00113.1.

"Comment Period Extended on Polar Bear Extinction Threat." Environment News Service, October 2, 2007. http://www.ens-newswire.com/ens/oct2007/2007-10-02-091.asp (no longer available).

Commoner, Barry. *Making Peace with the Planet.* New York: Pantheon, 1990.

Cone, Marla. "Dozens of Words for Snow, None for Pollution." *Mother Jones*, January–February 2004. http://www.hartford-hwp.com/archives/27b/059.html.

Cone, Marla. *Silent Snow: The Slow Poisoning of the Arctic.* New York: Grove Press, 2005.

Connor, Steve. "Global Warming Is Blamed for First Collapse of a Caribbean Coral Reef." *The Independent* (London), May 4, 2000, 12.

Connor, Steve. "Britain Could Become as Cold as Moscow." *The Independent* (London), June 21, 2001, 14.

Connor, Steve. "Malaria Could Become Endemic Disease in U.K." *The Independent* (London), September 12, 2001, 14.

Connor, Steve. "Strangers in the Seas; Exotic Marine Species Are Turning Up Unexpectedly in the Cold Waters of the North Atlantic." *The Independent* (London), August 5, 2002, 12–13.

Connor, Steve. "Peat Bog Gases Accelerate Global Warming." *The Independent* (London), July 8, 2004, 9.

Connor, Steve. "Meltdown: Arctic Wildlife Is on the Brink of Catastrophe; Polar Bears Could Be Decades from Extinction." *The Independent* (London), November 11, 2004, n.p. (LEXIS).

Conover, Ted. "Capitalist Roaders." *The New York Times Sunday Magazine*, July 2, 2006. http://www.nytimes.com/2006/07/02/magazine/02china.html.

"Contrails Linked to Temperature Changes." Environment News Service, August 8, 2002. http://ens-news.com/ens/aug2002/2002-08-08-09.asp#anchor4 (no longer available).

Cook, A. J., et al. "Retreating Glacier Fronts on the Antarctic Peninsula over the Past Half-Century." *Science* 308 (April 22, 2005): 541–544.

Cook, Edward R., et al. "Long-Term Aridity Changes in the Western United States." *Science* 306 (November 5, 2004): 1015–1018.

Cooke, Robert. "Waters Reflect Weather Trend; Study Finds Warming Effects." *Newsday*, December 18, 2003, A2.

Copeland, Larry, and William M. Welch. "'Walls of Water' Hit with Little Warning." *USA Today*, June 14, 2010, 1A, 5A, 6A.

Copland, L., D. R. Mueller, and L. Weir. "Rapid Loss of the Avies Ice Shelf, Ellesmere Island, Canada." *Geophysical Research Letters* 34 (2007): L21501.

"COP 21: Amazon Fire Destroying Rare Forest Home of Uncontacted Tribe." Survival International press release by e-mail, December 1, 2015, by e-mail. http://www.survivalinternational.org/news/11033.

"Corals 'Could Survive a More Acidic Ocean.'" University News. University of Western Australia, April 2, 2015. http://www.news.uwa.edu.au/201204024487/international/corals-could-survive-more-acidic-ocean.

"Corals Inherit Love for Heat." *Nature* 523 (July 2, 2015): 8.

Cordalis, D., and D. B. Suagee. "The Effects of Climate Change on American Indian and Alaska Native Tribes." *Natural Resources and Environment* 22(3) (Winter 2008): 45–49.

Corkery, Michael. "As Coal's Future Grows Murkier, Banks Pull Financing." *The New York Times*, March 21, 2016. http://www.nytimes.com/2016/03/21/business/dealbook/as-coals-future-grows-murkier-banks-pull-financing.html.

Coumou, Dim, Jascha Lehmann, and Johanna Beckmann. "The Weakening Summer Circulation in the Northern Hemisphere Mid-Latitudes." *Science* 348 (April 17, 2015): 324–327.

Cowen, Robert C. "Into the Cold? Slowing Ocean Circulation Could Presage Dramatic—and Chilly—Climate Change." *Christian Science Monitor*, September 26, 2002, 14.

Cowen, Robert C. "One Large, Overlooked Factor in Global Warming: Tropical Forest Fires." *Christian Science Monitor*, November 7, 2002, 14.

"Cow Skulls and Dust: Drought Grips Argentina." *Omaha World-Herald*, January 26, 2009, 3A.

Cox, John D. "More Greenland Ice Melting Faster." Discovery News, March 24, 2010. http://news.Discovery.com/earth/more-greenland-ice-melting-faster.html.

Cox, P. M., et al. "Amazonian Forest Dieback under Climate-Carbon Cycle Projections for the 21st Century." *Theoretical and Applied Climatology* 78 (2004): 137–156.

Cox, Stan. "Your Air Conditioner Is Making the Heat Wave Worse." *Washington Post*, July 22, 2016. https://www.washingtonpost.com/posteverything/wp/2016/07/22/your-air-conditioner-is-making-the-heat-wave-worse/?wpisrc=nl_headlines&wpmm=1.

Coy, Peter. "The Hydrogen Balm? Author Jeremy Rifkin Sees a Better, Post-Petroleum World." *Businessweek*, September 30, 2002, 83.

Cozzetto, K., et al. "Climate Change Impacts on the Water Resources of American Indians and Alaska Natives in the U.S." *Climatic Change* 120 (2013): 569–584.

Cramb, Auslan. "Highland River Salmon 'On Verge of Extinction.'" *Daily Telegraph* (London), July 15, 2002, 7.

Crilly, Rob. "2050 to Be Good Year for Scottish Wine; Global Warming Will Bring Grapes North." *The Herald* (Glasgow), November 20, 2002, 11.

Crucifix, Michael. "Earth's Narrow Escape from a Big Freeze." *Nature* 529 (January 14, 2016): 162–163.

Crutzen, Paul J. "Biographical." *The Nobel Prizes 1995*, ed. Tore Frängsmyr. Stockholm: Nobel Foundation, 1996. nobelprize.org/nobel_prizes/chemistry/laureates/1995/crutzen-bio.html.

Crutzen, Paul J. "The Antarctic Ozone Hole, a Human-Caused Chemical Instability in the Stratosphere: What Should We Learn from It?" Pp. 1–11 in Lennart O. Bengtsson and Claus U. Hammer, eds., *Geosphere-Biosphere Interactions and Climate*. Cambridge, UK: Cambridge University Press, 2001.

Crutzen, Paul J. "Albedo Enhancement by Stratospheric Sulfur Injections: A Contribution to Resolving Policy Dilemma?" *Climatic Change* 77 (2006): 211–220.

Crutzen, Paul J., and M. O. Andreae. "Biomass Burning in the Tropics: Impact on Atmospheric Chemistry and Biogeochemical Cycles." *Science* 250 (1990): 1669–1678.

Crutzen, Paul J., and John W. Birks: "The Atmosphere after a Nuclear War: Twilight at Noon." *Ambio* 11(2–3) (1982): 114–125.

Crutzen, Paul J., et al. "N_2O Release from Agro-Biofuel Production Negates Global Warming Reduction by Replacing Fossil Fuels." *Atmospheric Chemistry and Physics* 8 (2008): 389–395.

Cunningham, Stuart A., et al. "Temporal Variability of the Atlantic Meridional Overturning Circulation at 26.5°N." *Science* 317 (2007): 935–937. doi: 10.1126/science.1141304.

Curry, Andrew. "Racing the Thaw." *Science* 346 (10 October 10, 2014): 157–159.

Curry, Ruth, Bob Dickson, and Igor Yashayaev. "A Change in the Freshwater Balance of the Atlantic Ocean over the Past Four Decades." *Nature* 426 (December 18, 2003): 826–829.

Curry, Ruth, and Cecilie Mauritzen. "Dilution of the Northern North Atlantic Ocean in Recent Decades." *Science* 308 (June 17, 2005): 1772–1774.

Daigle, J. J., and D. Putnam. "The Meaning of a Changed Environment: Initial Assessment of Climate Change Impacts in Maine—Indigenous Peoples." Pp. 35–38 in G. L. Jacobson, et al., eds., *Maine's Climate Future: An Initial Assessment.* Orono: University of Maine Press, 2009.

Dalton, Alastair. "Ice Pack Clue to Climate-Change Effects." *The Scotsman* (Edinburgh), October 18, 2001, 7.

Dalton, Rex. "How Aircraft Emissions Contribute to Warming." *Nature*, December 21, 2009. http://www.nature.com/news/2009/091221/full/news.2009.1157.html.

Davenport, Carol. "Industry Awakens to Threat of Climate Change." *The New York Times*, January 24, 2014. http://www.nytimes.com/2014/01/24/science/earth/threat-to-bottom-line-spurs-action-on-climate.html?hp&_r=0.

Davenport, Carol. "Miami Finds Itself Ankle-Deep in Climate Change Debate," *The New York Times*, May 8, 2014. http://www.nytimes.com/2014/05/08/us/florida-finds-itself-in-the-eye-of-the-storm-on-climate-change.html.

Davenport, Carol. "Optimism Faces Grave Realities at Climate Talks." *The New York Times*, December 1, 2014. http://www.nytimes.com/2014/12/01/world/climate-talks.html.

Davenport, Carol. "Philippines Pushes Developing Countries to Cut Their Emissions." *The New York Times*, December 8, 2014. http://www.nytimes.com/2014/12/09/world/americas/philippines-pushes-developing-countries-to-cut-their-emissions.html.

Davenport, Carol. "E.P.A. to Set New Limits on Airplane Emissions." *The New York Times*, June 2, 2015. http://www.nytimes.com/2015/06/03/business/energy-environment/epa-to-set-new-limits-on-airplane-emissions.html.

Davenport, Carol. "The Marshall Islands Are Disappearing; Rising Seas Are Claiming a Vulnerable Nation." *The New York Times*, December 2, 2015. http://www.nytimes.com/interactive/2015/12/02/world/The-Marshall-Islands-Are-Disappearing.html.

Davenport, Carol. "As Wind Power Lifts Wyoming's Fortunes, Coal Miners Are Left in the Dust." *The New York Times*, June 19, 2016.

Davenport, Carol, and Laurie Goldstein. "Pope Francis Steps Up Campaign on Climate Change, to Conservatives' Alarm." *The New York Times*, April 27, 2015. http://www.nytimes.com/2015/04/28/world/europe/pope-francis-steps-up-campaign-on-climate-change-to-conservatives-alarm.html.

Davenport, Carol, and Campbell Robertson. "Resettling the First American 'Climate Refugees.'" *The New York Times*, May 2, 2016. http://www.nytimes.com/2016/05/03/us/resettling-the-first-american-climate-refugees.html.

Davey, Monica. "Balmy Weather May Bench a Baseball Staple." *The New York Times*, July 11, 2007. http://www.nytimes.com/2007/07/11/us/11ashbat.html.

Davey, Monica. "Death, Havoc, and Heat Mar Chicago Race." *The New York Times*, October 8, 2007, A1, A14.

Davidson, Helen. "Pacific Islanders Blockade Newcastle Coal Port to Protest Rising Sea Levels." *The Guardian* (U.K.), October 17, 2014. http://www.theguardian.com/environment/2014/oct/17/pacific-islanders-blockade-newcastle-coal-port-to-protest-rising-sea-levels.

Davidson, Keay. "Media Goofed on Antarctic Data; Global Warming Interpretation Irks Scientists." *San Francisco Chronicle*, February 4, 2002, A8.

Davidson, Keay. "Going to Depths for Evidence of Global Warming; Heating Trend in North Pacific Baffles Researchers." *San Francisco Chronicle*, March 1, 2004, A4.

Davis, Curt H., et al. "Snowfall-Driven Growth in East Antarctic Ice Sheet Mitigates Recent Sea-Level Rise." *Science* 308(2) (June 24, 2005): 1898–1901.

Davis, Julie Hirschfeld. "Obama Recasts Climate Change as Peril with Far-Reaching Effects." *The New York Times*, May 20, 2015. http://www.nytimes.com/2015/05/21/us/obama -recasts-climate-change-as-a-more-far-reaching-peril.html.

Davis, Neil. *Permafrost: A Guide to Frozen Ground in Transition*. Fairbanks: University of Alaska Press, 2001.

Davis, Robert E., et al. "Changing Heat-Related Mortality in the United States." *Environmental Health Perspectives*, July 23, 2003. doi: 10.1289/ehp.6336.

Daviss, Bennett. "Green Sky Thinking: Could Maverick Technologies Turn Aviation into an Eco-Success Story? Yes, but Time Is Running Out." *New Scientist*, February 24, 2007, 32–38.

"Deadly Tornado Hits New Zealand." *Washington Post*, December 6, 2012. http://www .washingtonpost.com/world/asia_pacific/tornado-rips-through-northeast-of-tokyo-dozens-of-people-injured/2012/05/06/gIQAV4Jw4T_print.html (no longer available).

Dean, Cornelia. "Study Warns of Threat to Coasts from Rising Sea Levels." *The New York Times*, January 17, 2009. http://www.nytimes.com/2009/01/17/science/earth/17sea.html.

Dean, Cornelia. "Rising Acidity Is Threatening Food Web of Oceans, Science Panel Says." *The New York Times*, January 31, 2009. http://www.nytimes.com/2009/01/31/science/earth/31ocean.html.

Dean, Cornelia. "Sea's Rise May Prove the Greater in Northeast." *The New York Times*, May 28, 2009. http://www.nytimes.com/2009/05/28/science/earth/28warming.html

Dean, Cornelia. "El Niño Variant Is Linked to Hurricanes in Atlantic." *The New York Times*, July 2, 2009. http://www.nytimes.com/2009/07/03/science/earth/03hurricane.html.

Dean, Cornelia. "An 'Increase' in Big Storms May Just Be Better Detection." *The New York Times*, August 12, 2009. http://www.nytimes.com/2009/08/13/science/earth/13atlantic .html.

De'ath, Glenn, Janice M. Louygh, and Katharina E. Fabricius. "Declining Coral Calcification on the Great Barrier Reef." *Science* 323 (January 2, 2009): 116–119.

"Death Toll in Australian Bushfires Climbs to 84." Associated Press in *The New York Times*, February 9, 2009. http://www.nytimes.com/2009/02/09/world/asia/09australia.html.

"Decade-Long Grassroots Campaign Shuts Two Chicago Coal Plants." Environment News Service, March 2, 2012. http://ens-newswire.com/2012/03/02/decade-long-grassroots -campaign-shuts-two-chicago-coal-plants/.

DeConto, Robert M., and David Pollard. "Contribution of Antarctica to Past and Future Sea-Level Rise." *Nature* 531 (March 31, 2016): 591–597.

Del Genio, Anthony, Mao-Sung Tao, and Jeffery Jonas. "Will Moist Convection Be Stronger in a Warmer World Climate?" *Geophysical Research Letters* 34(16) (August 17, 2007): 16703.

"The Demise of the Warm Blob." NASA Earth Observatory, February 16, 2016. http:// earthobservatory.nasa.gov/IOTD/view.php?id=87513&src=eoa-iotd.

Denby, David. "Review, An Inconvenient Truth." *The New Yorker*, June 19, 2006, 23.

Derbyshire, David. "Global Warming Fails to Boost Butterfly Visitors." *Daily Telegraph* (London), November 1, 2001, 13.

Derbyshire, David. "Baffled Bumble Bee Lured Out Early by Changing Climate." *Daily Telegraph* (London), March 12, 2004, 15.

Derksen, C., and R. Brown. "Spring Snow Cover Extent Reductions in the 2008–2012 Period Exceeding Climate Model Projections." *Geophysical Research Letters* 39(19) (October 16, 2012). http://onlinelibrary.wiley.com/doi/10.1029/2012GL053387/abstract.

Deschamps, Pierre, et al. "Ice-Sheet Collapse and Sea-Level Rise at the Bølling Warming 14,600 Years Ago." *Nature* 483 (March 29, 2012): 559–564.

Dessler, Andrew C., and Steven C. Sherwood. "A Matter of Humidity." *Science* 323 (February 20, 2009): 1020–1021.

Deutsch, Claudia. "Companies Giving Green an Office." *The New York Times*, July 3, 2007. http://www.nytimes.com/2007/07/03/business/03sustain.html.

Deutsch, Claudia. "A Threat So Big, Academics Try Collaboration." *The New York Times*, December 25, 2007. http://www.nytimes.com/2007/12/25/business/25sustain.html.

Dickens, Gerald R. "A Methane Trigger for Rapid Warming?" *Science* 299 (February 14, 2003): 1017. Review of James P. Kennett et al., *Methane Hydrates in Quaternary Climate Change: The Clathrate Gun Hypothesis*. Washington, DC: American Geophysical Union, 2002.

Dickens, Gerald R. "Global Change: Hydrocarbon-Driven Warming." *Nature* 429 (June 3, 2004): 513–515.

Dickson, B., et al. "Rapid Freshening of the Deep North Atlantic Ocean over the Past Four Decades." *Nature* 416 (April 25, 2002): 832–836.

Diffenbaugh, Noah S., and Christopher B. Field. "A Wet Winter Won't Save California." *The New York Times*, September 18, 2015. http://www.nytimes.com/2015/09/19/opinion/a-wet-winter-wont-save-california.html.

"Diminished Snow Pack in the Pacific Northwest." NASA Earth Observatory, May 17, 2015. http://earthobservatory.nasa.gov/IOTD/view.php?id=85887&src=eoa-iotd.

Dionne, E. J., Jr. "Gore's Energy Oomph." *Washington Post*, July 18, 2008. http://www.washingtonpost.com/wp-dyn/content/article/2008/07/17/AR2008071701840_pf.html.

Dixson, Danielle L., Philip L. Munday, and Geoffrey P. Jones. "Ocean Acidification Disrupts the Innate Ability of Fish to Detect Predator Olfactory Cues." *Ecology Letters*, November 16, 2009. http://www3.interscience.wiley.com/journal/122684940/abstract?CRETRY=1&SRETRY=0] doi: 10.1111/j.1461-0248.2009.01400.x.

Dobb, Edwin. "The New Oil Landscape." *National Geographic*, March 2013, 28–59.

Domack, Eugene, et al. "Stability of the Larsen B Ice Shelf on the Antarctic Peninsula During the Holocene Epoch." *Nature* 436 (August 4, 2005): 681–685.

Doney, Scott C. "Plankton in a Warmer World." *Nature* 444 (December 7, 2006): 695–696.

Donn, Jeff. "New England's Brilliant Autumn Sugar Maples—and Their Syrup—Threatened by Warmth." Associated Press, September 23, 2002 (LEXIS).

Donnelly, Jeffrey P., and Jonathan D. Woodruff. "Intense Hurricane Activity over the Past 5,000 Years Controlled by El Niño and the West African Monsoon." *Nature* 447 (May 24, 2007): 465–468.

Doran, Peter. "Cold, Hard Facts." *The New York Times*, July 27, 2006. http://www.nytimes.com/2006/07/27/opinion/27doran.html.

Doran, Peter T., et al. "Antarctic Climate Cooling and Terrestrial Ecosystem Response." *Nature* 415 (January 30, 2002): 517–520.

Dorell, Oren. "Report: Tropical Storm Activity at 30-Year Low." *USA Today*, November 12, 2008, 3A.

Dorell, Oren. "Like Cloud Lifted, Drought Eases." *USA Today*, November 12, 2009, A-3.

Dorell, Oren. "Scouts Ambushed by Heat, Storm." *USA Today*, July 26, 2010, A1.

Dorrepaal, Ellen, et al. "Carbon Respiration from Subsurface Peat Accelerated by Climate Warming in the Subarctic." *Nature* 460 (July 30, 2009): 616–619.

D'Ortenzio, Eric, et al. "Evidence of Sexual Transmission of Zika Virus." *New England Journal of Medicine* 374 (June 2, 2016): 2195–2198.

"Dow Jones Plans Large Solar Installation on N.J. Campus." *USA Today*, April 12, 2010, 2B.

Doyle, Alister. "Methane Bubbles." Reuters in Australian Broadcasting Corp., March 5, 2010. http://www.abc.net.au/science/articles/2010/03/05/2837443.htm.

Drajem, Mark. "Recent Heat Waves Caused by Global Warming, Hansen Says." Bloomberg Business News. August 06, 2012. http://www.businessweek.com/printer/articles/301255?type=bloomberg (no longer available).

Draper, Robert. "Australia's Dry Ruin." *National Geographic*, April 2009, 34–59.

"Drier, Warmer Springs in U.S. Southwest Stem from Human-Caused Changes in Winds." NASA Earth Observatory, August 19, 2008. http://earthobservatory.nasa.gov/Newsroom/MediaAlerts/2008/2008081927359.html (no longer available).

"Drought, Excessive Heat Ruining Harvests in Western Europe." *Daytona Beach News-Journal* (Florida), August 5. 2003, 3A.

"Drought Forces Barcelona to Ship in Drinking Water." *Wall Street Journal*, May 14, 2008, A13.

"Drought in Iraq." NASA Earth Observatory, June 4, 2008. http://earthobservatory.nasa.gov/IOTD/view.php?id=8797&eocn=image&eoci=related_image.

"Drought Shrinking São Paulo Reservoirs." NASA Earth Observatory, October 23, 2014. http://earthobservatory.nasa.gov/IOTD/view.php?id=84564&src=eoa-iotd.

Dube, Francine. "North America's Growing Season 12 Days Longer." *National Post* (Canada), September 5, 2001. http://www.nationalpost.com (no longer available).

Duplessy, J. C., D. M. Roche, and M. Kageyama. "The Deep Ocean during the Last Interglacial Period." *Science* 316 (April 6, 2007): 89–91.

Durack, Paul J., Susan E. Wiffels, and Richard J. Matear. "Ocean Salinities Reveal Strong Global Water Cycle Intensification during 1950 to 2000." *Science* 336 (April 27, 2012): 455–458.

Dutrieux, Pierre, et al. "Strong Sensitivity of Pine Island Ice-Shelf Melting to Climatic Variability." *Science* 343 (January 10, 2014): 174–178.

Dutton, A., and K. Lambeck. "Ice Volume and Sea Level during the Last Interglacial." *Science* 337 (July 13, 2012): 216–219.

Dutton, A., et al. "Sea Level Rise Due to Polar Ice-Sheet Mass Loss during Past Warm Periods." *Science* 349 (July 10, 2015). doi: 10.1126/science.aaa4019.

"Each Degree of Warming Will Raise Sea Levels 7.5 Feet." Environment News Service, July 15, 2013. http://ens-newswire.com/2013/07/15/each-degree-of-warming-will-raise-sea-levels-7-5-feet/.

"Earth's Lakes Are Warming Faster Than Its Air." *Science*, December 18, 2015, 1449.

Eddy, Melissa. "Germany May Offer Model for Reining in Fossil Fuel Use." *The New York Times*, December 3, 2015. http://www.nytimes.com/2015/12/04/world/europe/germany-may-offer-model-for-reining-in-fossil-fuel-use.html.

Edenhofer, Ottmar. "King Coal and the Queen of Subsidies." *Science* 349 (September 18, 2015): 1286–1287.

Edgerton, Lynne T., and the Natural Resources Defense Council. *The Rising Tide: Global Warming and World Sea Levels*. Washington, DC: Island Press, 1991.

Egan, Timothy. "Alaska, No Longer So Frigid, Starts to Crack, Burn, and Sag." *The New York Times*, June 16, 2002, A1.

Egan, Timothy, "Now, in Alaska, Even the Permafrost Is Melting," *The New York Times*, June 16, 2002, A1.

Egan, Timothy. "On Hot Trail of Tiny Killer in Alaska." *The New York Times*, June 25, 2002, F1.

Egan, Timothy. "The Race to Alaska Before It Melts." *The New York Times*, June 26, 2005. http://www.nytimes.com/2005/06/26/travel/the-race-to-alaska-before-it-melts.html.

Egan, Timothy. "Seattle on the Mediterranean." *The New York Times*, July 3, 2015. http://www.nytimes.com/2015/07/04/opinion/seattle-on-the-mediterranean.html.

Eggert, David. "Shareholders File Global Warming Resolutions at Ford, GM." Associated Press, December 11, 2002 (LEXIS).

Ehrlich, Greterl. "Rotten Ice: Traveling by Dogsled in the Melting Arctic." *Harper's*, April 2015, 41–50.

Eilperin, Juliet. "Warming Tied to Extinction of Frog Species." *Washington Post*, January 12, 2006, A1. http://www.washingtonpost.com/wp-dyn/content/article/2006/01/11/AR2006011102121_pf.html.

Eilperin, Juliet. "Antarctic Ice Sheet Is Melting Rapidly; New Study Warns of Rising Sea Levels." *Washington Post*, March 3, 2006, A1.

Eilperin, Juliet. "Growing Acidity of Oceans May Kill Corals." *Washington Post*, July 5, 2006, A1. http://www.washingtonpost.com/wp-dyn/content/article/2006/07/04/AR2006070400772_pf.htmla.

Eilperin, Juliet. "U.S., China Got Climate Warnings Toned Down." *Washington Post*, April 7, 2007, A5. http://www.washingtonpost.com/wp-dyn/content/article/2007/04/06/AR2007040600291.html.

Eilperin, Juliet. "Clues to Rising Seas Are Hidden in Polar Ice." *Washington Post*, July 16, 2007, A6. http://www.washingtonpost.com/wp-dyn/content/article/2007/07/15/AR2007071500882_pf.html.

Eilperin, Juliet. "Carbon Is Building Up in Atmosphere Faster Than Predicted." *Washington Post*, September 26, 2008, A2. http://www.washingtonpost.com/wp-dyn/content/article/2008/09/25/AR2008092503989_pf.html.

Eilperin, Juliet. "Study Ties Tree Deaths to Change in Climate." *Washington Post*, January 23, 2009, A8. http://www.washingtonpost.com/wp-dyn/content/article/2009/01/22/AR2009012202473_pf.html.

Eilperin, Juliet. "Geo-Engineering Sparks International Ban, First-Ever Congressional Report." *Washington Post*, October 29, 2010. http://www.washingtonpost.com/wp-dyn/content/article/2010/10/29/AR2010102906365_pf.html.

Eilperin, Juliet, and Jason Samenow. "Greenland Glacier Loses Large Mass of Ice." *Washington Post*, July 17, 2012. http://www.washingtonpost.com/national/health-science/greenland-glacier-loses-large-mass-of-ice/2012/07/17/gJQAf5CQsW_print.html.

Eilperin, Juliet, and Mary Beth Sheridan. "New Data Show Rapid Arctic Ice Decline." *Washington Post*, April 7, 2009, A3. http://www.washingtonpost.com/wp-dyn/content/article/2009/04/06/AR2009040601634_pf.html.

Eligon, John. "After Drought, Rains Plaguing Midwest Farms." *The New York Times*, June 9, 2013. http://www.nytimes.com/2013/06/10/us/after-drought-rains-plaguing-midwest-farms.html.

Elliott, Valerie. "Polar Bears Surviving on Thin Ice." *The Times* (London), October 30, 2003, n.p. (LEXIS)

Ellison, Christopher R. W., Mark R. Chapman, and Ian R. Hall. "Surface and Deep Ocean Interactions during the Cold Climate Event 8200 Years Ago." *Science* 312 (June 30, 2006): 1929–1932.

Elsner, James B. "Tempests in Time." *Nature* 447 (June 7, 2007): 647–648.

Elsner, James B., James P. Kossin, and Thomas H. Jagger. "The Increasing Intensity of the Strongest Tropical Cyclones." *Nature* 455 (September 4, 2008): 92–95.

Emanuel, K. A. "An Air-Sea Interaction Theory for Tropical Cyclones. Part I: Steady-State Maintenance." *Journal of the Atmospheric Sciences* 43 (1986): 585–604.

Emanuel, K. A. "The Dependence of Hurricane Intensity on Climate." *Nature* 326(2) (April 1987): 483–485.

Emanuel, K. A. "The Maximum Intensity of Hurricanes." *Journal of the Atmospheric Sciences* 45 (April 1, 1988): 1143–1156. http://www.aos.wisc.edu/~aos718/basic%20dynamics/ mpi.emanuel.pdf.

Emanuel, K. A. "Toward a General Theory of Hurricanes." *American Scientist* 76 (July–August 1988): 370–379. ftp://texmex.mit.edu/pub/emanuel/PAPERS/Amer _sci.pdf.

Emanuel, K. A. "Thermodynamic Control of Hurricane Intensity." *Nature* 401 (October 14, 1999): 665–669.

Emanuel, K. A. "Increasing Destructiveness of Tropical Storms over the Past 30 Years." *Nature* 436 (August 4, 2005): 686–688.

Emanuel, K. A. "Downscaling CMIP5 Climate Models Shows Increased Tropical Cyclone Activity over the 21st Century." *Proceedings of the United States National Academy of Sciences* 110(30) (July 2013): 12219–12224.

Emanuel, Kerry. "Downscaling CMIP5 Climate Models Shows Increased Tropical Cyclone Activity over the 21st Century." *Proceedings of the United States National Academy of Sciences* 110(30) (July 2013): 12219–12224.

Emanuel, Kerry A. "Increasing Destructiveness of Tropical Storms over the Past 30 Years." *Nature* 436 (August 4, 2005): 686–688.

"Energy Is Another Victim of Drought." *Omaha World-Herald*, May 1, 2015, 3A.

English, Andrew. "Feeding the Dragon: How Western Car-Makers Are Ignoring Ecological Dangers in Their Rush to Exploit a Wide-Open Market." *Daily Telegraph* (London), October 30, 2004, 1.

Environment America. *When It Rains, It Pours: Global Warming and the Rising Frequency of Extreme Precipitation in the United States*. Boston: Environment America, 2007. http:// www.environmentamerica.org/reports/ame/when-it-rains-it-pours-global-warming-and -rising-frequency-extreme-precipitation-united.

Epstein, Paul R. *Climate, Ecology, and Human Health*. December 18, 1998. http:// www.iitap .iastate.edu/ gccourse/ issues/ health/ health.html (no longer available).

Epstein, Paul R. "Profound Consequences: Climate Disruption, Contagious Disease, and Public Health." *Native Americas* 16(3–4) (Fall–Winter 1999): 64–67.

Epstein, Paul R., and Dan Ferber. *Changing Planet, Changing Health: How the Climate Crisis Threatens Our Health and What We Can Do about It*. Berkeley: University of California Press, 2011.

Epstein, Paul R., et al. "Biological and Physical Signs of Climate Change: Focus on Mosquito-Borne Diseases." *Bulletin of the American Meteorological Society* 79, Part 1 (1998): 409–417.

Erbacher, Jochen, et al. "Increased Thermohaline Stratification as a Possible Cause for an Ocean Anoxic Event in the Cretaceous Period." *Nature* 409 (January 18, 2001): 325–327.

Erdbrink, Thomas. "Its Great Lake Shriveled, Iran Confronts Crisis of Water Supply." *The New York Times*, January 30, 2014. http://www.nytimes.com/2014/01/31/world/middlee ast/its-great-lake-shriveled-iran-confronts-crisis-of-water-supply.html.

Erdbrink, Thomas. "Scarred Riverbeds and Dead Pistachio Trees in a Parched Iran." *The New York Times*, August 19, 2015. http://www.nytimes.com/2015/12/19/world/middle east/scarred-riverbeds-and-dead-pistachio-trees-in-a-parched-iran.html.

Erdman, Jon. "Storm in Italy May Have Set New World Record for 24-Hour Snowfall." Weather.com. http://www.weather.com/storms/winter/news/world-snow-record-italy-24 -hour-march-2015 (no longer available).

Erickson, Jim. "Boulder Team Sees Obstacle to Saving Ozone Layer; 'Rocks' in Arctic Clouds Hold Harmful Chemicals." *Rocky Mountain News* (Denver), February 9, 2001, 37A.

"Ethanol Production Threatens Plains States with Water Scarcity." Environment News Service, September 21, 2007. http://www.ens-newswire.com/ens/sep2007/2007-09-21-091. asp (no longer available).

Eunjung Cha, Ariana. "China Embraces Nuclear Future; Optimism Mixes with Concern as Dozens of Plants Go Up." *Washington Post*, May 29, 2007, D1.

"Europe and Pacific Northwest Face Record Heat." NASA Earth Observatory, July 12, 2015. http://earthobservatory.nasa.gov/IOTD/view.php?id=86204.

"Europe's Heat Wave Toll Tops 19,000." *Omaha World-Herald*, September 26, 2003, 2.

Evan, Amato. "Climate Science: Aerosols and Atlantic Aberrations." *Nature* 484 (April 12, 2012): 170–171.

Evan, Amato T., et al. "Arabian Sea Tropical Cyclones Intensified by Emissions of Black Carbon and Other Aerosols." *Nature* 479 (November 3, 2011): 94–97.

Evans-Pritchard, Ambrose. "Dutch Have Only Years before Rising Seas Reclaim Land: Dikes No Match against Global Warming Effects." *Daily Telegraph* (London), October 30, 2004, 1.

"Expert Fears Warming Will Doom Bears." *Times-Colonist* (Victoria, BC), January 5, 2003, C8.

"The Extended Reach of Australian Drought." *Nature* 483 (March 1, 2012): 8.

"Extreme Rainstorms a New Texas Trend." Environment News Service, December 3, 2007. http://www.ens-newswire.com/ens/dec2007/2007-12-04-095.asp (no longer available).

"Extreme Weather Events Signal Global Warming to World's Meteorologists." Environment News Service, August 17, 2010. http://www.ens-newswire.com/ens/aug2010/2010-08 -17-01.html.

"ExxonMobil Agrees to Report Carbon Stranded Asset Risk." Environment News Service, March 24, 2014. http://ens-newswire.com/2014/03/24/exxonmobil-agrees-to-report -carbon-stranded-asset-risk/.

Ezer, Tal, et al. "Gulf Stream's Induced Sea Level Rise and Variability along the U.S. Mid-Atlantic Coast." *Journal of Geophysical Research* 118(2) (February 2013): 685–697. doi: 10.1002.

Faerber, Fritz. 2010. *Oil Spill Threatens Native American "Water" Village*. National Geographic. Online film. http://news.nationalgeographic.com/news/2010/06/100608-us-oil-gulf -indians-video/. Accessed May 16, 2012.

Fahey, D. W., et al. "The Detection of Large HNO_3-Containing Particles in the Winter Arctic Stratosphere." *Science* 291 (February 9, 2001): 1026–1031.

Fahrenthold, David A. "Eco-Bills Come Due at Bay's Beaches Region Pays Dearly for Climate Change in Erosion, Abatement." *Washington Post*, March 19, 2009, A1. http://www .washingtonpost.com/wp-dyn/content/article/2009/03/18/AR2009031804178.html.

Fahrenthold, David A. "Scientists Say Mountaintop Mining Should Be Stopped." *Washington Post*, January 8, 2010, A3. http://www.washingtonpost.com/wp-dyn/content/arti cle/2010/01/07/AR2010010702530_pf.html.

Fahy, Declan. "Nature Charts Its Own Change: Irish Researchers Are Finding Signs of Climate Change in Trees and Bird Species." *Irish Times*, September 13, 2001, n.p. (LEXIS).

"Failure to Manage Global Warming Would Cripple World Economy." Environment News Service, October 13 and October 30, 2006. http://www.ens-newswire.com/ens/oct2006/2006-10-30-06.asp (no longer available).

Fairless, Daemon. "Renewable Energy: Energy Go-Round." *Nature* 447 (June 28, 2007): 1046–1048.

Falbel, Gerald. "Proposal for Reduction in Global Warming and Climate Change." July 25, 2015. http://www.freelunchclimatechangesolution.com.

Fallows, James. "Turn Left at Cloud 109." *New York Times Sunday Magazine*, November 21, 1999, 84–89.

Fang, Janet. "Wildlife Service Plans for a Warmer World." *Nature* 464 (March 18, 2010): 332–333.

Fargione, Joseph, et al. "Land Clearing and the Biofuel Carbon Debt." *Science* 319 (February 29, 2008): 1235–1238.

Faris, Stephan. "Ice Free." *New York Times Sunday Magazine*, July 27, 2008. http://www.nytimes.com/2008/07/27/magazine/27wwln-phenom-t.html.

Farman, J. C., B. G. Gardiner, and J. D. Shanklin. "Large Losses of Total Ozone Reveal Seasonal $ClPO_x/NO_x$ Interaction." *Nature* 315 (1985): 207–210.

Fears, Darryl. "Scientists Say California Hasn't Been This Dry in 500 Years." *Washington Post*, September 14, 2015. http://www.washingtonpost.com/news/energy-environment/wp/2015/09/14/scientists-say-its-been-500-years-since-california-was-this-dry/?wpmm=1&wpisrc=nl_evening.

Feder, Barnaby J. "The Showhouse That Sustainability Built." *The New York Times*, March 26, 2008. http://www.nytimes.com/2008/03/26/business/businessspecial2/26boulder.html.

Fedorov, Alexey V., Christopher M. Brierley, and Kerry Emanuel. "Tropical Cyclones and Permanent El Niño in the early Pliocene Epoch." *Nature* 463 (February 25, 2010): 1066–1070.

Fedorov, A.V., et al. "The Pliocene Paradox (Mechanisms for a Permanent El Niño)." *Science* (June 9, 2006): 1485–1489.

Feely, Richard A., et al. "Impact of Anthropogenic CO_2 on the $CaCO_3$ System in the Oceans." *Science* 305 (July 16, 2004): 362–366.

Feely, Richard A., et al. "Evidence for Upwelling of Corrosive 'Acidified' Water onto the Continental Shelf." *Science* 320 (June 13, 2008): 1490–1492.

Feldmann, Johannes, and Anders Levermann. "Collapse of the West Antarctic Ice Sheet after Local Destabilization of the Amundsen Basin." *Proceedings of the National Academy of Sciences* 112(46) (November 2015): 201512482.

Feresin, Emiliano. "Europe Looks to Draw Power from Africa." *Nature* 450 (November 29, 2007): 595.

Feulner, G., and S. Rahmstorf. "On the Effect of a New Grand Minimum of Solar Activity on the Future Climate on Earth." *Geophysical Research Letters* 37 (2010): L05707. doi: 10.1029/2010GL042710.

"Few Canadian Glaciers Left by 2100." *Nature* 520 (April 9, 2015): 134.

Field, Michael. "Dying Pacific Breadfruit New Sign of Looming Disaster." *Agence France-Presse*, December 1, 2002 (in LEXS).

"Fifth of World's Corals Dead." Discovery Channel.com. December 10, 2008. http://dsc.discovery.com/news/2008/12/10/coral-reef-death.html.

Filkins, Dexter. "The End of Ice: Exploring a Himalayan Glacier." *The New Yorker*, April 4, 2016, 59–65.

"Findings." *Harper's*, February, 2017, 96.

Finn, Peter. "In Balmy Europe, Feverish Choruses of 'Let It Snow.'" *Washington Post*, December 20, 2006, A1. http://www.washingtonpost.com/wp-dyn/content/article/2006/12/19/AR2006121901681_pf.html.

Finnegan, Seth, et al. "Paleontological Baselines for Evaluating Extinction Risk in the Modern Oceans." *Science* 348 (May 1, 2015): 567–570.

"Fire an Integral, but Unappreciated, Part of Global Warming." Environment News Service, April 23, 2009. http://www.ens-newswire.com/ens/apr2009/2009-04-23-095.asp.

Fischer, Erich M., and Reto Knutti. "Anthropogenic Contribution to Global Occurrence of Heavy-Precipitation and High-temperature Extremes." *Nature Climate Change*, April 27, 2015. http://www.nature.com/nclimate/journal/vaop/ncurrent/full/nclimate2617.html.

"Fished to the Point of Ruin, North Sea Cod Stocks So Low as to Spell Disaster." *The Herald* (Glasgow, Scotland), November 7, 2000, 18.

Fiske, Molly Hennessy, "Bayou on the Brink Again." *Los Angeles Times*, August 4, 2010. http://articles.latimes.com/2010/aug/04/nation/la-na-grand-bayou-20100804.

Fitter, A. H., and R. S. R. Fitter. "Rapid Changes in Flowering Time in British Plants." *Science* 296 (May 31, 2002): 1689–1691.

Flannery, Tim. *The Weather Makers: How Man Is Changing the Climate and What It Means for Life on Earth*. New York: Atlantic Monthly Press, 2005.

"Flash Floods Wreak Havoc." *Omaha World-Herald*, May 25, 2015, 7A.

Fleck, John. "Dry Days, Warm Nights." *Albuquerque Journal*, December 28, 2003, B1.

Folger, Tim. "Viking Weather: The Changing Face of Greenland." *National Geographic*, June 2010, 48–63.

Forcada, Jaume, and Joseph Iván Hoffman. "Climate Change Selects for Heterozygosity in a Declining Fur Seal Population." *Nature* 511 (July 24, 2014): 462–465.

Forero, Juan. "As Andean Glaciers Shrink, Water Worries Grow." *The New York Times*, November 24, 2002, A3.

Foster, Taryn, et al. "Ocean Acidification Causes Structural Deformities in Juvenile Coral Skeletons." *Science Advances* 2(2) (February 19, 2016): e1501130. doi: 10.1126/sciadv.1501130.

Fountain, Henry. "Observatory: Threat to Rice Crops." *The New York Times*, December 12, 2000, F5.

Fountain, Henry. "Observatory: Rice and Warm Weather." *The New York Times*, June 29, 2004, F1.

Fountain, Henry. "More Acidic Ocean Hurts Reef Algae as Well as Corals." *The New York Times*, January 8, 2008, D3.

Fountain, Henry. "A Storm Study Ties Warming to Ozone Loss." *The New York Times*, July 27, 2012, A1, A14.

Fountain, Henry. "Climate Aids in Study Face Big Obstacles." *The New York Times*, January 17, 2014. http://www.nytimes.com/2014/01/17/science/earth/climate-aids-in-study-face-big-obstacles.html.

Fountain, Henry. "For Already Vulnerable Penguins, Study Finds Climate Change Is Another Danger." *The New York Times*, January 29, 2014. http://www.nytimes.com/2014/01/30/science/earth/climate-change-taking-toll-on-penguins-study-finds.html.

Fountain, Henry. "Corralling Carbon before It Belches from Stack." *The New York Times*, July 21, 2014. http://www.nytimes.com/2014/07/22/science/corralling-carbon-before-it-belches-from-stack.html.

Fountain, Henry. "Sierra Nevada Snow Won't End California's Thirst." *The New York Times*, April 12, 2016. http://www.nytimes.com/2016/04/12/science/california-snow-drought-sierra-nevada-water.html.

Fountain, Henry. "Scientists See Push from Climate Change in Louisiana Flooding." *The New York Times*, September 7, 2016. http://www.nytimes.com/2016/09/08/science/global -warming-louisiana-flooding.html.

Fountain, Henry, and Justin Gillis. "Typhoon in Philippines Casts Long Shadow over U.N. Talks on Climate Treaty." *The New York Times*, November 11, 2013. http://www.nytimes. com/2013/11/12/world/asia/typhoon-in-philippines-casts-long-shadow-over-un-talks -on-climate-treaty.html.

Fowler, A. M., and K. J. Hennessy. "Potential Impacts of Global Warming on the Frequency and Magnitude of Heavy Precipitation." *Natural Hazards* 11 (1995): 283. https://link .springer.com/article/10.1007/BF00613411. doi: 10.1007/BF00613411.

Fox, Douglas. "Could East Antarctica Be Headed for Big Melt?" *Science* 328 (June 25, 2010): 1630–1631.

Fox, Douglas. "Antarctica Undercut," *National Geographic*, January 2012, 35.

Fox, Douglas. "Polar Research: Trouble Bares Its Claws." *Nature* 492 (December 2012): 170–172.

Francis, Jennifer A., and Stephen J. Vavrus. "Evidence Linking Arctic Amplification to Extreme Weather in Mid-latitudes." *Geophysical Research Letters*, 39 (March 17, 2012): L06801. http://marine.rutgers.edu/~francis/pres/Francis_Vavrus_2012GL051000_pub .pdf.

Francis, Justin. Responsibletravel.com. United Kingdom. Accessed April 30, 2006. http:// www.responsibletravel.com/Copy/Copy101993.htm (no longer available).

Frank, Thomas. "Planes Fly More, Emit Less Greenhouse Gas." *USA Today*, May 9, 2008, 1B.

Freeman, C., C. D. Evans, and D. T. Monteith. "Export of Organic Carbon from Peat Soils." *Nature* 412 (August 23, 2001): 785.

Freeman, C., et al. "Export of Dissolved Carbon from Peatlands under Elevated Carbon Dioxide Levels." *Nature* 430 (July 8, 2004): 195–198.

Fried, Jeremy S., Margaret S. Torn, and Evan Mills. "The Impact of Climate Change on Wild-fire Severity." *Climatic Change* 64 (May 2004): 169–191.

Friedman, Thomas. "Learning to Speak Climate." *The New York Times*, August 6, 2008. http://www.nytimes.com/2008/08/06/opinion/06friedman.html.

Freidman, Thomas L. "Is It Weird Enough Yet?" *The New York Times*, September 14, 2011, A29.

Fritz, Angela. "Freak Storm Pushes North Pole 50 Degrees above Normal to Melting Point." *Washington Post*, December 30, 2015, n.p.

Fromer, Jacob, and Edward Wong. "An Epic Downpour Wipes Away a Capital's Sheen." *The New York Times*, July 27, 2012, A10.

Frosch, Dan. "Citing Need for Assessments, U.S. Freezes Solar Energy Projects." *The New York Times*, June 27, 2008, A13.

"Fueling Jets with Animal Fat." Environment News Service, July 18, 2007. http://www.ens -newswire.com/ens/jul2007/2007-07-18-09.asp#anchor7 (no longer available).

Fukasawa, Masao, et al. "Bottom Water Warming in the North Pacific Ocean." *Nature* 427 (February 26, 2004): 825–827.

Fukuda, Misao, et al. "Climate Change Is Associated with Male:Female Ratios of Fetal Deaths and Newborn Infants in Japan." *Fertility and Sterility*, September 2014. http://www.fert stert.org/article/S0015-0282%2814%2901840-8/fulltext.

Funk, McKenzie. "Cold Rush: The Coming Fight for the Melting North." *Harper's*, September 2007, 54–55.

Funk, McKenzie. *Windfall: The Booming Business of Global Warming*. New York: Penguin Press, 2014.

Gaarder, Nancy. "State Enjoying 'Exceptional' Warmth." *Omaha World-Herald*, December 4, 2001, A1, A2.

Gaarder, Nancy. "Come Autumn, Ragweed Will Stick around Longer." *Omaha World-Herald*, March 6, 2011. http://www.livewellnebraska.com/apps/pbcs.dll/article?AID=/20110306/LIVEWELL01/703069909/1161 (no longer available).

Gaarder, Nancy. "Harvests Will Take Hit from Heat's Effect on Pollen." *Omaha World-Herald*, July 25, 2011, A1, A2.

Gaarder, Nancy. "Experts Unravel Causes of Punishing Heat Wave." *Omaha World-Herald*, June 29, 2012, 1A, 8A.

Gaarder, Nancy. "Pines Burned Out of Pine Ridge." *Omaha World-Herald*, January 7, 2013, A1, A2.

Gaarder, Nancy. "An Ag 'Battleground:' Federal Climate Report Says Midlands at Risk of Experiencing More Extremes." *Omaha World-Herald*, February 4, 2013, D1, D4.

Gaarder, Nancy. "Arctic Warming Blamed for Last Year's Heat and This Year's Chill." *Omaha World-Herald*, March 30, 2013. http://www.omaha.com/article/20130330/NEWS/130339998/1685#arctic-warming-blamed-for-last-year-s-heat-and-this-year-s-chill (no longer available).

Gaarder, Nancy. "Heavy Storm Hurdles through Sidney, Neb." *Omaha World-Herald*, September 11, 2013, 6B.

Gaarder, Nancy, David Hendee, and Andrew J. Nelson. "Once-in-a-Thousand-Years Rain." *Omaha World-Herald*, May 8, 2015, A-1, A-4.

Gabriel, Trip, and Carol Davenport. "Activists Push Fracking onto Agenda." *The New York Times*, April 5, 2016, A16.

Galloway, Elaine, and Chloe Rhodes. "Warm Spell Brings Early Start to Hay-Fever Misery." *Evening Standard* (London), April 14, 2003, 16.

Ganopolski, A., R. Winkelmann, and H. J. Schellnhuber. "Critical Insolation: CO_2 Relation for Diagnosing Past and Future Glacial Inception." *Nature* 529 (January 14, 2016): 200–203.

Gao, Y., et al. "Sources and Pathways of 90Sr in the North Atlantic-Arctic Region: Present Day and Global Warming." *Journal of Environmental Radioactivity* 5 (May 2009): 375–395.

Garcia, Rolando R. "Atmospheric Science: An Arctic Ozone Hole?" *Nature* 478 (October 27, 2011): 462–463.

Gardiner, Beth. "Report: Extreme Weather on the Rise, Likely to Get Worse." Associated Press Worldstream, International News, London. February 27, 2003 (LEXIS).

Gardner, T. A., et al. "Long-Term Region-Wide Declines in Caribbean Corals." *Science* 301 (July 18, 2003): online edition. http://science.sciencemag.org/content/301/5635/958.

Garfin, Gregg, et al. *Assessment of Climate Change in the Southwestern United States: A Report Prepared for the National Climate Assessment*. Washington, DC: Island Press, 2013. http://swccar.org/sites/all/themes/files/SW-NCA-color-FINALweb.pdf.

Gartner, John. "Climbers Bring Climate Change from Mountaintop to Laptops." Environment News Service, July 31, 2007. http://www.ens-newswire.com/ens/jul2007/2007-07-31-04.asp (no longer available).

Gaskill, Melissa, and Michael Wines. "At Least 3 Are Killed and 12 Are Missing as Storms Ravage Texas and Oklahoma." *The New York Times*, May 25, 2015, A9.

"Gators Spotted in Mississippi River Backwaters at Memphis." *Daytona Beach News-Journal*, May 13, 2006.

Gearheard, S., et al. "Linking Inuit Knowledge and Meteorological Station Observations to Understand Changing Wind Patterns at Clyde River, Nunavut." *Climatic Change*, 100 (2010), 267–294. doi: 10.1007/s10584-009-9587-1.

Gelbspan, Ross. *The Heat Is On: The High Stakes Battle over Earth's Threatened Climate*. Reading, MA: Addison-Wesley, 1997.

Gelling, Peter. "Forest Loss in Sumatra Becomes a Global Issue." *The New York Times*, December 6, 2007. http://www.nytimes.com/2007/12/06/world/asia/06indo.html.

George, Jane. "Global Warming Threatens Nunavut's National Parks." *Nunatsiaq News*, May 19, 2000. http://www.nunatsiaq.com/archives/nunavut000531/nvt20519_18.html.

"Georgia Judge Yanks Coal Power Permit on Climate Concerns." Environment News Service, June 30, 2008. http://www.ens-newswire.com/ens/jun2008/2008-06-30-091.asp (no longer available).

Georgiou, Lucy, et al. "PH Homeostasis during Coral Calcification in a Free Ocean CO_2 Enrichment (FOCE) Experiment, Heron Island Reef Flat, Great Barrier Reef." *Proceedings of the National Academy of Sciences*, October 6, 2015. http://www.pnas.org/content/early/2015/10/01/1505586112.full.pdf?sid=26b9f3e1-4ff7-4247-821f-e806352a8708.

Gertner, Jon. "The Secrets in Greenland's Ice Sheet." *The New York Times Sunday Magazine*, November 15, 2015. http://www.nytimes.com/2015/11/15/magazine/the-secrets-in-greenlands-ice-sheets.html.

Gething, Peter W., et al. "Climate Change and the Global Malaria Recession." *Nature* 465 (May 20, 2010): 342–345

Gewin, Virginia. "North Pacific 'Blob' Stirs Up Fisheries Management." *Nature* 524 (August 27, 2015): 396.

"Giant Squid Film Team Makes Spectacular Catch." September 14, 2002. Agence France Presse, September 14, 2002.

Giardina, C., and M. Ryan. "Evidence that Decomposition Rates of Organic Carbon in Mineral Soil Do Not Vary with Temperature. *Nature* 404 (2000): 858–861.

Gibbs, Walter. "Scientists Back Off Theory of a Colder Europe in a Warming World." *The New York Times*, May 15, 2007. http://www.nytimes.com/2007/05/15/science/earth/15cold.html.

Gibbs, Walter, and Sarah Lyall. "Gore Shares Peace Prize for Climate Change Work." *The New York Times*, October 13, 2007, A1, A13.

Gilbert, Jonathan. "Argentina Struggles to Control a Major Outbreak of Dengue." *The New York Times*, February 18, 2016, A10.

Giles, Jim. "The Outlook for Amazonia Is Dry." *Nature* 442 (August 17, 2006): 726–727.

Giles, Jim. "How Much Will It Cost to Save the World?" *Nature* 444 (November 2, 2006): 6–7.

Gill, Richardson Benedict. *The Great Maya Droughts: Water, Life, and Death*. Albuquerque: University of New Mexico Press, 2000.

Gill, Victoria. "Climate Change Link to Lizard Extinction." British Broadcasting Corp., May 14, 2010. http://news.bbc.co.uk/2/hi/science_and_environment/10113949.stm.

Gille, Sarah T. "Warming of the Southern Ocean since the 1950s." *Science* 295 (February 15, 2002): 1275–1277.

Gillet-Chaulet, Fabien, and Gael Durand. "Ice-Sheet Advance in Antarctica." *Nature* 467 (October 14, 2010): 794–795.

Gillett, Nathan P., and David W. J. Thompson. "Simulation of Recent Southern Hemisphere Climate Change." *Science* 302 (October 10, 2003): 273–275.

Gillis, Justin. "In Weather Chaos, a Case for Global Warming." *The New York Times*, August 14, 2010. http://www.nytimes.com/2010/08/15/science/earth/15climate.html.

Gillis, Justin. "A Warming Planet Struggles to Feed Itself." *The New York Times*, June 4, 2011. http://www.nytimes.com/2011/06/05/science/earth/05harvest.html.

Gillis, Justin. "The Threats to a Crucial Canopy." *The New York Times*, October 1, 2011. http://www.nytimes.com/2011/10/01/science/earth/01forest.html.

Gillis, Justin. "As Permafrost Thaws, Scientists Study the Risks." *The New York Times*, December 16, 2011. http://www.nytimes.com/2011/12/17/science/earth/warming-arctic-permafrost-fuels-climate-change-worries.html.

Gillis, Justin. "Rising Sea Levels Seen as Threat to Coastal U.S." *The New York Times*, March 13, 2012. http://www.nytimes.com/2012/03/14/science/earth/study-rising-sea-levels-a-risk-to-coastal-states.html.

Gillis, Justin. "Global Warming Makes Heat Waves More Likely, Study Finds." *The New York Times*, July 10, 2012. http://www.nytimes.com/2012/07/11/science/earth/global-warming-makes-heat-waves-more-likely-study-finds.html.

Gillis, Justin. "Study Finds More of Earth Is Hotter and Says Global Warming Is at Work." *The New York Times*, August 6, 2012. http://www.nytimes.com/2012/08/07/science/earth/extreme-heat-is-covering-more-of-the-earth-a-study-says.html.

Gillis, Justin. "Satellites Show Sea Ice in Arctic Is at a Record Low." *The New York Times*, August 27, 2012. http://www.nytimes.com/2012/08/28/science/earth/sea-ice-in-arctic-measured-at-record-low.html.

Gillis, Justin. "West Antarctica Warming Faster Than Thought, Study Finds." *The New York Times*, December 23, 2012.

Gillis, Justin. "How High Could the Tide Go?" *The New York Times*, January 21, 2013. http://www.nytimes.com/2013/01/22/science/earth/seeking-clues-about-sea-level-from-fossil-beaches.html.

Gillis, Justin. "Climate Maverick to Retire from NASA." *The New York Times*, April 1, 2013. http://www.nytimes.com/2013/04/02/science/james-e-hansen-retiring-from-nasa-to-fight-global-warming.html.

Gillis, Justin. "In Sign of Warming, 1,600 Years of Ice in Andes Melted in 25 Years." *The New York Times*, April 4, 2013. http://www.nytimes.com/2013/04/05/world/americas/1600-years-of-ice-in-perus-andes-melted-in-25-years-scientists-say.html.

Gillis, Justin. "Rebuilding the Shores, Increasing the Risks." *The New York Times*, April 8, 2013. http://www.nytimes.com/2013/04/09/science/earth/rebuilding-our-shores-increasing-the-risks.html.

Gillis, Justin. "Timing a Rise in Sea Level." *The New York Times*, August 12, 2013. http://www.nytimes.com/2013/08/13/science/timing-a-rise-in-sea-level.html.

Gillis, Justin. "Climate Change Seen Posing Risk to Food Supplies." *The New York Times*, November 1, 2013. http://www.nytimes.com/2013/11/02/science/earth/science-panel-warns-of-risks-to-food-supply-from-climate-change.html?_r=0.

Gillis, Justin. "A Jolt to Complacency on Food Supply." *The New York Times*, November 11, 2013. http://www.nytimes.com/2013/11/12/science/earth/warning-on-global-food-supply.html.

Gillis, Justin. "In New Jersey Pines, Trouble Arrives on Six Legs." *The New York Times*, December 1, 2013. http://www.nytimes.com/2013/12/02/science/earth/in-new-jersey-pines-trouble-arrives-on-six-legs.html.

Gillis, Justin. "The Flood Next Time." *The New York Times*, January 13, 2014. http://www.nytimes.com/2014/01/14/science/earth/grappling-with-sea-level-rise-sooner-not-later.html.

Gillis, Justin. "U.N. Says Lag in Confronting Climate Woes Will Be Costly." *The New York Times*, January 16, 2014. http://www.nytimes.com/2014/01/17/science/earth/un-says-lag-in-confronting-climate-woes-will-be-costly.html.

Gillis, Justin. "Freezing January for Easterners Was Not Felt Round the World." *The New York Times*, February 20, 2014. http://www.nytimes.com/2014/02/21/science/earth/more-bite-left-to-winter-but-it-hasnt-been-as-bad-as-you-think.html.

Gillis, Justin. "Panel's Warning on Climate Risk: Worst Is Yet to Come." *The New York Times*, March 30, 2014. http://www.nytimes.com/2014/03/31/science/earth/panels-warning-on-climate-risk-worst-is-yet-to-come.html.

Gillis, Justin. "Climate Efforts Falling Short, U.N. Panel Says." *The New York Times*, April 13, 2014, A1.

Gillis, Justin. "U.S. Climate Has Already Changed, Study Finds, Citing Heat and Floods." *The New York Times*, May 6, 2014. http://www.nytimes.com/2014/05/07/science/earth/climate-change-report.html?hp.

Gillis, Justin. "Looks Like Rain Again. And Again." *The New York Times*, May 13, 2014. http://www.nytimes.com/2014/05/13/science/looks-like-rain-again-and-again.html?ref=science.

Gillis, Justin. "Bipartisan Report Tallies High Toll on Economy from Global Warming." *The New York Times*, June 24, 2014. http://www.nytimes.com/2014/06/24/science/report-tallies-toll-on-economy-from-global-warming.html.

Gillis, Justin. "Greenhouse Gas Emissions Are Growing, and Growing More Dangerous, Draft of U.N. Report Says." *The New York Times*, August 26, 2014. http://www.nytimes.com/2014/08/27/science/earth/greenhouse-gas-emissions-are-growing-and-growing-more-dangerous-draft-of-un-report-says.html.

Gillis, Justin. "Sun and Wind Alter Global Landscape, Leaving Utilities Behind." *The New York Times*, September 13, 2014. http://www.nytimes.com/2014/09/14/science/earth/sun-and-wind-alter-german-landscape-leaving-utilities-behind.html.

Gillis, Justin. "U.N. Panel Issues Its Starkest Warning Yet on Global Warming." *The New York Times*, November 2, 2014. http://www.nytimes.com/2014/11/03/world/europe/global-warming-un-intergovernmental-panel-on-climate-change.html.

Gillis, Justin, "A Tricky Transition from Fossil Fuel." *The New York Times*, November 10, 2014. http://www.nytimes.com/2014/11/11/science/earth/denmark-aims-for-100-percent-renewable-energy.html

Gillis, Justin. "2014 Breaks Heat Record, Challenging Global Warming Skeptics." *The New York Times*, January 1, 2015. http://www.nytimes.com/2015/01/17/science/earth/2014-was-hottest-year-on-record-surpassing-2010.html.

Gillis, Justin. "A Climate Change Safety Net Frays." *The New York Times*, March 24, 2015, D3.

Gillis, Justin. "Climate Change Threatens to Kill Off More Aspen Forests by 2050s, Scientists Say." *The New York Times*, March 31, 2015. http://www.nytimes.com/2015/03/31/science/earth/climate-change-threatens-to-kill-off-more-aspen-forests-by-2050s-scientists-say.html.

Gillis, Justin. "New Study Links Weather Extremes to Global Warming." *The New York Times*, April 27, 2015. http://www.nytimes.com/2015/04/28/science/new-study-links-weather-extremes-to-global-warming.html.

Gillis, Justin. "Naomi Oreskes, a Lightning Rod in a Changing Climate." *The New York Times*, June 15, 2015. http://www.nytimes.com/2015/06/16/science/naomi-oreskes-a-lightning-rod-in-a-changing-climate.html?_r=0

Gillis, Justin. "Climate Change Intensifies California Drought, Scientists Say." *The New York Times*, August 20, 2015. https://www.nytimes.com/2015/08/21/science/climate-change-intensifies-california-drought-scientists-say.html.

Gillis, Justin. "Study Predicts Antarctica Ice Melt If All Fossil Fuels Are Burned." *The New York Times*, September 12, 2015. http://www.nytimes.com/2015/09/12/science/climate-study-predicts-huge-sea-level-rise-if-all-fossil-fuels-are-burned.html.

Gillis, Justin. "2015 Likely to Be Hottest Year Ever Recorded." *The New York Times*, October 22, 2015. http://www.nytimes.com/2015/10/22/science/2015-likely-to-be-hottest-year-ever-recorded.html.

Gillis, Justin. "2015 Was Hottest Year in Recorded History, Scientists Say." *The New York Times*, January 20, 2016. http://www.nytimes.com/2016/01/21/science/earth/2015-hottest-year-global-warming.html.

Gillis, Justin. "Seas Are Rising at Fastest Rate in Last 28 Centuries." *The New York Times*, February 22, 2016. http://www.nytimes.com/2016/02/23/science/sea-level-rise-global-warming-climate-change.html.

Gillis, Justin. "Scientists Warn of Perilous Climate Shift within Decades, Not Centuries." *The New York Times*, March 22, 2016. http://www.nytimes.com/2016/03/23/science/global-warming-sea-level-carbon-dioxide-emissions.html.

Gillis, Justin. "Climate Model Predicts West Antarctic Ice Sheet Could Melt Rapidly." *The New York Times*, March 31, 2016. http://www.nytimes.com/2016/03/31/science/global-warming-antarctica-ice-sheet-sea-level-rise.html.

Gillis, Justin, and Kenneth Chang. "Scientists Warn of Rising Oceans from Polar Melt." *The New York Times*, May 12, 2014. http://www.nytimes.com/2014/05/13/science/earth/collapse-of-parts-of-west-antarctica-ice-sheet-has-begun-scientists-say.html.

Gillis, Justin, and Joanna M. Foster. "Weather Runs Hot and Cold, So Scientists Look to the Ice." *The New York Times*, March 28, 2012. http://www.nytimes.com/2012/03/29/science/earth/arctic-sea-ice-eyed-for-clues-to-weather-extremes.html.

Gillis, Justin, and Henry Fountain. "Global Warming Cited as Wildfires Increase in Fragile Boreal Forest." *The New York Times*, May 10, 2016. http://www.nytimes.com/2016/05/11/science/global-warming-cited-as-wildfires-increase-in-fragile-boreal-forest.html

Gillis, Justin, and Somini Sengupta. "Limited Progress Seen Even as More Nations Step Up on Climate." *The New York Times*, September 28, 2015. http://www.nytimes.com/2015/09/28/world/limited-progress-seen-even-as-more-nations-step-up-on-climate.html.

Gilmour, James P., Luke D. Smith, Andrew J. Heyward, et al. "Recovery of an Isolated Coral Reef System Following Severe Disturbance." *Science* 340 (April 5, 2013): 69–71.

Gilmour, James P., et al. "Recovery of an Isolated Coral Reef System Following Severe Disturbance." *Science* 340 (April 5, 2013): 69–71.

"GISS 2007 Temperature Analysis." Goddard Institute for Space Studies. January 15, 2008. http://www.columbia.edu/~jeh1/mailings/20080114_GISTEMP.pdf.

Gjerdrum, Carina, et al. "Tufted Puffin Reproduction Reveals Ocean Climate Variability." *Proceedings of the National Academy of Sciences* 100(16) (August 5, 2003): 9377–9382.

"Glacial Retreat Seen Worldwide." Environment News Service, May 30, 2002. http://ens-news.com/ens/may2002/2002-05-30-09.asp#anchor2 (no longer available).

"Glacier Ice in Everest Region Could Vanish by 2100." Environment News Service, June 3, 2015. http://ens-newswire.com/2015/06/03/glacier-ice-in-everest-region-could-vanish-by-2100/.

"Glaciers and Ice Caps Quickly Melting into the Seas." Environment News Service, July 20, 2007. http://www.ens-newswire.com/ens/jul2007/2007-07-20-03.asp.

Glantz, Michael H. *Climate Affairs: A Primer*. Washington, DC: Island Press, 2003.

Gleckler, Peter J., et al. "Industrial-Era Global Ocean Heat Uptake Doubles in Recent Decades." *Nature Climate Change*, January 18, 2016. doi: 10.1038/nclimate2915.

Glick, Daniel. "The Heat Is On: Geosigns." *National Geographic*, September 2004, 12–33.

Glick, Patricia. *Global Warming: The High Costs of Inaction*. San Francisco: Sierra Club, 1998.

"Global Aviation Industry Vows to Halve CO$_2$ Emissions by 2050." Environment News Service, October 15, 2009. http://www.ens-newswire.com/ens/oct2009/2009-10-15-01.asp.

"Global Climate Shift Feeds Spreading Deserts." Environment News Service, June 17, 2002. http://ens-news.com/ens/jun2002/2002-06-17-03.asp.

"Globally, Year 2010 Ranked Second Warmest on Record; The Summer of 2010: A World-Wide Heat Wave." Environment News Service, June 28, 2011. http://www.ens-newswire.com/ens/jun2011/2011-06-28-01.html.

"Global Oil and Gas Methane Reduction Opportunities." Environmental Defense Fund, April 2015. http://www.edf.org/climate/rhodium-group-report-global-oil-gas-methane-emissions?_ga=1.219423705.1787517823.1427851088.

"Global Sea Levels Could Soon Rise 20 Feet as Climate Warms." Environment News Service, July 13, 2015. http://ens-newswire.com/2015/07/12/global-sea-levels-could-soon-rise-20-as-climate-warms/.

"Global Warming Blamed for Heavy Snowstorms, Record Floods." Environment News Service, March 2, 2011. http://www.ens-newswire.com/ens/mar2011/2011-03-02-02.html.

"Global Warming Blamed for Rising Sea Levels." *Omaha World-Herald*, November 25, 2001, 20A.

"Global Warming Could Hamper Ocean Sequestration." Environment News Service, December 4, 2002. http://ens-news.com/ens/dec2002/2002-12-04-09.asp (no longer available).

"Global Warming Could Speed Up." *Nature* 519 (March 12, 2015): 132.

"Global Warming Forecast to Delay Ozone Layer Recovery." Environment News Service, February 6, 2009. http://www.ens-newswire.com/ens/feb2009/2009-02-06-02.asp (no longer available).

"Global Warming Linked to Increase in Kidney Stones." Environment News Service, May 19, 2008. https://www.sciencedaily.com/releases/2008/05/080515072740.htm.

"Global Warming Means More Snow for Great Lakes Region." Ascribe Newswire, November 4. 2003 (LEXIS).

"Global Warming Sticker Shock." Environment News Service, May 23, 2008. http://www.ens-newswire.com/ens/may2008/2008-05-23-01.asp (no longer available).

"Global Warming Would Foster Spread of Dengue Fever into Some Temperate Regions." *Science Daily*, March 10, 1998.

Global Wind Energy Council. *Global Wind Statistics, 2014*. February 10, 2015. http://www.gwec.net/wp-content/uploads/2015/02/GWEC_GlobalWindStats2014_FINAL_10.2.2015.pdf. Accessed April 18, 2015.

"Global Wind Power Generated Record Year in 2006." Environment News Service, February 12, 2007.

Goddard Institute for Space Studies. "GISS Surface Temperature Analysis." Washington, DC: National Aeronautics and Space Administration, 2007.

Goes, Joaquim I., et al. "Warming of the Eurasian Landmass Is Making the Arabian Sea More Productive." *Science* 308 (April 22, 2005): 545–547.

"Going Hot and Cold in February." NASA Earth Observatory, March 10, 2015. http://earthobservatory.nasa.gov/IOTD/view.php?id=85460&src=eoa-iotd.

Goldman, Erica. "Even in the High Arctic, Nothing Is Permanent." *Science* 297 (August 30, 2002): 1493–1494.

Golledge, N. R., et al. "The Multi-Millennial Antarctic Commitment to Future Sea-Level Rise." *Nature* 526 (October 15, 2015): 421–425.

Gonzalez, Patrick, et al. "Aboveground Live Carbon Stock Changes of California Wildland Ecosystems, 2001–2010." *Forest Ecology and Management* 348 (July 15, 2015): 68–77. http://www.sciencedirect.com/science/article/pii/S0378112715001796.

Goodall, Chris. *How to Live a Low-Carbon Life: The Individual's Guide to Stopping Climate Change*. London: Earthscan Publications, 2007.

Goodall, Chris. Analysis of 27 footnoted sources in Bjorn Lomborg's *Cool It!* No date. http://www.carboncommentary.com/2007/10/15/29. Accessed October 19, 2007 (no longer available).

Goode, Erica. "Cod's Failure to Recover Is Linked to Warming Gulf of Maine." *The New York Times*, October 30, 2015, A18.

Goodell, Jeff. "Our Black Future." *The New York Times*, June 23, 2006. http://www.nytimes.com/2006/06/23/opinion/23goodell.html.

Goodell, Jeff. *How to Cool the Planet: Geoengineering and the Audacious Quest to Fix Earth's Climate*. Boston: Houghton-Mifflin Harcourt, 2011.

Goodman, J. David. "Australia Floods Show No Signs of Retreating." *The New York Times*, December 31, 2010. http://www.nytimes.com/2011/01/01/world/asia/01australia.html.

Gordon, Anita, and David Suzuki. *It's a Matter of Survival*. Cambridge, MA: Harvard University Press, 1991.

Gordon, Susan. "U.S. Pacific Northwest Gets Reduced Supply of Snow, Climate Study Says." *News-Tribune* (Tacoma). February 7, 2003, n.p. (LEXIS).

Gore, Al. "A Generational Challenge to Repower America." July 17, 2008. http://www.realclearpolitics.com/articles/2008/07/al_gores_energy_speech.html.

Goreau, Thomas J., et al. "Elevated Sea-Surface Temperatures Correlate with Caribbean Coral Reef Beaching." Pp. 225–262 in Richard A. Geyer, ed., *A Global Warming Forum: Scientific, Economic, and Legal Overview*. Boca Raton, FL: CRC Press, 1993.

Gorman, James. "A Tiny Horse That Got Even Tinier as the Planet Heated Up." *The New York Times*, February 23, 2012. http://www.nytimes.com/2012/02/24/science/sifrhippus-the-first-horse-got-even-tinier-as-the-planet-heated-up.html.

Goswami, B. N., et al. "Increasing Trend of Extreme Rain Events over India in a Warming Environment." *Science* 314 (December 1, 2006): 1442–1445.

Gough, Myles. "Study Shows Flash Flooding Risks Increase as Peak Downpours Intensify." UNSW Newsroom, June 9, 2015. https://newsroom.unsw.edu.au/news/science-tech/study-shows-flash-flooding-risks-increase-peak-downpours-intensify.

Gough, Robert. "Stress on Stress: Global Warming and Aquatic Resource Depletion." *Native Americas* 16(3–4) (Fall–Winter 1999): 46–48. http://nativeamericas.aip.cornell.edu (no longer available).

Government of Nunavut (GN). *Inuit Qaujimajangit Hilap Alanguminganut (Inuit Knowledge of Climate Change: A Sample of Inuit Experiences of Climate Change in Nunavut)*. Baker Lake and Arviat, Nunavut, January–March 2001. Government of Nunavut, Department of Sustainable Development, Environmental Protection Services. http://climatechangenunavut.ca/en/resources/news/inuit-qaujimajatuqangit-and-climate-change.

Grabherr, G., M. Gottfried, and H. Pauli. "Climate Effects on Mountain Plants." *Nature* 339: 448–451.

Grady, Denise. "Managing Planet Earth: On an Altered Planet, New Diseases Emerge as Old Ones Re-Emerge." *The New York Times*, August 20, 2002, F2.

Gramling, Carolyn. "Arctic Impact: Is the Melting Arctic Really Bringing Frigid Winters to North America and Eurasia?" *Science* 347 (February 20, 2015): 818–821.

Grant, Christine. "Swelling Seas Eating Away at Country's Monuments." *The Scotsman* (Edinburgh), December 24, 2001, 5.

Grant, Paul M. "Hydrogen Lifts Off—With a Heavy Load: The Dream of Clean, Usable Energy Needs to Reflect Practical Reality." *Nature* 424 (July 10, 2003): 129–130.

Gray, Josh M., et al. "Direct Human Influence on Atmospheric CO_2 Seasonality from Increased Cropland Productivity." *Nature* 515 (November 20, 2014): 398–401.

Gray, William M. "Hurricanes and Hot Air." *Wall Street Journal*, July 26, 2006, A12.

Greenaway, Norma. "Disaster Toll from Weather Up Tenfold: Droughts, Floods Need More Damage Control, Report Says." *Edmonton Journal*, February 28, 2003, A5.

"A Greener Bottom Line?" *Forbes*, November 7, 2012. http://www.forbes.com/sites/erik amorphy/2012/10/07/green-accounting-a-greener-bottom-line/.

"Greenland Melt Ponds." NASA Earth Observatory. March 21, 2013. http://earthobservatory .nasa.gov/IOTD/view.php?id=80677&src=eoa-iotd.

"Greenland's Ice Is Growing Darker." NASA Earth Observatory. January 10, 2012. http:// earthobservatory.nasa.gov/IOTD/view.php?id=76916&src=eoa-iotd.

Greenstone, Michael. "If We Dig Out All Our Fossil Fuels, Here's How Hot We Can Expect It to Get." *The New York Times*, April 9, 2015. http://www.nytimes.com/2015/04/09/upshot/ if-we-dig-out-all-our-fossil-fuels-heres-how-hot-we-can-expect-it-to-get.html.

"Greens Want Global Warming Examined in Bushfire Inquiry." Australian Associated Press, January 21, 2003 (LEXIS).

Gregg, Watson W., and Margarita E. Conkright. "Decadal Changes in Global Ocean Chlorophyll." *Geophysical Research Letters* 29(15) (2002): 10. doi: 1029/2002GL014689.

Gregory, Angela. "Fear of Rising Seas Drives More Tuvaluans to New Zealand." *New Zealand Herald*, February 19, 2003, n.p. (LEXIS).

Gregory, Jonathan M., Philippe Huybrechts, and Sarah C. B. Raper. "Threatened Loss of the Greenland Ice Sheet." *Nature* 428 (April 8, 2004): 616.

Griscom-Little, Amanda. "Detroit Takes Charge." *Outside*, April 2007, 60.

"A Growing Rift in Antarctic Ice." NASA Earth Observatory, October 19, 2012. http:// earthobservatory.nasa.gov/IOTD/view.php?id=79440&src=eoa-iotd.

"Growth in China's CO_2 Emissions Double Previous Estimates." Environment News Service, March 11, 2008. http://www.ens-newswire.com/ens/mar2008/2008-03-11-01.asp.

"Growth of Solar in the Gobi Desert." NASA Earth Observatory, June 17, 2015. http:// earthobservatory.nasa.gov/IOTD/view.php?id=86060&src=eoa-iotd.

Gruber, Nicolas, et al. "Rapid Progression of Ocean Acidification in the California Current System." *Science* 337 (July 13, 2012): 220–223.

Gugliotta, Guy. "In Antarctica, No Warming Trend; Scientists Find Temperatures Have Gotten Colder in Past Two Decades." *Washington Post*, January 14, 2002, A2.

Gugliotta, Guy. "Warming May Threaten 37 Per Cent of Species by 2050." *Washington Post*, January 8, 2004, A1. http://www.washingtonpost.com/wp-dyn/articles/A63153-2004Jan7.html (no longer available).

Gunter, G. "An Example of Oyster Production Decline with a Change in the Salinity Characteristics of an Estuary—Delaware Bay, 1800–1973." *Proceedings of the National Shellfish Association* 65 (1974): 3–13.

Haarsma, R. J., J. F. B. Mitchell, and C. A. Senior. "Tropical Disturbances in a G[lobal] C[limate] M[odel]." *Climate Dynamics* 8 (1993): 247–257.

Haigh, Joanna D. "Climate Variability and the Influence of the Sun." *Science* 294 (December 7, 2001): 2109–2111.

Hakim, Danny. "Several States Likely to Follow California on Car Emissions." *The New York Times*, June 11, 2004, C4.

Häkkinen, Sirpa, and Peter B. Rhines. "Decline of Subpolar North Atlantic Circulation During the 1990s." *Science* 304 (April 23, 2004): 555–559.

Hale, Ellen. "Seas Create Real Water Hazard; Changing Climate at Root of Erosion That's Putting Links Courses in Jeopardy." *USA Today*, July 18, 2001, 3C.

"Half U.S. Climate Warming Due to Land Use Changes." Environment News Service, May 28, 2003. http://ens-news.com/ens/may2003/2003-05-28-01.asp (no longer available).

Hall, Carl T. "Ocean Tells the Story: Earth Is Heating Up; Human Activity, Not Variables in Nature, Cited as Culprit." *San Francisco Chronicle*, April 29, 2005, A1.

Hallam, Anthony, and Paul Wignall. *Mass Extinctions and Their Aftermath*. Oxford, UK: Oxford University Press, 1997.

Hall-Spencer, Jason M., et al. "Volcanic Carbon Dioxide Vents Show Ecosystem Effects of Ocean Acidification." *Nature* 454 (July 3, 2008): 96–99.

Halweil, Brian. "The Irony of Climate." *World Watch*, March–April 2005, 18–23.

Hamilton, Clive. *Earthmasters: The Dawn of the Age of Climate Engineering*. New Haven, CT: Yale University Press, 2014.

Hamilton, Clive. "The Risks of Climate Engineering." *The New York Times*, February 12, 2015, A27.

Hand, Eric. "Acid Oceans Cited in Earth's Worst Die-off." *Science* 348 (April 10, 2015): 165–166.

Hand, Eric. "Record Ozone Hole May Open over Arctic in the Spring." *Science*, February 10, 2016, 650. http://www.sciencemag.org/news/2016/02/record-ozone-hole-may-open -over-arctic-spring.

Hann, Judith. "Spring Wakes Early, but Will Autumn Lie in Late Again? What Will Tomorrow's World Look Like?" United Kingdom Woodland Trust, 2002. http://www.woodland -trust.org.uk/news/subindex.asp?aid=328 (no longer available).

Hanna, E., et al. "Increased Runoff from the Greenland Ice Sheet: A Response to Global Warming." *Journal of Climate* 21 (January 15, 2008): 331–341.

Hanna, E., et al. "Greenland Blocking Index 1851–2015: A Regional Climate Change Signal." *International Journal of Climatology* (May 2016), JOC-15-0742.R1.

Hansen, Bogi, William R. Turrell, and Svein Sterhus. "Decreasing Overflow from the Nordic Seas into the Atlantic Ocean through the Faroe Bank Channel since 1950." *Nature* 411 (June 21, 2001): 927–930.

Hansen, James E. "Defusing the Global Warming Time Bomb." *Scientific American* 290(3) (March 2004): 68–77.

Hansen, James E. "Is There Still Time to Avoid 'Dangerous Anthropogenic Interference' with Global Climate? A Tribute to Charles David Keeling." Paper delivered to the American Geophysical Union, San Francisco, December 6, 2005. http://www.columbia.edu/~jeh1/ keeling_talk_and_slides.pdf.

Hansen, James E. "The Threat to the Planet." *New York Review of Books*, July 2006, 12–16.

Hansen, James E. "Declaration of James E. Hansen." *Green Mountain Chrysler-Plymouth-Dodge-Jeep, et al., Plaintiffs v. Thomas W. Torti, Secretary of the Vermont Agency of Natural Resources, et al., Defendants*. Case Nos. 2:05-CV-302 and 2:05-CV-304, Consolidated. United States District Court for the District of Vermont. August 14, 2006. http://www.giss.nasa.gov/~dcain/ recent_papers_proofs/vermont_14aug20061_textwfigs.pdf (no longer available).

Hansen, James E. "Political Interference with Government Climate Change Science." *Testimony of James E. Hansen, 4273 Durham Road, Kintnersville, PA, to Committee on Oversight and Government Reform United States House of Representatives*, March 19, 2007.

Hansen, James E. Personal communication, April 12, 2007.

Hansen, James E. "Coal Trains of Death." James Hansen's e-mail list, July 23, 2007.

Hansen, James E. "Game Over for the Climate." *The New York Times*, May 9, 2012. http:// www.nytimes.com/2012/05/10/opinion/game-over-for-the-climate.html.

Hansen, James E. "Climate Change Is Here—and Worse Than We Thought." *Washington Post*, August 3, 2012. http://www.washingtonpost.com/opinions/climate-change-is-here—and-worse-than-we-thought/2012/08/03/6ae604c2-dd90-11e1-8e43-4a3c4375504a_print.html.

Hansen, James E. "Making Things Clearer: Exaggeration, Jumping the Gun, and The Venus Syndrome." April 15, 2013. http://www.columbia.edu/~jeh1/mailings/2013/20130415_Exaggerations.pdf.

Hansen, James E. "An Old Story, but Useful Lessons." September 26, 2013. http://www.columbia.edu/~jeh1/mailings/2013/20130926_PTRSpaperDiscussion.pdf.

Hansen, James E. Personal correspondence, July 27, 2015.

Hansen, James E., et al. "Climate Impact of Increasing Atmospheric Carbon Dioxide." *Science* 213 (1981), 957–966.

Hansen, James E., et al. "Climate Sensitivity: Analysis of Feedback Mechanisms." Pp. 130–163 in *Geophysical Monograph 29 Maurice Ewing Vol. 5*. Washington, DC: American Geophysical Union, 1984.

Hansen, James E., et al. "Earth's Energy Imbalance: Confirmation and Implications." *Science* 308 (June 3, 2005): 1431–1435.

Hansen, James E., et al. "Climate Change and Trace Gases." *Philosophical Transactions of the Royal Society A*, July 15, 2007. http://rsta.royalsocietypublishing.org/content/365/1856/1925.

Hansen, James E., et al. "Target Atmospheric CO_2: Where Should Humanity Aim?" *Open Atmospheric Science Journal* 2 (2008): 217–231.

Hansen, James E., et al. "Global Temperature and Europe's Frigid Air." Draft paper, December 10, 2010. 20101211_TemperatureandEurope(3).pdf.

Hansen, James E., et al. "Global Surface Temperature Change." *Reviews of Geophysics*, December 14, 2010. http://onlinelibrary.wiley.com/doi/10.1029/2010RG000345/abstract.

Hansen, James E., et al. "Global Temperature in 2014 and 2015." January 16, 2015. http://csas.ei.columbia.edu/2015/01/16/global-temperature-in-2014-and-2015/.

Hansen, James E., et al. "Ice Melt, Sea Level Rise and Superstorms: Evidence from Paleoclimate Data, Climate Modeling, and Modern Observations That 2 Degrees C. Global Warming Is Highly Dangerous." *Atmospheric Chemistry and Physics Discussions* 15 (July 2015): 20059–20179. www.atmos-chem-phys-discuss.net/15/20059/2015/. doi: 10.5194/acpd-15-20059-2015.

Hansen, James E., et al. "Ice Melt, Sea Level Rise, and Superstorms: Evidence from Paleoclimate Data, Climate Modeling, and Modern Observations That 2°C Global Warming Could Be Dangerous." *Atmospheric Chemistry and Physics* 16(3) (March 22, 2016), 3761–3812. doi: 10.5194/acp-16-3761-2016.

Hansen, James E., and Makiko Sato. "Paleoclimate Implications for Human-Made Climate Change." New York: NASA Goddard Institute for Space Studies and Columbia University Earth Institute, January 18, 2011. http://www.columbia.edu/~jeh1/mailings/2011/20110118_MilankovicPaper.pdf.

Hansen, James E., and Makiko Sato. "Paleoclimate Implications for Human-Made Climate Change." In A. Berger, F. Mesinger, and D Šijački, eds., *Climate Change at the Eve of the Second Decade of the Century: Inferences from Paleoclimate and Regional Aspects: Proceedings of Milutin Milankovitch 130th Anniversary Symposium*, July 2011. New York: Springer.

Hansen, James E., and Makiko Sato. *Update of Greenland Ice Sheet Mass Loss: Exponential?* December 26, 2012. http://www.columbia.edu/~jeh1/mailings/2012/20121226_GreenlandIceSheetUpdate.pdf.

Hansen, James E., Makiko Sato, and R. Ruedy. "The Missing Climate Forcing." *Philosophical Transactions of the Royal Society of London B* 352 (1997): 231–240.

Hansen, James E., Makiko Sato, and Reto Ruedy. "Perceptions of Climate Change: The New Climate Dice." New York: NASA Goddard Institute for Space Studies and Columbia University Earth Institute, 2012. http://globalwarming-sowhat.com/warm—cool-/loaded -temps-dice-since.pdf.

Hansen, James E., Makiko Sato, and Reto Ruedy. "Perception of Climate Change." *Proceedings of the National Academy of Sciences*, August 6, 2012. http://www.pnas.org/content/ early/2012/07/30/1205276109.abstract. doi:10.1073/pnas.1205276109.

Hansen, Kathryn. "Researchers Warm Up to Melt's Role in Greenland Ice Loss." NASA.gov, April 27, 2008. https://www.nasa.gov/centers/goddard/news/topstory/2008/greenland _speedup.html.

Hansen, Terri. "Climate Disruptions Hitting More and More Tribal Nations." Indian Country Today Media Network, May 7, 2014. http://indiancountrytodaymedianetwork. com/2014/05/07/climate-disruptions-hitting-more-and-more-tribal-nations-154747.

Harden, Blaine. "Tree-Planting Drive Seeks to Bring a New Urban Cool; Lower Energy Costs Touted as Benefit." *Washington Post*, September 4, 2006; A1. http://www.washingtonpost. com/wp-dyn/content/article/2006/09/03/AR2006090300926_pf.html.

Harden, Blaine. "Air, Water Powerful Partners in Northwest." *Washington Post*, March 21, 2007, A3. http://www.washingtonpost.com/wp-dyn/content/article/2007/03/20/AR20070 32001634_pf.html.

Harden, Blaine, and Juliet Eilperin. "On the Move to Outrun Climate Change; Self-Preservation Forcing Wild Species, Businesses, Planning Officials to Act." *Washington Post*, November 25, 2006, A3.

Harper, J., et al. "Greenland Ice-Sheet Contribution to Sea-Level Rise Buffered by Meltwater Storage in Firn." *Nature* 491 (November 8, 2012): 240–243.

"Harper's Index." *Harper's*, September 2006, 17.

Harris, Gardiner. "Borrowed Time on Disappearing Land: Facing Rising Seas, Bangladesh Confronts the Consequences of Climate Change." *The New York Times*, March 28, 2014. http://www.nytimes.com/2014/03/29/world/asia/facing-rising-seas-bangladesh-confron ts-the-consequences-of-climate-change.html.

Harris, Gardiner. "Coal Rush in India Could Tip Balance on Climate Change." *The New York Times*, November 17, 2014. http://www.nytimes.com/2014/11/18/world/coal-rush-in-india-could-tip-balance-on-climate-change.html.

Hartmann, D. L. "Pacific Sea Surface Temperature and the Winter of 2014." *Geophysical Research Letters*, March 19, 2015, 42. http://onlinelibrary.wiley.com/doi/10.1002 /2015GL063083/abstract?campaign=wlytk-41855.5282060185.

"Harvard Medical School: Pollen Production—and Allergies—May Rise Significantly over Next 50 Years." ScienceDaily, March 22, 2002. http://www.sciencedaily.com/releases/ 2002/03/020322075343.htm.

"Harvard Student Arrested in Fossil Fuel Divestment Protest." Environment News Service, May 2, 2014. http://ens-newswire.com/2014/05/02/ameriscan-may-2-2014/.

Harvell, C. D., et al. "Emerging Marine Diseases—Climate Links and Anthropogenic Factors." *Science* 285(3) (September 3, 1999): 1505–1510.

Harvey, Chelsea. "What Scientists Just Discovered in Greenland Could Be Making Sea-Level Rise Even Worse." *Washington Post*, January 4, 2015. https://www.washingtonpost .com/news/energy-environment/wp/2016/01/04/what-scientists-discovered-in -greenland-could-be-making-sea-level-rise-even-worse/?wpmm=1&wpisrc=nl _headlines.

Harvey, Chelsea. "Scientists Confirm There's Enough Fossil Fuel on Earth to Entirely Melt Antarctica." *Washington Post*, September 11, 2015. http://www.washingtonpost.com/news/energy-environment/wp/2015/09/11/scientists-confirm-theres-enough-fossil-fuel-on-earth-to-melt-all-of-antarctica/?wpmm=1&wpisrc=nl_headlines.

Harvey, Chelsea. "Dominoes Fall: Vanishing Arctic Ice Shifts Jet Stream, Which Melts Greenland Glaciers." *Washington Post*, May 2, 2016. https://www.washingtonpost.com/news/energy-environment/wp/2016/05/02/dominoes-fall-vanishing-arctic-ice-shifts-jet-stream-melts-greenland-glaciers/?utm_term=.a8421ba10c66.

Harvey, Chelsea. "How Climate Disasters Can Drive Violent Conflict around the World." *Washington Post*, July 25, 2016. https://www.washingtonpost.com/news/energy-environment/wp/2016/07/25/how-climate-disasters-can-drive-violent-conflict-around-the-world/?wpisrc=nl_headlines&wpmm=1.

Harvey, Fiona. "Arctic May Have No Ice in Summer by 2070, Warns Climate Change Report." *Financial Times* (London), November 2, 2004, 1.

Harwood, David M. Personal correspondence, February 25, 2016.

Hay, S. I., et al. "Climate Change and the Resurgence of Malaria in the East African Highlands." *Nature* 425 (February 21, 2002): 905–909.

Hayes, David J., and James H. Stock. "The Real Cost of Coal." *The New York Times*, March 24, 2015. http://www.nytimes.com/2015/03/24/opinion/the-real-cost-of-coal.html?emc=edit_th_20150324&nl=todaysheadlines&nlid=35795487.

"Health Risks." *National Geographic*, November 2015, 96–97.

Healy, Jack. "Starving Sea Lions Washing Ashore by the Hundreds in California." *The New York Times*, March 13, 2015. http://www.nytimes.com/2015/03/13/us/starving-sea-lions-washing-ashore-by-the-hundreds-in-california.html.

Healy, Patrick. "Warming Waters: Lobstermen on Cape Cod Blame Light Hauls on Higher Ocean Temperatures." *Boston Globe*, August 30, 2002, B1.

Hearn, Kelly. "Huge Man-Made Algae Swarm Devoured—Bad for Climate?" *National Geographic News*, March 27, 2009. http://news.nationalgeographic.com/news/2009/03/090327-iron-seeding.html.

"Heat Fuels Fire at Fort McMurray." NASA Earth Observatory, May 7, 2016. http://earthobservatory.nasa.gov/NaturalHazards/view.php?id=87992.

"Heat Intensifies Siberian Wildfires." NASA Earth Observatory, August 2, 2013. http://earthobservatory.nasa.gov/IOTD/view.php?id=81736&src=eoa-iotd.

"Heat Wave Hits Thailand, India." NASA Earth Observatory, May 4, 2016. http://earthobservatory.nasa.gov/IOTD/view.php?id=87981&src=eoa-iotd.

"Heat Wave Stifles Australia." NASA Earth Observatory, January 12, 2014. http://earthobservatory.nasa.gov/IOTD/view.php?id=82790&src=eoa-iotd.

"Heavy Rains Threaten Flood-Prone Venice." *The Straits Times* (Singapore), June 8, 2002, n.p. (LEXIS).

Hedin, Lars O. "Biogeochemistry: Signs of Saturation in the Tropical Carbon Sink." *Nature* 519 (March 19, 2015): 295–296.

Heimann, Martin. "How Stable Is the Methane Cycle?" *Science* 327 (March 5, 2010): 1211–1212.

Hellmer, Hartmut H., et al. "Twenty-First-Century Warming of a Large Antarctic Ice-Shelf Cavity by a Redirected Coastal Current." *Nature* 485 (May 10, 2012): 225–228.

Hendee, David. "Peril Is Seen to State's Water Table." *Omaha World-Herald*, October 22, 2006, A1, A2.

Henderson, Casper. "The Other CO_2 Problem." *New Scientist*, August 5, 2006, 28–33.

Henderson, Mark. "Past Ten Summers Were the Hottest in 500 Years." *Times* (London), March 5, 2004, 10.

Henry, Tom. "Global Warming Grips Greenland, Leaving Lasting Mark." *Toledo Blade*, October 12, 2008. http://www.toledoblade.com/apps/pbcs.dll/article?AID=2008810109858 (no longer available).

Herrick, Thaddeus. "The New Texas Wind Rush: Oil Patch Turns to Turbines, as Ranchers Sell Wind Rights; A New Type of Prospector." *Wall Street Journal*, September 23, 2002, B1, B3.

Hertsgaard, Mark. "Harvesting a Climate Disaster." *The New York Times*, September 12, 2012. http://www.nytimes.com/2012/09/13/opinion/the-farm-bill-should-help-the-planet-not-just-crops.html.

Hesselbo, Stephen P., et al. "Massive Dissociation of Gas Hydrate during a Jurassic Oceanic Anoxic Event." *Nature* 406 (July 27, 2000): 392–395.

Hickman, Martin. "The Prince of Emissions." *The Independent* (London), April 1, 2006, 1.

Highfield, Roger. "Winter Floods 'Five Times More Likely.'" *Daily Telegraph* (London), January 31, 2002, 8.

Hillman, Mayer, and Tina Fawcett. *The Suicidal Planet: How to Prevent Global Climate Catastrophe*. New York: St. Martin's Press/Thomas Dunne Books, 2007.

Hinrichs, Kai-Uwe, Laura R. Hmelo, and Sean P. Sylva. "Molecular Fossil Record of Elevated Methane Levels in Late Pleistocene Coastal Waters." *Science* 299 (February 21, 2003): 1214–1217.

Hirsch, Jerry. "Damage to Coral Reefs Mounts, Study Says; Broad Survey Cites Human Causes Such as Overfishing and Pollution; Reefs Are a Key Indicator of the Health of Oceans, Scientists Say." *Los Angeles Times*, August 26, 2002, 14.

Hoag, Hannah. "Degree-by-Degree Breakdown of Climate Effects Published." *Nature* 466 (July 22, 2010): 425.

Hoag, Hannah. "How Cities Can Beat the Heat." *Nature* 524 (August 27, 2015): 402–404. http://www.nature.com/news/how-cities-can-beat-the-heat-1.18228.

Hoegh-Guldberg, O., et al. "Is Coral Bleaching Really Adaptive?" *Nature* 415 (February 7, 2001): 601–602.

Hoegh-Guldberg, Ove, et al. "Coral Reefs under Rapid Climate Change and Ocean Acidification." *Science* 318 (December 14, 2007): 1737–1742.

Hoffman, Ian. "Global Warming Killing Crops." *Daily Democrat*, March 17, 2007. http://www.dailydemocrat.com/article/ZZ/20070317/NEWS/703179724.

Hoffman, Ian. "Iron Curtain over Global Warming; Ocean Experiment Suggests Phytoplankton May Cool Climate." *Daily Review* (Hayward, CA), April 17, 2004, n.p. (LEXIS).

Hogan, Treacy. "Still Raining in Costa del Ireland." *Belfast Telegraph*, November 26, 2002, n.p. (LEXIS).

Holland, Jennifer S. "The Acid Threat: As CO_2 Rises, Shelled Animals May Perish." *National Geographic*, November 2001, 110–111.

Holleley, Clare E., et al. "Sex Reversal Triggers the Rapid Transition from Genetic to Temperature-Dependent Sex." *Nature* 523 (July 2, 2015): 79–82. http://www.nature.com/nature/journal/v523/n7558/full/nature14574.html?message-global=remove.

Holly, Chris. "Sea-Level Rise Seen as Key Global Warming Threat." *The Energy Daily* 32:36 (February 25, 2004), n.p. (LEXIS).

Homer-Dixon, Thomas. "The Tar Sands Disaster." *The New York Times*, March 31, 2013. http://www.nytimes.com/2013/04/01/opinion/the-tar-sands-disaster.html

Horovitz, Bruce. "Can Eateries Go Green, Earn Green? " *USA Today*, May 16, 2008, 1B, 2B.

Hotz, Robert Lee. "Greenland's Ice Sheet Is Slip, Sliding Away." *Los Angeles Times*, June 24, 2006, n.p.

Houghton, John. *Global Warming: The Complete Briefing*. Cambridge, UK: Cambridge University Press, 1997.

Houghton, J. T., et al. *Climate Change 2001: The Scientific Basis*. Cambridge, UK: Cambridge University Press, 2001.

Houghton, R. A., et al. "Annual Fluxes of Carbon from Deforestation and Regrowth in the Brazilian Amazon." *Nature* 403 (January 20, 2000): 301–304.

Houlder, Vanessa. "Rise Predicted in Aviation Carbon Dioxide Emissions." *Financial Times* (London), December 16, 2002, 2.

Howard, Touché. "University of Texas Study Underestimates National Methane Emissions at Natural Gas Production Sites Due to Instrument Sensor Failure." *Energy Science and Engineering*, August 4, 2015. http://onlinelibrary.wiley.com/doi/10.1002/ese3.81/abstract. doi: 10.1002/ese3.81.

Howat, I. M., et al. "Rapid Retreat and Acceleration of Helheim Glacier, East Greenland." *Geophysical Research Letters* 32 (2005): L22502. doi: 10.1029/2005GL024737.

Howden, S. Mark, et al. "Adapting Agriculture to Climate Change." *Proceedings of the National Academy of Sciences* 104 (December 11, 2007): 19691–19696.

"How Widespread Was the Australian Heatwave?" NASA Earth Observatory, January 23, 2013. http://earthobservatory.nasa.gov/IOTD/view.php?id=80232&src=eoa-iotd.

Hoyos, C. D., et al. "Deconvolution of the Factors Contributing to the Increase in Global Hurricane Intensity" *Science* 312 (April 7, 2006): 94–97.

Hu, A., et al. "Transient Response of the MOC and Climate to Potential Melting of the Greenland Ice Sheet in the 21st Century." *Geophysical Research Letters* 36 (May 29, 2009). doi: 10.1029/2009GL037998.

Hu, Feng Sheng, et al. "Cyclic Variation and Solar Forcing of Holocene Climate in the Alaskan Subarctic." *Science* 301 (September 26, 2003): 1890–1893.

Hubbard, Russell. "UNL Report Casts Doubt on Use of Corn-Crop Leftovers for Ethanol." *Omaha World-Herald*, April 29, 2014. http://www.omaha.com/article/20140420/MONEY/140429949/1697#unl-report-casts-doubt-on-use-of-corn-crop-leftovers-for-ethanol (no longer available).

Huber, Brian T., Kenneth G. MacLeod, and Scott L. Wing. *Warm Climates in Earth History*. Cambridge, UK: Cambridge University Press, 2000.

"Hudson Bay Ice-Free by 2050, Scientists Say." CBC News, May 14, 2001. http://www.cbc.ca/news/canada/hudson-bay-ice-free-by-2050-scientists-say-1.286276.

Hudson, Kris. "Whose Beach Is This, Anyway?" *Wall Street Journal*, December 12, 2007, B1, B8.

Huey, Raymond, Jonathan B. Losos, and Craig Moritz. "Are Lizards Toast?" *Science* 328 (May 14, 2010): 832–833.

Hughes, T. J., J. L. Fastook, and G. H. Denton. *Climatic Warming and the Collapse of the West Antarctic Ice Sheet*. Orono: University of Maine Press, 1979.

Hughes, T. P., et al. "Climate Change, Human Impacts, and the Resilience of Coral Reefs." *Science* 301 (August 15, 2003): 929–933. http://science.sciencemag.org/content/301/5635/929.

Hughes, Trevor. "It's a New Dawn for Solar Industry—Mainly." *USA Today*, January 14, 2016, 6B.

Hulme, Mike. *Can Science Fix Climate Change? A Case against Climate Engineering*. Cambridge, UK: Polity Press, 2014.

"Human Health." *National Geographic*, January 2016, n.p. https://www.scribd.com/doc/293386356/National-Geographic-USA-January-2016.

Human, Katy. "Disappearing Arctic Ice Chills Scientists; A University of Colorado Expert on Ice Worries That the Massive Melting Will Trigger Dramatic Changes in the World's Weather." *Denver Post*, October 5, 2004, B2.

Human, Katy, and Kim McGuire. "Ozone Decline Stuns Scientists." *Denver Post*, March 2, 2005, A8.

Hume, Stephen. "A Risk We Can't Afford: The Summer of Fire and the Winter of the Deluge Should Prove to the Naysayers That If We Wait Too Long to React to Climate Change We'll Be in Grave Peril." *Vancouver Sun*, October 23, 2003, A-13.

"Hunger, Water Scarcity Displaces Thousands of Afghans." *The New York Times*, June 4, 2008. http://www.nytimes.com/reuters/world/international-afghan-displacement.html (no longer available).

Huq, Saleemul. "Climate Change and Bangladesh." *Science* 294 (November 23, 2001):1617.

"Ice a Scarce Commodity on Arctic Rinks: Global Warming Blamed for Shortened Hockey Season." *Financial Post* (Canada), January 7, 2003, A3.

"Ice Loss in the Canadian Arctic Archipelago." NASA Earth Observatory, June 1, 2011. http://earthobservatory.nasa.gov/IOTD/view.php?id=50726&src=eoa-iotd.

"Ice Sheet of Greenland Melting Away at Faster Pace." *Omaha World-Herald*, January 20, 2008, 18A.

Idso, Sherwood B. *Carbon Dioxide: Friend or Foe?* Tempe, AZ: IBR Press, 1982.

Idso, S. B., R. C. Balling Jr., and R .S. Cerveny. "Carbon Dioxide and Hurricanes: Implications of Northern Hemispheric Warming for Atlantic/Caribbean Storms." *Meteorology and Atmospheric Physics* 42 (1990): 259–63.

Iglesias-Rodriguez, M. Debora, et al. "Phytoplankton Calcification in a High-CO_2 World." *Science* 320 (April 18, 2008): 336–340.

Immerzeel, Walter W., et al. "Climate Change Will Affect the Asian Water Towers." *Science* 328 (June 11, 2010): 1382–1385.

Imtiaz, Saba, and DeClan Walsh. "Karachi Heat Wave Death Toll Tops 650 during Ramadan Fast." *The New York Times*, June 23, 2015. http://www.nytimes.com/2015/06/24/world/asia/pakistan-says-more-than-600-have-died-in-heat-wave.html.

"Increasing Antarctic Sea Ice Extent Linked to the Ozone Hole." NASA Earth Observatory and British Antarctic Survey, April 21, 2009. https://www.bas.ac.uk/media-post/increasing-antarctic-sea-ice-extent-linked-to-the-ozone-hole/.

"India Blames Climate Change for Heat Wave Deaths." Environment News Service. June 2, 2015. http://ens-newswire.com/2015/06/02/india-blames-climate-change-for-heat-wave-deaths/.

"India Faces Deadly Heat Wave." NASA Earth Observatory, June 5, 2015. http://earthobservatory.nasa.gov/IOTD/view.php?id=85986&src=eoa-iotd.

"India Orders Some Areas to Cook at Night." *Omaha World-Herald*, April 30, 2016, 7A.

"Indigenous Peoples, Lands, and Resources." Chapter 12, *National Climate Assessment*, U.S. Global Change Research Program, 2014. http://nca2014.globalchange.gov/report/sectors/indigenous-peoples.

"Indonesia Fights Climate Threat That Rises from the Ground Up." *Omaha World-Herald*, November 26, 2009, A15.

Ingham, John. "Stingers Thrive as the Country Gets Warmer; Invasion of the Scorpions." *Daily Express* (U.K.), June 18, 2004, 40.

Inhofe, James M. *The Greatest Hoax: How the Global Warming Conspiracy Threatens Our Future.* New York: WND Books, 2012.

Inman, Mason. "Hot, Flat, Crowded—And Preparing for the Worst." *Science* 326 (October 30, 2009): 662–663.

Innis, Michelle. "Warming Oceans May Threaten Krill, a Cornerstone of the Antarctic Eco-system." *The New York Times*, October 19, 2015. http://www.nytimes.com/2015/10/20/science/australia-antarctica-krill-climate-change-ocean.html?_r=0.

Innis, Michelle. "Record Heat Puts Australia at Risk of Intense Fire Season." *The New York Times*, November 20, 2015. http://www.nytimes.com/2015/11/21/world/australia/australia-fires-record-temperatures.html.

Innis, Michelle. "Climate-Related Death of Coral around World Alarms Scientists." *The New York Times*, April 9, 2016. http://www.nytimes.com/2016/04/10/world/asia/climate-related-death-of-coral-around-world-alarms-scientists.html.

"Insects Feast under High CO2." *Nature* 519 (March 12, 2015): 133.

"Intense Fires in Northern Canada." NASA Earth Observatory, June 3, 2015. http://earthobservatory.nasa.gov/IOTD/view.php?id=85972&src=eoa-iotd.

Intergovernmental Panel on Climate Change. *Climate Change 2007: Synthesis Report.* IPCC Fourth Assessment Report (AR4). 2007. https://www.ipcc.ch/publications_and_data/publications_ipcc_fourth_assessment_report_synthesis_report.htm.

Intergovernmental Panel on Climate Change. United Nations. *Climate Change 2014: Synthesis Report.* Rajendra K. Pachauri and Leo Meyer, eds. https://www.ipcc.ch/pdf/assessment-report/ar5/syr/SYR_AR5_FINAL_full.pdf.

"Investors Sizing Up 'Carbon Footprints.'" *Omaha World-Herald*, September 30, 2007, D1.

"Iron Link to CO$_2$ Reductions Weakened." Environment News Service, April 10, 2003. http://ens-news.com/ens/apr2003/2003-04-10-09.asp#anchor8.

Isom, Hannah. "Ozone Hole Healing Could Cause Further Climate Warming." University of Leeds, U.K. January 25, 2010. http://www.eurekalert.org/pub_releases/2010-01/uol-ohh012210.php.

"Italy to Build World's First Hydrogen-Fired Power Plant." Environment News Service, December 18, 2006. http://www.ens-newswire.com/ens/dec2006/2006-12-18-05.asp (no longer available).

Ivins, Erik R. "Ice Sheet Stability and Sea Level." *Science* 324 (May 15, 2009): 888–889.

Izrael, Yu. "Climate Change Impact Studies: The IPCC Working Group II Report." Pp. 83–86 in J. Jager and H. L. Ferguson, eds., *Climate Change: Science, Impacts, and Policy.* Proceedings of the Second World Climate Conference. Cambridge, UK: Cambridge University Press, 1991.

Jackson, Derrick Z. "Sweltering in a Winter Wonderland." *Boston Globe*, December 5, 2001, A23.

Jaiser, R., et al. "Impact of Sea Ice Cover Changes on the Northern Hemisphere Atmospheric Winter Circulation." *Tellus* 64 (2012): 11595. http://www.tellusa.net/index.php/tellusa/article/view/11595.

Jakob, Michael, and Jerome Hilaire. "Climate Science: Unburnable Fossil-Fuel Reserves." *Nature* 517 (January 8, 2015): 150–152. doi: 10.1038/517150a.

Jans, Nick. "Living with Oil: The Real Price." Pp. 147–161 in Steven Kazlowski, *The Last Polar Bear: Facing the Truth of a Warming World.* Seattle: Braided River, 2008.

Jardine, Kevin. "The Carbon Bomb: Climate Change and the Fate of the Northern Boreal Forests." Ontario, Canada: Greenpeace International, 1994. http://dieoff.org/page129.htm.

Jarvis, Lisa. "Kindling for Climate Change." *Chemical and Engineering News*, August 17, 2009.

Jawad, Adil. "Pakistan's Largest City Thirsts for a Water Supply." Associated Press. August 24, 2014. http://www.benningtonbanner.com/ci_26396810/pakistans-largest-city-thirsts-water-supply (no longer available).

Jeffries, M. O. "Ellesmere Island Ice Shelves and Ice Islands." *Satellite Image Atlas of Glaciers of the World: Glaciers of North America—Glaciers of Canada.* J147-J164. March 7, 2002 (approved for publication). https://pubs.usgs.gov/pp/p1386j/iceshelves/iceshelves-hires .pdf.

"Jellyfish Attack Destroys Salmon." British Broadcasting Corporation, November 21, 2007. http://news.bbc.co.uk/2/hi/uk_news/northern_ireland/7106631.stm.

"Jellyfish Close Scottish Nuclear Power Plant." Environment News Service, June 30, 2011. http://ens-newswire.com/2011/07/01/jellyfish-close-scottish-nuclear-power-plant/.

"Jellyfish Swarms Invade Ecosystems Out of Balance." Environment News Service, December 16, 2008. http://www.ens-newswire.com/ens/dec2008/2008-12-16-01.asp (no longer available).

Jenkins, Mark. "True Colors: The Changing Face of Greenland." *National Geographic*, June 2010, 36–43.

Jenkyns, Hugh C., et al. "High Temperatures in the Late Cretaceous Arctic Ocean." *Nature* 432 (December 16, 2004): 888–892.

Jessepe, Lorraine. "Alaskan Native Communities Facing Climate-Induced Relocation." Indian Country Today Media Network, June 21, 2012. http://indiancountrytodaymedianetwork. com/article/alaskan-native-communities-facing-climate-induced-relocation-119615.

Jin, F.-F., J. Boucharel, and I.-I. Lin. "Eastern Pacific Tropical Cyclones Intensified by El Niño Delivery of Subsurface Ocean Heat." *Nature* 516 (December 4, 2014): 82–85.

Johannessen, Ola M., et al. "Recent Ice-Sheet Growth in the Interior of Greenland." *Science* 310 (November 11, 2005): 1013–1016.

Johansen, Bruce E. "Pristine No More: The Arctic, Where Mother's Milk Is Toxic." *The Progressive*, December 2000, 27–29.

Johansen, Bruce E. "Arctic Heat Wave." *The Progressive*, October 2001, 18–20.

Johansen, Bruce E. *The Dirty Dozen: Toxic Chemicals and the Earth's Future*. Westport, CT: Praeger, 2003.

Johansen, Bruce. "Global Warming Approaching." *Z Magazine*, April 1, 2005. https://zcomm .org/zmagazine/global-warming-approaching-by-bruce-johansen/.

Johansen, Bruce E. "The Paul Revere of Global Warming." *The Progressive*, August 2006, 26–28.

Johansen, Bruce E. "Scandinavia Gets Serious about Global Warming." *The Progressive*, July 2007, 22–25.

Johansen, Bruce E. "Global Warming, 'Thermal Inertia,' and Tomorrow's News." *Nebraska Report*, October 2007, 7.

Johansen, Bruce E. "Hansen Battles Climatic Self-Deception World-Wide." *The Nebraska Report*, October 2008, 7.

Johansen, Bruce E. *The Encyclopedia of Global Warming Science and Technology*. Santa Barbara, CA: Greenwood Press, 2009.

Johansen, Bruce E. "Has Our Nasty Winter Ended Global Warming? We Should Be So Lucky." *The Nebraska Report*, April 2010. http://nebraskansforpeace.org/nasty-winter -global-warming (no longer available).

Johansen, Bruce E. *Resource Exploitation in Native North America: A Plague upon the Peoples*. Santa Barbara, CA: Praeger, 2016.

Johnson, Bob. *Carbon Nation: Fossil Fuels in the Making of American Culture*. Lawrence: University Press of Kansas, 2014.

Johnson, Kirk. "Alaska Looks for Answers in Glacier's Summer Flood Surges." *The New York Times*, July 23, 2013. http://www.nytimes.com/2013/07/23/us/alaska-looks-for-answers -in-glaciers-summer-flood-surges.html.

Johnson, Kirk. "As Alaska Warms, the Iditarod Adapts." *The New York Times*, March 6, 2016. http://www.nytimes.com/2016/03/07/us/as-alaska-warms-the-iditarod-adapts.html.

Johnson, Tim. "As Globe Warms, Melting Glaciers Revealing More Than Bare Earth." McClatchy Newspapers, June 19, 2015. http://www.mcclatchydc.com/news/nation -world/world/article25186561.html.

Johnston, David Cay. "Some Need Hours to Start Another Day at the Office." *Omaha World-Herald*, Feb. 6, 2000, 1G.

Joling, Dan. "Study: Polar Bears May Turn to Cannibalism." Associated Press, June 14, 2006 (LEXIS).

Joling, Dan. "Walruses Abandon Ice for Alaska Shore." *Washington Post*, October 4, 2007. http://www.washingtonpost.com/wp-dyn/content/article/2007/10/04/AR200710 0402299_pf.html (no longer available).

Joling, Dan. "Thousands of Pacific Walruses Die; Global Warming Blamed." Associated Press, December 14, 2007 (LEXIS).

Jolly, David. "$1,200 a Pound, Truffles Suffer in Heat." *The New York Times*, December 20, 2012. http://www.nytimes.com/2012/12/21/business/global/is-climate-change-shrinking -the-luxury-truffle-crop.html.

Jolly, David. "Heat Waves in Europe Will Increase, Study Finds." *The New York Times*, December 8, 2014. http://www.nytimes.com/2014/12/09/world/europe/global-warming-to -make-european-heat-waves-commonplace-by-2040s-study-finds.html.

Jolly, W. Matt, et al. "Climate-Induced Variations in Global Wildfire Danger from 1979 to 2013." *Nature Communications* 6 (7537) (July14, 2015). http://www.nature.com/ ncomms/2015/150714/ncomms8537/full/ncomms8537.html.

Jones, Nicola. "Rising Tide: Researchers Struggle to Project How Fast, How High, and How Far the Oceans Will Rise." *Nature* 501 (September 19, 2013). http://www.nature.com/ news/climate-science-rising-tide-1.13749.

Joughin, Ian, W. Abdalati, and M. Fahnestock. "Large Fluctuations in Speed on Greenland's Jakobshavn Isbrae Glacier." *Nature* 432 (December 2, 2004): 608–610.

Joughin, Ian, Benjamin E. Smith, and Brooke Medley. "Marine Ice Sheet Collapse Potentially Under Way for the Thwaites Glacier Basin, West Antarctica." *Science* 344 (May 16, 2014): 735–738.

Joyce, Christopher. "'Alarming' Amazon Droughts May Have Global Fallout." National Public Radio, February 7, 2011. http://www.npr.org/2011/02/07/133462608/alarming-amazon -droughts-may-have-global-fallout?ft=1&f=1007.

"Jumbo Squid Has a Message for Us: Changing Global Patterns Are Going to Bring Different Species into Our Waters." *Times-Colonist* (Victoria, BC), October 8, 2004, A10.

"Jumbo Squid Invade San Diego Shores, Spook Divers." Associated Press in New Zealand Herald, July 17, 2009 (LEXIS).

Kafarowski, Joanna. "Cloutier, Shiela-Watt." *Encyclopedia of the Arctic*. London: Routledge, 2007. http://www.routledge-ny.com/ref/arctic/watt.html (no longer available). Accessed November 10, 2007.

Kahn, Joseph, and Jim Yardley. "As China Roars, Pollution Reaches Deadly Extremes." *The New York Times*, August 26, 2006. http://www.nytimes.com/2007/08/26/world/asia/ 26china.html.

Kalkstein, Laurence S. "Direct Impacts in Cities." *Lancet* 342 (December 4, 1993): 1397–1400.

Kalnay, Eugenia, and Ming Cai. "Impact of Urbanization and Land-Use Change on Climate." *Nature* 423 (May 29, 2003): 528–531.

Kanter, James. "Across the Atlantic, Slowing Breezes." *The New York Times*, March 7, 2007. http://www.nytimes.com/2007/03/07/business/businessspecial2/07europe.html.

Kanter, James. "E.U. Considers Emission Fines for Chinese and Indian Airlines." *The New York Times*, May 16, 2013. http://www.nytimes.com/2013/05/17/business/global/17iht -emit17.html.

Kaplow, Larry. "Solar Water Heaters: Israel Sets Standard for Energy; Cutting Dependence: Jerusalem's Alternative Energy Use a Lesson for United States." *Atlanta Journal Constitution*, August 5, 2001, 1P.

Karl, Thomas R., Neville Nicholls, and Jonathan Gregory. "The Coming Climate: Meteorological Records and Computer Models Permit Insights into Some of the Broad Weather Patterns of a Warmer World." *Scientific American* 276 (1997): 79–83. http://www .scientificamerican.com/0597issue/0597karl.htm (no longer available).

Karl, Thomas R., Richard W. Knight, and Bruce Baker. "The Record-Breaking Global Temperatures of 1997 and 1998: Evidence for an Increase in the Rate of Global Warming." *Geophysical Research Letters* 27 (March 1, 2000): 719–722.

Karl, Thomas R., N. Nicholls, and J. Gregory. "The Coming Climate." *Scientific American* 276 (1997): 79–83.

Karoly, David J. "Ozone and Climate Change." *Science* 302 (October 10, 2003): 236–237.

Katz, Miriam E., et al. "The Source and Fate of Massive Carbon Input during the Latest Paleocene Thermal Maximum." *Science* 286 (November 19, 1999): 1531–1533.

Katz, Richard F., and M. Grae Worster. "Stability of Ice-Sheet Grounding Lines." *Proceedings of the Royal Society A*, January 13, 2010. http://rspa.royalsocietypublishing.org/content/ early/2010/01/13/rspa.2009.0434.abstract. doi: 10.1098/rspa.2009.0434.

Kaufman, Darrell S., et al. "Recent Warming Reverses Long-Term Arctic Cooling." *Science* 325 (September 4, 2009): 1236–1239.

Kaufman, Leslie. "Front-Line City in Virginia Tackles Rise in Sea." *The New York Times*, November 25, 2010. http://www.nytimes.com/2010/11/26/science/earth/26norfolk.html.

Kaufman, Marc. "Warming Arctic Is Taking a Toll." *Washington Post*, April 15, 2006, A7. http:// www.washingtonpost.com/wp-dyn/content/article/2006/04/14/AR2006041401368_pf.html.

Kaufman, Marc. "Southwest May Get Even Hotter, Drier; Report on Warming Warns of Droughts." *Washington Post*, (April 6, 2007, A-3,i). http://www.washingtonpost.com/wp -dyn/content/article/2007/04/05/AR2007040501180_pf.html.

Kaufman, Marc. "Decline in Snowpack Is Blamed on Warming." *Washington Post*, February 1, 2008, A1. http://www.washingtonpost.com/wp-dyn/content/article/2008/01/31/AR20 08013101868_pf.html.

Kaufman, Marc. "Perennial Arctic Ice Cover Diminishing, Officials Say." *Washington Post*, March 19, 2008, A3. http://www.washingtonpost.com/wp-dyn/content/article/2008/ 03/18/AR2008031802903_pf.html.

Kazlowski, Steven, et al. *The Last Polar Bear: Facing the Truth of a Warming World.* Seattle: Braided River Books, 2008.

Keeling, Charles D., and Timothy P. Whorf. "The 1,800-Year Oceanic Tidal Cycle: A Possible Cause of Rapid Climate Change." *Proceedings of the National Academy of Sciences of the United States of America* 97(8) (April 11, 2000): 3814–3819.

Keen, Judy. "Neighbors at Odds over Noise from Wind Turbines." *USA Today*, November 4, 2008, A3.

Keenlyside, N.S., et al. "Advancing Decadal-Scale Climate Prediction in the North Atlantic Sector." *Nature* 453 (May 1, 2008): 84–88.

Keith, David. *A Case for Climate Engineering.* Cambridge, MA: MIT/Boston Review, 2013.

Kelleher, Lynne. "Look Who's Here: Tropical Fish Warming to Waters around Ireland." *Sunday Mirror* (London), October 20, 2002, 15.

Kelley, Colin, et al. "Climate Change in the Fertile Crescent and Implications of the Recent Syrian Drought." *Proceedings of the National Academy of Sciences* 112(8) (March 2, 2015). doi: 10.1073/pnas.1421533112.

Kelly, Mick. "Halting Global Warming." Pp. 83–112 in Jeremy Leggett, ed., *Global Warming: The Greenpeace Report*. New York: Oxford University Press, 1990.

"Ken Caldeira." Carnegie Institution for Science, Department of Global Ecology, Stanford University. No date. http://globalecology.stanford.edu/labs/caldeiralab/Caldeira_bio.html. Accessed February 25, 2015.

Kenneally, Christine. "The Inferno." *The New Yorker*, October 26, 2009, 46–53.

Kennett, Douglas J., et al. "Development and Disintegration of Maya Political Systems in Response to Climate Change." *Science* 338 (November 9, 2012): 788–791.

Kennett, James P., et al. "Carbon Isotopic Evidence for Methane Hydrate Instability during Quaternary Interstadials." *Science* 288 (Apr 7, 2000): 128–133.

Kennett, James P., et al. *Methane Hydrates in Quaternary Climate Change: The Clathrate Gun Hypothesis*. Washington, DC: American Geophysical Union, 2003.

Kennett, James P., et al. "Development and Disintegration of Maya Political Systems in Response to Climate Change." *Science* 338 (November 9, 2012): 788–791.

Kerr, Jeremy T., et al. "Climate Change Impacts on Bumblebees Converge across Continents." *Science* 349 (July 10, 2015): 177–180.

Kerr, Richard A. "West Antarctica's Weak Underbelly Giving Way?" *Science* 281 (July 24, 1998): 499–500.

Kerr, Richard A. "Deep Chill Triggers Record Ozone Hole." *Science* 282 (October 16, 1998): 391.

Kerr, Richard A. "Oceanography: Has a Great River in the Sea Slowed Down?" *Science* 286 (November 5, 1999): 1061–1062.

Kerr, Richard A. "A Smoking Gun for an Ancient Methane Discharge." *Science* 286 (November 19, 1999): 1465.

Kerr, Richard A. "Globe's 'Missing Warming' Found in the Ocean." *Science* 287 (March 24, 2000): 2126–2127.

Kerr, Richard A. "A Variable Sun Paces Millennial Climate." *Science* 294 (November 16, 2001): 1431–1432.

Kerr, Richard A. "A Single Climate Mover for Antarctica." *Science* 296 (May 3, 2002): 825–826.

Kerr, Richard A. "A Warmer Arctic Means Change for All." *Science* 297 (August 30, 2002): 1490–1492.

Kerr, Richard A. "European Climate: Mild Winters Mostly Hot Air, Not Gulf Stream." *Science* 297 (September 27, 2002): 2202.

Kerr, Richard A. "Climate Change: Sea Change in the Atlantic." *Science* 303 (January 2, 2004): 35.

Kerr, Richard A. "Looking Back for the World's Climatic Future." *Science* 312 (June 9, 2005): 1456–1457.

Kerr, Richard A. "Atlantic Mud Shows How Melting Ice Triggered an Ancient Chill." *Science* 312 (June 30, 2006): 1860.

Kerr, Richard A. "Pollute the Planet for Climate's Sake?" *Science* 314 (October 20, 2006): 401–403.

Kerr, Richard A. "False Alarm: Atlantic Conveyor Belt Hasn't Slowed Down After All." *Science* 314 (November 17, 2006): 1064.

Kerr, Richard A. "Is Battered Arctic Sea Ice Down For the Count?" *Science* 318 (October 5, 2007): 33–34.

Kerr, Richard A. "How Urgent Is Climate Change?" *Science* 318 (November 23, 2007): 1230–1231.

Kerr, Richard A. "Climate Tipping Points Come in from the Cold." *Science* 319 (January 11, 2008): 153.

Kerr, Richard A. "More Climate Wackiness in the Cretaceous Supergreenhouse?" *Science* 319 (January 11, 2008): 145.

Kerr, Richard A. "Galloping Glaciers of Greenland Have Reined Themselves in." *Science* 323 (January 23, 2009): 458.

Kerr, Richard A, "The Many Dangers of Greenhouse Acid." *Science* 323 (January 23, 2009): 459.

Kerr, Richard A. "Arctic Summer Sea Ice Could Vanish Soon but Not Suddenly." *Science* 323 (March 27, 2009): 1655.

Kerr, Richard A. "Antarctic Glacier Off Its Leash." *Science* 327 (January 22, 2010): 409.

Kerr, Richard A. "Models Foresee More-Intense Hurricanes in the Greenhouse." *Science* 327 (January 22, 2010): 399.

Kerr, Richard A. "Ocean Acidification Unprecedented, Unsettling." *Science* 328 (June 18, 2010): 1500–1501.

Kerr, Richard A. "'Arctic Armageddon' Needs More Science, Less Hype." *Science* 329 (August 6, 2010): 620–621.

Kerr, Richard A. "The Greenhouse Is Making the Water-Poor Even Poorer." *Science* 336 (April 27, 2012): 405.

Kerr, Richard A. "Ice-Free Arctic Sea May Be Years, Not Decades, Away." *Science* 337 (September 28, 2012): 1591.

Kerr, Richard A. "Experts Agree Global Warming Is Melting the World Rapidly." *Science* 338 (November 30, 2012a): 1138.

Kerr, Richard A. "An Oil Gusher in the Offing, but Will It Be Enough? *Science* 338 (November 30, 2012b): 1139.

Keys, David. "Global Warming: Methane Threatens to Repeat Ice-Age Meltdown." *The Independent* (London), June 16, 2001, n.p. (LEXIS).

Khan, Shfaqat Abbas, et al. "Spread of Ice Mass Loss into Northwest Greenland Observed by GRACE and GPS." *Geophysical Research Letters* 37 (March 23, 2010): L06501. http://www.agu.org/pubs/crossref/2010/2010GL042460.shtml (no longer available). doi: 10.1029/2010GL042460.

Khan, Shfaqat Abbas, et al. "Sustained Mass Loss of the Northeast Greenland Ice Sheet Triggered by Regional Warming." *Nature Climate Change*, March 2014. doi: 10.1038/nclimate2161.

Khatiwala, S., F. Primeau, and T. Hall. "Reconstruction of the History of Anthropogenic CO_2 Concentrations in the Ocean." *Nature* 462 (November 19, 2009): 346–349.

Khazendar, Ala, et al. "The Evolving Instability of the Remnant Larsen B Ice Shelf and Its Tributary Glaciers." *Earth and Planetary Science Letters* 419(1) (June 2015): 199–210. http://www.sciencedirect.com/science/article/pii/S0012821X1500151X.

Kiesecker, Joseph M., Andrew R. Blaustein, and Lisa K. Belden. "Complex Causes of Amphibian Population Declines." *Nature* 410 (April 5, 2001): 681–684.

Kim, Jim Yong. "Make Climate Change a Priority." *Washington Post*, January 24, 2013. http://www.washingtonpost.com/opinions/make-climate-change-a-priority/2013/01/24/6c5c2b66-65b1-11e2-9e1b-07db1d2ccd5b_print.html.

Kim, Seon Tae, et al. "Response of El Niño Sea Surface Temperature Variability to Greenhouse Warming." *Nature Climate Change*, August 2014. http://www.researchgate.net/

publication/264424604_Response_of_El_Nio_sea_surface_temperature_variability_to _greenhouse_warming.

Kinnard, Christophe, et al. "Reconstructed Changes in Arctic Sea Ice over the Past 1,450 Years." *Nature* 479 (November 24, 2011): 509–512.

Kintisch, Eli. "Hot Times for the Cretaceous Oceans." *Science* 311 (February 24, 2006): 1095.

Kintisch, Eli. "As the Seas Warm." *Science* 313 (August 11, 2006): 776–779.

Kintisch, Eli. "Making Dirty Coal Plants Cleaner." *Science* 317 (July 13, 2007): 184–186.

Kintisch, Eli. "Light-Splitting Trick Squeezes More Electricity Out of Sun's Rays." *Science* 317 (August 3, 2007): 583–584.

Kintisch, Eli. "Rules for Ocean Fertilization Could Repel Companies." *Science* 322 (November 7, 2008): 835.

Kintisch, Eli. *Hack the Planet: Science's Best Hope—or Worst Nightmare—for Averting Climate Catastrophe.* New York: Wiley, 2010.

Kintisch, Eli. "Hansen's Retirement from NASA Spurs Look at His Legacy." *Science* 340 (May 3, 2013): 540–541.

Kintisch, Eli. "U.S. Carbon Plan Relies on Uncertain Capture Technology." *Science* 341 (September 27, 2013): 1438–1439.

Kintisch, Eli. "Dr. Cool." *Science* 342 (October 18, 2013): 307–309.

Kintisch, Eli. "Into the Maelstrom." *Science* 344 (April 18, 2014): 250–253.

Kintisch, Eli. "'Sea Butterflies' Are a Canary for Ocean Acidification." *Science* 344 (May 9, 2014): 569.

Kintisch, Eli. "New Solution to Carbon Pollution?" *Science* 352 (June 10, 2016): 1262–1263.

Kintisch, Eli. "Sea Ice Retreat Said to Accelerate Greenland Melting." *Science* 352 (June 17, 2016): 1377.

Kintisch, Eli, and Erik Stokstad. "Ocean CO_2 Studies Look Beyond Coral." *Science* 319 (February 22, 2008): 1029.

Kirchner, Stephanie. "Be Aggressive about Passive [Solar Power]." *Time*, April 9, 2007. http:// www.time.com/time/printout/0,8816,1603747,00.html (no longer available).

Kirk-Davidoff, Daniel, Daniel P. Schrag, and James G. Anderson. "On the Feedback of Stratospheric Clouds on Polar Climate." *Geophysical Research Letters* 29(11) (2002): 14659–14663.

Kjær, Kurt H., et al. "Aerial Photographs Reveal Late–20th-Century Dynamic Ice Loss in Northwestern Greenland." *Science* 337(August 3, 2012): 569–573.

Klein, Allison, Fredrick Kunkle, and Tim Craig. "Torrential Rains Inundate D.C. Region: 3 Killed, Roads and Schools Closed." *Washington Post*, September 8, 2011. http://www .washingtonpost.com/local/tropical-storm-lee-inundates-dc-region/2011/09/08/gIQA 7OHIDK_print.html.

Klein, Naomi. *This Changes Everything: Capitalism and the Climate.* New York: Simon & Schuster, 2014.

Kleiven, Helga, et al. "Reduced North Atlantic Deep Water Coeval with the Glacial Lake Agassiz Freshwater Outburst." *Science* 3129 (January 4, 2008): 60–64.

Klug, Edward C. "Global Warming: Melting Down the Facts about This Overheated Myth." CFACT Briefing Paper Number 105, November 1997. http://www.cfact.org/IssueArchive /greenhouse.bp.n97.txt (no longer available).

Knoblauch, Jessica A. "Have It Your (the Sustainable) Way." *EJ* (*Environmental Journalism*), Spring 2007, 28–30, 46.

Knorr, W., et al. "Long-Term Sensitivity of Soil Carbon Turnover to Warming." *Nature* 433 (January 20, 2005): 298–301.

Knutson, Thomas R., et al. "Impact of CO_2-Induced Warming on Hurricane Intensities as Simulated in a Hurricane Model with Ocean Coupling." *Journal of Climate* 14 (2001): 2458–2469.

Knutson, Thomas R., and Robert E. Tuleya. "Impact of CO_2-Induced Warming on Simulated Hurricane Intensity and Precipitation: Sensitivity to the Choice of Climate Model and Convective Parameterization." *Journal of Climate* 17(18) (September 15, 2004): 3477–3495.

Knutson, Thomas R., Robert E. Tuleya, and Y. Kurihara. "Simulated Increase in Hurricane Intensities in a CO_2-Warmed Climate." *Science* 279 (February 13, 1998), 1018–1020.

Knutson, Thomas R., et al. "Impact of CO_2-Induced Warming on Hurricane Intensities as Simulated in a Hurricane Model with Ocean Coupling." *Journal of Climate* 14 (2001): 2458–2469.

Knutson, Thomas R., et al. "Simulated Reduction in Atlantic Hurricane Frequency under Twenty-First-Century Warming Conditions." *Nature GeoScience* online, May 18, 2008. http://www.nature.com/ngeo/journal/vaop/ncurrent/abs/ngeo202.html. doi: 10.1038/ngeo202.

Knutti, R., et al. "Strong Hemispheric Coupling of Glacial Climate through Freshwater Discharge and Ocean Circulation." *Nature* 430 (August 19, 2004): 851–856.

Koch, A., and W. Peden, eds. *The Life and Selected Writings of Thomas Jefferson.* New York: Random House, 1944.

Koch, Wendy. "Major Builders Offer Affordable 'Net Zero' Homes." *USA Today*, April 20, 2011, B1.

Koch, Wendy. "Solar Energy Saw Boom in 2010." *USA Today*, June 16, 2011, 3A.

Koch, Wendy. "Climate Change Linked to More Pollen, Allergies, Asthma." *USA Today*, May 31, 2013. http://www.usatoday.com/story/news/nation/2013/05/30/climate-chan ge-allergies-asthma/2163893/.

Koch, Wendy. "Wildfire Smoke Becoming a Serious Health Hazard." *USA Today*, October 25, 2013. http://www.usatoday.com/story/news/nation/2013/10/24/wildfires-smoke -climate-change-harm-health/3173165/?utm.

Koch, Wendy. "Rising Sea Levels Torment Norfolk, Va., and Coastal U.S." *USA Today*, December 18, 2013, 1A, 2A. http://www.usatoday.com/story/news/nation/2013/12/17/sea-leve l-rise-swamps-norfolk-us-coasts/3893825/.

Koch, Wendy. "Would Keystone Pipeline Unload 'Carbon Bomb' or Job Boom?" *USA Today*, March 10, 2014. http://www.usatoday.com/story/news/nation/2014/03/01/keyston exls-myths-debunked/5651099/.

Koch, Wendy. "Greenland's Ice Decline Accelerates." *USA Today*, March 17, 2014, 4A.

Koch, Wendy. "Miami Is One of USA's Top Hot Spots for Climate Change. *USA Today*, May 7, 2014. http://www.usatoday.com/story/news/nation/2014/05/07/miami-south -florida-climate hot-spots/8803849/.

Kolbert, Elizabeth. "The Darkening Sea: What Carbon Emissions Are Doing to the Oceans." *The New Yorker*, November 20, 2006, 66–75. http://www.newyorker.com/maga zine/2006/11/20/the-darkening-sea.

Kolbert, Elizabeth. "Don't Drive, He Said." *The New Yorker*, May 7, 2007, 23–24.

Kolbert, Elizabeth. "Running on Fumes." *The New Yorker*, November 5, 2007, 87–90.

Kolbert, Elizabeth. "Unconventional Crude: Canada's Synthetic-fuels Boom." *The New Yorker*, November 12, 2007, 46–51.

Kolbert, Elizabeth. "Testing the Climate." *The New Yorker*, December 24 and 31, 2007, 43–44.

Kolbert, Elizabeth. "The Island in the Wind." *The New Yorker*, July 7, 2008. http://www
.newyorker.com/magazine/2008/07/07/the-island-in-the-wind.

Kolbert, Elizabeth. "Hosed: Is There a Quick Fix for the Climate?" *The New Yorker*, November
16, 2009, 75–77.

Kolbert, Elizabeth. "The Acid Sea." *National Geographic*, April 2011, 100–121.

Kolbert, Elizabeth. "The Big Heat." *The New Yorker*, July 23, 2012, 19–20.

Kolbert, Elizabeth. "The Weight of the World: Can Christina Figures Persuade Humanity
to Save Itself?" *The New Yorker*, August 24, 2015, 24–30.

Kolbert, Elizabeth. "Unsafe Climates." *The New Yorker*, December 7, 2015, 23–24.

Kolbert, Elizabeth. "The Siege of Miami: As Temperatures Rise, So Will Sea Levels." *The New
Yorker*, December 21 and 28, 2015, 42–50.

Kolbert, Elizabeth. "Unnatural Selection: What Will It Take to Save the World's Reefs and
Forests?" *The New Yorker*, April 18, 2016, 22–28.

Kolbert, Elizabeth. "A Song of Ice: What Happens When a Country Starts to Melt?" *The
New Yorker*, October 24, 2016, 50–61.

Kondratyev, Kirill, Vladimir F. Krapivin, and Costas A. Varotsos. *Global Carbon Cycle and
Climate Change*. Berlin: Springer/Praxis, 2004.

Konviser, Bruce I. "Glacier Lake Puts Global Warming on the Map." *Boston Globe*, July 16,
2002, C1.

Kopp, Robert E. "Paleoclimate: Tahitian Record Suggests Antarctic Collapse." *Nature* 483
(March 29, 2012): 549–550.

Kopp, Robert E., et al. "Probabilistic Assessment of Sea Level During the Last Interglacial
Stage." *Nature* 462 (December 17, 2009): 863–867.

Kopp, Robert E., et al. "Temperature-Driven Global Sea-Level Variability in the Common
Era." *Proceedings of the National Academy of Sciences*, February 22, 2016. doi: 10.1073/
pnas.1517056113.

Korhonen, H., et al. "Aerosol Climate Feedback Due to Decadal Increases in Southern Hemisphere Wind Speeds." *Geophysical Research Letters*, January 27, 2010 (online). doi:
10.1029/2009GL041320.

Körner, Christian, and David Basler. "Phenology under Global Warming." *Science* 327
(March 19, 2010): 1461–1462.

Krajick, Kevin. "Arctic Life, on Thin Ice." *Science* 291 (January 19, 2001): 424–425.

Krajick, Kevin. "Tracing Icebergs for Clues to Climate Change." *Science* 292 (June 22, 2001):
2244–2245.

Kramer, Andrew E. "Russians and Their Crops Wilt under Heat Wave." *The New York Times*,
July 19, 2010. http://www.nytimes.com/2010/07/20/world/europe/20russia.html.

Kramer, Andrew E. "Warming Revives Dream of Sea Route in Russian Arctic." *The New York
Times*, October 18, 2011. http://www.nytimes.com/2011/10/18/business/global/warming
-revives-old-dream-of-sea-route-in-russian-arctic.html.

Kramer, Andrew E., and Andrew C. Revkin. "Arctic Shortcut Beckons Shippers as Ice Thaws."
The New York Times, September 11, 2009. http://www.nytimes.com/2009/09/11/science/
earth/11passage.html.

Krauss, Clifford. "Bear Hunting Caught in Global Warming Debate." *The New York Times*,
May 27, 2006. http://www.nytimes.com/2006/05/27/world/americas/27bears.html.

Krauss, Clifford. "Move Over, Oil, There's Money in Texas Wind." *The New York Times*, February 23, 2008. http://www.nytimes.com/2008/02/23/business/23wind.html.

Krauss, Clifford. "Coal Miners Struggle to Survive in an Industry Battered by Layoffs and
Bankruptcy." *The New York Times*, July 17, 2015. http://www.nytimes.com/2015/07/18/

business/energy-environment/coal-miners-struggle-to-survive-in-an-industry-battered -by-layoffs-and-bankruptcy.html.

Krauss, Clifford, and Keith Bradsher. "China's Global Search for Energy." *The New York Times*, May 22, 2014. http://www.nytimes.com/2014/05/22/business/international/ chinas-global-search-for-energy.html?hp&_r=0.

Krauss, Clifford, et al. "As Polar Ice Turns to Water, Dreams of Treasure Abound." *The New York Times*, October 10, 2005. http://www.nytimes.com/2005/10/10/science/10arctic.html.

Kristof, Nicholas. "For Pacific Islanders, Global Warming Is No Idle Threat." *The New York Times*, March 2, 1997. http://sierraactivist.org/library/990629/islanders.html (no longer available).

Krueger, Andrew, and John Myers. "Duluth Mayor Declares State of Emergency Due to Flooding." *Duluth News Tribune*, June 20, 2012. http://www.duluthnewstribune.com/ event/article/id/234914/.

Kruger, Tim. "Stratospheric Folly." *Nature* 508 (April 24, 2014): 457.

Krugman, Paul. "The Sum of All Ears." *The New York Times*, January 29, 2007, A23.

Krugman, Paul. "Wind, Sun, and Fire." *The New York Times*, February 1, 2016, A21.

Krupnik, I., and D. Jolly, eds. 2002. *The Earth Is Faster Now: Indigenous Observations of Arctic Environmental Change*. Fairbanks, AK: Arctic Research Consortium of the United States. http://www.arcus.org/publications/eifn.

Kump, Lee R. "What Drives Climate?" *Nature* 408 (December 7, 2000): 651–652.

Kump, Lee R. "Chill Taken Out of the Tropics." *Nature* 413 (October 4, 2001): 470–471.

Kunzig, Robert. "Drying of the West." *National Geographic*, February 2008, 90–113.

Kurz, Werner A., et al. "Global Climate Change: Disturbance Regimes and Biospheric Feedbacks of Temperate and Boreal Forests." Pp. 119–133 in George M. Woodwell and Fred T. MacKenzie, eds., *Biotic Feedbacks in the Global Climate System: Will the Warming Feed the Warming?* New York: Oxford University Press, 1995.

Kurz, Werner A., et al. "Mountain Pine Beetle and Forest Carbon Feedback to Climate Change." *Nature* 452 (April 24, 2008): 987–991.

Laghari, Javaid R. "Melting Glaciers Bring Energy Uncertainty." *Nature* 502 (October 30, 2013): 617–618.

Laidler, G. J., et al. "Travelling and Hunting in a Changing Arctic: Assessing Inuit Vulnerability to Sea Ice Change in Igloolik, Nunavut." *Climatic Change* 94 (2009): 363–397. doi: 10.1007/s10584-008-9512-z./

"Lake Mead's Level Hits Landmark Low." *Omaha World-Herald*, May 1, 2015, 3A.

Lam, Linda, Nick Wiltgen, and Jon Erdman Wilt. "Lake-Effect Snow Recap: Up to 88 Inches of Snow Buries Parts of Western New York, Including the Buffalo Southtowns." Weather .com, November 21, 2014. http://www.weather.com/storms/winter/news/lake-effect -snow-significant-lake-erie-lake-ontario-20141115.

Landauer, Robert. "Big Changes in Our China Suburb." *Sunday Oregonian*, October 20, 2002, F4.

"Land Clearances Turned up the Heat on Australian Climate." *New Scientist*, May 16, 2009. http://www.newscientist.com/article/mg20227084.700-land-clearances-turned-up-the -heat-on-australian-climate.html?DCMP=OTC-rss&nsref=environment.

Landsea, Christopher W. "A Climatology of Intense (or Major) Atlantic Hurricanes." *Monthly Weather Review* 121 (1993): 1703–1713.

Landsea, Christopher W. "NOAA: Report on Intensity of Tropical Cyclones", August 12, 1999. Miami, FL: NOAA Hurricane Research Division. http://www.aoml.noaa.gov/ hrd/tcfaq/tcfaqG.html#G3.

Landsea, Christopher W., et al. "Downward Trends Atlantic Hurricanes During the Past Five Decades." *Geophysical Research Letters* 23 (1996): 1697–1700.

Landsea, Christopher W., et al. "Impact of Duration Thresholds on Atlantic Tropical Cyclone Counts." *Journal of Climate*, August 2009. http://ams.allenpress.com/perlserv/?request=get-abstract&doi=10.1175%2F2009JCLI3034.1. doi: 10.1175/2009JCLI3034.1.

Larsen, C. F., et al. "Surface Melt Dominates Alaska Glacier Mass Balance." *Geophysical Research Letters*, June 4, 2015. http://onlinelibrary.wiley.com/doi/10.1002/2015GL064349/abstract.

"Last Call for Larsen B." NASA Earth Observatory, June 9, 2015. http://earthobservatory.nasa.gov/IOTD/view.php?id=86002&src=eoa-iotd.

Latif, M., et al. "Is the Thermohaline Circulation Changing?" *Journal of Climate* 19 (18) (September 15, 2006): 4631–4637.

Laukaitis, Algis J. "Talmage Man: No Kiddo . . . It's an Armadillo!" *Lincoln Journal Star*, August 18, 2005, A1.

Lavelle, Marianne. "Good Gas, Bad Gas." *National Geographic*, December 2012, 90–109.

Lavelle, Marianne. "Collapse of New England's Iconic Cod Tied to Climate Change." *Science* Online, October 29, 2015. http://news.sciencemag.org/climate/2015/10/collapse-new-england-s-iconic-cod-tied-climate-change.

Lavers, Chris. *Why Elephants Have Big Ears.* New York: St. Martin's Press, 2000.

Lawrence, D. M., and A. G. Slater. "A Projection of Severe Near-Surface Permafrost Degradation during the 21st Century." *Geophysical Research Letters* 32 (2005): 1–5.

Lawton, R. O., et al. "Climatic Impact of Tropical Lowland Deforestation on Nearby Montane Cloud Forests." *Science* 294 (October 19, 2001): 584–587.

Lazaroff, Cat. "Climate Change Threatens Global Biodiversity." Environment News Service, February 7, 2001. http://ens-news.com/ens/feb2002/2002L-02-07-06.html (no longer available).

Lazaroff, Cat. "Land Use Rivals Greenhouse Gases in Changing Climate." Environment News Service, October 2, 2002. http://ens-news.com/ens/oct2002/2002-10-02-06.asp (no longer available).

Lazo, Alejandro. "A Shorter Link Between the Farm and Dinner Plate." *Washington Post*, July 29, 2007, A1. http://www.washingtonpost.com/wp-dyn/content/article/2007/07/28/AR2007072801255.html?wpisrc=newsletter.

Lea, David W. "The 100,000-Year Cycle in Tropical SST, Greenhouse Forcing, and Climate Sensitivity." *Journal of Climate* 17(11) (June 1, 2004): 2170–2179. https://courses.seas.harvard.edu/climate/seminars/pdfs/Lea_JofClimate2004.pdf.

Leake, Jonathan. "Fiery Venus Used to Be Our Green Twin." *Sunday Times* (London), December 15, 2002, 11.

"Leaking Underground CO_2 Storage Could Contaminate Drinking Water." EurekaAlert! November 11, 2010. http://www.eurekalert.org/pub_releases/2010-11/du-luc111110.php.

Lean, Geoffrey. "Quarter of World's Corals Destroyed." *The Independent* (London), January 7, 2001, 7.

Lean, Geoffrey. "We Regret to Inform You that the Flight to Malaga Is Destroying the Planet; Air Travel Is Fast Becoming One of the Biggest Causes of Global Warming." *The Independent* (London), August 26, 2001, 23.

Lean, Geoffrey. "Hot Summer Sparks Global Food Crisis." *The Independent* (London), August 31, 2003, 4.

Lean, Geoffrey. "Worst U.S. Drought in 500 Years Fuels Raging California Wildfires." *The Independent* (London), July 25, 2004, 20.

Lean, Geoffrey. "Global Warming Will Redraw Map of the World." *The Independent* (London), November 7, 2004, 8.

Lean, Geoffrey. "The Big Thaw: Global Disaster Will Follow If the Ice Cap on Greenland Melts." *The Independent* (London), November 20, 2005. http://arizonaenergy.org/ News_06/News_Feb06/The%20Big%20Thaw—%20Global%20Disaster%20Will%20 Follow%20If%20the%20Ice%20Cap%20on%20Greenland%20Melts.htm.

Lean, J., J. Beer, and R. Bradley. "Reconstruction of Solar Irradiance Since 1610: Implications for Climate Change." *Geophysical Research Letters* 22 (1995): 3195–3198.

Leatherman, Stephen P. "Coastal Land Loss in the Chesapeake Bay Region: An Historical Analog Approach to Global Change Analysis." Pp. 17–27 in Jurgan Schmandt and Judith Clarkson, eds., *The Regions and Global Warming: Impacts and Response Strategies*. New York: Oxford University Press, 1992.

Ledford, Heidi. "Malaria May Not Rise as World Warms." *Nature* 465 (May 20, 2010):280.

Lee, Xuhui, et al. "Observed Increase in Local Cooling Effect of Deforestation at Higher Latitudes." *Nature* 479 (November 17, 2011): 384–387.

Leggett, Jeremy, ed. *Global Warming: The Greenpeace Report*. New York: Oxford University Press, 1990.

Leggett, Jeremy. *The Carbon War: Global Warming and the End of the Oil Era*. New York: Routledge, 2001.

Leggett, Karby. "In Rural China, General Motors Sees a Frugal but Huge Market: It Bets Tractor Substitute Will Look Pretty Good to Cold, Wet Farmers." *Wall Street Journal*, January 16, 2001, A19.

Leidig, Michael, and Roya Nikkhah. "The Truth about Global Warming—It's the Sun That's to Blame." *Sunday Telegraph* (London), July 18, 2004, 5.

Le Maho, Yvon. "Conservation: After the Ice." *Nature* 468 (December 23, 2010): 1034–1035.

Lemonick, Michael D. "East Coast Faces Rising Seas from Slowing Gulf Stream." Weather .com. February 13, 2013. http://www.climatecentral.org/news/east-coast-faces-rising -seas-from-slowing-gulf-stream-15587.

Levermann, Anders, et al. "The Multimillennial Sea-Level Commitment of Global Warming." *Proceedings of the National Academy of Sciences* 110(34) (August 20, 2013): 13745– 13750. http://www.pnas.org/content/110/34/13745.full.pdf.

Levitt, Steven D., and Stephen J. Dubner. *SuperFreakonomics: Global Cooling, Patriotic Prostitutes, and Why Suicide Bombers Should Buy Life Insurance*. New York: William Morrow, 2009.

Levitus, Sydney, et al. "Warming of the World Ocean." *Science* 287 (2000): 2225–2229.

Levitus, Sydney, et al. "Anthropogenic Warming of Earth's Climate System." *Science* 292 (April 13, 2001): 267–270.

Libellaug, Henrik Pryser. "Lightning Strike Kills More Than 300 Reindeer in Norway." *The New York Times*, August 29, 2016. http://www.nytimes.com/2016/08/30/world/europe/ hardangervidda-norway-lightning-reindeer.html.

Lighthill, J., et al. "Global Climate Change and Tropical Cyclones." *Bulletin of the American Meteorological Society* 75(1994):2147–2157.

"Linda." Review of Inhofe, *The Greatest Hoax*, Amazon.com. December 24, 2014. http:// www.amazon.com/The-Greatest-Hoax-Conspiracy-Threatens/product-reviews /1936488493/ref=cm_cr_dp_synop?ie=UTF8&showViewpoints=0&sortBy=bySubmis sionDateDescending#R3PEM77NQ445GW (no longer available).

Linden, Eugene. "The Big Meltdown: As the Temperature Rises in the Arctic, It Sends a Chill around the Planet." *Time* 156(10), September 4, 2000. http://www.time.com/time/mag azine/articles/0,3266,53418,00.html (no longer available).

Lindroth, A., A. Grelle, and A. S. Moren. "Long-Term Measurements of Boreal Forest Carbon Balance Reveal Large Temperature Sensitivity." *Global Change Biology* 4 (April 1998), 443–450.

Little, Mark G., and Robert B. Jackson. "Potential Impacts of Leakage from Deep CO_2 Geo-Sequestration on Overlying Freshwater Aquifers." *Environmental Science and Technology* 44 (2010): 9225–9232. http://pubs.acs.org/doi/abs/10.1021/es102235w.

Liu, Chong, et al. "Artificial Photosynthesis Steps Up." *Science* 352 (June 3, 2016): 1210–1213.

Liu, Chong, et al. "Water Splitting–Biosynthetic System with CO2 Reduction Efficiencies Exceeding Photosynthesis." *Science* 352 (June 3, 2016): 1210–1213.

Liu, Zhu, et al. "Reduced Carbon Emission Estimates from Fossil Fuel Combustion and Cement Production in China." *Nature* 524 (August 20, 2015): 335–338.

"Livermore Scientists Find Global Ocean Warming Has Doubled in Recent Decades." Lawrence Livermore National Laboratory, January 18, 2016. https://www.llnl.gov/news/livermore-scientists-find-global-ocean-warming-has-doubled-recent-decades.

Lizza, Ryan. "The President and the Pipeline." *The New Yorker*, September 16, 2013, 38–51.

Ljunggren, David. "Effects of Global Warming Clear in Canada Arctic." Environmental News Network, April 20, 2000. http://www.enn.com/enn-subsciber-news-archive/2000/04/04202000/reu_arctwarm_12170.asp (no longer available).

Logan, Tracey. "Alaska's Biggest Tundra Fire Sparks Climate Warning." *New Scientist*, July 30, 2009. http://www.newscientist.com/article/dn17537-alaskas-biggest-tundra-fire-sparks-climate-warning.html?DCMP=OTC-rss&nsref=environment.

"Loggerhead Turtle Nesting Down by Half Since 1998." Environment News Service, November 8, 2007. http://www.ens-newswire.com/ens/nov2007/2007-11-08-094.asp

Lohr, Steve. "Global Warming Is a Challenge for Economic Policy, Too." *The New York Times*, November 12, 2006. http://www.nytimes.com/2006/12/12/business/worldbusiness/12iht-climate.3877222.html.

Lomborg, Bjorn. *The Skeptical Environmentalist*. Cambridge, UK: Cambridge University Press, 2001.

Lomborg, Bjorn. *Cool It! The Skeptical Environmentalist's Guide to Global Warming*. New York: Knopf (Random House), 2007.

Lomborg, Bjorn. "The Poor Need Cheap Fossil Fuels." *The New York Times*, December 3, 2013. http://www.nytimes.com/2013/12/04/opinion/the-poor-need-cheap-fossil-fuels.html.

Long, Stephen P., et al. "Food for Thought: Lower-Than-Expected Crop Yield Stimulation with Rising CO_2 Concentrations." *Science* 312 (June 30, 2006): 1918–1921.

"Longer, More Frequent Fire Seasons." NASA Earth Observatory, July 28, 2015. http://earthobservatory.nasa.gov/IOTD/view.php?id=86268&src=eoa-iotd.

"Long-Term Global Warming Trend Continues." NASA Earth Observatory. January 16, 2013 http://earthobservatory.nasa.gov/IOTD/view.php?id=80167&src=eoa-iotd.

"Lopsided Warming North to South." *Nature* 487 (July 5, 2012): 9.

"Loss of Arctic Ice Opens New Cable Route." *Omaha World-Herald*, January 22, 2010, A4.

"Lotion in the Ocean." *National Geographic*, June 2015, 10.

Lovett, Ian. "A 'Wall of Mud' in California, and Warnings to Heed El Niño." *The New York Times*, October 17, 2015. http://www.nytimes.com/2015/10/17/us/a-wall-of-mud-in-california-and-warnings-to-heed-el-nino.html.

Lovett, Ian, and Jennifer Medina. "Fires in West Have Residents Gasping on the Soot Left Behind." *The New York Times*, September 9, 2015. http://www.nytimes.com/2015/09/10/us/fires-in-west-leave-residents-gasping-on-the-soot-left-behind.html.

Lovins, Amory. "More Profit with Less Carbon." *Scientific American* 74 (September 2005): 76–83.

Loya, Wendy M., and Paul Grogan. "Carbon Conundrum on the Tundra." *Nature* 431 (September 23, 2004): 406–407.

Lublin, Joann S. "Environmentalism Sprouts Up on Corporate Boards." *Wall Street Journal*, August 11, 2006, B6.

Lucht, Wolfgang, et al. "Climatic Control of the High-Latitude Vegetation Greening Trend and Pinatubo Effect." *Science* 296 (May 31, 2002): 1687–1689.

Luhnow, David, and Geraldo Samor. "As Brazil Fills Up on Ethanol, It Weans Off Energy Imports." *Wall Street Journal*, January 9, 2006, A1, A8.

Luterbacher, Jurg, et al. "European Seasonal and Annual Temperature Variability, Trends, and Extremes Since 1500." *Science* 303 (March 5, 2004): 1499–1503.

Lyall, Sarah. "At Risk From Floods, but Looking Ahead with Floating Houses." *The New York Times*, April 3, 2007. http://query.nytimes.com/gst/fullpage.html?res=9B0CE6DB1 F30F930A35757C0A9619C8B63.

Lyall, Sarah. "Warming Revives Flora and Fauna in Greenland." *The New York Times*, October 28, 2007. http://www.nytimes.com/2007/10/28/world/europe/28greenland.html.

Lydersen, Kari. "Sea-Ice Melt Imperils Walruses, and Economy Based on Them." *Washington Post*, August 29, 2008. http://www.washingtonpost.com/wp-dyn/content/article/2008 /08/28/AR2008082803489.html.

Lyman, John M., et al. "Robust Warming of the Global Upper Ocean." *Nature* 465 (May 20, 2010): 334–337.

Lynas, Mark. *High Tide: The Truth about Our Climate Crisis*. New York: Picador/St. Martins, 2004.

Lynas, Mark. "Fly and Be Damned." *New Statesman* (London), April 3, 2006, 12–15. http:// www.newstatesman.com/node/164067.

Lynas, Mark. *Six Degrees: Our Future on a Hotter Planet*. London: Fourth Estate (HarperCollins), 2007.

Lynch-Stieglitz, Jean, et al. "Atlantic Meridional Overturning Circulation during the Last Glacial Maximum." *Science* 316 (April 6, 2007): 66–69.

Macalister, Terry. "Shell Chief Delivers Global Warming Warning to Bush in His Own Back Yard." *The Guardian* (U.K.), March 12, 2003, 19.

Macdougall, Doug. *Frozen Earth: The Once and Future Story of Ice Ages*. Berkeley: University of California Press, 2004.

Macey, Richard. "Climate Change Link to Clearing." *Sydney Morning Herald*, June 29, 2004.

Machguth, Horst, et al. "Greenland Meltwater Storage in Firn Limited by Near-Surface Ice Formation." *Nature Climate Change* (2016). http://www.nature.com/nclimate/journal/ vaop/ncurrent/full/nclimate2899.html. doi: 10.1038/nclimate2899.

Mack, Michelle C., et al. "Ecosystem Carbon Storage in Arctic Tundra Reduced by Long-Term Nutrient Fertilization." *Nature* 432 (September 23, 2004): 440–443.

Macken, Julie. "The Double-Whammy Drought." *Australian Financial Review*, May 4, 2004, 61.

MacKinnon, Morag. "Flash Floods Kill Five as Torrential Rains Hit Australia's East Coast." Reuters, May 2, 2015. http://www.reuters.com/article/2015/05/02/us-australia-storm -idUSKBN0NN03D20150502.

MacLeod, Calum. "China, on Cusp as Superpower, Wrestles Plant." *USA Today*, August 10, 2010, 6A.

Madigan, Nick. "South Florida Faces Ominous Prospects from Rising Waters." *The New York Times*, November 11. 2013. http://www.nytimes.com/2013/11/11/us/south-florida-faces-ominous-prospects-from-rising-waters.html.

Magill, Bobby. "The Navajo Nation's Shifting Sands of Climate Change." Climate Central, May 28, 2014. http://www.climatecentral.org/news/navajo-nation-climate-change-17326.

Mahapatra, Dhananjay. "Small Diesel Car Buyers to Pay One-Time Pollution Tax: SC [Supreme Court]." *Times of India*, December 17, 2015, 1.

Mahtab, Fasih Uddin. "The Delta Regions and Global Warming: Impact and Response Strategies for Bangladesh." Pp. 28–43 in Jurgan Schmandt and Judith Clarkson, eds., *The Regions and Global Warming: Impacts and Response Strategies*. New York: Oxford University Press, 1992.

"Maine: Shrimp Season Is Called Off." *The New York Times*, December 4, 2013, A21.

"Maine's Lobster Bubble Is Threatening to Burst, Thanks to Global Warming and Canada." quartz.com. http://qz.com/112273/maines-lobster-bubble-is-threatening-to-burst-thanks-to-global-warming-and-canada/. Accessed August 30, 2013.

"Major Iceberg Cracks Off Pine Island Glacier." NASA Earth Observatory, November 15, 2013. http://earthobservatory.nasa.gov/IOTD/view.php?id=82392&src=eoa-iotd.

Malakoff, David. "Researchers Struggle to Assess Responses to Ocean Acidification." *Science* 338 (October 5, 2012): 27–28.

Malcolm, Jay R., and Adam Markham. Global Warming and Terrestrial Biodiversity Decline: A Modeling Approach. Report Prepared for the World Wildlife Fund, July 2000. http://www.wwf.se/source.php/1117008/figures.pdf.

"Maldives Government Dives for Climate Change." Associated Press in *The New York Times*, October 17, 2009. http://www.nytimes.com/aponline/2009/10/17/world/AP-AS-Maldives-Underwater-Cabinet.html (no longer available).

Maldonado, J. K., et al. "The Impact of Climate Change on Tribal Communities in the U.S.: Displacement, Relocation, and Human Rights." *Climatic Change* 120 (2013): 601–614.

Malhi, Yadvinder, et al. "Climate Change, Deforestation, and the Fate of the Amazon." *Science* 319 (January 11, 2008): 169–172.

Mallaby, Sebastian. "Carbon Policy That Works." *Washington Post*, July 23, 2007, A17. http://www.washingtonpost.com/wpdyn/content/article/2007/07/22/AR2007072200884_pf.html (no longer available).

Malone, Thomas F., Edward D. Goldberg, and Walter H. Munk. "Roger Randall Dougan Revelle, 1909–1991." No date. Washington, DC: National Academy of *Sciences*. http://www.nap.edu/readingroom/books/biomems/rrevelle.html.

Maloney, Peter. "Solar Projects Draw New Opposition." *The New York Times*, September 24, 2008, H2.

Mann, Michael E., Raymond S. Bradley, and Michael K. Hughes. "Global-Scale Temperature Patterns and Climate Forcing over the Past Six Centuries." *Nature* 392 (April 23, 1998): 779–787.

Manney, Gloria L., et al. "Unprecedented Arctic Ozone Loss in 2011." *Nature* 478 (October 27, 2011): 469–475.

Marchese, Anthony J., et al. "Methane Emissions from United States Natural Gas Gathering and Processing." *Environmental Science and Technology Letters*, August 18, 2015. http://pubs.acs.org/doi/abs/10.1021/acs.est.5b02275.

Marcott, Shaun A., et al. "A Reconstruction of Regional and Global Temperature for the Past 11,300 Years." *Science* 339 (March 8, 2012): 1198–1201.

"Marine Life on a Warming Planet" (Editorial). *The New York Times*, December 2, 2012. http://www.nytimes.com/2012/12/03/opinion/marine-life-on-a-warming-planet.html.

Marris, Emma, and Daemon Fairless. "Wind Farms' Deadly Reputation Hard to Shift." *Nature* 447 (May 10, 2007): 126.

Marshall, Jessica. "Solar Energy: Springtime for the Artificial Leaf." *Nature* 510 (June 4, 2014): 22–24. http://www.nature.com/news/solar-energy-springtime-for-the-artificial-leaf-1.15341.

Martens, Pim. "How Will Climate Change Affect Human Health?" *American Scientist* 87(6) (November–December 1999): 534–541.

Martens, Willem J. M., Theo H. Jetten, and Dana A. Focks. "Sensitivity of Malaria, Schistosomiasis, and Dengue to Global Warming." *Climatic Change* 35 (1997): 145–156.

Martin, Andrew. "In Eco-Friendly Factory, Low-Guilt Potato Chips." *The New York Times*, November 15, 2007, A1, A22.

Martin, Richard. "A Big Leap for an Artificial Leaf: A New System for Making Liquid Fuel from Sunlight, Water, and Air Is a Promising Step for Solar Fuels." *MIT Technology Review*, June 7, 2016. https://www.technologyreview.com/s/601641/a-big-leap-for-an-artificial-leaf/.

Mascarelli, Amanda. "Climate-Change Adaptation: Designer Reefs." *Nature* 508 (April 24, 2014): 444–446. http://www.nature.com/news/climate-change-adaptation-designer-reefs-1.15073.

Mason, John, Jack A. Bailey, and Ardea London. "Doomsday for Butterflies as Britain Warms Up; Dozens of Native Species at Risk of Extinction as Habitats Come under Threat." *The Independent* (London), September 29, 2002, 12.

Masood, Salman. "Starved for Energy, Pakistan Braces for a Water Crisis." *The New York Times*, February 13., 2015. http://www.nytimes.com/2015/02/13/world/asia/pakistan-braces-for-major-water-shortages.html.

"Mass Extinction of Ocean Species Soon to Be 'Inevitable.'" Environment News Service, June 21, 2011. http://www.ens-newswire.com/ens/jun2011/2011-06-21-01.html (no longer available).

Mathews-Amos, Amy, and Ewann A. Berntson. *Turning Up the Heat: How Global Warming Threatens Life in the Sea*. World Wildlife Fund and Marine Conservation Biology Institute, 1999. http://www.worldwildlife.org/news/pubs/wwf_ocean.htm.

Matter, Juerg M., et al. "Rapid Carbon Mineralization for Permanent Disposal of Anthropogenic Carbon Dioxide Emissions." *Science* 352 (June 10, 2016): 1312–1314.

Matthews, H. Damon, and Ken Caldeira. "Stabilizing Climate Requires Near-Zero Emissions." *Geophysical Research Letters* 35 (February 27, 2008): L04705. doi: 10.1029/2007 GL032388.

Maue, Ryan. "Global Tropical Storm Activity." 2008. http://policlimate.com/tropical/.

May, Wilhelm, Reinhard Voss, and Erich Roeckner. "Changes in the Mean and Extremes of the Hydrological Cycle in Europe under Enhanced Greenhouse Gas Conditions in a Global Time-slice Experiment." Pp. 1–30 in Martin Beniston, *Climatic Change: Implications for the Hydrological Cycle and for Water Management*. Dordrecht, Germany: Kluwer Academic Publishers, 2002.

Maynard, Roger. "Climate Change Bringing More Floods to Australia." *The Straits Times* (Singapore), March 14, 2001, 17.

McCarthy, Michael. "Climate Change Provides Exotic Sea Life with a Warm Welcome to Britain." *The Independent* (London), January 24, 2002, 13.

McCarthy, Michael. "'Rainforests of the Sea' Ravaged; Overfishing and Pollution Kill 80 Per Cent of Coral on Caribbean Reefs." *The Independent* (London), July 18, 2003, 3.

McCarthy, Michael. "The Four Degrees: How Europe's Hottest Summer Shows Global Warming Is Transforming Our World." *The Independent* (London), December 8, 2003, 3.

McCarthy, Michael. "Countdown to Global Catastrophe." *The Independent* (London), January 24, 2005, 1.

McCord, Joel. "Marshes in Decay Haunt the Bay." *Baltimore Sun*, December 6, 2000, 1B.

McCrea, Steve. "Air Travel: Eco-Tourism's Hidden Pollution." *Earth Times* (San Diego), August 1996. http://www.sdearthtimes.com/et0896/et0896s13.html.

McCulloch, Malcolm, et al. "Coral Resilience to Ocean Acidification and Global Warming Through pH Up-Regulation." *Nature Climate Change* 2(2012):623–627. http://www.nature.com/nclimate/journal/v2/n8/full/nclimate1473.html.

McEwen, Bill. "The West's Dying Forests." *The New York Times*, August 2, 2004, A16.

McFarling, Usha Lee. "Fear Growing over a Sharp Climate Shift." *Los Angeles Times*, July 13, 2001, A1.

McFarling, Usha Lee. "Scientists Now Fear 'Abrupt' Global Warming Changes." *Los Angeles Times*, December 12, 2001, A30.

McFarling, Usha Lee. "Glacial Melting Takes Human Toll; Avalanche in Russia and Other Disasters Show That Global Warming Is Beginning to Affect Areas Much Closer to Home." *Los Angeles Times*, September 25, 2002, A4.

McFarling, Usha Lee. "A Tiny 'Early Warning' of Global Warming's Effect; The Population of Pikas, Rabbit-Like Mountain Dwellers, Is Falling, a Study Finds." *Los Angeles Times*, February 26, 2003, A17.

McGlade, Christophe, and Paul Ekins. "The Geographical Distribution of Fossil Fuels Unused When Limiting Global Warming to 2°C." *Nature* 517 (January 8, 2015): 187–190. doi: 10.1038/nature14016.

McGrath, Ben. "The White Wall: In the Iditarod, Alaska's Mushing Dynasties Confront the Elements—and a Generational Divide." *The New Yorker*, April 22, 2013, 81–95.

McGrath, Ben. "Drill, Baby, Drill." *The New Yorker*, December 2, 2013, 24.

McGrath, Susan. "On Thin Ice." *National Geographic*, July 2011, 64–75.

McGuire, Bill. *Surviving Armageddon: Solutions for a Threatened Planet.* New York: Oxford University Press, 2005.

McKibben, Bill. *The End of Nature.* New York: Random House, 1989.

McKibben, Bill. "Welcome to the Climate Crisis: How to Tell Whether a Candidate Is Serious About Combating Global Warming." *Washington Post*, May 27, 2006, A25. http://www.washingtonpost.com/wp-dyn/content/article/2006/05/26/AR2006052601549_pf.html.

McKibben, Bill. *Earth: Making a Life on a Tough New Planet.* New York: Times Books, 2010.

McKibben, Bill. "A Link Between Climate Change and Joplin Tornadoes? Never!" *Washington Post*, May 23, 2011. http://www.washingtonpost.com/opinions/a-link-between-climate-change-and-joplin-tornadoes-never/2011/05/23/AFrVC49G_print.html.

McKibben, Bill. "Obama's Catastrophic Climate-Change Denial." *The New York Times*, May 13, 2015. http://www.nytimes.com/2015/05/13/opinion/obamas-catastrophic-climate-change-denial.html?hp&action=click&pgtype=Homepage&module=c-column-top-span-region®ion=c-column-top-span-region&WT.nav=c-column-top-span-region&_r=0.

McKie, Robin. "Dying Seas Threaten Several Species." *The Observer* (London), December 2, 2001, 14.

McKie, Robin. "Decades of Devastation Ahead as Global Warming Melts the Alps: A Mountain of Trouble as Matterhorn Is Rocked by Avalanches." *The Observer* (London), July 20, 2003, 18.

McLean, Kirsty Galloway. *Advance Guard: Climate Change Impacts, Adaptation, Mitigation and Indigenous Peoples*. U.N. University Institute of Advanced Studies. Traditional Knowledge Initiative. Traditional Knowledge Initiative. Darwin, Australia, 2010. http://archive.ias .unu.edu/resource_centre/UNU_Advance_Guard_Compendium_2010_final_web.pdf.

McMahon, Bucky. "Relocate! Relocate! Relocate!" *GQ*, December 2009, 296–306.

McMillan, Malcolm, et al. "A High-Resolution Record of Greenland Mass Balance." *Geophysical Research Letters* 43 (July 9, 2016): 7002–7010.

McNeil, Ben I., and Tristan P. Sasse. "Future Ocean Hypercapnia Driven by Anthropogenic Amplification of the Natural CO_2 Cycle." *Nature* 529(January 21, 2016):383–386.

McNulty, Sheila. "U.S. Power Generation Answer Is Blowing in the Wind." *Financial Times* (London), April 24, 2007, 7.

McPhaden, M. J., and D. Zhang. "Slowdown of the Meridional Overturning Circulation in the Upper Pacific Ocean." *Nature* 415 (February 6, 2002): 603–607.

McWilliams, Brendan. "Study of Plants Confirms Global Warming." *Irish Times*, November 1, 2001, 26.

Meagher, John. "Look What the Changing Climate Dragged in . . ." *Independent* (Dublin), July 9, 2004, n.p. (LEXIS).

"Mediterranean Drying." *Nature* 479 (November 17, 2011): 270.

Medvigy, David, et al. "Simulated Changes in Northwest U.S. Climate in Response to Amazon Deforestation." *Journal of Climate* 26 (2013): 9115–9136.

Meehl, Gerald A., and Claudia Tebaldi. "More Intense, More Frequent, and Longer Lasting Heat Waves in the 21st Century." *Science* 305 (August 13, 2004): 994–997.

Meek, James. "Wildflowers Study Gives Clear Evidence of Global Warming." *The Guardian* (U.K.), May 31, 2002, 6.

Meek, James. "Global Warming Gives Pests Taste for Life in London." *The Guardian* (U.K.), October 8, 2002, 6.

Meier, Mark F., and Mark B. Dyurgerov. "Sea-Level Changes: How Alaska Affects the World." *Science* 297 (July 19, 2002): 350–351.

"Melting Andean Glaciers Could Leave 30 Million High and Dry." Environment News Service, April 28, 2008. http://www.ens-newswire.com.

"Melting Antarctic Glacier Could Raise Sea Level." Reuters, July 24, 1998. http://benetton. dkrz.de:3688/homepages/georg/kimo/0254.html (no longer available).

"Melting Faster." *Science* 316(1) (May 18, 2007): 955.

"Melting Ice May Release Frozen Influenza Viruses." Environment News Service, November 27, 2006. http://www.ens-newswire.com/ens/nov2006/2006-11-27-09.asp#anchor3 (no longer available).

"Melting on the Greenland Ice Cap, 2008." NASA Earth Observatory, February 24, 2009. http://earthobservatory.nasa.gov/IOTD/view.php?id=37215.

"Melting Planet: Species Are Dying Out Faster Than We Have Dared Recognize, Scientists Will Warn This Week." *The Independent* (London), October 2, 2005. http://news.inde pendent.co.uk/world/environment/article316604.ece (no longer available).

Menzel, A., and P. Fabian. "Growing Season Extended in Europe." *Nature* 397 (1999): 659–662.

Merchant, Brian. "The Nation's Top Climate Scientist Predicts an 'Ice-Free, Human-Free' Planet." Motherboard, April 17, 2013. http://motherboard.vice.com/blog/the-nations -top-climate-scientist-predicts-an-ice-free-human-free-planet (no longer available).

Merzer, Martin. "Study: Global Warming Likely Making Hurricanes Stronger." *Miami Herald*, August 1, 2005, n.p. (LEXIS).

"Metro Vancouver Air Quality Comparable to Beijing: Health Authorities Advise Caution Due to Smoke from Hundreds of Wildfires Across B.C." CBC News, July 6, 2015. http://www.cbc.ca/news/canada/british-columbia/metro-vancouver-air-quality-comparable-to-beijing-1.3140735.

Meuvret, Odile. "Global Warming Could Turn Siberia into Disaster Zone: Expert." Agence France Presse, October 2, 2003.

"Mild Weather Zaps Sap of Maple-Syrup Farmers." *USA Today*, March 26, 2010, 4A.

Miles, B. W. J., et al. "Rapid, Climate-Driven Changes in Outlet Glaciers on the Pacific Coast of East Antarctica." *Nature* 500 (August 29, 2013): 563–566.

Miles, Paul. "Fiji's Coral Reefs Are Being Ruined by Bleaching." *Daily Telegraph* (London), June 2, 2001, 4.

Miller, Kenneth G., et al. "A Geological Perspective on Sea-Level Rise and Its Impacts along the U.S. Mid-Atlantic Coast." *Earth's Future* 1(1) (December 5, 2013). http://onlinelibrary.wiley.com/doi/10.1002/2013EF000135/abstract.

Miller, T., et al. "Impact of Global Climate Change on Urban Infrastructure." In J. B. Smith and D. A. Tirpak, eds., *Potential Effects of Global Climate Change on the United States: Appendix H, Infrastructure.* Washington, DC: U.S. Environmental Protection Agency, 1989.

Miller, Webb, et al. "Polar and Brow Bear Genomes Reveal Ancient Admixture and Demographic Footprints of Past Climate Change." *Proceedings of the National Academy of Sciences* 109(36) (September 4, 2012): E2382–E2390. doi: 10.1073/pnas.1210506109.

Milly, P. C., et al. "Increasing Risk of Great Floods in a Changing Climate." *Nature* 415 (January 30, 2002): 514–517.

Milmo, Cahal, and Elizabeth Nash. "Fish Farms Push Atlantic Salmon Towards Extinction." *The Independent* (London), June 1, 2001, 11.

Min, Seung-Ki, et al. "Human Contribution to More-Intense Precipitation Extremes." *Nature* 470 (February 17, 2011): 378–381.

Minard, Anne. "Methane 'Hot Spot' Seen from Space Hovers over Four Corners." Indian Country Today Media Network, October 28, 2014. http://indiancountrytodaymedianetwork.com/2014/10/28/methane-hot-spot-seen-space-hovers-over-four-corners-157560.

Minnis, Patrick, et al. "Contrails, Cirrus Trends, and Climate." *Journal of Climate*, April 5, 2004, 1671–1685.

Mitchell, John G. "Down the Drain: The Incredible Shrinking Great Lakes." *National Geographic*, September 2002, 34–51.

"MIT: Climate Change Odds Much Worse Than Thought." NASA Earth Observatory, May 19, 2009. http://earthobservatory.nasa.gov/Newsroom/view.php?id=38818&src=eoa-manews.

Mitrovica, Jerry X., Natalya Gomez, and Peter U. Clark. "The Sea-Level Fingerprint of West Antarctic Collapse." *Science* 323 (February 6, 2009): 753.

Mitrovica, Jerry X., and Maureen E. Raymo. "Collapse of Polar Ice Sheets during the Stage 11 Interglacial." *Nature* 483 (March 22, 2012): 453–456.

Mittelstaedt, Martin. "World Faces Perpetual Food Crisis: Study." *Toronto Globe and Mail*, January 8, 2009. http://www.theglobeandmail.com/servlet/story/RTGAM.20090108.wclimate0108/BNStory/International/home (no longer available).

Mohan, Vizhwa. " 'Historic' Climate Deal Done, Meets India's Key Concerns." *Sunday Times of India*, December 13, 2015, A1.

Monaghan, Andrew J., et al. "Insignificant Change in Antarctic Snowfall since the International Geophysical Year." *Science* 313 (August 11, 2006): 827–831.

Monastersky, Richard. "The Long Goodbye: Alaska's Glaciers Appear to Be Disappearing before Our Eyes. Are They a Sign of Things to Come?" *New Scientist*, April 14, 2001, 30–32.

Monastersky, Richard. "Seabed Scars Raise Questions over Carbon-Storage Plan." *Nature* 504 (December 19, 2013): 339–340. http://www.nature.com/news/seabed-scars-raise -questions-over-carbon-storage-plan-1.14386.

Monbiot, George. "We Are All Killers: Until We Stop Flying." *The Guardian* (U.K.), February 28, 2006. http://www.monbiot.com/archives/2006/02/28/we-are-all-killers/.

"Monsoon Wet Spells Get Wetter." *Nature* 509(May 1, 2014):11.

"Monster Iceberg Heads into Antarctic Waters." Agence France Presse, October 22, 2002 (LEXIS).

Montaigne, Fen. "The Ice Retreat: Global; Warming and the Adelie Penguin." *The New Yorker*, December 21 and 28, 2009, 72–82.

Montaigne, Fen. *Fraser's Penguins: A Journey to the Future of Antarctica*. New York: Henry Holt, 2010.

Moon, T., et al. "Twenty-First-Century Evolution of Greenland Outlet Glacier Velocities." *Science* 336 (May 4, 2012): 576–578.

Mooney, Chris. "Global Warming Is Now Slowing Down the Circulation of the Oceans— with Potentially Dire Consequences." *Washington Post*, March 24, 2015. http://www .washingtonpost.com/news/energy-environment/wp/2015/03/23/global-warming-is- now-slowing-down-the-circulation-of-the-oceans-with-potentially-dire-consequences/.

Mooney, Chris. "Climate Change Could Shrink Mount Everest's Glaciers by 70 Percent, Study Finds." *Washington Post*, May 27, 2015. http://www.washingtonpost.com/news/energy -environment/wp/2015/05/27/climate-change-could-shrink-glaciers-in-the-mount-ever est-region-by-70-percent-study-finds/.

Mooney, Chris. "Giant Earthquakes Are Shaking Greenland—and Scientists Just Figured Out the Disturbing Reason Why." *Washington Post*, June 25, 2015. http://www.washing tonpost.com/news/energy-environment/wp/2015/06/25/giant-earthquakes-are-shaking -greenland-and-scientists-just-figured-out-the-disturbing-reason-why/?wpisrc=nl_head lines&wpmm=1.

Mooney, Chris. "Alaska's Terrifying Wildfire Season and What It Says about Climate Change." *Washington Post*, July 26, 2015. http://www.washingtonpost.com/news/energy-environment /wp/2015/07/26/alaskas-terrifying-wildfire-season-and-what-it-says-about-climate-cha nge/?wpisrc=nl_headlines&wpmm=1.

Mooney, Chris. "Scientists Say a Dramatic Worldwide Coral Bleaching Event Is Now Under- way." *Washington Post*, October 8, 2015. http://www.washingtonpost.com/news/energy -environment/wp/2015/10/08/scientists-say-a-dramatic-worldwide-coral-bleaching-event -is-now-underway/?wpmm=1&wpisrc=nl_headlines.

Mooney, Chris. "What the Earth Will Be Like in 10,000 Years, According to Scientists." *Wash- ington Post*, February 10, 2016. https://www.washingtonpost.com/news/energy-environ ment/wp/2016/02/08/what-the-earth-will-be-like-in-10000-years-according-to -scientists/.

Mooney, Chris. "We Had All Better Hope These Scientists Are Wrong about the Planet's Future." *Washington Post*, March 22, 2016. https://www.washingtonpost.com/news/energy -environment/wp/2016/03/22/we-had-all-better-hope-these-scientists-are-wrong-about -the-planets-future/?wpmm=1&wpisrc=nl_evening.

Mooney, Chris. "World Bank: The Way Climate Change Is Really Going to Hurt Us Is through Water." *Washington Post*, May 3, 2016. https://www.washingtonpost.com/news/energy -environment/wp/2016/05/03/world-bank-the-way-climate-change-is-really-going-to -hurt-us-is-through-water/?wpmm=1&wpisrc=nl_evening.

Mooney, Chris, and Jason Samenow. "The North Pole Is an Insane 36 degrees Warmer Than Normal as Winter Descends." *Washington Post*, November 17, 2016. https://www.washingtonpost.com/news/energy-environment/wp/2016/11/17/the-north-pole-is-an-insane-36-degrees-warmer-than-normal-as-winter-descends/?utm_term=.4c41bc26bbbe&wpisrc=nl_headlines&wpmm=1.

Mooney, Chris, and Jody Warrick. "Research Casts Alarming Light on Decline of West Antarctic Glaciers." *Washington Post*, December 5, 2014. http://www.washingtonpost.com/national/health-science/research-casts-alarming-light-on-decline-of-west-antarctic-ice-sheets/2014/12/04/19efd3e4-7bbe-11e4-84d4-7c896b90abdc_story.html?wpisrc=nl-headlines&wpmm=1.

Moore, Molly. "In Northern France, Warming Presses Fall Grape Harvest into Summertime." *Washington Post*, September 2, 2007, A1. http://www.washingtonpost.com/wp-dyn/content/article/2007/09/01/AR2007090101360_pf.html.

Moore, Sam, et al. "Deep Instability of Deforested Tropical Peatlands Revealed by Fluvial Organic Carbon Fluxes." *Nature* 493 (January 31, 2013): 660–663.

"More Cyclones for Hawaiian Islands." *Nature* 497 (May 16, 2013): 290.

"More Rain in Ozone's Absence." *Nature* 498 (June 6, 2013): 9.

"More Smoke Can Mean Less Rain." NASA Earth Observatory. September 24, 2015. http://earthobservatory.nasa.gov/IOTD/view.php?id=86672&src=eoa-iotd.

Morello, Lauren. "Summer Storms Bolster Arctic Ice." *Nature* 500 (August 29, 2013): 512.

Morison, James, et al. "Changing Arctic Ocean Freshwater Pathways." *Nature* 481 (January 5, 2012): 66–70.

Moritz, Craig, and Rosa Agudo. "The Future of Species under Climate Change: Resilience or Decline?" *Science* 341 (August 2, 2013): 504–508.

Morrison, John, and Alex Sink. "The Climate Change Peril That Insurers See." *Washington Post*, September 27, 2007, A25. http://www.washingtonpost.com/wp-dyn/content/article/2007/09/26/AR2007092602070_pf.html.

Morton, Joseph. "Ethanol Rises above Partisanship in Iowa." *Omaha World-Herald*, November 11, 2013, 1A, 3A.

Morton, Oliver. "Is This What It Takes to Save the World?" *Nature* 447 (2007): 132–136.

Moss, Stephen. "Casualties." *The Guardian* (U.K.), April 26, 2001, 18.

"Most Serious Greenhouse Gas Is Increasing, International Study Finds." *Science Daily*, April 27, 2001. http://www.sciencedaily.com/releases/2001/04/010427071254.htm.

Mouawad, Jad. "Deal on Aviation Emissions Sets Can't-Miss Goals." *The New York Times*, February 16, 2016. http://www.nytimes.com/2016/02/16/business/energy-environment/a-hollow-agreement-on-aviation-emissions.html.

Mouawad, Jad. "Wind Power Grows 39 Percent for the Year." *The New York Times*, January 25, 2010, B1, B6.

Mouawad, Jad, and Diane Cardwell. "Farm Waste and Animal Fats Will Help Power a United Jet." *The New York Times*, June 30, 2015. http://www.nytimes.com/2015/06/30/business/energy-environment/farm-waste-and-animal-fats-will-help-power-a-united-jet.html.

Mouginot, J. E., E. Rignot, and B. Scheuchl. "Sustained Increase in Ice Discharge from the Amundsen Sea Embayment, West Antarctica, from 1973 to 2013." *Geophysical Research Letters* 41(5) (March 16, 2014): 1576–1584.

Mouginot, J., et al. "Fast Retreat of Zachariæ Isstrøm, Northeast Greenland." *Science*. Published online November 15, 2015. http://www.sciencemag.org/content/early/2015/11/11/science.aac7111.abstract?sid=47d39afa-a6a5-415e-918c-0735fab5c92f. doi: 10.1126/science.aac7111.

Moy, Andrew D., et al. "Reduced Calcification in Modern Southern Ocean Planktonic Foraminifera." *Nature GeoScience*, March 8, 2009 (online). doi: 10.1038/ngeo460.

Moyer, Justin W. "Drought-Stricken California's San Joaquin Valley Is Sinking, NASA Says." *Washington Post*, August 20, 2015. http://www.washingtonpost.com/news/morning-mix/wp/2015/08/20/drought-stricken-californias-san-joaquin-valley-is-sinking-nasa-says/?wpmm=1&wpisrc=nl_headlines.

"Much Deforestation Driven by Population, Poverty." *Global Futures Bulletin 84* (May 15, 1999). http://www.gsreport.com/articles/art000149.html.

Mudelsee, Mandred, et al. "No Upward Trends in the Occurrence of Extreme Floods in Central Europe." *Nature* 425 (September 11, 2003): 166–169.

Mufson, Steven. "Coal Rush Reverses, Power Firms Follow." *Washington Post*, September 4, 2007, D1. http://www.washingtonpost.com/wp-dyn/content/article/2007/09/03/AR2007090301119_pf.html.

Mufson, Steven. "Siphoning Off Corn to Fuel Our Cars." *Washington Post*, April 30, 2008, A1. http://www.washingtonpost.com/wp-dyn/content/article/2008/04/29/AR2008042903092_pf.html.

Mufson, Steven. "Power-Sector Emissions of China to Top U.S." *Washington Post*, August 27, 2008, D1. http://www.washingtonpost.com/wp-dyn/content/article/2008/08/26/AR2008082603096_pf.html.

Mufson, Steven. "The Car of the Future—But at What Cost?" *Washington Post*, November 25, 2008, A1. http://www.washingtonpost.com/wp-dyn/content/article/2008/11/24/AR2008112403211_pf.html.

Mufson, Steven. "China Steps Up, Slowly but Surely." *Washington Post*, October 24, 2009. http://www.washingtonpost.com/wp-dyn/content/article/2009/10/23/AR2009102304075_pf.html.

Muller, Phillip. "Pacific Islands' Deadly Threat from Climate Change." *Washington Post*, May 30, 2013. http://www.washingtonpost.com/opinions/pacific-islands-face-a-deadly-threat-from-climate-change/2013/05/30/86ff1956-c7a9-11e2-9245-773c0123c027_print.html.

Muller, Richard A. "The Truth about Tornadoes." *The New York Times*, November 20, 2013. http://www.nytimes.com/2013/11/21/opinion/the-truth-about-tornadoes.html.

Mulvaney, Robert, et al. "Recent Antarctic Peninsula Warming Relative to Holocene Climate and Ice-Shelf History." *Nature* 489 (September 6, 2012): 141–144.

Muñoz, Nicolas J., et al. "Adaptive Potential of a Pacific Salmon Challenged by Climate Change." *Nature Climate Change* 5 (February 2015): 163–166. doi: 10.1038/nclimate2473.

Munro, Margaret. "Global Warming Affecting Squirrels' Genes, Study Finds: Research in Yukon: 'Phenomenal Change' Seen as Rodents Breed Earlier." *National Post* (Canada), February 12, 2003, A2.

Munro, Margaret. "Puffin Colony Threatened by Warming: A Few Degrees Can Be Devastating. Thousands of Triangle Island Chicks Die When Heat Drives Off Their Favoured Fish." *Montreal Gazette*, July 15, 2003, A12.

Munro, Margaret. "Earth's 'Big Burp' Triggered Warming: Prehistoric Release of Methane a Cautionary Tale for Today." *Edmonton Journal*, June 3, 2004, A10.

Murphy, Kim. "Front-Row Exposure to Global Warming; Climate: Engineers Say Alaskan Village Could Be Lost as Sea Encroaches." *Los Angeles Times*, July 8, 2001, A1.

Murray, Tavi. "Greenland's Ice on the Scales." *Nature* 443 (September 21, 2006): 277–278.

Murray, T., et al. "Reverse Glacier Motion during Iceberg Calving and the Cause of Glacial Earthquakes." *Science* (online), June 26, 2015. http://science.sciencemag.org/content/early/2015/06/24/science.aab0460. doi: 10.1126/science.aab0460.

Mydans, Seth. "Reports from Four Fronts in the War on Warming." *The New York Times*, April 3, 2007. http://www.nytimes.com/2007/04/03/science/earth/03clim.html.

Mydans, Seth. "Vietnam Finds Itself Vulnerable If Sea Rises." *The New York Times*, September 24, 2009. http://www.nytimes.com/2009/09/24/world/asia/24delta.html.

Myers, Samuel S., et al. "Increasing CO_2 Threatens Human Nutrition." *Nature* 510 (June 5, 2014): 139–142.

Nagourney, Adam. "As California Drought Enters 4th Year, Conservation Efforts and Worries Increase." *The New York Times*, March 17, 2015. http://www.nytimes.com/2015/03/18/us/as-california-drought-enters-4th-year-conservation-efforts-and-worries-increase.html.

Naik, Gautam. "Global Warming May Be Spurring Allergy, Asthma." *Wall Street Journal*, May 3, 2007, A1, A13. https://www.wsj.com/articles/SB117815689129890415.

Naik, Gautam. "Arctic Becomes Tourism Hot Spot, but Is It Cool?" *Wall Street Journal*, September 24, 2007, A1, A12.

Naik, Gautam. "Warmer Climate Gives Cheer to Makers of British Bubbly." *Wall Street Journal*, May 11, 2010, A1, A18.

Naik, Gautam, and Geraldo Samor. "Drought Spotlights Extent of Damage in Amazon Basin." *Wall Street Journal*, October 21, 2005, A12.

Naish, T., et al. "Orbitally Induced Oscillations in the East Antarctic Ice Sheet at the Oligocene/Miocene Boundary." *Nature* 413 (October 18, 2001): 719–723.

Naish, T., et al. "Obliquity-Paced Pliocene West Antarctic Ice Sheet Oscillations." *Nature* 457 (March 19, 2009): 322–325.

Najar, Nida. "More Rescues as Rain Hits South India." *The New York Times*, December 4, 2015, A14.

Najar, Nida. "Streets, If Not the Air, Clear Out as Delhi Tests Car Restrictions." *The New York Times*, January 5, 2016, A4.

"NASA Data Show Some African Drought Linked to Warmer Indian Ocean." NASA Earth Observatory, August 5, 2008. http://earthobservatory.nasa.gov/Newsroom/NasaNews/2008/2008080527314.html (no longer available).

"NASA Finds Warmer Ocean Speeding Greenland Glacier Melt." NASA Earth Observatory, February 16, 2009. http://earthobservatory.nasa.gov/Newsroom/view.php?id=42840&src=eoa-nnews (no longer available).

"NASA: Recent Mideast Drought Worst in 900 Years." *Omaha World-Herald*, March 4, 2016, 7A.

"NASA Researchers Find Snowmelt in Antarctica Creeping Inland." NASA Earth Observatory, September 20, 2007. https://www.nasa.gov/centers/goddard/news/topstory/2007/antarctic_snowmelt.html.

"NASA Study Finds Carbon Emissions Could Dramatically Increase Risk of U.S. Megadroughts." NASA press release, February 12, 2015. http://www.nasa.gov/press/2015/february/nasa-study-finds-carbon-emissions-could-dramatically-increase-risk-of-us/#.VN3kU8Ysou0.

"NASA Study Finds World Warmth Edging Ancient Levels." NASA, September 25, 2006. http://www.giss.nasa.gov/research/news/20060925.

"NASA Study Links Severe Storm Increases, Global Warming." NASA Earth Observatory, December 19, 2008. http://earthobservatory.nasa.gov/Newsroom/view.php?id=36309.

"NASA Study Says Climate Adds Fuel to Asian Wildfire Emissions." NASA Earth Observatory, April 30, 2009. http://earthobservatory.nasa.gov/Newsroom/view.php?id=38433&src=eoa-nnews.

"NASA Study Shows Role of Melt in Arctic Sea Ice Loss." NASA press release, November 9, 2010. http://www.jpl.nasa.gov/news/news.php?feature=2811.

"NASA: Vast Areas of West Antarctica Melted in 2005." Environment News Service, May 28, 2007. http://www.ens-newswire.com/ens/may2007/2007-05-28-09.asp#anchor1 (no longer available).

National Academy of Sciences. *Policy Implications of Greenhouse Warming.* Washington, DC: National Academy Press, 1991.

National Climate Assessment. Washington, DC: U.S. Government Printing Office, January 2013. http://nca2014.globalchange.gov/.

National Science Foundation. "Jellyfish Gone Wild!!" Arlington, VA: NSF, no date. https://www.nsf.gov/news/special_reports/jellyfish/textonly/index.jsp.

"The Navajo Memory Complements Science in Study of Climate Change." USGS Newsroom. October 21, 2011. http://www.usgs.gov/newsroom/article.asp?ID=3007&from=rss#.VAb2Wmv29vZ (no longer available).

Navarro, Mireya. "Environment Blamed in Western Tree Deaths." *The New York Times,* January 23, 2009. http://www.nytimes.com/2009/01/23/us/23trees.html.

Navarro, Mireya. "New York Is Lagging as Seas and Risks Rise, Critics Warn." *The New York Times,* September 10, 2012. http://www.nytimes.com/2012/09/11/nyregion/new-york-faces-rising-seas-and-slow-city-action.htm (no longer available).

Nepstad, Daniel C. *Report from the Amazon.* Fallmouth, MA: Woods Hole Research Center, May 1988. http://terra.whrc.org/science/tropfor/fire/report2.htm (no longer available).

Nesmith, Jeff. "Antarctic Glacier Melt Increases Dramatically." *Atlanta Journal-Constitution,* September 22, 2004, 9A.

"New Ice Island at Pine Island Glacier." NASA Earth Observatory, July 28, 2013, http://earthobservatory.nasa.gov/IOTD/view.php?id=81674&src=eoa-iotd.

"New Melt Record for Greenland Ice Sheet." NASA Earth Observatory, January 21, 2011. http://www.eurekalert.org/pub_releases/2011-01/ccon-nmr012111.php.

"New Research Links Global Warming to Wildfires Across the West." *Omaha World-Herald,* November 5, 2004, 11A.

"New Study: Heat Is Being Stored Beneath the Ocean Surface." NASA Earth Observatory, July 10, 2015. http://earthobservatory.nasa.gov/IOTD/view.php?id=86184&src=eoa-iotd.

"New Wave of Bleaching Hits Coral Reefs Worldwide." Environment News Service, October 29, 2002. http://ens-news.com/ens/oct2002/2002-10-29-19.asp#anchor1 (no longer available).

"New York Lakes Fail to Freeze." Environment News Service, March 21, 2002. http://ens-news.com/ens/mar2002/2002L-03-21-09.html#anchor1.

Nicholls, Neville. "The Changing Nature of Australian Droughts." *Climatic Change* 63 (2004): 323–336.

Nickels, S., et al. *Unikkaaqatigiit: Putting the Human Face on Climate Change: Perspectives from Inuit in Canada.* Quebec City, Quebec: Inuit Tapiriit Kanatami, Nasivvik Centre for Inuit Health and Changing Environments at Université Laval and the Ajunnginiq Centre at the National Aboriginal Health Organization, Ottawa, 2005. https://www.itk.ca/sites/default/files/unikkaaqatigiit01_0.pdf.

Nieves, Veronica, Josh K. Willis, and William C. Patzert. "Recent Hiatus Caused by Decadal Shift in Indo-Pacific Heating." *Science,* July 9, 2015. doi: 10.1126/science.aaa4521.

Nijhuis, Michelle. "Forest Fires: Burn Out." *Nature* 489 (September 20, 2012): 352–354. http://www.nature.com/news/forest-fires-burn-out-1.11424.

Nijhuis, Michelle. "Can Coal Ever Be Clean?" *National Geographic,* April 2014, 28–61.

Nikiforuk, Andrew. "A Forest Threatened by Keystone XL." *The New York Times,* November 18, 2014. http://www.nytimes.com/2014/11/18/opinion/a-forest-threatened-by-keystone-xl.html.

Nisbet, E. G. "The End of the Ice Age." *Canadian Journal of Earth Sciences* 27(1) (1990): 148–157.

Nisbet, E. G., and J. Chappellaz. "Shifting Gear, Quickly." *Science* 324 (April 24, 2009): 477–478.

"No Magic Fix for Carbon." *Nature* 509 (May 1, 2014): 7.

Normile, Dennis. "Some Coral Bouncing Back from El Niño." *Science* 288 (May 12, 2000): 941–942.

Normile, Dennis. "Hard Summer for Corals Kindles Fears for Survival of Reefs." *Science* 329 (August 27, 2010): 1001.

Normile, Dennis. "Clues to Supertyphoon's Ferocity Found in the Western Pacific." *Science* 342 (November 29, 2013): 1027.

North, Gerald R., Jurgen Schmandt, and Judith Clarkson. *The Impact of Global Warming on Texas.* Austin: University of Texas Press, 1995.

"Northeast Storms Growing More Fierce." *USA Today*, April 6, 2010, 3A.

"North Pole Had Sub-Tropical Seas Because of Global Warming." Agence France Presse, September 7, 2004 (LEXIS).

"Northwest Passage Nearly Open." NASA Earth Observatory, August 27, 2007. https://archive.org/details/npseaice_amsre_2007234.

Nosengo, Niccola. "Venice Floods: Save Our City!" *Nature* 424 (August 7, 2003): 608–609.

"No Sugar-Beet Answer." *Omaha World-Herald*, April 7, 2007, 6B.

Nowak, Rachel. "The Continent That Ran Dry." *The New Scientist*, June 16, 2007, 8–11.

Nowakowski, Kelsey. "When Food Is Fuel." *National Geographic*, October 2014, 18–19.

Nunes, Flavia, and Richard D. Norris. "Abrupt Reversal in Ocean Overturning during the Palaeocene/Eocene Warm Period." *Nature* 439 (January 5, 2006): 60–63.

Nuttall, Nick. "Global Warming 'Will Turn Rainforests into Deserts.'" *The Times* (London). November 3, 1998. http://bonanza.lter.uaf.edu/~davev/nrm304/glbxnews.htm (no longer available).

Nuttall, Nick. "Climate Change Lures Butterflies Here Early." *The Times* (London), May 24, 2000, n.p.

Nuttall, Nick. "Coral Reefs 'On the Edge of Disaster.'" *The Times* (London), October 25, 2000, n.p. (LEXIS).

Nyberg, Johan, et al. "Low Atlantic Hurricane Activity in the 1970s and 1980s Compared to the Past 270 Years," *Nature* 447 (June 7, 2007): 698–701.

"A OBLS Personer: Bjorn Lomborg." Danmarks Radio. June 12, 2007. http://www.dr.dk/obls/personer-lomborg.html (no longer available).

O'Brien, S. T., B. P. Hayden, and H. H. Shugart. "Global Climatic Change, Hurricanes, and a Tropical Forest." *Climatic Change* 22 (1992): 175–190.

"Ocean Acidification (Effects on Marine Animals: Fish)—Summary." *CO₂ Science*. Ocean Acidification Database. November 14, 2014. http://www.co2science.org/subject/o/summaries/acidfish.php.

"Ocean Warming Underestimated." *Nature* 514 (October 9, 2014): 142–143.

O'Connell, Sanjida. "Power to the People." *Times* (London), May 20, 2002, n.p. (LEXIS).

Ogle, Andy. "Squirrels Get Squirrelier Earlier: Climate Change to Blame. Breeding Season in Yukon Advances 18 Days in Decade." *Montreal Gazette*, February 12, 2003, A12.

O'Harra, Doug. "Marine Parasite Infects Yukon River King Salmon; Fish Are Left Inedible; Scientists Study Overall Impacts." *Anchorage Dispatch News*, January 28, 2004, A1.

"Oldest Arctic Sea Ice Is Disappearing." NASA Earth Observatory, March 1, 2012. http://earthobservatory.nasa.gov/IOTD/view.php?id=77270&src=eoa-iotd.

Oliver, Thomas A., and Stephen R. Palumbi. "Distributions of Stress-Resistant Coral Symbionts Match Environmental Patterns at Local but Not Regional Scales." *Marine Ecology Progress Series*, March 2009, 93–103. doi: 10.3354/meps07871.

O'Malley, Brendan. "Global Warming Puts Rainforest at Risk." *Courier-Mail* (Cairns, Australia), July 24, 2003, 14.

O'Neill, Graeme. "The Heat Is On." *Sunday Herald-Sun* (Sydney, Australia), March 31, 2002.

"One in Five ExxonMobil Shareholders Want Climate Action." Environment News Service. May 28, 2003. http://ens-news.com/ens/may2003/2003-05-28-09.asp#anchor3 (no longer available).

Onishi, Normimitsu, and Malia Wollanjan. "Severe Drought Grows Worse in California." *The New York Times*, January 17, 2014. http://www.nytimes.com/2014/01/18/us/as-californias-drought-deepens-a-sense-of-dread-grows.html.

Onyanga-Omara, Jane. "At Least 16 Killed in French Riviera Flash Floods." *USA Today*, October 5, 2015, 6A.

Oppenheimer, Michael, and Robert H. Boyle. *Dead Heat: The Race against the Greenhouse Effect*. New York: Basic Books, 1990.

O'Reilly, Catherine M., et al. "Climate Change Decreases Aquatic Ecosystem Productivity of Lake Tanganyika, Africa." *Nature* 424 (August 14, 2003): 766–768.

Oreskes, Naomi, and Erik M. Conway. *Merchants of Doubt*. New York: Bloomsbury Press, 2010.

Orr, James C., et al. "Anthropogenic Ocean Acidification over the Twenty-First Century and Its Impact on Calcifying Organisms." *Nature* 437 (September 29, 2005): 681–686.

Orsi, Alejandro H., et al. "Cooling and Ventilating the Abyssal Ocean." *Geophysical Research Letters* 2(15) (August 1, 2001): 2923–2926.

"Our Changing Climate." National Climate Assessment. United States. 2014. http://nca2014.globalchange.gov/highlights/report-findings/our-changing-climate.

Overbye, Dennis. "Looking to Mars to Help Understand Changing Climates." *The New York Times*, December 8, 2014. http://www.nytimes.com/2014/12/09/science/looking-to-mars-to-help-understand-changing-climates.html.

"Over 80 Percent of Indonesia's Coral Reefs under Threat." *Jakarta Post*, September 13, 2001, n.p. (LEXIS).

Overpeck, Jonathan T., et al. "Paleoclimatic Evidence for Future Ice-Sheet Instability and Rapid Sea-Level Rise." *Science* 311 (March 24, 2006): 1747–1750.

Overpeck, Jonathan, and Bradley Udall. "Dry Times Ahead." *Science* (June 25, 2010): 1642–1643.

Overpeck, J. T., et al. "Arctic System on Trajectory to New, Seasonally Ice-Free State." *Eos* 86(34) (August 23, 2005): 309, 312.

Owen, David. "Where the River Runs Dry." *The New Yorker*, May 25, 2015, 52–63.

"Ozone Hole Fans African Heat." *Nature* 502 (October 17, 2013): 275.

"Ozone Hole 2013." NASA Earth Observatory. October 26, 2013. http://earthobservatory.nasa.gov/IOTD/view.php?id=82235&src=eoa-iotd.

"Ozone Hole 2015." NASA Earth Observatory. October 27, 2015. http://earthobservatory.nasa.gov/IOTD/view.php?id=86869&src=eoa-iotd.

"Ozone Loss Reaches New Record." Environment News Service, October 2, 2006. http://www.ens-newswire.com/ens/oct2006/2006-10-02-01.asp (no longer available).

Pacala, S., and R. Socolow. "Stabilization Wedges: Solving the Climate Problem for the Next 50 Years with Current Technologies." *Science* 305 (August 13, 2004): 968–972.

"Pacific Coast Turning More Acidic." Earth Observatory. May 22, 2008. http://earthobservatory.nasa.gov/Newsroom/MediaAlerts/2008/2008052226903.html (no longer available).

Padma, T. V. "Himalayan Plants Seek Cooler Climes." *Nature* 512 (August 28, 2014): 359. http://www.nature.com/news/himalayan-plants-seek-cooler-climes-1.15771.

Pagani, Mark, et al. "The Role of Carbon Dioxide during the Onset of Antarctic Glaciation." *Science* 334 (December 2, 2011): 1261–1264.

Page, Susan E., et al. "The Amount of Carbon Released from Peat and Forest Fires in Indonesia during 1997." *Nature* 420 (November 7, 2002): 61–65.

Pal, Jeremy S., and Elfatih A.B. Eltahir. "Future Temperature in Southwest Asia Projected to Exceed a Threshold for Human Adaptability." *Nature Climate Change*, November 2015. http://www.nature.com/nclimate/journal/vaop/ncurrent/full/nclimate2833.html.

Pala, Christopher. "Warming May Not Swamp islands." *Science* 345 (August 1, 2014): 496–497.

Pala, Christopher. "Corals Tie Stronger El Niños to Climate Change." *Science* 354 (December 9, 2016): 1210.

Pall, Pardeep, et al. "Anthropogenic Greenhouse Gas Contribution to Flood Risk in England and Wales in Autumn 2000." *Nature* 470 (February 17, 2011): 382–385.

Palmer, Brian. "The Bearded Ladies: Climate Change May Eliminate Male Bearded Dragons." Natural Resources Defense Council, July 2015. http://www.onearth.org/earthwire/climate-change-bearded-dragons-sex-determination.

Palmer, M. A., et al. "Mountaintop Mining Consequences." *Science* 327 (January 8, 2010): 148–149.

Palmer, T. N., and J. Ralsanen. "Quantifying the Risk of Extreme Seasonal Precipitation Events in a Changing Climate." *Nature* 415 (January 30, 2002): 512–514.

Paolo, Fernando S., Helen A. Fricker, and Laurie Padman. "Volume Loss from Antarctic Ice Shelves Is Accelerating." *Science* (online), March 26, 2015. http://science.sciencemag.org/content/early/2015/03/25/science.aaa0940. doi: 10.1126/science.aaa0940.

Papiez, Chelsie. "Climatic Change in the Quileute and Hoh Nations of Coastal Washington." Pp. 68–88 in Zoltan Grossman and Alan Parker, eds., *Asserting Native Resilience: Pacific Rim Indigenous Nations Face the Climate Crisis*. Corvallis, OR: Oregon State University Press, 2012.

Paraskevas, Joe. "Glaciers in the Canadian Rockies Shrinking to Their Lowest Level in 10,000 Years." *National Post* (Canada), December 4, 2003, A8.

Park, Haeyoun, Damien Cave, and Wilson Andrews. "After Years of Drought, Wildfires Rage in California." *The New York Times*, July 15, 2015. http://www.nytimes.com/interactive/2015/07/15/us/california-fire-season-drought.html?hp&action=click&pgtype=Homepage&module=second-column-region®ion=top-news&WT.nav=top-news.

Parker, Cindy L., and Steven M. Shapiro. *Climate Chaos: Your Health at Risk: What You Can Do to Protect Yourself and Your Family*. Westport, CT: Praeger, 2008.

Parker, Laura. "Treading Water." *National Geographic*, February 2015, 106–128.

Parkinson, C. L., and D. J. Cavalieri. "Antarctic Sea Ice Variability and Trends, 1979–2010." *The Cryosphere* 6 (August 15, 2012): 871–880.

Parkinson, Giles. "Citicorp Says the 'Age of Renewables' Has Begun'" *Renew Economy*, March 27, 2014. http://reneweconomy.com.au/2014/citigroup-says-the-age-of-renewables-has-begun-69852.

Parmesan, Camille, et al. "Poleward Shifts in Geographical Ranges of Butterfly Species Associated with Regional Warming." *Nature* 399 (June 10, 1999): 579–583.

Parry, Martin. *Climate Change and World Agriculture*. London: Earthscan, 1990.

Parry, Martin, and Zhang Jiachen. "The Potential Effect of Climate Changes on Agriculture." Pp. 279–289 in J. Jager and H. L. Ferguson, *Climate Change: Science, Impacts, and Policy*. Proceedings of the Second World Climate Conference. Cambridge, UK: Cambridge University Press, 1991.

Paterson, W. S. B., and N. Reeh. "Thinning of the Ice Sheet in Northwest Greenland over the Past Forty Years." *Nature* 414 (November 1, 2001): 60–62.

Paulson, Henry M. "The Coming Climate Crash: Lessons for Climate Change in the 2008 Recession." *The New York Times*, June 21, 2014. http://www.nytimes.com/2014/06/22/opinion/sunday/lessons-for-climate-change-in-the-2008-recession.html.

Payne, A. J., et al. "Recent Dramatic Thinning of Largest West Antarctic Ice Stream Triggered by Oceans." *Geophysical Research Letters* 31(23) (December 9, 2004): L23401. Paper no. 10.1029/2004GL021284.

Paytan, Adina, et al. "Methane Transport from the Active Layer to Lakes in the Arctic Using Toolik Lake, Alaska, as a Case Study." *Proceedings of the National Academy of Sciences* March 9, 2015. http://www.pnas.org/content/early/2015/03/05/1417392112.abstract?sid=6bb09f2b-6806-46e4-981e-ade09ad31498.

PBS News Hour. "Native Lands Wash Away as Sea Levels Rise." Louisiana Public Broadcasting (co-producer), June 1, 2012. http://www.pbs.org/newshour/updates/climate-change/jan-june12/louisianacoast_05-30.html (no longer available).

Pearce, Fred. "Nature Plants Doomsday Devices." *The Guardian* (U.K.), November 25, 1998. http://go2.guardian.co.uk/science/912000568-disast.html.

Pearce, Fred. "Failing Ocean Current Raises Fears of Mini Ice Age." *New Scientist*, November 30, 2005. https://www.newscientist.com/article/dn8398-failing-ocean-current-raises-fears-of-mini-ice-age/.

Pearce, Fred. *With Speed and Violence: Why Scientists Fear Tipping Points in Climate Change.* Boston: Beacon Press, 2007.

Pearce, Fred. "But Here's What They Didn't Tell Us." *New Scientist*, February 10–16, 2007, 6–9.

Pearson, Paul N., et al. "Warm Tropical Sea-Surface Temperatures in the Late Cretaceous and Eocene Epochs." *Nature* 413 (October 4, 2001): 481–487.

Pederson, Gregory T., et al. "The Unusual Nature of Recent Snowpack Declines in the North American Cordillera." *Science* 15 (July 2011): 332–335.

Pegg, J. R. "The Earth Is Melting, Arctic Native Leader Warns." Environment News Service, September 16, 2004. http://www.ens-newswire.com/ens/sep2004/2004-09-16-10.asp.

Pegg, J. R. "Warming Climate Adds to U.S. Flood Fears." Environment News Service, July 2, 2008. http://www.ens-newswire.com/ens/jul2008/2008-07-02-10.asp (no longer available).

Pelly, David F. *The Sacred Hunt: A Portrait of the Relationship Between Seals and Inuit.* Vancouver, BC: Greystone, 2001.

Peng, Shaobing, et al. "Rice Yields Decline with Higher Night Temperature from Global Warming." *Proceedings of the National Academy of Sciences* 101(27) (July 6, 2004): 9971–9975.

Pennisi, Elizabeth. "Calcification Rates Drop in Australian Reefs." *Science* 323 (January 2, 2009): 27.

Pennisi, Elizabeth. "Western U.S. Forests Suffer Death by Degrees." *Science* 323 (January 23, 2009): 447.

Perlman, David. "Decline in Oceans' Phytoplankton Alarms Scientists." *San Francisco Chronicle*, October 6, 2003, A6.

Perlman, David. "Shrinking Glaciers Evidence of Global Warming; Differences Seen by Looking at Photos from 100 Years Ago." *San Francisco Chronicle*, December 17, 2004, A18.

Perry, Michael. "Global Warming Devastates World's Coral Reefs." Reuters, November 26, 1998. http://www.gsreport.com/articles/art000023.html (no longer available).

Pershing, Andrew J., et al. "Slow Adaptation in the Face of Rapid Warming Leads to Collapse of the Gulf of Maine Cod Fishery." *Science* 350 (November 13, 2015): 745.

Peters, Robert L. "Effects of Global Warming on Biological Diversity." Pp. 82–95 in Edwin Abrahamson, ed., *The Challenge of Global Warming*. Washington, DC: Island Press, 1989.

Peterson, Bruce J., et al. "Increasing River Discharge to the Arctic Ocean." *Science* 298 (December 13, 2002): 2171–2173.

Petit, Charles W. "Polar Meltdown: Is the Heat Wave on the Antarctic Peninsula a Harbinger of Global Climate Change?" *U.S. News and World Report*, February 28, 2000, 64–74.

Petit, Charles. "In the Rockies, Pines Die and Bears Feel It." *The New York Times*, January 30, 2007. http://www.nytimes.com/2007/01/30/science/30bear.html.

Petrenko, Vasilii V., et al. "$^{14}CH_4$ Measurements in Greenland Ice: Investigating Last Glacial Termination CH_4 Sources." *Science* 324 (April 24, 2009): 506–508.

Petrillo, Lisa. "Turning the Tide in Venice." Copley News Service April 28, 2003 (LEXIS).

"Phew, What a Scorcher—and It's Going to Get Worse." Agence France Presse, December 1, 2004 (LEXIS).

"Philippines Climate Negotiator Vows Hunger Strike." Environment News Service, November 12, 2013. http://ens-newswire.com/2013/11/12/philippines-climate-negotiator-vows -hunger-strike/.

"Philippine Typhoon Leaves up to 10,000 Dead in Tacloban City." Associated Press in Canadian Broadcasting Corp., November 11, 2013. http://www.cbc.ca/news/world/philip pine-typhoon-leaves-up-to-10-000-dead-in-tacloban-city-1.2421580.

Phillips, Oliver L., et al. "Drought Sensitivity of the Amazon Rainforest." *Science* 323 (March 6, 2009): 1344–1347.

Phrampus, Benjamin J., and Matthew J. Hornbach. "Recent Changes to the Gulf Stream Causing Widespread Gas Hydrate Destabilization." *Nature* 490 (October 25, 2012): 527–530.

Pianin, Eric. "A Baltimore without Orioles? Study Says Global Warming May Rob Maryland, Other States of Their Official Birds." *Washington Post*, March 4, 2002, A3.

Pianin, Eric. "Study Fuels Worry over Glacial Melting; Research Shows Alaskan Ice Mass Vanishing at Twice Rate Previously Estimated." *Washington Post*, July 19, 2002, A14.

Pielke, Roger. "Land Use Changes and Climate Change." *Philosophical Transactions: Mathematical, Physical & Engineering Sciences* (Journal of the Royal Society of London), August 2002.

Pilford, Nicholas W., et al. "Migratory Response of Polar Bears to Sea Ice Loss: To Swim or Not to Swim." *Ecography*, April 14, 2016. http://onlinelibrary.wiley.com/doi/10.1111/ ecog.02109/abstract.

Pilkey, Orrin H., and Andrew G. Cooper. "Society and Sea Level Rise." *Science* 303 (March 19, 2004): 1781–1782.

Pilkey, Orrin H., Linda Pilkey-Jarvis, and Keith C. Pilkey. *Retreat from a Rising Sea: Hard Choices in an Age of Climate Change*. New York: Columbia University Press, 2016.

Pimm, Stuart, and Jeff Harvey. "No Need to Worry about the Future." *Nature* 414 (November 8, 2001): 149.

Pincock, Steve. "Showdown in a Sun-burnt Country." *Nature* 450 (November 15, 2007): 336–338.

"Pine Beetles Changing Rocky Mountain Air Quality, Weather," Environment News Service, October 1, 2008. http://www.ens-newswire.com/ens/oct2008/2008-10-01-091.asp (no longer available).

Pleven, Liam, Ian McDonald, and Karen Richardson. "As Hurricane Season Starts, Disaster Insurance Runs Short." *Wall Street Journal*, July 10, 2006, A-1, A-8.

Plumer, Brad. "CBO: Carbon Capture Efforts Aren't Going So Well." *Washington Post*, July 2, 2012. https://www.washingtonpost.com/pb/news/wonk/wp/2012/07/02/cbo-carbon

-capture-efforts-arent-going-so-well/?outputType=accessibility&nid=menu_nav_accessib ilityforscreenreader.

Poggioli, Sylvia. "Venice Struggling with Increased Flooding." National Public Radio Morning Edition, November 29, 2002 (LEXIS).

Pohl, Otto. "New Jellyfish Problem Means Jellyfish Are Not the Only Problem." *The New York Times*, May 21, 2002, F3.

Polakovic, Gary. "Deforestation Far Away Hurts Rain Forests, Study Says; Downing Trees on Costa Rica's Coastal Plains Inhibits Cloud Formation in Distant Peaks." *Los Angeles Times*, October 19, 2001, A1.

Polidoro, Beth, and Kent Carpenter. "Dynamics of Coral Reef Recovery." *Science* 340 (April 5, 2013): 34–35.

Pollack, Henry. *A World Without Ice*. London: Avery/Penguin, 2009.

Pollard, David, and Robert M. DeConto. "Modelling West Antarctic Ice Sheet Growth and Collapse through the Past Five Million Years." *Nature* 457 (March 19, 2009): 329–332.

Pollard, David, Robert M. DeConto, and Richard B. Alley. "Potential Antarctic Ice Sheet Retreat Driven by Hydrofracturing and Ice Cliff Failure." *Earth and Planetary Science Letters* 412 (February 15, 2015): 112–121.

"Polynyas and the Pine Island Glacier, Antarctica." NASA Earth Observatory, November 18, 2011. http://earthobservatory.nasa.gov/IOTD/view.php?id=76437&src=eoa-iotd.

Porter, Eduardo. "Old Forecast of Famine May Yet Come True." *The New York Times*, April 1, 2014. https://www.nytimes.com/2014/04/02/business/energy-environment/a-200-year -old-forecast-for-food-scarcity-may-yet-come-true.html.

Portner, Hans O., and Rainer Knust. "Climate Change Affects Marine Fishes through the Oxygen Limitation of Thermal Tolerance." *Science* 315 (January 5, 2007): 95–97.

"Portugal Celebrates Massive Solar Plant." *The New York Times*, March 28, 2007. http:// www.nytimes.com/aponline/technology/AP-Portugal-Solar-Power-Plant.html (no longer available).

Post, Eric, et al. "Ecological Consequences of Sea-Ice Decline." *Science* 341 (August 2, 2013): 519–524.

Potter, Thomas D., and Bradley R. Colman, eds. *Handbook of Weather, Climate, and Water: Dynamics, Climate, Physical Meteorology, Weather Systems, and Measurements*. Hoboken, NJ: Wiley-InterScience, 2003.

Pounds, J. Alan. "Climate and Amphibian Decline." *Nature* 410 (April 5, 2001): 639–640.

Pounds, J. Alan, and Karen L. Masters. "Amphibian Mystery Misread." *Nature* 462 (November 5, 2009): 38–39.

Pounds, J. Alan, and Robert Puschendorf. "Clouded Futures." *Nature* 427 (January 8, 2004): 107–108.

Pounds, J. Alan, et al. "Widespread Amphibian Extinctions from Epidemic Disease Driven by Global Warming." *Nature* 439 (January 12, 2006): 161–167.

Powell, James Lawrence. *The Inquisition of Climate Science*. New York: Columbia University Press, 2011.

Povoledo, Elisabetta. "Scientists and Religious Leaders Discuss Climate Change at Vatican." *The New York Times*, April 29, 2015. http://www.nytimes.com/2015/04/29/world/europe/ scientists-and-religious-leaders-discuss-climate-change-at-vatican.html.

Powlson, David. "Will Soil Amplify Climate Change?" *Nature* 433 (January 20, 2005): 204–205.

"Predicting the Indian Monsoon." *Nature* 484 (April 19, 2012): 291.

"Prehistoric Extinction Linked to Methane." *Omaha World-Herald*, July 27, 2000, 9.

Prein, Andreas F., et al. "The Future Intensification of Hourly Precipitation Extremes." *Nature Climate Change*, December 2016. http://www.nature.com/nclimate/journal/vaop/ncur rent/full/nclimate3168.html.

"Prepare Farms for the Future: Scientists Must Work Closely with Farmers to Ensure That Agriculture Can Stand Up to the Ravages of Climate Change." *Nature* 523 (July 23, 2015): 381. http://www.nature.com/news/prepare-farms-for-the-future-1.18018.

Prigg, Mark. "Despite All the Heavy Rain, That Was the Hottest June for 28 Years." *Evening Standard* (London), July 1, 2004, A9.

"Prince of Wales Warns Climate Crash Could Dwarf Financial Crisis." Environment News Service, May 24, 2011. http://www.ens-newswire.com/ens/may2011/2011-05-24-01.html.

Pringle, Heather. "Did Pulses of Climate Change Drive the Rise and Fall of the Maya?" *Science* 338 (November 9, 2012): 730–731.

Pritchard, H. D., et al. "Antarctic Ice-Sheet Loss Driven by Basal Melting of Ice Shelves." *Nature* 484 (April 26, 2012): 502–505.

Putkonen, J. K, and G. Roe. "Rain-on-Snow Events Impact Soil Temperatures and Affect Ungulate Survival." *Geophysical Research Letters* 30(4) (2003): 1188–1192.

Qiu, Jane. "Measuring the Meltdown." *Nature* 468 (November 11, 2010): 141–142.

Qiu, Jane. "Slip Sliding Away." *Science* 338 (November 16, 2012): 881.

Qiu, Jane. "Winds of Change." *Science* 338 (November 16, 2012): 879–881.

Qiu, Jane. "Himalayan Ice Can Fool Climate Studies." *Science* 347 (March 27, 2015): 1404–1405.

Qiu, Jane. "Trouble in Tibet: Rapid Changes in Tibetan Grasslands Are Threatening Asia's Main Water Supply and the Livelihood of Nomads." *Nature* 529 (January 14, 2016): 142–145. http://www.nature.com/news/trouble-in-tibet-1.19139.

Quadfasel, D. "Oceanography: The Atlantic Heat Conveyor Slows." *Nature* 438 (December 1, 2005): 565–566.

Quammen, David. "Planet of Weeds: Tallying the Losses of Earth's Animals and Plants." *Harper's*, October 1998, 57–69.

"Rachel's No. 466: Climate and Infectious Disease." Rachel's Democracy and Health News. November 2, 1995. Environmental Research Foundation, Annapolis, Maryland. https:// rachelsdemocracyandhealthnews.wordpress.com/2013/10/27/rachels-466-climate-and -infectious-disease-part-1/.

Radford, Tim. "As the World Gets Hotter, Will Britain Get Colder?" *The Guardian* (U.K.), June 21, 2001, 3.

Radford, Tim. "Ten Key Coral Reefs Shelter Much of Sea Life: American Association Scientists Identify Vulnerable Marine 'Hot Spots' with the Richest Biodiversity on Earth." *The Guardian* (U.K.), February 15, 2002, 12.

Radford, Tim. "85 Per Cent of Alaskan Glaciers Melting at 'Incredible Rate.'" *The Guardian* (U.K.), July 19, 2002, 9.

Radford, Tim. "2020: The Drowned World." *The Guardian* (U.K.), September 11, 2004, 10.

Radowitz, John von. "Calmer Sun Could Counteract Global Warming." Press Association, October 5, 2003 (LEXIS).

Rahmstorf, Stefan. "Thermohaline Circulation: The Current Climate." *Nature* 421 (February 13, 2003): 699.

Rahmstorf, Stefan. "A Semi-Empirical Approach to Projecting Future Sea-Level Rise." *Science* 215 (January 19, 2007): 368–370.

Rahmstorf, Stefan, Grant Foster, and Anny Cazenave. "Comparing Climate Projections to Observations up to 2011." *Environmental Research Letters* 7(4) (2012). http://iopscience.iop.org/1748-9326/7/4/044035/article.

Rahmstorf, Stefan, et al. "Exceptional Twentieth-Century Slowdown in Atlantic Ocean over Turning Circulation." *Nature Climate Change*, March 23, 2015. http://www.nature.com/articles/nclimate2554.epdf.

Ramanathan, V. "Observed Increases in Greenhouse Gases and Predicted Climatic Changes." Pp. 239–247 in Edwin Abrahamson, ed., *The Challenge of Global Warming*. Washington, DC: Island Press, 1989.

Ramanathan, V., and Y. Feng. "On Avoiding Dangerous Anthropogenic Interference with the Climate System: Formidable Challenges Ahead." *Proceedings of the National Academy of Sciences* 105(38) (September 23, 2008): 14245–14250.

Randall, C. E., et al. "Stratospheric Effects of Energetic Particle Precipitation in 2003–2004." *Geophysical Research Letters* 32 (March 2, 2005): L05802. doi: 10.1029/2004GL022003, 2005.

Randall, Tom. "Fossil Fuels Just Lost the Race against Renewables: This Is the Beginning of the End." *Bloomberg Business*, April 14, 2015. http://www.bloomberg.com/news/articles/2015-04-14/fossil-fuels-just-lost-the-race-against-renewables.

Range, Stacey. "Climatologists Say Midlands in Dust Bowl-Like Drought." *Omaha World-Herald*, January 12, 2000, 18.

Rangel, Thiago F. "Amazonian Extinction Debts." *Science* 337 (July 13, 2012): 162–163.

"The Rapid Melt of Greenland" *Nature* 491 (November 8, 2012): 163.

"Rare Tropical Cyclone Strikes Somalia." Environment News Service, November 12, 2013. http://earthobservatory.nasa.gov/IOTD/view.php?id=82377&src=eoa-iotd.

Ravilious, Kate. "The Arctic Sea a Stagnant Soup?" *New Scientist*, August 6, 2009. http://www.newscientist.com/article/mg20327204.800-arctic-ocean-may-be-polluted-soup-by-2070.html.

Ravishankara, A. R. "Water Vapor in the Lower Stratosphere." *Science* 337 (August 17, 2012): 809–810.

Réale, Denis, et al. "Genetic and Plastic Responses of a Northern Mammal to Climate Change." *Proceedings of the Royal Society B*, March 22, 2003. doi: 10.1098/rspb.2002.2224.

"Recent Warming of Arctic May Affect World-Wide Climate." National Aeronautics and Space Administration Press Release, October 23, 2003. http://www.gsfc.nasa.gov/topstory/2003/1023esuice.html (no longer available).

"Record Low Arctic Sea Ice Extent for January." NASA Earth Observatory, February 8, 2011. http://earthobservatory.nasa.gov/IOTD/view.php?id=49132&src=eoa-iotd.

"Record Ozone Hole May Open over Arctic in the Spring." *Science*, February 12, 2016, 650. http://www.sciencemag.org/news/2016/02/record-ozone-hole-may-open-over-arctic-spring.

"Record Rainfall Floods India." *The New York Times*, July 28, 2005, A12.

"Records Fall in 2015 Cyclone Season." NASA Earth Observatory, December 5, 2015. http://earthobservatory.nasa.gov/IOTD/view.php?id=87092&src=eoa-iotd.

Redsteer, M. H., et al. "Disaster Risk Assessment Case Study: Recent Drought on the Navajo Nation, Southwestern United States." Chapter 3 in *Global Assessment. Report on Disaster Risk Reduction*, 2010. http://www.preventionweb.net/english/hyogo/gar/2011/en/what/drought.html.

Redsteer, M. H., R. C. Bogle, and J. M. Vogel. "Monitoring and Analysis of Sand Dune Movement and Growth on the Navajo Nation, Southwestern United States." 2011a. Fact Sheet Number 3085. Reston, VA: U.S. Geological Survey.

Redsteer, M. H., K. B. Kelley, and H. Francis. "Increasing Vulnerability to Drought and Climate Change on the Navajo Nation." 2011b. Paper GC43B-0928, delivered at American Geophysical Union annual meeting, December 5–9, 2011, San Francisco.

Redsteer, M. H., et al. "Unique Challenges Facing Southwestern Tribes." Pp. 385–404 in G. Garfin et al., eds., *Assessment of Climate Change in the Southwest United States: A Report Prepared for the National Climate Assessment.* Washington, DC: Island Press, 2013. http://www.swcarr.arizona.edu/sites/default/files/ACCSWUS_Ch17.pdf.

Reece, Steven Y., et al. "Wireless Solar Water Splitting Using Silicon-Based Semiconductors and Earth-Abundant Catalysts." *Science* 334 (November 4, 2011): 645–648.

Reed, Nicholas. "Mild Winter Stirs Wildlife to Early Thoughts of Love." *Vancouver Sun*, February 12, 2003, B1.

Reed, Stanley. "Shift to Lower-Carbon Energy Is Too Slow, Report Warns." *The New York Times*, November 9, 2015. http://www.nytimes.com/2015/11/10/business/energy-environment/shift-to-lower-carbon-energy-is-too-slow-report-warns.html.

Regaldo, Antonio. "The Ukukus Wonder Why a Sacred Glacier Melts in Peru's Andes." *Wall Street Journal*, June 17, 2005, A1, A10.

Reilly, Michael. "Tibetan Glaciers Melting at Stunning Rate." Discovery.com. November 24, 2008. http://tibet.net/2008/11/tibetan-glaciers-melting-at-stunning-rate/.

"Report Says Oceans Hit by Carbon Dioxide Use." *Omaha World-Herald*, July 17, 2004, 5A.

"Researchers Warm Up to Melt's Role in Greenland Ice Loss." NASA Earth Observatory, April 17, 2008. http://earthobservatory.nasa.gov/Newsroom/NasaNews/2008/2008041726522.html (no longer available).

"Retreat of Jakobshavn Glacier, Greenland." NASA Earth Observatory, June 11, 2014. http://earthobservatory.nasa.gov/IOTD/view.php?id=83837&src=eoa-iotd.

Revelle, R., and H. E. Suess. "Carbon Dioxide Exchange between Atmosphere and Ocean and the Question of an Increase of Atmospheric CO_2 during the Past Decades." *Tellus* 9 (1957), 18–27.

Revkin, Andrew C. "Antarctic Test Raises Hope on a Global-Warming Gas." *The New York Times*, October 12, 2000, A18.

Revkin, Andrew C. "A Message in Eroding Glacial Ice: Humans Are Turning Up the Heat," *The New York Times*, February 19, 2001, A1.

Revkin, Andrew C. "Forecast for a Warmer World: Deluge and Drought." *The New York Times*, August 28, 2002, A10.

Revkin, Andrew. "An Icy Riddle as Big as Greenland." *The New York Times*, June 8, 2004. www.nytimes.com/learning/teachers/featured_articles/20040609wednesday.html.

Revkin, Andrew. "Antarctic Glaciers Quicken Pace to Sea; Warming Is Cited." *The New York Times*, September 24, 2004, A24.

Revkin, Andrew. "Global Warming Is Expected to Raise Hurricane Intensity." *The New York Times*, September 30, 2004, A20.

Revkin, Andrew C. "'An Inconvenient Truth': Al Gore's Fight against Global Warming." *The New York Times*, May 22, 2006. http://www.nytimes.com/2006/05/22/movies/22gore.html.

Revkin, Andrew C. "Studies Portray Tropical Arctic in Distant Past." *The New York Times*, June 1, 2006. http://www.nytimes.com/2006/06/01/science/earth/01climate.html.

Revkin, Andrew C. "NASA's Goals Delete Mention of Home Planet." *The New York Times*, July 22, 2006, A1, A10.

Revkin, Andrew C. "Climate Experts Warn of More Coastal Building." *The New York Times*, July 25, 2006, D2.

Revkin, Andrew. "Study Links Tropical Ocean Warming to Greenhouse Gases." *The New York Times*, September 12, 2006. http://www.nytimes.com/2006/09/12/science/12ocean.html.

Revkin, Andrew C. "Gore Calls for Immediate Freeze on Heat-Trapping Gas Emissions." *The New York Times*, September 19, 2006. http://www.nytimes.com/2006/09/19/washington/19gore.html.

Revkin, Andrew C. "Budgets Falling in Race to Fight Global Warming." *The New York Times*, October 30, 2006. http://www.nytimes.com/2006/10/30/business/worldbusiness/30energy.html.

Revkin, Andrew C. "Arctic Ice Shelf Broke Off Canadian Island." *The New York Times*, December 30, 2006. http://www.nytimes.com/2006/12/30/science/earth/30ice.html.

Revkin, Andrew. "Carbon-Neutral Is Hip, but Is It Green?" *The New York Times*, April 29, 2007. http://www.nytimes.com/2007/04/29/weekinreview/29revkin.html.

Revkin, Andrew C. "Climate Panel Reaches Consensus on the Need to Reduce Harmful Emissions." *The New York Times*, May 4, 2007. http://www.nytimes.com/2007/05/04/science/04climate.html.

Revkin, Andrew C. "Analysis Finds Large Antarctic Area Has Melted." *The New York Times*, May 16, 2007. http://www.nytimes.com/2007/05/16/science/earth/16melt.html.

Revkin, Andrew C. "Study Finds Hurricanes Frequent in Some Cooler Periods." *The New York Times*, May 24, 2007. http://www.nytimes.com/2007/05/24/science/earth/24storm.html.

Revkin, Andrew C. "Cyclone Nears Iran and Oman." *The New York Times*, June 6, 2007. http://www.nytimes.com/2007/06/06/world/middleeast/06storm.html.

Revkin, Andrew C. "Arctic Update: Resilient Bears, Shrinking Ice." *The New York Times*, December 13, 2007. http://dotearth.blogs.nytimes.com/2007/12/12/arctic-update-resilient-bears-vanishing-ice/index.html.

Revkin, Andrew C. "In Greenland, Ice and Instability." *The New York Times*, January 8, 2008. http://www.nytimes.com/2008/01/08/science/earth/08gree.html.

Revkin, Andrew C. "In a New Climate Model, Short-Term Cooling in a Warmer World." *The New York Times*, May 1, 2008. http://www.nytimes.com/2008/05/01/science/earth/01climate.html (no longer available).

Revkin, Andrew C. "Tropical Warming Tied to Flooding Rains." *The New York Times*, August 8, 2008. http://www.nytimes.com/2008/08/08/science/earth/08rain.html.

Revkin, Andrew C. "Arctic Ocean Ice Retreats Less Than Last Year." *The New York Times*, December 30, 2008. http://www.nytimes.com/2008/09/17/science/earth/17ice.html.

Revkin, Andrew C. "Gore, Will, Climate, and Complexity." *The New York Times*, February 25, 2009. http://dotearth.blogs.nytimes.com/2009/02/25/gore-and-will-and-climate-and-the-press.

Revkin, Andrew C. "Caribbean Reefs Face Severe Summer Threat." *The New York Times*, July 22, 2009. http://dotearth.blogs.nytimes.com/2009/07/22/carribean-reefs-face-severe-summer-threat/?hpw.

Revkin, Andrew C. "Walruses Suffer Substantial Losses as Sea Ice Erodes." *The New York Times*, October 3, 2009. http://www.nytimes.com/2009/10/03/science/earth/03walrus.html.

Revkin, Andrew C., and Matthew L. Wald. "Solar Power Wins Enthusiasts but Not Money." *The New York Times*, July 16, 2007. http://www.nytimes.com/2007/07/16/business/16solar.html.

Rex, Markus, et al. "Arctic Ozone Loss and Climate Change." *Geophysical Research Letters* 31 (March 10, 2004). http://www.eurekalert.org/pub_releases/2004-03/agu-ajh031004.php.

Reyes, Alberto V., et al. "South Greenland Ice-Sheet Collapse during Marine Isotope Stage 11." *Nature* 510 (June 26, 2014): 525–528. doi: 10.1038/nature13456.

Reynolds, James. "Earth Is Heading for Mass Extinction in Just a Century." *The Scotsman*, June 18, 2003, 6.

Rice, Doyle. "Hemingway's Famous Icy Peaks Are Thawing Fast." *USA Today*, November 3, 2009, 7D.

Rice, Doyle. "'Weather Whiplash' Swamps Midwest." *USA Today*, April 22, 2013, 3A.

Rice, Doyle. "Storm Warning: Climate Change to Spawn More Hurricanes." *USA Today*, July 8, 2013. http://www.usatoday.com/story/weather/2013/07/08/climate-change-glo bal-warming-hurricanes/2498611/.

Rice, Doyle. "Arctic Gets a Break in 2013, but Warming Still Ongoing." *USA Today*, December 13, 2013, 3A.

Rice, Doyle. "When It Rains, It Pours This Winter." *USA Today*, February 5, 2015, 4A.

Rice, Doyle. "Buried in Boston? Blame It on Climate Change—Maybe." *USA Today*, February 10, 2015. http://www.usatoday.com/story/weather/2015/02/09/northeast-snowsto rms-climate-change-global-warming/23133913/.

Rice, Doyle. "Megadrought May Plague Parts of USA." *USA Today*, February 13, 2015. http://www.usatoday.com/story/weather/2015/02/12/western-plains-drought-climate-change/23298093/.

Rice, Doyle. "Wildfires Decimating Forest Service's Budget." *USA Today*, August 6, 2015, 3A.

Rice, Doyle. "Study: Risk of Wildfires Rises." *USA Today*, August 28, 2015, 4A.

Rice, Doyle. "Study: Ocean Acidification Stunting Growth of Coral Reefs." *USA Today*, February 24, 2016. http://www.usatoday.com/story/weather/2016/02/24/ocean-acidificati on-coral-reefs/80857670/.

Rich, Nathaniel. "The Most Ambitious Environmental Lawsuit Ever." *The New York Times Magazine*, October 5, 2014. http://www.nytimes.com/interactive/2014/10/02/magazine/ mag-oil-lawsuit.html.

Richardson, Franci. "Sharks Take the Bait: Experts—Sightings in Maine an 'Unusual Circumstance.'" *Boston Herald*, August 11, 2002, 3.

Ridley, Jeff K., and Helene T. Hewitt. "A Mechanism for Lack of Sea Ice Reversibility in the Southern Ocean." *Geophysical Research Letters*, December 1, 2014. doi: 10.1002 /2014GL062167.

Riebesell, U., et al. "Enhanced Biological Carbon Consumption in a High CO_2 Ocean." *Nature* 450 (November 22, 2007): 545–548.

Rifkin, Jeremy. *The Hydrogen Economy: The Creation of the Worldwide Energy Web and the Redistribution of Power on Earth.* New York: Jeremy P. Tarcher/Putnam, 2002.

Rignot, E. J. "Fast Recession of a West Antarctic Glacier." *Science* 281 (July 24, 1998): 549–551.

Rignot, E. J., et al. "Accelerated Ice Discharge from the Antarctic Peninsula Following the Collapse of Larsen B Ice Shelf." *Geophysical Research Letters* 31(18) (September 22, 2004). doi: 10.1029/2004GL020697.

Rignot, Eric, et al. "Recent Antarctic Ice Mass Loss from Radar Interferometry and Regional Climate Modelling." *Nature GeoScience* 1 (2008): 106–110.

Rignot, E. J., et al. "Widespread, Rapid Grounding Line Retreat of Pine Island, Thwaites, Smith, and Kohler Glaciers, West Antarctica from 1992 to 2011." *Geophysical Research Letters*, May 2014. doi: 10.1002/2014GL060140.

Rinne, Tim. "Local View: City Dwellers Can Grow Food in a Risky Climate." *Lincoln Journal Star* (Nebraska), November 30, 2014. http://journalstar.com/news/opinion/editorial/co lumnists/local-view-city-dwellers-can-grow-food-in-a-risky/article_b5198714-8021 -54c8-aa89-9cb09d42b186.html.

"Rising Seas Could Swamp One in 10 People by 2100." Environment News Service, March 10, 2009. http://www.ens-newswire.com/ens/mar2009/2009-03-10-03.asp (no longer available).

"Rising Temperatures: A Month Versus a Decade." NASA Earth Observatory, October 24, 2014. http://earthobservatory.nasa.gov/IOTD/view.php?id=84589&src=eoa-iotd.

"Rising Tide: Who Needs Essex Anyway." *The Guardian* (U.K.), June 12, 2003, 4.

Ritchie, Kathy. "Dune and Gloom." *Arizona Highways*, 2014. http://www.arizonahighways. com/extras/dune.asp (no longer available).

Ritter, Malcolm. "This Is Winter? Much of Nation Basked in Warm January," Associated Press, February 3, 2006 (LEXIS).

Roach, John. "Where Did Global Warming Go? The Deep Ocean, Experts Say." NBC News. April 11, 2013. http://science.nbcnews.com/_news/2013/04/11/17708881-where-did -global-warming-go-the-deep-ocean-experts-say?lite.

Robbins, Jim. "Snow in July is a Mixed Blessing for the Northern Rockies." *The New York Times*, July 2, 2008, A9, A10.

Robbins, Jim. "Bark Beetles Kill Millions of Acres of Trees in West." *The New York Times*, November 18, 2008. http://www.nytimes.com/2008/11/18/science/18trees.html.

Robbins, Jim. "Some See Beetle Attacks on Western Forests as a Natural Event." *The New York Times*, July 7, 2009, D3.

Robbins, Jim. "Moose Die-Off Alarms Scientists." *The New York Times*, October 14, 2013. http://www.nytimes.com/2013/10/15/science/earth/something-is-killing-off-the-moose. html

Robbins, Jim. "Deforestation and Drought." *The New York Times*, October 9, 2015. http:// www.nytimes.com/2015/10/11/opinion/sunday/deforestation-and-drought.html.

Robel, Alexander. "The Long Future of Antarctic Melting." *Nature* 526 (October 15, 2015): 327–328.

Roberts, Callum M., et al. "Marine Biodiversity Hotspots and Conservation Priorities for Tropical Reefs." *Science* 295 (February 15, 2002): 1280–1284.

Roberts, Paul. *The End of Oil: The Edge of a Perilous New World.* Boston: Houghton-Mifflin, 2004.

Robertson, Campbell. "Iraq, a Land between Two Rivers, Suffers as One of Them Dwindles." *The New York Times*, July 14, 2009, AS1, S8.

Robertson, Campbell, and Kim Severson. "South Assesses the Toll after a Deadly Barrage of Tornadoes." *The New York Times*, April 28, 2011. http://www.nytimes.com/2011/04/29/ us/29storm.html.

Robinson, Eugene. "Feeling the Heat." *Washington Post*, July 2, 2012. http://www.washington post.com/opinions/eugene-robinson-feeling-the-heat/2012/07/02/gJQANNZGJW_ print.html.

Robock, Alan. "Whither Geo-Engineering?" *Science* 320 (May 30, 2008): 1166–1167.

Roe, Nicholas. "Show Me a Home Where the Reindeer Roam." *Times* (London), November 10, 2001, n.p. (LEXIS).

Rogers, David J., and and Sarah E. Randolph. "The Global Spread of Malaria in a Future, Warmer World." *Science* 289 (September 8, 2000): 1763–1766.

Rogers, Paul. "Solar Energy Heats Up." *Omaha World-Herald*, October 15, 2006, 1RE, 2RE.

Rohling, E. J., et al. "High Rates of Sea-Level Rise during the Last Interglacial Period." *Nature GeoScience* (December 16, 2007). http://www.nature.com/ngeo/journal/v1/n1/full/ngeo.2007.28.html. doi: 10.1038/ngeo.2007.28.

Rohter, Larry. "Antarctica, Warming, Looks Ever More Vulnerable." *The New York Times*, January 25, 2005. http://www.nytimes.com/2005/01/25/science/earth/25ice.html.

Rohter, Larry. "A Record Amazon Drought, and Fear of Wider Ills." *The New York Times*, December 11, 2005. http://www.nytimes.com/2005/12/11/international/americas/11amazon.html.

Rohter, Larry. "With Big Boost from Sugar Cane, Brazil Is Satisfying Its Fuel Needs." *The New York Times*, April 10, 2006. http://www.nytimes.com/2006/04/10/world/americas/10brazil.html.

Rohter, Larry. "Brazil, Alarmed, Reconsiders Policy on Climate Change." *The New York Times*, July 31, 2007. http://www.nytimes.com/2007/07/31/world/americas/31amazon.html.

Rojas, Rick. "Drought's Extremes Tallied at Record-Low Lake Mead." *The New York Times*, May 5, 2015. http://www.nytimes.com/2015/05/05/us/droughts-extremes-can-be-measured-at-record-low-lake-mead.html.

Rolfe, Christopher. "Comments on the British Columbia Greenhouse Gas Action Plan." West Coast Environmental Law Association. April 17, 1996. http://www.wcel.org/wcelpub/11026.html (no longer available).

Romero, Simon. "Swallowing Rain Forest, Cities Surge in Amazon." *The New York Times*, November 24, 2012. http://www.nytimes.com/2012/11/25/world/americas/swallowing-rain-forest-brazilian-cities-surge-in-amazon.html.

Romero, Simon. "Taps Start to Run Dry in Brazil's Largest City; São Paulo Water Crisis Linked to Growth, Pollution and Deforestation." *The New York Times*, February 17, 2015. http://www.nytimes.com/2015/02/17/world/americas/drought-pushes-sao-paulo-brazil-toward-water-crisis.html.

Romero, Simon. "Alarm Spreads in Brazil over a Virus and a Surge in Malformed Infants." *The New York Times*, December 30, 2015. http://www.nytimes.com/2015/12/31/world/americas/alarm-spreads-in-brazil-over-a-virus-and-a-surge-in-malformed-infants.html.

Romm, Joseph. *Hell and High Water: Global Warming—the Solution and the Politics—and What We Should Do.* New York: William Morrow, 2007.

Romm, Joseph. "Desertification: The Next Dust Bowl." *Nature* 478 (October 27, 2011): 450–451.

Rosenbloom, Stephanie. "Giant Retailers Look to Sun for Energy Savings." *The New York Times*, August 11, 2008. http://www.nytimes.com/2008/08/11/business/11solar.html.

Rosenthal, Elisabeth. "As the Climate Changes, Bits of England's Coast Crumble." *The New York Times*, May 4, 2007. http://www.nytimes.com/2007/05/04/world/europe/04erode.html.

Rosenthal, Elisabeth. "Likely Spread of Deserts to Fertile Land Requires Quick Response, U.N. Report Says." *The New York Times*, June 28, 2007. http://www.nytimes.com/2007/06/28/world/28deserts.html.

Rosenthal, Elisabeth. "Studies Deem Biofuels a Greenhouse Threat." *The New York Times*, February 8, 2008. http://www.nytimes.com/2008/02/08/science/earth/08wbiofuels.html.

Rosenthal, Elizabeth. "Europe Turns Back to Coal, Raising Climate Fears." *The New York Times*, April 23, 2008. http://www.nytimes.com/2008/04/23/world/europe/23coal.html.

Rosenthal, Elisabeth. "Water Is New Battleground in Drying Spain." *The New York Times*, June 3, 2008, A1, A12.

Rosenthal, Elisabeth. "China Increases Lead as Biggest Carbon Dioxide Emitter." *The New York Times*, June 14, 2008. http://www.nytimes.com/2008/06/14/world/asia/14china.html.

Rosenthal, Elisabeth. "Stinging Tentacles Offer Hint of Oceans' Decline." *The New York Times*, August 3, 2008. http://www.nytimes.com/2008/08/03/science/earth/03jellyfish.html.

Rosenthal, Elisabeth. "In German Suburb, Life Goes On without Cars." *The New York Times*, May 12, 2009. http://www.nytimes.com/2009/05/12/science/earth/12suburb.html.

Rosenthal, Elisabeth. "An Amazon Culture Withers as Food Dries Up." *The New York Times*, July 25, 2009. http://www.nytimes.com/2009/07/25/science/earth/25tribe.html.

Rosenthal, Elisabeth. "To Cut Global Warming, Swedes Study Their Plates." *The New York Times*, October 23, 2009. http://www.nytimes.com/2009/10/23/world/europe/23degrees.html.

Rosenthal, Elisabeth. "Turtles Are Casualties of Warming in Costa Rica." *The New York Times*, November 14, 2009. http://www.nytimes.com/2009/11/14/science/earth/14turtles.html.

Rosenthal, Elisabeth. "In Bolivia, Water and Ice Tell of Climate Change." *The New York Times*, December 14, 2009. http://www.nytimes.com/2009/12/14/science/earth/14bolivia.html.

Rosenthal, Elisabeth. "Britain Curbing Airport Growth to Aid Climate." *The New York Times*, July 2, 2010. http://www.nytimes.com/2010/07/02/science/earth/02runway.html.

Rosenthal, Elisabeth. "Portugal Gives Itself a Clean-Energy Makeover." *The New York Times*, August 10, 2010, A1, A8.

Rosenthal, Elisabeth. "Nations That Debate Coal Use Export It to Feed China's Need." *The New York Times*, November 21, 2010. http://www.nytimes.com/2010/11/22/science/earth/22fossil.html.

Rosenthal, Elisabeth. "Race Is On as Ice Melt Reveals Arctic Treasures." *The New York Times*, September 18, 2012. http://www.nytimes.com/2012/09/19/science/earth/arctic-resources-exposed-by-warming-set-off-competition.html.

Rosenthal, Elisabeth. "A Melting Greenland: Perils Against Potential." *The New York Times*, September 23, 2012. http://www.nytimes.com/2012/09/24/science/earth/melting-greenland-weighs-perils-against-potential.html.

Rosner, Hillary. "The Bug's That Eating the Woods." *National Geographic*, April 2015, 96–115.

Ross, D., et al. "Obliquity-Paced Pliocene West Antarctic Ice Sheet Oscillations." *Nature* 457 (March 19, 2009): 322–325.

Rowan, Rob. "Thermal Adaptation in Reef Coral Symbionts." *Nature* 430 (August 12, 2004): 742.

Rowland, Sherwood, and Mario Molina. "Stratospheric Sink for Chlorofluoromethanes: Chlorine Atom-Catalyzed Destruction of Ozone." *Nature* 249 (June 28, 1974): 810–812.

Rubin, Daniel. "Venice Sinks as Adriatic Rises." Knight-Ridder News Service, July 1, 2003, n.p. (LEXIS).

Rubin, Josh. "Toronto's Blooming Warm; Gardens, Golfers Spring to Life as Record High Nears." *Toronto Star*, December 5, 2001, B2.

Rudolf, John Collins. "The Warming of Greenland." *The New York Times*, January 16, 2007. http://www.nytimes.com/2007/01/16/science/earth/16gree.html.

Ruff, Joe. "U[nion] P[acific] Target of Global Warming Protest." *Omaha World-Herald*, June 17, 2010. http://www.omaha.com/article/20100617/MONEY/100619732 (no longer available).

Ruhlmann, Carsten, et al. "Warming of the Tropical Atlantic Ocean and Slowdown of Thermohaline Circulation during the Last Deglaciation." *Nature* 402 (December 2, 1999): 511–514.

Running, Steven W. "Is Global Warming Causing More, Larger Wildfires?" *Science* 313 (August 18, 2006): 927–928.

Russell, Sabin. "Glaciers on Thin Ice; Expert Says Melting to Be Faster Than Expected." *San Francisco Chronicle*, February 17, 2002, A4.

Rustin, Susanna. "A Life in Writing: Bill McKibben." *The Guardian* (U.K.), December 3, 2010. http://www.theguardian.com/books/2010/dec/06/bill-mckibben-interview.

Ruttimann, Jacqueline. "Oceanography: Sick Seas." *Nature* 442 (August 31, 2006): 978–980.

Ryall, Julian. "Tokyo Plans City Coolers to Beat Heat." *Times* (London), August 11, 2002, 20.

Ryan, Siobhain. "National Icons Feel the Heat." *Courier Mail* (Australia), February 4, 2002, 1.

Ryskin, G. "Methane-Driven Oceanic Eruptions and Mass Extinctions." *Geology* 31 (2003): 737–740.

Sabadini, Roberto. "Ice Sheet Collapse and Sea Level Change." *Science* 295 (March 29, 2002): 2376–2377.

Sabine, Christopher L., et al. "The Oceanic Sink for Anthropogenic CO_2." *Science* 305 (July 16, 2004): 367–371.

"Saharan Dust Has Chilling Effect on North Atlantic." NASA Earth Observatory Press Release, December 14, 2007. http://earthobservatory.nasa.gov/Newsroom/NasaNews/2007/2007121425986.html.

Sallenger, Asbury, Jr., Kara S. Doran, and Peter A. Howd. "Hotspot of Accelerated Sea-Level Rise on the Atlantic Coast of North America." *Nature Climate Change* 2(2012): 884–888.

Samenow, Jason. "'Potentially Catastrophic' Category 5 Hurricane Patricia to Slam West Coast of Mexico Friday." *Washington Post*, October 22, 2015. https://www.washingtonpost.com/news/capital-weather-gang/wp/2015/10/22/extremely-dangerous-category-4-hurricane-patricia-to-slam-west-coast-of-mexico-friday/?wpmm=1&wpisrc=nl_headlines.

Sample, Ian. "Warming Hits 'Tipping Point.'" *The Guardian* (U.K.), August 18, 2005, 1. http://www.guardian.co.uk/guardianweekly/story/0,12674,1550685,00.html.

Samuelson, Robert J. "The Retreat of 'Peak Oil.'" *Washington Post*, June 15, 2015. http://www.washingtonpost.com/opinions/the-retreat-of-peak-oil/2015/06/14/76a24ae4-1124-11e5-9726-49d6fa26a8c6_story.html.

Sanderson, Katharine. "Flights of Green Fancy." *Nature* 453 (May 15, 2008): 264–265.

Santer, B. D., et al. "Forced and Unforced Ocean Temperature Changes in Atlantic and Pacific Tropical Cyclogenesis Regions." *Proceedings of the National Academy of Sciences* 103(38) (September 19, 2006): 13905–13910.

Santos, Fernanda. "Nine Straight Days of 110 or More: That's Hot, Even for Phoenix." *The New York Times*, August 14, 2012. http://www.nytimes.com/2012/08/15/us/unrelenting-heat-keeps-torrid-grip-on-phoenix.html.

Santoso, Agus, et al. "Late-Twentieth-Century Emergence of the El Niño Propagation Asymmetry and Future Projections." *Nature* 504 (2013): 126–130.

Saraux, Claire, et al. "Reliability of Flipper-Banded Penguins as Indicators of Climate Change." *Nature* 469 (January 13, 2011): 203–206.

"Saskatchewan First Nations Evacuate 13,000, Declare Wildfire State of Emergency." Indian Country Today Media Network, July 6, 2015. http://indiancountrytodaymedianetwork.com/2015/07/06/saskatchewan-first-nations-evacuate-13000-declare-wildfire-state-emergency-160973.

"Satellite Data Reveals Extreme Summer Snowmelt in Northern Greenland, CCNY Professor Says." NASA Earth Observatory, October 8, 2008 (no longer available). http://earthobservatory.nasa.gov/Newsroom/MediaAlerts/2008/2008100827658.html.

"Satellite Images Show Chunk of Broken Antarctic Ice Shelf." April 16, 1998. http://www
.eurekalert.org/releases/brkantartice.html. (no longer available).

"Satellite Images Show Continued Breakup of Greenland's Largest Glaciers, Predict Disintegration in Near Future." NASA Earth Observatory, August 20, 2008. http://eartho
bservatory.nasa.gov/Newsroom/MediaAlerts/2008/2008082027361.html (no longer
available).

"Satellites Show Arctic Literally on Thin Ice." NASA Earth Observatory, April 7, 2009. http://
earthobservatory.nasa.gov/Newsroom/view.php?id=37803&src=eoa-nnews (no longer
available).

Saunders, Mark A., and Adam S. Lea. "Large Contribution of Sea Surface Warming to Recent
Increase in Atlantic Hurricane Activity." *Nature* 451 (January 31, 2008): 557–560.

Savill, Richard. "Tropical Fish Hooked on Channel Holidays." *The Telegraph* (London, U.K.),
August 12, 2004. http://www.telegraph.co.uk/news/uknews/1469221/Tropical-fish
-hooked-on-Channel-holidays.html.

"Saving Earth's Plant Diversity From Global Warming." Environment News Service, May 22,
2007. http://www.ens-newswire.com/ens/may2007/2007-05-22-02.asp (no longer
available).

Scambos, Ted. "Anatomy of an Ice Shelf's Demise." *Nature* 503 (November 28, 2013): 441.

Scambos, T. A., et al. "Glacier Acceleration and Thinning after Ice Shelf Collapse in the
Larsen B Embayment, Antarctica." *Geophysical Research Letters* 31(18) (September 22,
2004). doi: 10.1029/2004GL020670.

Schaeffer, Michiel, et al. "Long-Term Sea-Level Rise Implied by 1.5°C and 2°C Warming
Levels." *Nature Climate Change* 2 (December 2012): 867–870.

Schapiro, Mark. *Carbon Shock: A Tale of Risk and Calculus on the Front Lines of the Disrupted
Global Economy*. White River Junction, VT: Chelsea Green, 2014.

Schar, Christoph, and Gerd Jendritzky. "Hot News from Summer 2003." *Nature* 432 (December 2, 2004): 559–561.

Schiermeier, Quirin. "The Oresmen." *Nature* 421 (January 9, 2003): 109–110.

Schiermeier, Quirin. "Gas Leak: Global Warming Isn't a New Phenomenon." *Nature* 423
(June 12, 2003): 681–682.

Schiermeier, Quirin. "A Rising Tide: The Ice Covering Greenland Holds Enough Water to
Raise the Oceans Six Metres—and It's Starting to Melt." *Nature* 428 (March 11, 2004):
114–115.

Schiermeier, Quirin. "Researchers Seek to Turn the Tide on Problem of Acid Seas." *Nature*
430 (August 19, 2004): 820.

Schiermeier, Quirin. "Poles Lose Out as Ozone Levels Begin to Recover." *Nature* 437 (September 8, 2005): 179.

Schiermeier, Quirin. "On Thin Ice: The Arctic Is the Bellwether of Climate Change." *Nature*
441 (May 10, 2007): 146–147.

Schiermeier, Quirin. "Ocean Circulation Noisy, Not Stalling." *Nature* 448 (August 23, 2007):
844–845.

Schiermeier, Quirin. "Europe to Capture Carbon." *Nature* 451 (January 17, 2008): 232.

Schiermeier, Quirin. "Global Warming Blamed for Growth in Storm Intensity." *Nature*
455, September 3, 2008. http://www.nature.com/news/2008/080903/full/
news.2008.1079.html.

Schiermeier, Quirin. "Atmospheric Science: Fixing the Sky." *Nature* 460 (August 13, 2009):
792–796.

Schiermeier, Quirin. "Glaciers' Wane Not All Down to Humans." *Nature* 465 (June 10, 2010):
677.

Schiermeier, Quirin. "Increased Flood Risk Linked to Global Warming; Likelihood of Extreme Rainfall May Have Been Doubled by Rising Greenhouse Gas Levels." *Nature* 470 February 17, 2011, 316. http://www.nature.com/news/2011/110216/full/470316a.html.

Schiermeier, Quirin. "Earth's Acid Test." *Nature* 471 (March 9, 2011): 154–156. http://www.nature.com/news/2011/110309/full/471154a.html.

Schiermeier, Quirin. "Climate and Weather: Extreme Measures." *Nature* 477 (September 7, 2011): 148–149.

Schiermeier, Quirin. "Ice Loss Shifts Arctic Cycles." *Nature* 489 (September 13, 2012): 185–186.

Schiermeier, Quirin. "Greenland Defied Ancient Warming." *Nature* 493 (January 24, 2013): 459–460. http://www.nature.com/news/greenland-defied-ancient-warming-1.12265.

Schiermeier, Quirin. "Wild Weather Can Send Greenhouse Gases Spiralling." *Nature* 496 (April 11, 2013): 147. http://www.nature.com/news/wild-weather-can-send-greenhouse-gases-spiralling-1.12764.

Schiermeier, Quirin. "Global Warming Brews Weird Weather." *Nature* (online), April 27, 2015. http://www.nature.com/news/global-warming-brews-weird-weather-1.17407.

Schiermeier, Quirin. "Hunting the Godzilla El Niño." *Nature* 526 (October 22, 2015): 490–491. http://www.nature.com/news/hunting-the-godzilla-el-ni%C3%B1o-1.18591?WT.mc_id=SFB_NNEWS_1508_RHBox.

Schiermeier, Quirin. "Solar on the Steppe: Ukraine Embraces Renewables Revolution: Former Soviet Nation Bids for Independence from Russian Fossil Fuels." *Nature* 537 (September 29, 2016): 598. doi: 10.1038/537598a.

Schimel, David. "Climate Change and Crop Yields: Beyond Cassandra." *Science* 312 (June 30, 2006): 1889–1890.

Schimel, David, and David Baker. "Carbon Cycle: The Wildlife Factor." *Nature* 420 (November 7, 2002): 29–30.

Schleck, Dave. "High Fliers May Be Creating Clouds: Global Warming May Be Worsened by Contrails from Aircraft." *Montreal Gazette*, July 18, 2004, D6.

Schleussner, Carl-Friedrich, et al. "Armed-Conflict Risks Enhanced by Climate-Related Disasters in Ethnically Fractionalized Countries." *Proceedings of the National Association of Sciences* 113(30), July 25, 2016. http://www.pnas.org/content/early/2016/07/20/1601611113. doi: 10.1073/pnas.1601611113.

Schlossberg, Tatiana. "Warmer Winter Brings Forest-Threatening Beetles North." *The New York Times*, March 18, 2016. http://www.nytimes.com/2016/03/22/science/southern-pine-beetles-new-england-forests.html.

Schlossberg, Tatiana. "2016 Already Shows Record Global Temperatures." *The New York Times*, April 19, 2016. http://www.nytimes.com/2016/04/20/science/2016-global-warming-record-temperatures-climate-change.html.

Schmid, Randolph E. "Warming Climate Reduces Yield for Rice, One of World's Most Important Crops." Associated Press, June 28, 2004 (LEXIS).

Schmid, Randolph E. "Climate Change Could Devastate U.S. Wineries." Associated Press, July 10, 2006 (LEXIS).

Schmid, Randolph E. "State of the Arctic Warming, with Widespread Melting." Associated Press, November 16, 2006 (LEXIS).

Schmidhuber, Josef, and Francesco Tubiello. "Global Food Security under Climate Change." *Proceedings of the National Academy of Sciences* 104 (December 11, 2007): 19703–19708.

Schmidt, Charles. "As Threats to Corals Grow, Hints of Resilience Emerge." *Science* 340 (March 29, 2013): 1517–1519.

Schmittner, Andreas. "Decline of the Marine Ecosystem Caused by a Reduction in the Atlantic Overturning Circulation." *Nature* 434 (March 31, 2005): 628–633.

Schneider, Stephen H. "No Therapy for the Earth: When Personal Denial Goes Global." In Michael Aleksiuk and Thomas Nelson, eds., *Nature, Environment & Me: Explorations of Self in a Deteriorating World*. Montreal: McGill-Queens University Press, 2000.

Schneider, Stephen H., and R. S. Chen. "Carbon Dioxide Warming and Coastline Flooding: Physical Factors and Climatic Impact." *American Review of Energy* 5 (1980): 107–140.

Schneider, Stephen, and Terry Root, eds. *Wildlife Responses to Climate Change*. Washington, DC: Island Press, 2001.

Schneider, Stephen, et al. "Misleading Math about the Earth." *Scientific American*, January 2002. http://www.scientificamerican.com/article/misleading-math-about-the/.

Schofield, Oscar, et al. "How Do Polar Marine Ecosystems Respond to Rapid Climate Change?" *Science* (June 18, 2010): 1520–1523.

Schrope, Mark. "Successes in Fight to Save Ozone Layer Could Close Holes by 2050." *Nature* 408 (December 7, 2000): 627.

Schrope, Mark. "Global Warming: A Change of Climate for Big Oil." *Nature* 411 (May 31, 2001): 516–518.

Schwalm, Christopher, Christopher A. Williams, and Kevin Schaeffer. "Hundred-Year Forecast: Drought." *The New York Times*, August 12, 2012. http://www.nytimes.com /2012/08/12/opinion/sunday/extreme-weather-and-drought-are-here-to-stay.html.

Schwartz, John. "Rockefellers, Heirs to an Oil Fortune, Will Divest Charity from Fossil Fuels." *The New York Times*, September 21, 2014. http://www.nytimes.com/2014/09/22/ us/heirs-to-an-oil-fortune-join-the-divestment-drive.html.

Schwartz, John. "Study Finds Low Cost in Reducing Methane Emissions." *The New York Times*, April 22, 2015. http://www.nytimes.com/2015/04/22/world/americas/study-finds -low-cost-in-reducing-methane-emissions.html.

Schwartz, John. "Norway Will Divest from Coal in Push against Climate Change." *The New York Times*, June 5, 2015. http://www.nytimes.com/2015/06/06/science/norway-in-push -against-climate-change-will-divest-from-coal.html.

Schwartz, John. "Methane Leaks May Greatly Exceed Estimates, Report Says." *The New York Times*, August 4, 2015. http://www.nytimes.com/2015/08/05/science/methane-leaks -may-greatly-exceed-estimates-report-says.html.

Schwartz, John. "Methane Leaks in Natural-Gas Supply Chain Far Exceed Estimates, Study Says." *The New York Times*, August 18, 2015. http://www.nytimes.com/2015/08/ 19/science/methane-leaks-in-natural-gas-supply-chain-far-exceed-estimates-study-says .html.

Schwartz, John. "As Fires Grow, a New Landscape Appears in the West." *The New York Times*, September 21, 2015. http://www.nytimes.com/2015/09/22/science/as-fires-grow-a-new -landscape-appears-in-the-west.html.

Schwartz, John. "Deadly Heat Is Forecast in Persian Gulf by 2100." *The New York Times*, October 26, 2015. http://www.nytimes.com/2015/10/27/science/intolerable-heat-may -hit-the-middle-east-by-the-end-of-the-century.html.

Schwartz, John. "The Pacific Ocean Becomes a Caldron." *The New York Times*, November 2, 2015. http://www.nytimes.com/2015/11/03/science/global-warming-pacific-ocean-el -nino-blob.html.

Schwartz, John. "Climate Models May Overstate Clouds' Cooling Power, Research Says." *The New York Times*, April 8, 2016. http://www.nytimes.com/2016/04/08/science/climate -models-may-overstate-clouds-cooling-power-research-says.html.

Schwartzman, Stephen. "Reigniting the Rainforest: Fires, Development and Deforestation." *Native Americas* 16(3–4) (Fall–Winter 1999): 60–63.

"Scientists Find Heat-Tolerant Coral Reefs That May Resist Climate Change." NASA Earth Observatory, May 19, 2009. http://earthobservatory.nasa.gov/Newsroom/view.php?id =38819&src=eoa-manews (no longer available).

"Scientists Predict Future of Weather Extremes." Environment News Service, October 20, 2006. http://www.ens-newswire.com/ens/oct2006/2006-10-20-03.html.

"Scientists Report Large Ozone Loss." *USA Today*, April 6, 2000, 3A.

"Scientist: Warming Threatens Koalas." *Omaha World-Herald*, May 8, 2008, 7A.

Scott, Michon. "Rapid Retreat: Ice Shelf Loss Along Canada's Ellesmere Coast." NASA Earth Observatory, September 9, 2008. http://earthobservatory.nasa.gov/Study/Ellesmere.

Screen, James A., and Ian Simmonds. "The Central Role of Diminishing Sea Ice in Recent Arctic Temperature Amplification." *Nature* 464 (April 29, 2010): 1334–1337.

Seabrook, Charles. "Amphibian Populations Drop." *Atlanta Journal-Constitution*, October 15, 2004, 1C.

"Sea Creatures Adapt to Acid." *Nature* 520 (April 9, 2015): 134.

Seager, R., et al. "Is the Gulf Stream Responsible for Europe's Mild Winters?" *Quarterly Journal of the Royal Meteorological Society* 128 (2002): 2563–2586.

"Sea Ice Penguin Theory Sinks." *Nature* 472 (April 21, 2011): 263.

Searchinger, Timothy, et al. "Use of U.S. Croplands for Biofuels Increases Greenhouse Gases through Emissions from Land-Use Change." *Science* 319 (February 29, 2008): 1238–1240.

"Second Giant Solar Plant Planned for Arizona Electricity Users." Environment News Service, May 25, 2009. http://www.ens-newswire.com/ens/may2009/2009-05-25-091.html.

Secord, Ross, et al. "Evolution of the Earliest Horses Driven by Climate Change in the Paleocene–Eocene Thermal Maximum." *Science* 335 (February 24, 2012): 959–962. http://www.sciencemag.org/content/335/6071/959.

Seelye, Katharine. "Environmental Groups Gain as Companies Vote on Issues." *The New York Times*, May 29, 2003, C1.

Seelye, Katharine. "Regulators Approve First Offshore Wind Farm in U.S." *The New York Times*, April 28, 2010. http://www.nytimes.com/2010/04/29/us/29wind.html (no longer available).

Seelye, Kathleen. "Survivor of Dust Bowl Now Battles a Fiercer Drought." *The New York Times*, May 4, 2011. http://www.nytimes.com/2011/05/04/us/04dust.html.

Seetharaman, G. "Getting Past the Gridlock." *Economic Times*, December 13–19, 2015, 2–15.

Seibel, Brad A., and Patrick J. Walsh. "Potential Impacts of CO_2 Injection on Deep-Sea Biota." *Science* 294 (October 12, 2001): 319–320.

Seidel, Dian, et al. "Widening of the Tropical Belt in a Changing Climate." *Nature GeoScience*, December 2, 2007. http://www.nature.com/ngeo/journal/v1/n1/full/ngeo.2007.38. html. doi: 10.1038/ngeo.2007.38.

"Self-Regulating Coral Protect Themselves against Ocean Acidification." University News. University of Western Australia. October 6, 2015. http://www.news.uwa.edu. au/201510068035/research/self-regulating-coral-protect-themselves-against-ocean-aci dification.

Semenov, Vladimir A., and Lennart Bengtsson. *Modes of Wintertime Arctic Temperature Variability*. Hamburg, Germany: Max Planck Institut fur Meteorologie, 2003.

Semmens, Grady. "Ecologists See Disaster in Dwindling Water Supply." *Calgary Herald*, November 27, 2003, A14.

Sengupta, Somini. "Sea's Rise in India Buries Islands and a Way of Life." *The New York Times,* April 11, 2007. http://www.nytimes.com/2007/04/11/world/asia/11india.html.

Sengupta, Somini. "Glaciers in Retreat." *The New York Times,* July 17, 2007. http://www.nytimes.com/2007/07/17/science/earth/17glacier.html.

Senna, Mônica Carneiro Alves, Marcos Heil Costa, and Gabrielle Ferreira Pires. "Vegetation-Atmosphere-Soil Nutrient Feedbacks in the Amazon for Different Deforestation Scenarios." *Journal of Geophysical Research,* February 18, 2009. doi: 10.1029/2008JD010401.

Service, Robert F. "Solar Power: Can the Upstarts Top Silicon?" *Science* 319 (February 8, 2008): 718–720.

Service, Robert F. "New Catalyst Marks Major Step in the March Toward Hydrogen Fuel." *Science* 321 (August 1, 2008): 620.

Service, Robert F. "Artificial Leaf Turns Sunlight into a Cheap Energy Source." *Science* 332 (April 1, 2011): 25.

Service, Robert F. "Outlook Brightens for Plastic Solar Cells." *Science* 332 (April 15, 2011): 293.

Service, Robert F. "Rising Acidity Brings an Ocean of Trouble." *Science* 337 (July 13, 2012): 146–148.

Service, Robert. "Perovskite Solar Cells Keep On Surging." *Science* 344 (May 2, 2014): 458.

Service, Robert F. "Meager Snows Spell Trouble Ahead for Salmon." *Science* 348 (April 17, 2015): 268–269.

Sessolo, M., and H. Bolink. "Perovskite Solar Cells Join the Major League." *Science* 350 (November 20, 2015): 917.

"Severe Thunderstorms and Climate Change." NASA Earth Observatory, April 7, 2013. http://earthobservatory.nasa.gov/IOTD/view.php?id=80825&src=eoa-iotd.

Shaffer, Gary, et al. "Long-Term Ocean Oxygen Depletion in Response to Carbon Dioxide Emissions from Fossil Fuels." *Nature GeoScience* 2 (2009): 105–109.

Shakhova, Natalia, et al. "Extensive Methane Venting to the Atmosphere from Sediments of the East Siberian Arctic Shelf." *Science* 327 (March 5, 2010): 1946–1950.

"Shanghai Mulls Building Dam to Ward off Rising Sea Levels." Agence France Presse, February 9, 2004. http://www.spacedaily.com/2004/040209065658.wbolr45c.html.

Sharmila, S., et al. "Asymmetry in Space-Time Characteristics of Indian Summer Monsoon Intraseasonal Oscillations during Extreme Years: Role of Seasonal Mean State." *International Journal of Climate,* June 2014, 1–16.

Shea, J. M., et al. "Modelling Glacier Change in the Everest Region, Nepal Himalaya." *The Cryosphere,* 9 (May 27, 2015): 1105–1128. http://www.the-cryosphere.net/9/1105/2015/tc-9-1105-2015.html.

Shearer, Christine. *Kivalina: A Climate Change Story.* Chicago: Haymarket Books, 2011.

Shearer, C. "The Social Construction of Alaska Native Vulnerability to Climate Change." *Race, Gender & Class* 19(3–4) (2012): 61–79.

Shellenberger, Michael. "The Climate War Is Over." *USA Today,* December 3, 2015, 7A.

Shen, Shu-zhong, et al. "Calibrating the End-Permian Mass Extinction." *Science* 334 (December 9, 2011): 1367–1372.

Sheppard, Cate. "Climate Change Takes a Village as the Planet Warms, a Remote Alaskan Town Shows Just How Unprepared We Are." Huffington Post, December 14, 2014. http://www.huffingtonpost.com/2014/12/14/shishmaref-alaska-climate-change-relocation_n_6296516.html.

Sheppard, Charles R. C. "Predicted Recurrences of Mass Coral Mortality in the Indian Ocean." *Nature* 425 (September 18, 2003): 294–297.

Shi, Dalin, et al. "Effect of Ocean Acidification on Iron Availability to Marine Phytoplankton." *Science* 327 (February 5, 2010): 676–679.

Shields, Craig. "Maybe It Is Time to Stop Building Things with Tailpipes and Smokestacks." 2Green Energy, October 12, 2014. http://2greenenergy.com/2014/10/12/tailpipes-and -smokestacks/.

Shindell, Drew T., David Rind, and Patrick Lonergan. "Increased Polar Stratospheric Ozone Losses and Delayed Eventual Recovery Owing to Increasing Greenhouse-Gas Concentrations." *Nature* 392 (April 9, 1998): 589–592.

Shindell, D. T., and G. A. Schmidt. "Southern Hemisphere Climate Response to Ozone Changes and Greenhouse Gas Increases." *Geophysical Research Letters* 31 (2004): L18209. doi: 10.1029/2004GL020724.

"Shrinking Lakes on the Mongolian Plateau." NASA Earth Observatory. April 8, 2015. http:// earthobservatory.nasa.gov/IOTD/view.php?id=85665&src=eoa-iotd.

Siegel, David A., and Bryan A. Franz. "Oceanography: Century of Phytoplankton Change." *Nature* 466 (July 29, 2010): 569–571.

Siegel, Matt. "Record Heat Fuels Widespread Fires in Australia." *The New York Times*, January 9, 2013. http://www.nytimes.com/2013/01/10/world/asia/record-heat-fuels-widespread -fires-in-australia.html.

Siegel, Matt. "Report Blames Climate Change for Extremes in Australia." *The New York Times*, March 4, 2013. http://www.nytimes.com/2013/03/05/world/asia/australian-government -blames-climate-change-for-angry-summer.html.

Sierra Club. "Global Warming: The High Costs of Inaction." 1999. http://www.sierraclub .org/global-warming/resources/innactio.htm (no longer available).

Silva, Rebecca. "California Fires Fueled by Global Warming, Says Reports." HNGN (Headlines and Global News). May 19, 2014. http://www.hngn.com/articles/31623/20140519/ california-fires-fueled-by-global-warming-says-reports.htm.

Silver, Cheryl Simon, and Ruth S. DeFries. *One Earth, One Future: Our Changing Global Environment*. Washington, DC: National Academy Press, 1990.

Simon, Stephanie. "Aspen Trees Die across the West." *Wall Street Journal*, October 14, 2009, A3.

"Simulation Shows World without Ozone Layer." NASA Earth Observatory, March 18, 2009. http://earthobservatory.nasa.gov/Newsroom/view.php?id=37588&src=eoa-nnews (no longer available).

Sinervo, Barry, et al. "Erosion of Lizard Diversity by Climate Change and Altered Thermal Niches." *Science* 328 (May 14, 2010): 894–899.

Slivka, Kelly. "Burst of Melting in Greenland's Ice Sheet Gives Scientists a Rare Climatic Snapshot." *The New York Times*, July 25, 2012, A4.

Sluijs, Appy, et al. "Subtropical Arctic Ocean Temperatures during the Palaeocene/Eocene Thermal Maximum." *Nature* 441 (June 1, 2006): 610–613.

Smith, Craig S. "One Hundred and Fifty Nations Start Groundwork for Global Warming Policies." *The New York Times*, January 18, 2001, 7.

Smith, Felisa A. "Some Like It Hot." *Science* 335 (February 24, 2012): 924–925.

Smith, L. C., et al. "Disappearing Arctic Lakes." *Science* 308 (June 3, 2005): 1429.

Smith, Lewis. "Falling Numbers Silence Cuckoo's Call of Spring." *Times* (London), March 6, 2002, n.p. (LEXIS).

Smith, Lindsay N. "Storm Surge." *National Geographic*, September 2015, 6.

Smith, O. Glenn. "Harvest the Sun—From Space." *The New York Times*, July 23, 2008. http://www.nytimes.com/2008/07/23/opinion/23smith.html.

Smith, Rebecca. "Wind, Solar Power Gain Users." *Wall Street Journal*, January 18, 2008, A6.

Smith, Stephen. "Comin' Ah-Choo: Tepid Temperatures Speeding Allergy Season." *Boston Globe*, April 10, 2002, A1.

"Smoke Blankets British Columbia." NASA Earth Observatory, July 8, 2015. http://earthobservatory.nasa.gov/IOTD/view.php?id=86190&src=eoa-iotd.

Smucker, Philip. "Global Warming Sends Troops of Baboons on the Run: Rising Temperatures and Humans Encroaching on Grasslands are Endangering the Ethiopian Primates." *Christian Science Monitor*, June 15, 2001, 7.

"Snow Cover Extent Declines in the Arctic." NASA Earth Observatory, January 8, 2013. http://earthobservatory.nasa.gov/IOTD/view.php?id=80102&src=eoa-iotd.

Sokolov, A. P., et al. "Probabilistic Forecast for 21st Century Climate Based on Uncertainties in Emissions (without Policy) and Climate Parameters. *Journal of Climate*, June 2009. doi: 10.1175/2009JCLI2863.1.

Solanki, S. K., et al. "Unusual Activity of the Sun during Recent Decades Compared to the Previous 11,000 Years." *Nature* 431 (October 28, 2004): 1084–1086.

"Solar in the Sahara." NASA Earth Observatory, January 10, 2016. http://earthobservatory.nasa.gov/IOTD/view.php?id=87293&src=eoa-iotd.

Solie, Stacey. "Scientists Adopt Tiny Island as a Warming Bellwether." *The New York Times*, October 6, 2012. http://www.nytimes.com/2012/10/07/us/scientists-in-washington-state-adopt-tiny-island-as-climate-change-bellwether.html.

Solnit, Rebecca. "Power in Paris." *Harper's*, December 2015, 5–7.

"Soot Drives Greenland Melting." *Nature* 509 (May 29, 2014): 537.

Sorkin, Andrew Ross. "A New Tack in the War on Mining Mountains." *The New York Times*, March 10, 2015, B1, B4. http://www.nytimes.com/2015/03/10/business/dealbook/pnc-joins-banks-not-financing-mountaintop-coal-removal.html?ref=business.

Souder, William. "Global Warming and a Toad Species' Decline." *Washington Post*, April 9, 2001, n.p. (in LEXIS)

"Sound Reasons for Ethanol." *Omaha World-Herald*, November 14, 2014, 4B.

Spalding, Mark. *World Atlas of Coral Reefs*. Berkeley: University of California Press, 2001.

Spalding, Mark. "Coral Grief: Rising Temperatures, Pollution, Tourism, and Fishing Have All Helped to Kill Vast Stretches of Reef in the Indian Ocean." *The Guardian* (U.K.), September 12, 2001, 8.

Spears, Tom. "Antarctica Rides Global 'Heatwave:' Continent's Warm Coast Causes Concern." *Ottawa Citizen*, August 8, 2001, A-1.

Spector, Michael. "The Climate Fixers: Is There a Technological Solution to Global Warming?" *The New Yorker*, May 14, 2012. http://www.newyorker.com/magazine/2012/05/14/the-climate-fixers.

Spencer, Jane. "Big Firms to Press Suppliers on Climate." *Wall Street Journal*, October 9, 2007, A7.

Speth, James Gustave. *Red Sky at Morning: America and the Crisis of the Global Environment*. New Haven, CT: Yale University Press, 2004.

Spielhagen, Robert F., et al. "Enhanced Modern Heat Transfer to the Arctic by Warm Atlantic Water." *Science* 331 (January 28, 2011): 450–453.

Spinrad, Richard W., and Ian Boyd. "Our Deadened, Carbon-Soaked Seas." *The New York Times*, October 15, 2015. http://www.nytimes.com/2015/10/16/opinion/our-deadened-carbon-soaked-seas.html.

Spracklen, D. V., S. R. Arnold, and C. M. Taylor. "Observations of Increased Tropical Rainfall Preceded by Air Passage over Forests." *Nature* 489 (September 13, 2012): 282–285.

Sridhar, Venkataramana, et al. "Large Wind Shift on the Great Plains during the Medieval Warm Period." *Science* 313 (July 21, 2006): 345–347.

Sriver, Ryan L. "Climate Change: Man-Made Cyclones." *Nature* 479 (November 3, 2011): 50–51.

Srokosz, M. A., and H. L. Bryden. "Observing the Atlantic Meridional Overturning Circulation Yields a Decade of Inevitable Surprises." *Science* 348 (June 19, 2015): 1330. http://science.sciencemag.org/content/348/6241/1255575. doi: 10.1126/science.1255575.

Staba, David. "Snowstorm Blankets Buffalo, Killing at Least 3." *The New York Times*, October 14, 2006. http://www.nytimes.com/2006/10/14/nyregion/14storm.html.

Staba, David. "An Old Steel Mill Retools to Produce Clean Energy." *The New York Times*, May 22, 2007. http://www.nytimes.com/2007/05/22/nyregion/22wind.html.

Stark, Mike. "Assault by Bark Beetles Transforming Forests; Vast Swaths of West Are Red, Gray, and Dying; Drought, Fire Suppression, and Global, Warming Are Blamed." *Los Angeles Times*, October 6, 2002, B1.

Steffen, Konrad. "Greenland Melt Accelerating, According to Colorado University-Boulder Study." Press release, University of Colorado, December 11, 2007 via NASA Earth Observatory, December 25, 2007. http://www.news.uiuc.edu/news/07/1210nitrogen.html (no longer available).

Steffensen, Jørgen Peder, et al. "High-Resolution Greenland Ice Core Data Show Abrupt Climate Change Happens in Few Years." *Science* 321 (August 1, 2008): 680–684.

Steig, Eric J. "Brief but Warm Antarctic Summer." *Nature* 489 (September 6, 2012): 39–40.

Steig, Eric J., et al. "Warming of the Antarctic Ice-Sheet Surface Since the 1957 International Geophysical Year." *Nature* 457 (January 22, 2008): 459–462.

Stein, Mara Lemos. "Ticket to Nowhere?" *Wall Street Journal*, March 9, 2009, R9.

Steinhauer, Jennifer. "In California, Heat Is Blamed for 100 Deaths." *The New York Times*, July 28, 2006. http://www.nytimes.com/2006/07/28/us/28heat.html.

Steinman, David. *Safe Trip to Eden: 10 Steps to Save Planet Earth from Global Warming Meltdown*. New York: Thunder's Mouth Press, 2007.

Stephens, Tim. "Methane in Arctic Lake Traced to Groundwater from Seasonal Thawing." University of California Santa Cruz Newscenter, March 9, 2015. http://news.ucsc.edu/2015/03/arctic-methane.html.

Stern, Nicholas H. *A Blueprint for a Safer Planet: How to Manage Climate Change and Create a New Era of Progress and Prosperity*. London: The Bodley Head, 2009.

The Stern Review. www.sternreview.org.uk. Accessed September 26, 2014.

Sternberg, Steve. "July Heat Left Records Smoldering." *USA Today*, August 1, 2011, A1.

Stevens, William K. *The Change in the Weather: People, Weather, and the Science of Climate*. New York: Delacorte Press, 1999.

Stewart, James B. "Coal Industry Wobbles as Market Forces Slug Away." *The New York Times*, August 7, 2015. http://www.nytimes.com/2015/08/07/business/energy-environment/coal-industrty-wobbles-as-market-forces-slu-away.html (no longer available).

St. Fleur, Nicholas. "Climate Change Is Shrinking Where Bumblebees Range, Research Finds." *The New York Times*, July 9, 2015. http://www.nytimes.com/2015/07/10/science/bumblebees-global-warming-shrinking-habitats.html

Stiffler, Linda, and Robert McClure. "Effects Could Be Profound." *Seattle Post-Intelligencer*, November 13, 2003, A8.

Stirling, Ian, and Andrew Derocher. "Possible Impacts of Climatic Warming on Polar Bears." *Arctic* 46(3) (September 1993): 240–245.

Stocker, Thomas F., Reto Knutti, and Gian-Kasper Plattner. "The Future of the Thermohaline Circulation—A Perspective." Pp. 277–293 in Dan Seidov, Bernd J. Haupt, and Mark Maslin, eds., *The Oceans and Rapid Climate Change: Past, Present, and Future*. Washington, DC: American Geophysical Union, 2001.

Stokstad, Erik. "Global Survey Documents Puzzling Decline of Amphibians." *Science* 306 (October 15, 2004): 391.

Stoll, John D. "Visions of the Future: What Will the Car of Tomorrow Look Like? Perhaps Nothing Like the Car of Today." *Wall Street Journal*, April 17, 2006, R8.

"Storm with 106-mph Gusts Slams Flooded Britain." Associated Press, February 12, 2014. http://www.foxnews.com/world/2014/02/12/storm-with-106-mph-gusts-slams-flooded-britain/.

Stott, Peter A., D.A. Stone, and M.R. Allen. "Human Contribution to the European Heatwave of 2003." *Nature* 432 (December 2, 2004): 610–613.

Straneo, Fiammetta, and Patrick Heimbach. "North Atlantic Warming and the Retreat of Greenland's Outlet Glaciers." *Nature* 504 (December 5, 2013): 36–43.

Strauss, Benjamin, and Robert Kopp. "Rising Seas, Vanishing Coastlines." *The New York Times*, November 24, 2012. http://www.nytimes.com/2012/11/25/opinion/sunday/rising-seas-vanishing-coastlines.html.

Stroeve, J. C., et al. "Tracking the Arctic's Shrinking Ice Cover: Another Extreme September Minimum in 2004." *Geophysical Research Letters* 32(4) (February 25, 2005): L04501. http://dx.doi.org/10.1029/2004GL021810.

Strong, Aaron, et al. "Ocean Fertilization: Time to Move On." *Nature* 461 (September 17, 2009): 347–348.

Struck, Doug. "'Rapid Warming' Spreads Havoc in Canada's Forests; Tiny Beetles Destroying Pines." *Washington Post*, March 1, 2006, A1. http://www.washingtonpost.com/wp-dyn/content/article/2006/02/28/AR2006022801772_pf.html.

Struck, Doug. "Icy Island Warms to Climate Change." *Washington Post*, June 7, 2007, A1. http://www.washingtonpost.com/wp-.

Stuart, Simon N., et al. "Status and Trends of Amphibian Declines and Extinctions Worldwide." *Science* 306 (December 3, 2004): 1783–1786.

Stuber, Nicola, et al. "The Importance of the Diurnal and Annual Cycle of Air Traffic for Contrail Radiative Forcing." *Nature* 441 (June 15, 2006): 864–867.

"Study Confirms Antarctica Warming." Environment News Service, September 6, 2006 (no longer available).

"Study Documents Widespread Extinction of Lizard Populations Due to Climate Change." NASA Earth Observatory, May 13, 2010. http://earthobservatory.nasa.gov/Newsroom/view.php?id=43996&src=eoa-manews (no longer available).

"Study Finds Climate Change a New Phenomenon in Planet's 11,000-Year Cooling Trend." Associated Press in CBS News. March 7, 2013. http://www.cbsnews.com/8301-205_162-57573132/study-finds-climate-change-a-new-phenomenon-in-planets-11000-year-cooling-trend.

"Study Reports Large-Scale Salinity Changes in Oceans; Saltier Tropical Oceans, Fresher Ocean Waters Near Poles Are Further Signs of Global Warming's Impacts on Planet." *Ascribe Newsletter*, December 17, 2003 (LEXIS).

"Study Reveals Increased River Discharge to Arctic Ocean; Finding Could Mean Big Changes to Global Climate." *Ascribe Newsletter*, December 12, 2002 (LEXIS).

"Study Says Glaciers Formed during a Very Warm Period." *The New York Times*, January 11, 2008. http://www.nytimes.com/2008/01/11/world/europe/11glacier.html.

Stukin, Stacie. "The Lean, Green Kitchen." *Vegetarian Times*, September 2007, 51–54.

Subramanian, Meera. *A River Runs Again: India's Natural World in Crisis.* New York: Public Affairs, 2025.

"Substantial Glacier Ice Loss in Central Asia's Largest Mountain Range." *Science Daily,* August 17, 2015. http://www.sciencedaily.com/releases/2015/08/150817132329.htm.

Sudarshan, V. "In Delhi, Just Waiting to Exhale." *The Hindu,* December 17, 2015, 11.

Sudetic, Chuck. "As the World Burns." *Rolling Stone,* September 2, 1999, 97–106, 129.

"Suffocation Suspected for Greatest Mass Extinction." NewScientist.com News Service, September 9, 2003. https://www.newscientist.com/article/dn4138-suffocation-suspected-for-greatest-mass-extinction/.

Suhr, Jim. "Armadillos Making Northern March." Associated Press, Illinois State Wire, September 16, 2006 (LEXIS).

"Summer Heat Wave in Europe Killed 35,000." United Press International, October 10, 2003 (LEXIS).

Sumner, Thomas. "No Stopping the Collapse of West Antarctic Ice Sheet." *Science* 344 (May 16, 2014): 683. http://www.nytimes.com/2008/01/08/science/earth/08gree.html.

Sun, Yadong, et al. "Lethally Hot Temperatures during the Early Triassic Greenhouse." *Science* 338 (October 19, 2012): 366–370.

Sunda, William G. "Iron and the Carbon Pump." *Science* 327 (February 5, 2010): 654–655.

"Sun's Heat Could Cut Fossil-Fuel Use." *Nature* 523 (July 23, 2015): 384.

Surendran, Aparna. "Fossil Fuel Cuts Would Reduce Early Deaths, Illness, Study Says." *Los Angeles Times,* August 17, 2001, A20.

Sutterley Tyler C., et al. "Mass Loss of the Amundsen Sea Embayment of West Antarctica from Four Independent Techniques." *Geophysical Research Letters,* December 5, 2014. doi: 10.1002/2014GL061940.

Svensen, Henrik, et al. "Release of Methane from a Volcanic Basin as a Mechanism for Initial Eocene Global Warming." *Nature* 429 (June 3, 2004): 542–545.

Svetsky, Benjamin. "How Al Gore Tamed Hollywood." *Entertainment Weekly,* July 21, 2006, 26–32.

"Swedish Bogs Flooding Atmosphere with Methane; Thawing Sub-Arctic Permafrost Increases Greenhouse Gas Emissions." American Geophysical Union. February 10, 2004. http://www.scienceblog.com/community/article2366.html (no longer available).

"Swinomish Climate Change Initiative." Pp. 133–145 in Zoltan Grossman and Alan Parker, eds., *Asserting Native Resilience: Pacific Rim Indigenous Nations Face the Climate Crisis.* Corvallis, OR: Oregon State University Press, 2012.

Swinomish Indian Tribal Community. *Swinomish Climate Change Initiative Climate Adaptation Action Plan.* La Conner, WA: Swinomish Indian Tribal Community Office of Planning and Community Development, 2010.

"Swinomish Study Social and Physical Impacts of Climate Change with EPA Grant." Indian Country Today Media Network, September 2, 2014. http://indiancountrytodaymedianetwork.com/2014/09/02/swinomish-study-social-and-physical-impacts-climate-change-epa-grant-156714.

Szabo, Liz. "Health Groups Say Obama Plan on Coal Will Save Lives." *USA Today,* August 4, 2015, 2A.

Tabuchi, Hiroko. "To Expand Offshore Power, Japan Builds Floating Windmills." *The New York Times,* October 24, 2013. http://www.nytimes.com/2013/10/25/business/international/to-expand-offshore-power-japan-builds-floating-windmills.html.

"Taking Stock of 2014 Fire Emissions." NASA Earth Observatory, August 11, 2015. http://earthobservatory.nasa.gov/IOTD/view.php?id=86388&src=eoa-iotd.

Tan, Ivy, Trude Storelvmo, and Mark D. Zelinka. "Observational Constraints on Mixed-Phase Clouds Imply Higher Climate Sensitivity." *Science* 352 (April 8, 2016): 224–227.

Tao, Shengli, et al. "Rapid Loss of Lakes on the Mongolian Plateau." *Proceedings of the National Academy of Sciences* 112 (7) (2015): L2281–L2286.

Tavernise, Sabrina. "Risk of Extreme Weather from Climate Change to Rise over Next Century, Report Says." *The New York Times*, June 22, 2015. http://www.nytimes.com/2015/06/23/science/risk-of-extreme-weather-from-climate-change-to-rise-over-next-century-report-says.html.

Tedesco, M., et al. "Extreme Snowmelt in Northern Greenland during Summer 2008." *Eos* 89(41) (October 7, 2008): 391.

Tedesco, M., et al. "The Role of Albedo and Accumulation in the 2010 Melting Record in Greenland." *Environmental Research Letters* 6(1) (January 21, 2011), n.p. http://iopscience.iop.org/article/10.1088/1748-9326/6/1/014005/meta;jsessionid=110ADC707B6345F40B3BE3E2C9A259C8.ip-10-40-1-105.

"10 Coastal Cities at Greatest Flood Risk as Sea Levels Rise." Environment News Service, September 3, 2013. http://ens-newswire.com/2013/09/03/10-coastal-cities-at-greatest-flood-risk-as-sea-levels-rise/.

Terris, Ben. "Jim Inhofe Is a Small-Plane-Flying, Global-Warming-Denying Senator. And Now He's Got a Gavel." *Washington Post*, January 9, 2015. http://www.washingtonpost.com/lifestyle/style/the-senates-top-climate-change-denier-is-flying-high-with-a-committee-chairmanship/2015/01/07/bf625fe0-9365-11e4-ba53-a477d66580ed_story.html.

Therrell, Matthew D., and Makayla J. Trotter. "Waniyetu Wówapi: Native American Records of Weather and Climate." *Bulletin of the American Meteorological Society* 92(5) (May 2011): 583–592. http://journals.ametsoc.org/doi/pdf/10.1175/2011BAMS3146.1.

"Thinning Ice More Fragile and Mobile." *Nature* 491 (November 15, 2012): 304.

"Thin Polar Bears Called Sign of Global Warming." Environment News Service, May 16, 2002. http://ens-news.com/ens/may2002/2002L-05-16-07.html (no longer available).

Thomas, Chris D., et al. "Extinction Risk from Climate Change." *Nature* 427 (January 8, 2004): 145–148.

Thomas, R. "Future Sea-Level Rise and Its Early Detection by Satellite Remote Sensing." In Vol. 4, *Effects of Changes in Atmospheric Ozone and Global Climate.* New York: United Nations Environment Program/United States Environmental Protection Agency, 1986.

Thomas, R., et al. "Accelerated Sea-Level Rise from West Antarctica." *Science* 306 (October 8, 2004): 255–258.

Thomas, William. "Salmon Dying in Hot Waters." Environment News Service, September 22, 1998. http://www.econet.apc.org/igc/en/hl/9809244985/hl11.html (no longer available).

Thompson, Andrea. "Growing Acid Problem Thins Shells of Ocean Creatures." Live Science, March 8, 2009. http://www.livescience.com/environment/090308-acidification-shells.html.

Thompson, Dan. "Experts Say California Wildfires Could Worsen with Global Warming." Associated Press, November 12, 2003 (LEXIS).

Thompson, David W. J., and Susan Solomon. "Interpretation of Recent Southern Hemisphere Climate Change." *Science* 296 (May 3, 2002): 895–899.

Thompson, Douglas. "Minnesota Tribes Collaborate to Save State's Disappearing Moose Population." Indian Country Today Media Network, February 11, 2015. http://indian countrytodaymedianetwork.com/2015/02/11/minnesota-tribes-collaborate-save-states -disappearing-moose-population-159151.

Thompson, L. G., et al. "Annually Resolved Ice Core Records of Tropical Climate Variability over the Past ~1800 Years." *Science* 340 (May 24, 2013): 945–950.

Thompson, Lonnie G., et al. "Glacier Loss on Kilimanjaro Continues Unabated." *Proceedings of the National Academy of Sciences.* Published online before print November 2, 2009. http://www.pnas.org/content/early/2009/10/30/0906029106.abstract. doi: 10.1073/pnas.0906029106.

Thorpe, N. 2000. "Contributions of Inuit Ecological Knowledge to Understanding the Impacts of Climate Change on the Bathurst Caribou Herd in the Kitikmeot Region, Nunavut." Unpublished master's thesis, School of Resource and Environmental Management. Project No. 268. Simon Fraser University, Burnaby, British Columbia.

Thorpe, R. B., et al. "Mechanisms Determining the Atlantic Thermohaline Circulation Response to Greenhouse Gas Forcing in a Non-Flux-Adjusted Coupled Climate Model." *Journal of Climate* 14 (July 15, 2001): 3102–3116.

Thurman, Judith. "In Fashion: Broad Stripes and Bright Stars: The Spring-Summer Men's Fashion Shows in Milan and Paris." *The New Yorker,* July 28, 2003, 78–82.

"Tibetan Glaciers Melting at Stunning Rate." November 24, 2008. Discovery.com. http://dsc.discovery.com/news/2008/11/24/tibet-glaciers-warming-02.html (no longer available).

Tidwell, Mike. *The Ravaging Tide: Strange Weather, Future Katrinas, and the Coming Death of America's Coastal Cities.* New York: Free Press, 2006.

"Time to Act on Nunavut Climate Change." Canadian Broadcasting Corporation, March 16, 2001. http://north.cbc.ca/cgi-bin/templates/view.cgi?/news/2001/03/16/16nunmoose (no longer available).

Tingley, Martin P., and Peter Huybers. "Recent Temperature Extremes at High Northern Latitudes Unprecedented in the Past 600 Years." *Nature* 496 (April 11, 2013): 201–205.

Tolbert, Margaret A., and Owen B. Toon. "Solving the P[olar] S[tratospheric] C[loud] Mystery." *Science* 292 (April 6, 2001): 61–63.

Tollefson, Jeff. "Heatwaves Blamed on Global Warming; Unusually High Frequency Points to Human Influence." *Nature* 488 (August 9, 2012): 143–144. http://www.nature.com/news/heatwaves-blamed-on-global-warming-1.11130.

Tollefson, Jeff. "Climate Science: A Line in the Sands." *Nature* 500 (August 8, 2013): 136–137. http://www.nature.com/news/climate-science-a-line-in-the-sands-1.135150 (no longer available).

Tollefson, Jeff. "China's Carbon Emissions Could Peak Sooner Than Forecast." *Nature* 531 (March 24, 2016): 425. http://www.nature.com/news/china-s-carbon-emissions-could -peak-sooner-than-forecast-1.19597.

Tomsho, Robert. "Wind Shift in Energy Debate." *Wall Street Journal,* June 19, 2008, A11.

Toner, Mike. "Huge Ice Chunk Breaks Off Antarctica." *Atlanta Journal-Constitution,* March 20, 2002, A1.

Toner, Mike. "Meltdown in Montana; Scientists Fear Park's Glaciers May Disappear Within 30 Years." *Atlanta Journal-Constitution,* June 30, 2002, 4A.

Toner, Mike. "Microscopic Ocean Life in Global Decline; Temperature Shifts a Cause or an Effect?" *Atlanta Journal-Constitution,* August 9, 2002, 3A.

Toner, Mike. "Arctic Ice Thins Dramatically, NASA Satellite Images Show." *Atlanta Journal-Constitution,* October 24, 2003, 1A.

Toniazzo, T., J. M. Gregory, and P. Huybrechts. "Climatic Impact of a Greenland Deglaciation and Its Possible Irreversibility." *Journal of Climate* 17(1) (January 1, 2004): 21–33.

"Tornado Rips through City Northeast of Tokyo, 1 Person Dead, Dozens Injured." Associated Press, in Yahoo News, May 5, 2012. http://news.yahoo.com/tornado-rips-city-northeast-tokyo-1-person-dead-103010197.html.

"Torrential Rains Flood Argentina, Overall Death Toll Rises to 52." Associated Press in *Washington Post*, April 3, 2013. http://www.washingtonpost.com/business/torrential-rains-flood-argentina-overall-death-toll-rises-to-31/2013/04/03/f2e584c8-9c6d-11e2-9219-51eb8387e8f1_print.html (no longer available).

"Torrential Rains Submerge Half of Philippine Capital, Trigger Landslide That Kills 8 People." Associated Press, August 7, 2012. https://www.washingtonpost.com/world/torrential-rains-submerge-half-of-philippine-capital/2012/08/07/29c94b48-e0a6-11e1-a19c-fcfa365396c8_gallery.html.

Torres, Christian, and Juliet Eilperin. "Mayor Bloomberg Gives $50 Million to Fight Coal-Fired Power Plants." *Washington Post*, July 21, 2011. http://www.washingtonpost.com/national/health-science/mayor-bloomberg-gives-50-million-to-fight-coal-fired-power-plants/2011/07/20/gIQAEKKURI_print.html.

Tosca, M. G., et al. "Human-Caused Fires Limit Convection in Tropical Africa: First Temporal Observations and Attribution." *Geophysical Research Letters* 42(15): August 5, 2015. http://onlinelibrary.wiley.com/doi/10.1002/2015GL065063/abstract#footer-support-info.

Toth, Lauren T., et al. "ENSO Drove 2500-Year Collapse of Eastern Pacific Coral Reefs." *Science* 337 (July 6, 2012): 81–84.

Townsend, Mark. "Monsoon Britain: As Storms Bombard Europe, Experts Say That What We Still Call 'Freak' Weather Could Soon Be the Norm." *The Observer* (London), August 11, 2002, 15.

Trapp, Robert J., et al. "Changes in Severe Thunderstorm Frequency during the 21st Century Caused by Anthropogenically Enhanced Global Radiative Forcing." *Proceedings of the National Academy of Sciences* 104(50) (December 4, 2007): 19719–19723.

Travis, Davis J., Andrew M. Carleton, and Ryan G. Lauritsen. "Climatology: Contrails Reduce Daily Temperature Range." *Nature* 418 (August 8, 2002): 593–594.

Treaster, Joseph B. "Gulf Coast Insurance Expected to Soar." *The New York Times*, September 24, 2005. http://www.nytimes.com/2005/09/24/business/24insure.html.

Trenberth, Kevin E. "The Ocean Is Warming, Isn't It?" *Nature* 465 (May 20, 2010): 304.

Trenberth, Kevin E., et al. "The Changing Character of Precipitation." *Bulletin of the American Meteorological Society* 84(9) (September 2003): 1205–1217.

"Tropical Storm Nears Hawaii, New Ocean Plan Predicts Worse." Environment News Service, July 29, 2013. http://ens-newswire.com/2013/07/29/tropical-storm-nears-hawaii-new-ocean-plan-predicts-worse/.

Tsuda, Atsushi, et al. "A Mesoscale Iron Enrichment in the Western Subarctic Pacific Induces a Large Centric Diatom Bloom." *Science* 300 (May 9, 2003): 958–961.

Tubiello, Francesco, Jean Francois Soussana, and S. Mark Howden. "Crop and Pasture Response to Climate Change." *Proceedings of the National Academy of Sciences* 104 (December 11, 2007): 19686–19690.

"Turkish Glaciers Shrink by Half." NASA Earth Observatory, July 2, 2015. http://earthobservatory.nasa.gov/IOTD/view.php?id=86140&src=eoa-iotd.

Turner J., et al. "Non-Annular Atmospheric Circulation Change Induced by Stratospheric Ozone Depletion and Its Role in the Recent Increase of Antarctic Sea Ice Extent." *Geophysical Research Letters* 36 (April 23, 2009): L08502. doi: 10.1029/2009GL037524.

"2011 Arctic Sea Ice." NASA Earth Observatory. October 4, 2011. http://earthobservatory. nasa.gov/IOTD/view.php?id=53108&src=eoa-iotd. Accessed October 5, 2011.

"2011 Global Temperatures." NASA Earth Observatory, January 20, 2012. http:// earthobservatory.nasa.gov/IOTD/view.php?id=76975&src=eoa-iotd.

"Typhoon Maysak Approaches the Philippines." NASA Earth Observatory, April 4, 2015. http://earthobservatory.nasa.gov/IOTD/view.php?id=85638&src=eoa-iotd.

Underwood, Emily. "Models Predict Longer, Deeper U.S. Droughts." *Science* 347 (February 13, 2015): 707.

U.N. Environment Program. "Global Outlook for Ice and Snow," 2007–2008. http://www .unep.org/geo/geo_ice/.

U.N. Environment Program. Migratory Species and Climate Change: Impacts of a Changing Environment on Wild Animals. Bonn, Germany: UNEP / CMS Convention on Migratory Species and DEFRA, 2006. http://www.cms.int/sites/default/files/document/ScC14 _Inf_09_Migratory_Species%26Climate_Change_E_0.pdf.

"Unexpected Ice Loss Detected in East Antarctica." NASA Earth Observatory, November 24, 2009. http://earthobservatory.nasa.gov/Newsroom/view.php?id=41455&src=eoa-nnews (no longer available).

"Unprecedented Climate Extremes Mark Decade 2001–2010." Environment News Service, July 17, 2013. http://ens-newswire.com/2013/07/17/unprecedented-climate-extremes -mark-decade-2001-2010/.

"Unstable Antarctica: What's Driving Ice Loss?" NASA Earth Observatory. December 15, 2010. http://www.nasa.gov/topics/earth/features/unstable-antarctica.html.

"UN Warns of Global Shift 'From a Fish to a Jellyfish Ocean.'" Environment News Service, May 30, 2013. http://ens-newswire.com/2013/05/30/un-warns-of-global-shift-from-a -fish-to-a-jellyfish-ocean/.

Unwin, Brian. "Tropical Birds and Exotic Sea Creatures Warm to Britain's Welcoming Waters." *The Independent* (London), August 20, 2001, 7.

Urban, Mark C. "Accelerating Extinction Risk from Climate Change." *Science* 348 (May 1, 2015): 571–573.

Urbina, Ian. "Perils of Climate Change Could Swamp Coastal Real Estate." *The New York Times*, November 24, 2016. http://www.nytimes.com/2016/11/24/science/global-warming-coastal-real-estate.html.

U.S. Department of the Interior. *The State of the Birds: 2010 Report on Climate Change*. Washington, DC: Fish and Wildlife Service.

U.S. Environmental Protection Agency. *Ecological Impacts from Climate Change: An Economic Analysis of Freshwater Recreational Fishing*, April 1995. Washington, DC: EPA.

Utton, Tim. "The Rat Rampage." *Daily Mail* (U.K.), January 22, 2003, n.p. (LEXIS).

"UV Radiation Linked to Deformed Amphibians." Environment News Service, June 2, 2002. http://www.ens-newswire.com/ens/jun2002/2002-06-21-09.html.

Våge, Kjetil, et al. "Surprising Return of Deep Convection to the Subpolar North Atlantic Ocean in Winter 2007–2008." *Nature GeoScience* 2(1) (January 2009). http://www.nature .com/ngeo/journal/v2/n1/abs/ngeo382.html.

Vance, Erik. "Letter from the Sea of Cortez: Emptying the World's Aquarium." *Harper's*, August 2013, 53–62.

Vanderkam, Laura. "Want to Save the Planet? Stay Home." *USA Today*, May 20, 2008, 9A.

Van Mantgem, Phillip J., et al. "Widespread Increase of Tree Mortality Rates in the Western United States." *Science* 323 (January 23, 2009): 521–524.

Vastag, Brian. "Warmer Still: Extreme Climate Predictions Appear Most Accurate, Report Says." *Washington Post*, November 8, 2012. http://www.washingtonpost.com/national/

health-science/warmer-still-extreme-climate-predictions-appear-most-accurate-study-says/2012/11/08/ebd075c6-29c7-11e2-96b6-8e6a7524553f_print.html.

Vastag, Brian, and Ed O'Keefe. "Obama to Visit Missouri Sunday; Deadly Tornadoes on the Rise." *Washington Post*, May 24, 2011. http://www.washingtonpost.com/national/environment/researchers-see-a-pattern-in-rise-of-deadly-tornadoes/2011/05/23/AFinz49G_print.html.

Vedantam, Shankar. "Glacier Melt Could Signal Faster Rise in Ocean Levels." *Washington Post*, February 17, 2006, A1.

Veizer, Jan, Yves Godderis, and Louis M. Francois. "Evidence for Decoupling of Atmospheric CO_2 and Global Climate during the Phanerozoic Eon." *Nature* 408 (December 7, 2000): 698–701.

Velicogna, Isabella, and John Wahr. "Measurements of Time-Variable Gravity Show Mass Loss in Antarctica." *Science* (online), March 2, 2006. doi: 10.1126/science.1123785.

Velicogna, Isabella, and John Wahr. "Acceleration of Greenland Ice Mass Loss in Spring 2004." *Nature* 443 (September 21, 2006): 329–331.

Vergano, Dan. "Sandy Revives Debate over Rising Sea Level." *USA Today*, November 28, 2012, 3A.

Vergano, Dan. "Kidney Stone Cases Could Heat Up." *USA Today*, July 15, 2008, A1.

Vergara, Walter. "The Impacts of Climate Change in Latin America." 2007. https://www.thefreelibrary.com/The+impacts+of+climate+change+in+Latin+America.-a0221761359.

Verschuren, Dirk. "Global Change: The Heat on Lake Tanganyika." *Nature* 424 (August 14, 2003): 731–732.

Vick, Karl, and Sonya Geis. "California Fires Continue to Rage." *Washington Post*, October 24, 2007, A1. http://www.washingtonpost.com/wp-dyn/content/article/2007/10/23/AR2007102300347_pf.html.

Victor, David G. *Climate Change: Debating America's Policy Options*. New York: Council on Foreign Relations, 2004.

Victor, David G., and James P. Leape. "Global Climate Agreement: After the Talks." *Nature* 527 (November 26, 2015): 439–441.

Vidal, John. "Antarctica Sends Warning of the Effects of Global Warming: Scientists Stunned as Ice Shelf Falls Apart in a Month." *The Guardian* (U.K.), March 20, 2002, 3.

Viles, Carson. "First Foods and Climate Change." Pacific Northwest Tribal Climate Change Project, University of Oregon. No date. http://tribalclimate.uoregon.edu/files/2010/11/firstfoods_climatechange_12-14-11_final.pdf. Accessed September 13, 2014.

Vincent, John, and Paul Brown. "Swoop to Conquer: Global Warming Brings Butterflies to Britain Earlier." *The Guardian* (U.K.), May 24, 2000, 9.

"Visualizing the 2012 Sea Ice Minimum." NASA Earth Observatory, September 27, 2012. http://earthobservatory.nasa.gov/IOTD/view.php?id=79256&src=eoa-iotd.

Vitello, Paul. "Hurricane Fears Cost Homeowners Coverage." *The New York Times*, October 16, 2007. http://www.nytimes.com/2007/10/16/nyregion/16insurance.html.

Voggesser, Garrit. "The Tribal Path Forward: Confronting Climate Change and Conserving Nature." *The Wildlife Professional*, Winter 2010.

Volk, Tyler. *CO_2 Rising: The World's Greatest Environmental Challenge*. Cambridge, MA: MIT Press, 2008.

"Vulnerability of Coastal Louisiana Tribes in a Climate Change Context." Natural Resources and Conservation Services (National Resources Conservation Service) Workshop. Tribes and Climate Change. Northern Arizona University. September 26, 2012. http://www4.nau.edu/tribalclimatechange/tribes/gulfcoast_lacoastal.asp.

Waddell, Lynn. "Rising Insurance Rates Push Florida Homeowners to Brink." *The New York Times*, June 29, 2006. http://www.nytimes.com/2006/06/29/us/29florida.html.

Wagner, Angie. "Debate over Causes Aside, Warm Climate's Effects Striking in the West." Associated Press, April 27, 2004 (LEXIS).

Wagner, Gernot, et al. "Energy Policy: Push Renewables to Spur Carbon Pricing." *Nature* 525 (September 3, 2015): 27–29. http://www.nature.com/news/energy-policy-push-renewables-to-spur-carbon-pricing-1.18260.

Wald, Matthew. "It's Free, Plentiful and Fickle." *The New York Times*, December 28, 2006. http://www.nytimes.com/2006/12/28/business/28wind.html.

Wald, Matthew. "What's So Bad about Big?" *The New York Times*, March 7, 2007. http://www.nytimes.com/2007/03/07/business/businessspecial2/07big.html.

Wald, Matthew. "Turning Glare into Watts." *The New York Times*, March 6, 2008. http://www.nytimes.com/2008/03/06/business/06solar.html.

Wald, Matthew L. "Mounting Costs Slow the Push for Clean Coal." *The New York Times*, May 30, 2008. http://www.nytimes.com/2008/05/30/business/30coal.html.

Wald, Matthew. "Two Large Solar Plants Planned in California." *The New York Times*, August 15, 2008. http://www.nytimes.com/2008/08/15/business/15solar.html.

Wald, Matthew. "Wind Energy Bumps into Power Grid's Limits." *The New York Times*, August 27, 2008. http://www.nytimes.com/2008/08/27/business/27grid.html.

Wald, Matthew L. "Despite Climate Concern, Global Study Finds Fewer Carbon Capture Projects." *The New York Times*, October 10, 2013. http://www.nytimes.com/2013/10/11/science/earth/study-finds-setbacks-in-carbon-capture-projects.html.

Walter, K. M., et al. "Methane Bubbling from Siberian Thaw Lakes as a Positive Feedback to Climate Warming. *Nature* 414 (September 7, 2006): 71–75. http://shadow.eas.gatech.edu/~kcobb/abrupt/walter06.pdf.

Walters, Pat. "Risk Takers: Ice Investigator." *National Geographic*, January 2013, 58–67.

Walther, Gian-Reto. "Weakening of Climatic Constraints with Global Warming and Its Consequences for Evergreen Broad-leaved Species." *Folia Geobotanica* 37 (2002): 129–139.

Walther, Gian-Reto, et al. "Ecological Responses to Recent Climate Change." *Nature* 416 (March 28, 2002): 389–395.

Walther, Gian-Reto. "Plants in a Warmer World." *Perspectives in Plant Ecology, Evolution, and Systematics* 6(3) (2003): 169–185.

Walther, Gian-Reto, et al. "Palms Tracking Climate Change." *Global Ecology and Biogeography*, May 5, 2007. doi: 10.1111/j.1466-8238.2007.00328.x.

"Warm Arctic, Chilly Mid-Latitudes." NASA Earth Observatory. April 3, 2013. http://earthobservatory.nasa.gov/IOTD/view.php?id=80804&src=eoa-iotd.

"Warmer Climate Hurting Birds, New IUCN Red List Shows." Environment News Service, May 19, 2008. http://www.ens-newswire.com/ens/may2008/2008-05-20-03.asp (no longer available).

"Warmest June on Record for Global Ocean Surface Temperature." Environment News Service, July 221, 2009. http://www.ens-newswire.com/ens/jul2009/2009-07-21-096.asp (no longer available).

"Warming Climate Jeopardizes California Fruit and Nut Crops." Environment News Service, July 23, 2009. http://www.ens-newswire.com/ens/jul2009/2009-07-23-095.html.

"Warming Climate Undermines World Food Supply." Environment News Service, December 3, 2007. http://www.ens-newswire.com/ens/dec2007/2007-12-03-05.asp (no longer available).

"Warming Could Submerge Three of India's Largest Cities: Scientist." Agence France Presse, December 6, 2003. http://www.spacedaily.com/2003/031206151721.ha5156ha.html.

"Warming Past, Present." *Omaha World-Herald*, September 4, 2009, 5A.

"Warming, Polluted Oceans Imperil 75 Percent of All Coral Reefs." Environment News Service, February 24, 2011. http://www.ens-newswire.com/ens/feb2011/2011-02-24-02.html.

"Warming Streams Could Wipe Out Salmon, Trout." Environment News Service, May 22, 2002. http://ens-news.com/ens/may2002/2002L-05-22-06.html (no longer available).

"Warming Trend Continues in 2014." World Meteorological Organization. February 2, 2015. https://www.wmo.int/media/content/warming-trend-continues-2014. "Warm Water and Strange Weather May Be Connected." NASA Earth Observatory, April 18, 2015. http://earthobservatory.nasa.gov/IOTD/view.php?id=85714&src=eoa-iotd.

"Warm Water and Strange Weather May Be Connected." NASA Earth Observatory, April 18, 2015. http://earthobservatory.nasa.gov/IOTD/view.php?id=85714&src=eoa-iotd.

"Warm Water Surging into Arctic Ocean." Environment News Service, September 27, 2006. http://www.ens-newswire.com/ens/sep2006/2006-09-27-01.asp (no longer available).

Warner, Melanie. "Is America Ready to Quit Coal?" *The New York Times*, February 15, 2009. http://www.nytimes.com/2009/02/15/business/15coal.html.

Warren, M. S., et al. "Rapid Responses of British Butterflies to Opposing Forces of Climate and Habitat Change." *Nature* 414 (November 1, 2001): 65–69.

Warrick, Joby. "CO_2 Levels in Atmosphere Rising at Dramatically Faster Rate, U.N. Report Warns." *Washington Post*, September 9, 2014. http://www.washingtonpost.com/national/health-science/co2-levels-in-atmosphere-rising-at-dramatically-faster-rate-un-report-warns/2014/09/08/3e2277d2-378d-11e4-bdfb-de4104544a37_story.html.

Warrick, Joby. "Greenhouse Gases Hit New Milestone, Fueling Worries about Climate Change." *Washington Post*, November 9, 2015. https://www.washingtonpost.com/national/health-science/greenhouse-gases-hit-new-milestone-fueling-worries-about-climate-change/2015/11/08/1d7c7ffc-8654-11e5-be39-0034bb576eee_story.html.

Warrick, Joby. "A Warming Globe Brings Invisible Threat: Disease." *Washington Post*, November 27, 2015. http://www.washingtonpost.com/sf/national/2015/11/27/disease/?wpmm=1&wpisrc=nl_headlines.

Wasko, Conrad, and Ashish Sharma. "Steeper Temporal Distribution of Rain Intensity at Higher Temperatures within Australian Storms." *Nature GeoScience* (June 2015), online letter. http://www.nature.com/ngeo/journal/vaop/ncurrent/full/ngeo2456.html.

"Watching the Ozone Hole Before and After the Montreal Protocol." NASA Earth Observatory. September 18, 2012. http://earthobservatory.nasa.gov/IOTD/view.php?id=79198&src=eoa-iotd.

"Waterfall-Like Rain Eases in Southwest Japan, but 28 Dead, Thousands of Homes Damaged." *Washington Post*, July 15, 2012. http://www.washingtonpost.com/world/asia_pacific/floods-in-southern-japan-kill-at-least-26-thousands-remain-cut-off/2012/07/16/gJQAxYWcnW_print.htm (no longer available).

"Water Levels Still Dropping Near São Paulo." NASA Earth Observatory, March 7, 2015. http://earthobservatory.nasa.gov/IOTD/view.php?id=85460&src=eoa-iotd.

Waters, Colin N., et al. "The Anthropocene Is Functionally and Stratigraphically Distinct from the Holocene." *Science* 351 (January 8, 2016). http://science.sciencemag.org/content/351/6269/aad2622. doi: 10.1126/science.aad2622.

Watson, A. J., et al. "Effect of Iron Supply on Southern Ocean CO_2 Uptake and Implications for Glacial Atmospheric CO_2." *Nature* 407 (October 12, 2000): 730–733.

Watson, Jeremy. "Plan to Hold Back Tides of Venice Runs into Flood of Opposition from Greens." *Scotland on Sunday*, December 30, 2001, 18.

Watson, Traci. "Companies Rethink Coal Plants." *USA Today*, March 9, 2009, 9A.

Watson, Traci. "Bumblebees Stung by Climate Change." *USA Today*, July 10, 2015, 4A.

Watt-Cloutier, Sheila. "Honouring Our Past, Creating Our Future: Education in Northern and Remote Communities." In Lynne Davis, Louise Lahache, and Marlene Castellano, eds., *Aboriginal Education: Fulfilling the Promise*. Vancouver: University of British Columbia Press, 2000.

Watt-Cloutier, Sheila. "Speech Notes for Sheila Watt-Cloutier, Chair, Inuit Circumpolar Conference." Conference of Parties to the United Nations Framework Convention on Climate Change, Milan, Italy, December 10, 2003. http://www.inuitcircumpolar.com/index.php?ID=242&Lang=En (no longer available).

Watt-Cloutier, Sheila. "Canada and Inuit: Addressing Global Environmental Challenges, Remarks by Sheila Watt-Cloutier, Chair, Inuit Circumpolar Conference at the Inaugural Environmental Protection Service Inuit Speaker Series." Ottawa, Ontario, January 16, 2004. http://www.inuitcircumpolar.com/index.php?ID=250&Lang=En.

Watt-Cloutier, Sheila. "The Climate Change Petition by the Inuit Circumpolar Conference to the Inter-American Commission on Human Rights: Presentation by Sheila Watt-Cloutier, Chair, Inuit Circumpolar Conference [at the] 11th Conference of Parties to the U.N. Framework Convention on Climate Change," December 7, 2005. Montreal. http://inuitcircumpolar.com/index.php?ID=318&Lang=En.

Watt-Cloutier, Sheila, and Terry Fenge. "Poisoned by Progress: Will Next Week's Negotiations in Bonn Succeed in Banning the Chemicals That Inuit Are Eating?" Press release, Inuit Circumpolar Conference, March 14, 2000. http://www.inuitcircumpolar.com/index.php?ID=250&Lang=En (no longer available).

Watts, Nick, et al. "Health and Climate Change: Policy Responses to Protect Public Health." *The Lancet*, June 23, 2015. http://www.thelancet.com/commissions/climate-change.

Weart, Spencer R. *The Discovery of Global Warming*. Cambridge, MA: Harvard University Press, 2003.

Weather.com. "Cyclone Chapala in the Arabian Sea Likely to Be Rare, Destructive Landfall in Yemen." November 3, 2015. http://www.weather.com/storms/hurricane/news/cyclone-chapala-yemen-oman-arabian-peninsula.

Weather.com. "February 2016 Was the Most Abnormally Warm Month Ever Recorded, Topping January 2016, NASA Says." March 14, 2016. https://weather.com/news/climate/news/record-warmest-february-global-2016.

Weatherhead, E., S. Gearheard, and R. G. Barr. "Changes in Weather Persistence: Insight from Inuit Knowledge." *Global Environmental Change* 20 (2010): 523–528.

Webb, Jason. "Mosquito Invasion as Argentina Warms." Reuters, 1998. http://ces.iisc.ernet.in/hpg/envis/doc98html/enarg117.html.

Webb, Jason. "Small Islands Say Global Warming Hurting Them Now." Reuters, 1998. http://bonanza.lter.uaf.edu/~davev/nrm304/glbxnews.htm (no longer available).

Webb, Jason. "World Forests Said Vulnerable to Global Warming." Reuters. November 4, 1998. http://bonanza.lter.uaf.edu/~davev/nrm304/glbxnews.htm (no longer available).

Webb, Jason. "World Temperatures Could Jump Suddenly." Reuters, November 4, 1998. http://bonanza.lter.uaf.edu/~davev/nrm304/glbxnews.htm.

Weinburg, Bill. "Hurricane Mitch, Indigenous Peoples, and Mesoamerica's Climate Disaster." *Native Americas* 16(3–4) (Fall–Winter 1999): 50–59. http://nativeamericas.aip.cornell.edu/fall99/fall99weinberg.html.

Weiner, Jonathan. *The Next One Hundred Years: Shaping the Fate of Our Living Earth.* New York: Bantam Books, 1990.

Weise, Elizabeth. "Warming Climate Makes Gardeners' Map Out-of-date." *USA Today,* April 24, 2008, A1.

Weise, Elizabeth. "Some Crops Migrate North with Warmer Temperatures." *USA Today,* September 17, 2013. http://www.usatoday.com/story/news/nation/2013/09/17/climate-change-agriculture-crops/2784561/.

Weiss, Rick. "Firms Seek Patents on 'Climate Ready' Altered Crops." *Washington Post,* May 13, 2008. http://www.washingtonpost.com/wp-dyn/content/article/2008/05/12/AR200 8051202919_pf.html.

Welch, Craig. "Global Warming Hitting Northwest Hard, Researchers Warn." *Seattle Times,* February 14, 2004. http://seattletimes.nwsource.com/html/localnews/2001857961 _warming14m.html (no longer available).

Welch, Craig. "Puget Sound Waters Becoming More Corrosive." *Seattle Times,* July 13, 2010, B1, B2.

Welch, Craig. "Sea Change: Scientists Fear Ocean Acidification Will Drive the Collapse of Alaska's Iconic Crab Fishery." *Seattle Times,* September 11, 2013a. http://apps.seattletimes .com/reports/sea-change/2013/sep/11/alaska-crab-industry/.

Welch, Craig. "Sea Change: The Pacific's Perilous Turn." *Seattle Times,* September 11, 2013b. http://apps.seattletimes.com/reports/sea-change/2013/sep/11/pacific-ocean-perilous-turn -overview/.

Welch, Ivo. "Why Divestment Fails." *The New York Times,* May 9, 2014. http://www.nytimes .com/2014/05/10/opinion/why-divestment-fails.html?hp&rref=opinion.

Wentz, Frank J., et al. "How Much More Rain Will Global Warming Bring?" *Science* 317 (July 2007): 233–235.

"West Antarctic Glacier Loss Appears Unstoppable, UCI-NASA Study Finds." UCI News. University of California–Irvine. December 2, 2014. http://news.uci.edu/press-releases/ west-antarctic-melt-rate-has-tripled-uc-irvine-nasa/West.

Westbrook, Graham K., et al. "Escape of Methane Gas from the Seabed along the West Spitsbergen Continental Margin." *Geophysical Research Letters* 36 (August 6, 2009): L15608. doi:10.1029/2009GL039191.

Westerling, A. L., et al. "Warming and Earlier Spring Increase Western U.S. Forest Wildfire Activity." *Science* 313 (August 18, 2006): 940–943.

"Western Wildfires Linked to Atlantic Ocean Temperatures." Environment News Service, January 3, 2007. http://www.ens-newswire.com/ens/jan2007/2007-01-03-09.asp# anchor2 (no longer available).

"Wetter, Drier with Warming." *Nature* 484 (May 3, 2012): 8.

Wheaton, Sarah. "Keystone XL Pipeline Fight Lifts Environmental Movement." *The New York Times,* January 24, 2014, A9. https://www.nytimes.com/2014/01/25/us/keystone-xl -pipeline-fight-lifts-environmental-movement.html.

"Where's the Snow?" First Stewards, May 8, 2014. http://www.firststewards.org/2014-press .html.

White, James C. "Do I Hear a Million?" *Science* 304 (June 11, 2004): 1609–1610.

White, Joseph R. "An Ecotopian View of Fuel Economy." *Wall Street Journal,* June 26, 2006, D4.

White, M. A., et al. "Extreme Heat Reduces and Shifts United States Premium Wine Production in the 21st Century." *Proceedings of the National Academy of Sciences* 103(30) (July 25, 2006): 11217–11222.

Whitfield, John. "Alaska's Climate: Too Hot to Handle." *Nature* 425 (September 25, 2003): 338–339.

Whitlock, Craig. "Cloudy Germany a Powerhouse in Solar Energy." *Washington Post*, May 5, 2007, A1. http://www.washingtonpost.com/wp-dyn/content/article/2007/05/04/AR200 7050402466_pf.html.

"Why Monitoring Emissions from Permafrost Matters." NASA Earth Observatory, April 12, 2016. http://earthobservatory.nasa.gov/IOTD/view.php?id=87850&src=eoa-iotd.

"Widespread Warmth Envelops Greenland." NASA Earth Observatory, May 18, 2016. http://earthobservatory.nasa.gov/IOTD/view.php?id=88048&src=eoa-iotd.

Wilber, Del Quentin. "U.S. Airlines under Pressure to Fly Greener." *Washington Post*, July 28, 2007, D1. http://www.washingtonpost.com/wp-dyn/content/article/2007/07/27/AR 2007072702256_pf.html.

"Wildfires Burn More U.S. Forest." *Nature* 538 (October 20, 2016): 292–293.

"Wildfires in Irkutsk." NASA Earth Observatory, May 21, 2014. http://earthobservatory.nasa .gov/IOTD/view.php?id=83702&src=eoa-iotd.

"Wildfires in Siberia." NASA Earth Observatory, July 12, 2012. http://earthobservatory.nasa .gov/IOTD/view.php?id=78515&src=eoa-iotd.

"Wildfires Set to Increase 50 Percent by 2050." NASA Earth Observatory, July 28, 2009. http://earthobservatory.nasa.gov/Newsroom/view.php?id=39650&src=eoa-manews (no longer available).

Wildlife Society, The. "Final Position Statement: Global Climate Change and Wildlife." http:// wildlife.org/wp-content/uploads/2014/05/global.climate.change.11.21.11.pdf.

Wilhite, Donald A. "Drought in the U.S. Great Plains." Pp. 743–758 in Thomas D. Potter and Bradley R. Colman, eds., *Handbook of Weather, Climate, and Water: Atmospheric Chemistry, Hydrology, and Societal Impacts.* Hoboken, NJ: Wiley InterScience, 2003.

Wilkin, Dwayne. "A Team of Glacial Ice Experts Say Mother Nature's Thermostat Has Kept the Eastern Arctic at About the Same Temperature since 1960." *Nunatsiaq News*, May 30, 1997. http://www.nunanet.com/~nunat/week/70530.html#7 (no longer available).

Wilkinson, Clive, et al. "Ecological and Socioeconomic Impacts of 1998 Coral Mortality in the Indian Ocean: An ENSO Impact and a Warning of Future Change?" *Ambio* 28(2) (March 1999): 188–196.

Wilkinson, Freddie. "Don't Climb Every Mountain." *The New York Times*, May 18, 2012, A27.

Will, George F. "Fuzzy Climate Math." *Washington Post*, April 12, 2007, A27. http://www .washingtonpost.com/wp-dyn/content/article/2007/04/11/AR2007041102109_pf.html.

Williams, A. P., et al. "Contribution of Anthropogenic Warming to California Drought during 2012–2014." *Geophysical Research Letters*, August 28, 2015. http://onlinelibrary. wiley.com/doi/10.1002/2015GL064924/abstract. doi: 10.1002/2015GL064924.

Williams, Frances. "Everest Hit by Effects of Global Warming." *Financial Times* (London), June 6, 2002, 2.

Williams, Gisela. "Resorts Prepare for a Future Without Skis." *The New York Times*, December 2, 2007. http://www.nytimes.com/2007/12/02/travel/02skiglobal.html.

Williams, Wendy, and Robert Whitcomb. *Cape Wind: Money, Celebrity, Class, Politics, and the Battle for Our Energy Future on Nantucket Sound.* New York: Public Affairs, 2007.

Wilmsen, Steven. "Critters Enjoy a Baby Boom; Mild Winter's Downside Is Proliferation of Vermin." *Boston Globe*, March 30, 2002, B1.

Wilson, Paul A., and Richard D. Norris. "Warm Tropical Ocean Surface and Global Anoxia during the Mid-Cretaceous Period." *Nature* 412 (July 26, 2001): 425–429.

Wilson, Rory P. "Animal Behavior: The Price Tag." *Nature* 469 (January 13, 2011): 164–165.

Wilson, Scott. "Warming Shrinks Peruvian Glaciers; Retreat of Andean Snow Caps Threatens Future for Valleys." *Washington Post*, July 9, 2001, A1.

Wiltgen, Nick. "Iraq [and] Iran Heat 125 Degrees; Feels Like Temperature Reaches 159." Weather.com. July 31, 2015. http://www.weather.com/news/news/iraq-iran-heat-middle-east-125-degrees.

Windeyer, Chris. "South Baffin Swelters in Winter Heat Wave." *Nunatsiaq News*, January 4, 2011. http://www.nunatsiaqonline.ca/stories/article/98789_south_baffin_swelters_in_winter_heat_wave/.

Wines, Michael. "Dinner Disappears, and African Penguins Pay the Price." *The New York Times*, June 4, 2007. http://www.nytimes.com/2007/06/04/world/africa/04robben.html.

Wines, Michael. "Worst Drought in Half Century Shrivels the Wheat Belt of China." *The New York Times*, February 25, 2009. http://www.nytimes.com/2009/02/25/world/asia/25drought.html.

Wines, Michael. "Colorado River Drought Forces a Painful Reckoning for States." *The New York Times*, January 5, 2014. http://www.nytimes.com/2014/01/06/us/colorado-river-drought-forces-a-painful-reckoning-for-states.html.

Wines, Michael. "Stanford to Purge $18 Billion Endowment of Coal Stock." *The New York Times*, May 7, 2014. http://www.nytimes.com/2014/05/07/education/stanford-to-purge-18-billion-endowment-of-coal-stock.html.

Wines, Michael. "Clean Mining a Deception in Kentucky, Groups Say." *The New York Times*, November 18, 2014. http://www.nytimes.com/2014/11/18/us/clean-mining-a-deception-in-kentucky-groups-say.html.

Wines, Michael. "Toxic Algae Outbreak Overwhelms a Polluted Ohio River." *The New York Times*, October 1, 2015. http://www.nytimes.com/2015/10/01/us/toxic-algae-outbreak-overwhelms-a-polluted-ohio-river.html.

Wines, Michael, and Jess Bidgood. "Waters Warm, and Cod Catch Ebbs in Maine." *The New York Times*, December 14, 2014. http://www.nytimes.com/2014/12/15/us/waters-warm-in-gulf-of-maine-and-cod-catch-ebbs.html.

Winkelmann, R., et al. "Increased Future Ice Discharge from Antarctica Owing to Higher Snowfall." *Nature* 492 (December 13, 2012): 239–242. doi: 10.1038/nature11616.

Winkelmann, Ricarda, et al. "Combustion of Available Fossil Fuel Resources Sufficient to Eliminate the Antarctic Ice Sheet." *Science Advances* 1(8) (September 11, 2015). e1500589. http://advances.sciencemag.org/content/1/8/e1500589. doi: 10.1126/sciadv.1500589.

"Winter Fog Becoming Rare in California." NASA Earth Observatory, June 5, 2014. http://earthobservatory.nasa.gov/IOTD/view.php?id=83806&src=eoa-iotd.

"Winter's Officially Here, Hits Flights. Worsens Pollution." *Sunday Times of India*, December 13, 2015, A1.

"Wintertime Disintegration of Wilkins Ice Shelf." NASA Earth Observatory. July 22, 2008. http://earthobservatory.nasa.gov/Newsroom/NewImages/images.php3?img_id=18095 (no longer available).

Wiseman, Paul. "Australia Pushes New Climate Plan." *USA Today*, September 6, 2007, 10A.

Witze, Alexandra. "Evidence Supports Warming Theory." *Omaha World-Herald*, January 12, 2000, 4.

Witze, Alexandra. "Climate Change: Losing Greenland." *Nature* 452 (April 17, 2008): 798–802.

Witze, Alexandra. "Drowned Tundra Emits More Carbon." *Nature*, August 4, 2009, online. http://www.nature.com/news/2009/090804/full/news.2009.778.html. doi: 10.1038/news.2009.778.

Witze, Alexandra. "Native Ecosystems Blitzed by Drought: California's Current Water Crisis Offers a Preview of What Climate Change Will Bring." *Nature* 512 (August 12, 2014): 121–122. http://www.nature.com/news/native-ecosystems-blitzed-by-drought-1.15707.

"WMO Issues: The Climate in Africa in 2013." World Meteorological Organization, February 3, 2015. https://public.wmo.int/en/media/news/wmo-issues-climate-africa-2013.

Wohlforth, Charles. *The Whale and the Supercomputer: On the Northern Front of Climate Change.* New York: North Point Press/Farrar, Strauss and Giroux, 2004.

Wong, Edward. "Beijing Takes Emergency Steps to Fight Smog." *The New York Times*, January 31, 2013. http://www.nytimes.com/2013/01/31/world/asia/beijing-takes-emergency -steps-to-fight-smog.html.

Wong, Edward. "Glut of Coal-Fired Plants Casts Doubts on China's Energy Priorities." *The New York Times*, November 12, 2015, A6.

Wong, Edward, and Chris Buckley. "Chinese Premier Vows Tougher Regulation on Air Pollution." *The New York Times*, March 16, 2015. http://www.nytimes.com/2015/03/16/ world/asia/chinese-premier-li-keqiang-vows-tougher-regulation-on-air-pollution.html.

Woodard, Colin. "Wind Turbines Sprout from Europe to U.S." *Christian Science Monitor*, March 14, 2001, 7.

Woodwell, George M. "The Effects of Global Warming." Pp. 116–132 in Jeremy Leggett, ed., *Global Warming: The Greenpeace Report*. New York: Oxford University Press, 1990.

Woodwell, G. M., et al. "Biotic Feedbacks in the Warming of the Earth." *Climatic Change* 40 (1998): 495–518.

Woody, Todd. "Alternative Energy Projects Stumble on a Need for Water." *The New York Times*, September 30, 2009. http://www.nytimes.com/2009/09/30/business/energy -environment/30water.html.

"World Economic Forum Feels Climate Change Pressure." Environment News Service, January 25, 2013. http://ens-newswire.com/2013/01/25/world-economic-forum-feels-clim ate-change-pressure/.

World Meteorological Organization. *The Global Climate 2001–2010: A Decade of Extremes*. 2013. http://library.wmo.int/pmb_ged/wmo_1119_en.pdf.

Wormworth, Janice. Bird Species and Climate Change: The Global Status Report. Fairlight, NSW, Australia: Climate Risk Pty Limited, n.d. https://www.wwf.or.jp/activities/lib/pdf _climate/environment/birdsFullReport.pdf.

"Worst Ice Year Kills Canadian Seals before Hunters Can." Environment News Service, March 26, 2010. http://www.ens-newswire.com/ens/mar2010/2010.

Wouters, B., et al. "Dynamic Thinning of Glaciers on the Southern Antarctic Peninsula." *Science* 348 (May 22, 2015): 899–903.

Wright, Shelly. *Our Ice Is Vanishing/Sikuvut Nunguliqtuq: A History of Inuit, Newcomers, and Climate Change*. Montreal: McGill-Queen's University Press, 2014.

Wuethrich, Bernice. "Lack of Icebergs Another Sign of Global Warming?" *Science* 285 (July 2, 1999): 37.

Wuethrich, Bernice. "How Climate Change Alters Rhythms of the Wild." *Science* 287 (February 4, 2000): 793–795.

Xu, Baiqing, et al. "Black Soot and the Survival of Tibetan Glaciers." *Proceedings of the National Academy of Sciences*. Published online before print December 8, 2009. doi: 10.1073/ pnas.0910444106.

Yardley, Jim. "China Says Rich Countries Should Take Lead on Global Warming." *The New York Times*, February 7, 2007. http://www.nytimes.com/2007/02/07/world/asia/07china .html.

Yardley, Jim. "China Releases Climate Change Plan." *The New York Times*, June 4, 2007. http://www.nytimes.com/2007/06/04/world/asia/04cnd-china.html.

Yavasli, D. D., C. J. Tucker, and K. A. Melocik. "Change in the Glacier Extent in Turkey during the Landsat Era." *Remote Sensing of Environment* 163 (June 15, 2015): 32–41. http://www.sciencedirect.com/science/article/pii/S003442571500098X.

Yeginsu, Ceylan. "In Turkey, Even Snow Can Be Tainted by Politics." *The New York Times*, February 20, 2015, A4.

Yoon, Carol Kaesuk. "Something Missing in Fragile Cloud Forest: The Clouds." *The New York Times*, November 20, 2001, F5.

Yue, Jia, et al. "Increasing Carbon Dioxide Concentration in the Upper Atmosphere Observed by SABER." *Geophysical Research Letters*, August 12, 2015. doi: 10.1002/2015GL064696.

"Zachariæ Isstrøm Glacier, Greenland." NASA Earth Observatory, December 4, 2015. http://earthobservatory.nasa.gov/IOTD/view.php?id=87086.

Zacharias, Yvonne. "Global Warming Exacerbates B.C. Wildfire Severity, Scientist Says." *Vancouver Sun*, July 6, 2015. http://www.vancouversun.com/technology/global+warming+e xacerbates+wildfire+severity+scientist+says/11192869/story.html.

Zachos, James C., et al. "Rapid Acidification of the Ocean during the Paleocene–Eocene Thermal Maximum." *Science* 308 (June 10, 2005): 1611–1615.

Zalasiewicz, Jan, and Mark Williams. *Ocean Worlds: The Story of Seas on Earth and Other Planets*. New York: Oxford University Press, 2016.

Zalasiewicz, Jan, et al. "The New World of the Anthropocene." *Environmental Science & Technology*, 2010, 44(7): 2228–2231.

Zeller, Tom, Jr. "America's Breadbasket Moves to Canada?" *The New York Times*, December 5, 2006. http://thelede.blogs.nytimes.com/2006/12/05/americas-breadbasket-moves-to -canada.

Zeng, Ning, et al. "Agricultural Green Revolution as a Driver of Increasing Atmospheric CO_2 Seasonal Amplitude." *Nature* 515 (November 20, 2014): 394–397.

Zhang, Keqi, Bruce C. Douglas, and Stephen P. Leatherman. "Global Warming and Coastal Erosion." *Climatic Change* 64(1 and 2) (May 2004): 41–58.

Zimmer, Carl. "Multitude of Species Face Climate Threat." *The New York Times*, April 4, 2011. http://www.nytimes.com/2011/04/05/science/earth/05climate.html.

Zimmer, Carl. "Study Finds Global Warming as Threat to 1 in 6 Species." *The New York Times*, May 1, 2015. http://www.nytimes.com/2015/05/05/science/new-estimates-for -extinctions-global-warming-could-cause.html.

Ziska, Lewis, et al. "Recent Warming by Latitude Associated with Increased Length of Ragweed Pollen Season in Central North America." *Proceedings of the National Academy of Sciences* 108(10) (February 22, 2011): 4248–4251.

Zremski, Jeremy. "A Chilling Forecast on Global Warming." *Buffalo News*, August 8, 2002, A1.

About the Author

Bruce E. Johansen is Fredrick W. Kayser Professor of Communication and Native American Studies at the University of Nebraska at Omaha, where he has been teaching and writing since 1982. He has written 47 published books, including *Eco-Hustle! Global Warming, Greenwashing, and Sustainability* (Praeger 2015). Johansen writes frequently about environmental subjects, including *The Encyclopedia of Global Warming Science and Technology* (2 vols., 2009), *Global Warming in the 21st Century* (3 vols., 2006), *The Global Warming Desk Reference* (2001), *The Dirty Dozen: Toxic Chemicals and the Earth's Future* (2003), and *Indigenous Peoples and Environmental Issues* (2004). He also writes as a journalist in several United States national forums, including the *Washington Post* and *The Progressive*, and has frequently had letters to the editor published in *The Atlantic, The New York Times, National Geographic*, and *Wall Street Journal*.

Index

Bold indicates volume numbers.